Dyna-Soar

Hypersonic Strategic
Weapons System

Compiled from the archives & edited
by Robert Godwin

The Editor would like to acknowledge the following people who provided unique insights for this book.

Wilfred Dukes (Bell Aerospace)
Neil Armstrong (NASA Dyna-Soar Pilot/Consultant)

Also the following people for helping with the search to find the material contained herein:

Richard Byron (Niagara Aerospace Museum)
Hugh Neeson (Niagara Aerospace Museum)
Bruce Hess (Wright-Patterson AFB)
Archie DiFante (Maxwell AFB)
Peter Merlin (Dryden Flight Research Center)
Stephen Garber, Colin Fries, Jane Odom, John Hargenrader, Nadine Andreassen (NASA HQ)
Shirley Jordan (US Department of Commerce)
Michael Shanklin (Oklahoma Omniplex)
Glen Swanson (Liftoff Books)
Jim Busby (Space Frontier Foundation)
John Zwez (Neil Armstrong Museum)
Thomas Lubbesmeyer (Boeing)
Gordon Woodcock (Boeing Retired)
Dennis Wingo (Skycorp)

For the photographs of the Boeing Models:
Jake Schultz (Boeing)

For Boeing Licensing:
Mary Kane
Eleanor Anderson

Computer Renderings by Tim McElyea (Media Fusion) and Rob Godwin

The following photos courtesy of Boeing Management Company. Used under license. All rights reserved
Text section - Page 350, pages 392-395, page 415, page 431 top, page 445 top,
Black & White section - Page 1 bottom right, pages 9 - 13, page 14 bottom left, page 15 top left
Color section - pages 4-6, page 7 top, pages 8-9, pages 14-16

We acknowledge the financial support of the Government of Canada through the
Book Publishing Industry Development Program for our publishing activities.
Published by Collector's Guide Publishing Inc., Box 62034,
Burlington, Ontario, Canada, L7R 4K2
Printed and bound in Canada
Dyna-Soar - Hypersonic Strategic Weapons System
by Robert Godwin
ISBN 1-896522-95-5
ISSN 1496-6921
Apogee Books Space Series
©2003 Apogee Books

Dyna-Soar

Hypersonic Strategic Weapons System

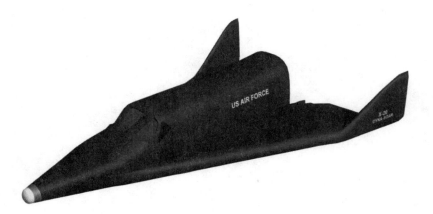

CONTENTS

Sänger Antipodal Bomber

Dry Weight 10 tons
Supporting Surface Area
Fuselage 80.8 m2
Wings 44.7 m2
Length 28m
Wing span 15m

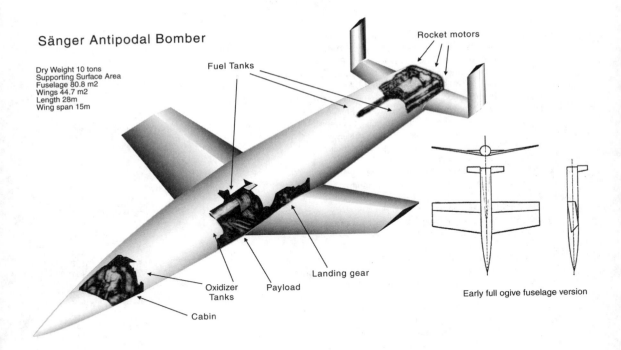

Starting point - The Sänger Antipodal Bomber
(Illustration by the Editor based on Sänger's original report)

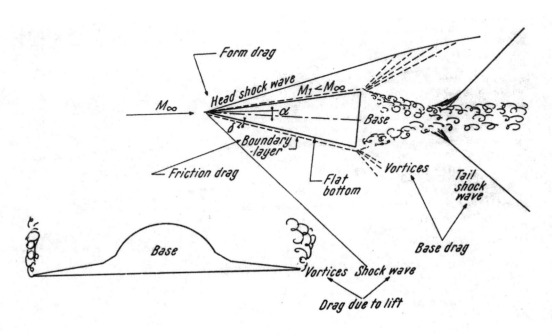

An energy dissipation diagram for a hypersonic vehicle
from 1954 by Bell Aircraft Engineer Krafft Ehricke

INTRODUCTION

Anyone who has followed this series of books may feel compelled to ask the question: Why the Dyna-Soar? They may well even ask: What is the Dyna-Soar?

The main reason for including this particular vehicle in our ongoing series lies in its remarkably foresighted design. Clearly the Dyna-Soar/X-20 was way ahead of its time (even though it was not completed) and that sort of advance in technology is expensive; consuming a sizeable portion of the military budget and employing nearly eight thousand people for over six years. In many ways it has been asserted that the story of the Dyna-Soar began in Austria in the mid-1930's. It was there in the University of Vienna that Dr Eugen Sänger, a rocket expert, first began to formulate the idea for a winged vehicle that could fly into space and arrive anywhere on Earth in a matter of a little over an hour. If it could be built, the so-called Sänger hypersonic space-plane would be a formidable weapon. On March 5th 1934 Sänger drafted a letter to the editor of the Journal of the British Interplanetary Society stating that he was, *"working on extremely fast aeroplanes, rocket-driven."* This was at a time when people who talked about spaceflight were usually considered cranks by the general public and the generous flow of information between rocket researchers was not yet obstructed by politicians or military security.

Just before the outbreak of World War II Sänger was drafted into the Herman Göring Institute and applied his considerable talents to the work undertaken there in various fields of aviation. During the war Sänger and his wife Irene Bredt sat down and wrote a four hundred page report which outlined the concept for the armed version of the space-plane. Now dubbed the antipodal bomber (due to its ability to deliver a warhead to opposite sides of the globe) the concept finally attracted the attention of those who were waging war. Copies of the report were mailed in total secrecy to a handful of the German military elite, including General Walter Dornberger who at the time was commander in charge of rocket research at Peenemünde.

Sänger's antipodal bomber was to have been pushed horizontally along an almost 2 mile track (sort of like *Fireball XL5* or the rocket on *When World's Collide*, for those of you who remember such things). The concept for horizontal launch had many advocates at the time because it was a method to reach escape velocity without fighting gravity. After the vehicle raced along the track it left behind a rocket-propelled sled and then subsequently shed its expended rockets and fuel tanks thus decreasing its overall mass. This method was obviously discarded because the vehicle must sustain a constant battle against the denser air at low altitude while it is reaching escape velocity. As anyone who has tried to beat the land speed record can testify, supersonic velocities at ground level are quite hazardous. However there is nothing fundamentally wrong with the principle, it's just that flying vertically and then gradually tipping over to horizontal once you reach the rarefied upper atmosphere is easier.

Another thing which Sänger realized was that the vehicle didn't necessarily need a large wing area. He intended to use the wide flat fuselage to provide a substantial amount of the required lift. Lifting bodies and flying wings were all the rage at the time. In England Armstrong-Whitworth were working on their AW-52 flying wing, in the USA Northrup were working towards their YB-49 and in Germany everyone from the Horten brothers to Messerschmitt were working on similar concepts that used unconventional arrangements for increasing lift. Consequently Sänger came up with a design that could have a relatively small wing arrangement with sharp leading edges and a wide flat-bottomed fuselage.

Some engineers have suggested that Sänger was too optimistic when it came to assessing the required thrust to make the vehicle fly. He proposed an engine with 200,000 pounds of thrust but given the overall weight of the vehicle he was suggesting, this may well have been inadequate.

Thankfully it was too late for the Germans to build this super-weapon, even if they could have surmounted the technical obstacles (they were actually contemplating using plain old steel for the fuselage), and so at the end of the war the Sänger-Bredt report fell into allied hands. Soviet leader Joseph Stalin was reported to have been so infatuated with the idea he allegedly dispatched a team which included his own son, to Paris, to find and "acquire" Sänger. The Soviet abduction team failed, and by this time the German rocket scientists who had surrendered to the Americans were already on their way to being fully integrated into the American missile program.

Sänger found himself working in Paris for the post-war French government and somehow was ignored by the contingent of Americans combing Europe for the cream of German aviation talent. He even applied to be taken to America but was dropped out of the running because, under Operation Paperclip, Eisenhower was only prepared to endorse just over a hundred Germans to come and work in America. This oversight ultimately led to Sänger finding himself working in Egypt for a different kind of Nasser! Sänger helped the Egyptians start their rocket program until the Israeli government, recognizing the threat that his ideas posed, demanded that the German government recall him to Germany, a request to which they subsequently acceded. His wife continued her work discreetly from her position at the Stuttgart institute while he spent the late years of his life as Professor of Space Flight Technology at West Berlin's Technological University. He died at the age of 58 and he certainly knew that his brain-child was beginning to take shape in America without him, since he had declined an invitation to come and join the team at Bell Aerospace, allegedly because the US government would not allow him to bring some of his staff with him.

Meanwhile Walter Dornberger found himself in America. After having his war record investigated by the British he had quickly relocated to Wright Patterson Air Force Base in Ohio before moving on to a consultancy position at Bell Aerospace in Niagara Falls New York. While he worked there he never forgot the Sänger report and by mid-1952 he was lobbying the United States Air Force to approve a series of studies on the feasibility of a long-range space bomber. On July 14th the NACA endorsed the study. Dornberger had made important strides in this direction during the last years of the war with his team at Peenemünde and so he believed that the technological hurdles could be overcome. One of the more difficult obstacles was convincing the American public (and thus the Congress and Senate) that man-in-space was not pure fantasy.

One of the first decisions Dornberger made was to recruit Krafft Ehricke away from the Army Ballistic Missile Division in Huntsville Alabama. Ehricke was an extraordinarily enthusiastic and talented engineer that had been naturalized into the United States under Operation Paperclip. It is interesting to note that between 1952, when Dornberger made his first proposal, and 1957 when the project was suddenly given the "go-ahead" that his old protégé, Wernher von Braun, had managed to launch an extremely well organized campaign (with the aid of Walt Disney, Collier's magazine and Viking publishing house) to influence the public that a space-faring future was imminent.

A small amount of money was allocated to Bell to pursue their research and a team of engineers were thrown into the deep waters of hypersonic flight. The obstacles were daunting. Requirements included entirely new materials and new ways of approaching the design of the vehicle. All of the disciplines had to be rethought since due to the unprecedented stresses and strains of hypersonic flight the standard approach of building a winged vehicle would be insufficient. The aerodynamic experts had to liaise with the structural and materials experts to try and fabricate something entirely revolutionary. This project brought about a strange confluence of old enemies as Bell assembled a team comprising the best of British (some from the Armstrong flying wing team), American and German engineering talent to continue the search for faster and better vehicles.

The project went through a variety of different names and acronyms, one of which was BoMi, short for Bomber Missile, then there was Brass Bell, a recon version and finally Project ROBO a rocket bomber which was investigated by Douglas, Convair and North American Aviation. In addition it enjoyed a series of speculative launch configurations which included everything from a single stage to orbit to an elaborate three-stage flyback arrangement. In its two stage configuration BoMi was to have a first stage approximately 120 feet in length with a wing span of some 60 feet and a lift-off weight of about 700,000 pounds. Initial discussions for propellants suggested the new hypergolic fuels of nitrogen tetroxide and UDMH. The first stage would have five engines and the second would have three, this was to make sub-orbital flights. A later orbital version would have been 25 feet longer and 240,000 pounds heavier, with the first stage almost entirely built of titanium. (Many years later titanium would be chosen for the United States hypersonic transport which never got off the ground for a similar reason – cost. Titanium is rare and the bulk of the world's resources are in Russia.) The second stage would almost certainly have employed an ablative thermal protection system (which was to have been sprayed, somewhat in similar fashion to the notorious pink X-15). The payload for the orbital BOMI would have been something in the order of 13,000 pounds.

Dornberger and Ehricke had even suggested a civilian version for long range transportation (see color section). They also planned a variant propelled by liquid hydrogen which made the second stage even larger. This may in turn have led to a three stage version called MX-2276 where the propellant tank was mounted between the flyback first stage and the orbital glider.

It was not long before investment questions were being raised about Bell's ability to cope with such a large program, however somehow the tide remained in Bell's favor and by 1954 a contract was issued which now proposed a large three-stage vehicle with pilots, orbital capability and a multi-tasking potential.

Two months after the launch of Sputnik the USAF invited contractors to begin work on what had now evolved into a fully militarized project. They would request a sort-of one-man space shuttle. A host of contractors were invited to get involved and indeed everyone from Martin, Boeing, Bell, Goodyear, North American, Lockheed, Douglas, Republic, Convair, Chance and even AVRO in Canada had some sort of orbital hypersonic vehicle on the drawing boards. Some of these companies were not directly involved with the Strategic Systems Division's request for proposals and just happened to be conducting their own research in the field. A.V.Roe in Canada were already immersed in their Arrow interceptor program and were looking ahead to Mach 3 variants, which inevitably led them further into the field of hypersonics. Indeed the chief structural engineer for Bell's BoMi program had been recruited from AVRO in 1951. By 1956 the United States government was telling the Canadian government that high speed aircraft were obsolete and that missiles were the way of the future, while simultaneously ramping up the budget for BoMi. Within a couple of years the Canadian government would buy into that myth and would decimate the Canadian aerospace industry by cancelling the entire Arrow interceptor program along with all of its attendant weaponry. Like-minded individuals in some departments of the US DOD would later push the same nonsense at home. It is interesting to note that Canada's Air Marshall Roy Slemon was at the X-20 roll-out in Las Vegas in September 1962. Slemon had doggedly defended the Arrow but to no avail.

Meanwhile, even though Bell had teamed with Martin to try and cope with the demands of such an enormous project, the hypersonic bomber missile contract would ultimately go to Boeing. Boeing would subsequently christen it the Dyna-Soar (for dynamic soaring) and then later the X-20.

Never before or since was such an expensive program subjected to so many changes in direction. The official Air Force history (page 349) reveals a labyrinth of bureaucratic legerdemain. It seems that the responsibility from the program was shifted on an almost monthly basis for six years. It began life as Sänger and Dornberger's antipodal bomber but ultimately ended up as a one man test vehicle. This lack of a defined military goal would ultimately be the death of the Dyna-Soar. However the people working on it had aspired to greater things over the course of its life. After the cancellation of the program Lockheed proposed a multi-stage flyback version which would have weighed 28,000 pounds and carried 12 people as well as 6,600 pounds of cargo.

Even the launch vehicle was never truly nailed down. Some advocates at Boeing wanted to use the Minuteman with solid propellant units to launch a smaller glider weighing 6,500 pounds. Before the contract was issued Boeing suggested the Atlas-Centaur but naturally Martin wanted to use their own Titan. Boeing got the glider but were told in April 1960 it was to be launched on sub-orbital trajectories by a Martin Titan I missile. The Titan I had only seen its first successful launch in February 1959 and it did not become operational as America's largest ICBM until April of 1962. It was 98 feet long and had a diameter which shrunk from 10 feet on the first stage to 8 feet on the second. It used perishable fuels (i.e. kerosene and liquid oxygen) and weighed in at 220,000lbs (without Dyna-Soar). The total thrust of the two stages was 380,000 lbs.

In January 1961 the proposed launcher for Dyna-Soar was upgraded to a Martin Titan II. This new ICBM was first contracted only six months earlier, in June of 1960, and didn't see its first successful flight until March 1962. It was five feet longer than the Titan I and the upper stage had been widened to make the missile a consistent 10 feet in diameter. The main improvement was the use of storable fuels, a combination of 50% Hydrazine and UDMH with Nitrogen Tetroxide as the oxidizer. Overall thrust was increased to 530,000 lbs but the hypergolic fuels required heavier tanks and so the Titan II weighed in at a hefty 330,000 lbs. The Titan II would go on to become the launcher for Project Gemini.

In July 1961 the Space Systems Division sent in their recommendation to use a version of the planned Phoenix booster called Space Launch System A388. This version of Phoenix would have been powered by solid-rockets on the first stage and the new LOX/LH$_2$ J2 engine for the second stage. Another possibility considered was von Braun's new Saturn C1 (NOVA) but was eliminated for cost and complexity reasons.

Finally in February 1962 the launcher for Dyna-Soar was upgraded once more for fully operational orbital flight by using a Titan III with its large strap-on solid rockets. The Titan III had been through a couple of early design configurations which used a Martin-Marietta upper transtage (Titan IIIA) and a Bell Agena D third stage (Titan IIIB) before being equipped with two enormous strap-on solid rockets developed by United Technologies. The boosters generated 1.3 million pounds of thrust each, creating a combined lifting thrust of 3.1 million pounds. This enormous booster had little application for delivering warheads (which were shrinking in size and weight) but was suitable for launching military payloads of up to 29,000 lbs into low earth orbit. It was perfect for Dyna-Soar. (Later the Titan IV would increase that payload to over 48,000 lbs.)

Other proposals included one from Martin which was a massive configuration called Boosted Arcturus and would have used a staggering conglomeration of seven Titans strapped together, able to send Dyna-Soar to the moon with 60,000 pounds of payload. Another money-saving version called Project Streamline went back to using the first stage of Wernher von Braun's new superbooster, the Saturn IB.

There were a variety of still more proposed configurations which all have only one thing in common, they all fully expected Dyna-Soar to sit on top and then fly home to a runway. It was to be the space equivalent of the Swiss Army knife, a tug-boat for Apollo - or a taxi to space-stations - or a bomber for the Air Force – or a spy-plane for the intelligence community.

Since all of these configurations didn't allow for any kind of escape tower NASA and the USAF began a program at Edward's Air Force Base in California to try and devise an abort procedure that would allow the crew to escape a bad launch. The experimental F5D aircraft was run through a series of extreme maneuvers by Neil Armstrong until a usable escape path was devised.

Next would have been a series of trials where Dyna-Soar would be dropped from a modified B-52, beginning in early 1965, and then make its first orbital launch in early 1966.

All this is very interesting but the Dyna-Soar was only half-built, and of course it never flew. On that basis it may not seem to merit the attention of an entire book. However, what can be found in the following pages are the details of an audacious program which cost the American taxpayer over $400 million between 1957 and 1963. Some of you may want to know where the money went.

The fact is, even though the Dyna-Soar was cancelled, the spin-offs in experience, materials science and technology were enormous and are still paying dividends today. 36 of the 66 projects were continued, including the pilot controlled booster which was finally used on the Saturn V. Some of the aerodynamic work ended up being used later in an unmanned program called Project Asset which successfully proved the aerodynamics of a similar structure when it flew six times between 1963 and 1965. Dyna-Soar was also the first successful use of Pulse Code Modulation and moreover incorporated a revolutionary inertial guidance system. The glider was to have been an entirely metal structure using the new René 41 alloy for the panels, zirconium for the nose and molybdenum for the hot spots. Because almost all of the effort was spent on figuring out how to bring the glider home and almost nothing was spent on its weapons capability it could be argued that the Dyna-Soar was just a dry-run for the Space Shuttle program. The money spent might therefore be put into that enormous pile where it would almost disappear.

What is particularly interesting are the implications of Dyna-Soar. If it had been seen successfully to its logical conclusion, the United States would have had a space shuttle when the Russians were still trying to launch a three-man pressurized sphere (which was so small they had to do it without the benefit of space-suits.) Apollo might have been accompanied into earth orbit by a sort of one-man companion for support and observation. The International Space Station would not currently be endangered with redundancy due to the lack of an adequate escape system. The enormous amounts of money used to deploy the B-2 (the stealth bomber shaped like a bat-wing that was partially derived from the Northrup flying wing) and the B-1 (variable geometry supersonic bomber) might have been unnecessary, the Dyna-Soar antipodal hypersonic space-bomber might have replaced both and been virtually unstoppable. Now this may seem like Monday-night quarter-backing but the fact is that today these very same plans are still on the table. The USAF recently announced *Project Falcon* hoping for a quicker way to accurately deliver a weapon to the world's hot-spots. NASA is searching for a life-boat under their *OSP* program and the Russians are threatening to build what is almost a clone of the Dyna-Soar, presumably based on their own research into such things, to take tourists into space. The problem in 1963 was that no one spent enough time explaining *why* the Air Force needed a manned space program. As early as September 1961 Secretary of Defense Robert McNamara questioned whether Dyna-Soar was the best expenditure of national resources.

Aside from the obvious technological anecdotes associated with the X-20 there are the human stories which involve characters like Dornberger, Sänger, the six pilots chosen to fly it (none of whom made it into orbit although two of them flew the X-15), Mercury astronauts Gus Grissom and Wally Schirra, X-15 pilots Milt Thompson, Bill Dana, Pete Knight and Neil Armstrong (who was one of six pilot/consultants brought into the program.) There is also General Curtis LeMay, an imposing figure in American military history, who urgently advocated the cause of Dyna-Soar, perhaps because he could see that its demise would spell the end of the manned military space program.

By July 1963 a decision had been made that NASA and the DOD would not indulge in any manned space programs without fully considering the other agency's needs. The Air Force announced their long held desire for a manned orbital space station that summer and a plan was evolving, which would use NASA's Gemini capsule on top of the Air Force' Titan 3, to shuttle men and equipment to that station. Dyna-Soar would have been capable of carrying an additional four people in the mid-section (allocated normally for equipment).

Running concurrent to the Dyna-Soar program was the Aerospace Plane program. It began around 1957 and shared many similarities with Dyna-Soar. The original design was to be single-stage to orbit but it evolved into a two-stage flyback version. The second stage of the Aerospace plane looked much like Dyna-Soar but it needed to be powered.

Dyna-Soar, meanwhile, had not originally been designed to be powered so could not carry a large fuel load. Now its mission was no longer as a suborbital glider but a full-fledged orbital vehicle. This meant it would need some kind of de-orbit engine and so the basic shape would have to be changed to accommodate the shifts in the centre of gravity caused by moving fuel. Suddenly versions of Dyna-Soar with engines began to appear in artist's renderings. Despite all these modifications what would prove fatal for Dyna-Soar was the success of Gemini, the lack of any solid reason from the Air Force to put a manned space craft into orbit and the fact that in 1959 the mandate had shifted from an orbital weapons platform to a suborbital research vehicle and then see-sawed back and forth for the next four years. Politics even intervened when the original landing site for the test flights, in Brazil, was suddenly not available. Consequently too much time was spent on how to get the thing back from space and not enough time figuring out what it would do when it got there. McNamara would not spend the money until some one could tell him what he was getting for it.

On November 25th 1963 the American nation watched in stunned silence as the funeral procession for President Kennedy was televised around the world. Just two weeks later despite the protestations of the Air Force Dyna-Soar was officially cancelled by Defense Secretary Robert McNamara and the new administration of Lyndon Johnson.

The Dyna-Soar story is a tale of high-stakes and high technology involving the leading players in the world of politics and aerospace. Some of the best-known heroes of the American space program played a hand in that story. It is in some ways America's parallel to the Canadian Avro Arrow story. The Arrow was way ahead of its time and when it was cancelled every working aircraft was destroyed for no logical reason. With Dyna-Soar it was a similar story. Today the only full-size mock-up of Dyna-Soar, built by Boeing, is lost, although a Martin X-23A version (which shows some similarity to the later X-24) built for the World's Fair still resides at the Oklahoma Omniplex.

This book looks at the story of Dyna-Soar through the official documentation that flowed between the Air Force, NASA and the contractors. There are a few interesting additions such as an article written by Dornberger years after Dyna-Soar was cancelled making a case for the validity of the aerospace plane concept and there is a copy of the bibliography that Sänger used when he prepared his original report. The details of the political machinations can be partially divined from reading these documents but the real narrative is still waiting to be told although an excellent effort was made by USAF Major Roy Houchin for his doctoral dissertation in 1995. This book however is just one chapter in an enormous story that is still being written.

In the end we can only imagine how the geopolitical climate might be different today had the X-20 Dyna-Soar flown in space.

Robert Godwin (Editor)

> **Date: Oct 1955   Title: Structural Design for Aerodynamic Heating**

An Introduction to Structural Design for Aerodynamic Heating (Extract)

By Wilfred Dukes and A. Schnitt - Bell Aircraft Corporation

WADC Technical Report 55-305 Pt II

Problem Boundaries

A. GENERAL

During the design of an airframe for subsonic or low supersonic speeds, the structure can be studied on the basis of local loading and section shape; other parameters generally have a secondary effect. Since the number of section shapes is quite limited, being usually flat or near flat panels, cylinders, or combinations of both, a general structural study can be made in simple terms.

When speeds are sufficient to cause aerodynamic heating, however, many new parameters are introduced, since the heating is a function of the complete flight history, the body shape, and the nature of the structure itself. The problem is broadened further because the presence of heating increases the number of possible types of structure, and the variety of structural materials, that must be considered.

If a comprehensive study of supersonic and hypersonic structure is to be made, with consideration of the many variables involved and the possible aircraft types and flight plans, structural arrangements and associated thermal parameters must be examined in a systematic manner. It is the purpose of this section of the report to make such an examination. In this way, only the practical range of thermal parameter values will be included in the structural study to follow and, conversely, the results of the study can be related to specific aircraft types.

By following this procedure, if it is possible to indicate the most suitable type of construction, the structural materials, the type of heat protection, and the unit structural weights for various types of future aircraft, the information will be of considerable assistance in preliminary structural design, and will also give direction to future research. In particular, it may avoid wasteful research directed at the attainment of impractical combinations of thermal parameter values.

This section begins with a discussion of probable future types of supersonic aircraft and the type of flight path each may use. The limitation of air-breathing engines and the efficient use of rocket power plants are discussed, since these form important practical limitations to the range of variables that must be considered. It is shown that each flight path may be divided generally into phases such as boost, glide, and cruise, and that considerations given to each phase are reasonably independent of the remainder of the flight. The factors controlling and limiting each phase of flight, such as required range, minimum energy expenditure during climb, and boost accelerations as limited by the crew, are proposed as the basis for computation of flight path relationships. The significant thermal parameters required for structural design at elevated temperatures are then given for each type of flight path and structural arrangement.

Theoretical methods of obtaining numerical values for the relationships between speed, altitude, time, and range for each flight path phase are developed on the basis of a number of simplifying assumptions. Such assumptions are necessary because the number of possible variables is large.

To include all effects is unnecessary for the purposes of this study, since they would only serve to obscure the important factors. Graphical results of the various flight path relationships are presented, based upon the theoretical considerations, and, from these, additional curves are produced to give values of the thermal parameters.

Because of the necessary simplifying assumptions made in the theoretical work discussed previously, additional support has been given to the calculated boundary values of the aerodynamic heating problem by the collection of statistical data from progress reports and final reports on existing and projected aircraft. Because much of this statistical information has a SECRET security classification, it is not included in this part of the report, but instead the comparisons with theory are made in the summary of problem boundaries in Section I of Part III.

B. FUTURE VEHICLE TYPES

Table II-1 summarizes those vehicles in the supersonic and hypersonic speed range to which it is expected that most attention will be directed in the next phase of airframe development. The table has been subdivided on the basis of the vehicle characteristics which are most significant in their effect on the vehicle design. These characteristics are the presence or absence of a crew, the type of power plant, and the particular flight path to be flown.

In selecting these vehicles it was presumed that for a considerable future time, the cost and complexity of high-speed flight could only be justified for the delivery of costly high-energy warheads. Such warheads would be used against large, important ground targets so that the primary vehicle for hypersonic flight is the medium or long-range bomber. Prestrike and poststrike reconnaissance is a necessary part of such a mission, and it is assumed that hypersonic reconnaissance vehicles will also be built. Finally, the use of such bombardment vehicles leads directly to interceptors of similar speed capabilities, as a countermeasure.

The presence of a crew within a vehicle generally requires that the vehicle be nonexpendable and that it be designed for repeated flights. The feature of recoverability usually requires lowspeed flight characteristics and sufficient wing area to support the aircraft during landing. A crew limits accelerations, particularly longitudinal, and may also necessitate larger safety factors for structural design.

The effects of power plant and flight path on the elevated temperature aspects of structural design are closely related, since the power plant is probably the most significant factor to be considered in establishing the flight path. Power plants considered in this study are the turbojet, the ramjet, and both liquid- and solid-propellant rockets. Each of these engines has its own characteristics and is best suited for a particular flight path.

TABLE II-1

SUMMARY OF VEHICLES DEFINING MAXIMUM ENVIRONMENTAL CONDITIONS

Vehicle Type		Unmanned Vehicles — Expendable — High Longitudinal Accelerations — Power Plant		Manned Vehicles — Repeated Flights — Limited Accelerations — Power Plant	
		Rocket	Air-breathing	Rocket	Air-breathing
Long- and Medium-Range Bombardment	Re-entry Vehicle	Ballistic trajectory. Extreme temperatures and heating rates. Very short exposure time.			
	Booster Vehicles	Moderate temperatures and heating rates. Very short exposure time.			
Long- and Medium-Range Bombardment	Bomber	Boost-glide flight path. High temperatures and heating rates. Moderate exposure time (max. 1-1/2 hours).	Climb-cruise flight path with turbojets. Low Mach numbers (3.0). Moderate temperatures and heating rates. Long exposure time (to 2-1/2 hours). Boost-cruise with ram-jets. Moderate Mach numbers (3.0 to 5.0) Moderate exposure time (1 to 1-1/2 hours).	Similar to unmanned vehicles, except speeds will be higher for same radius since range is required for return.	Same as for unmanned vehicles, except exposure times may be doubled for same radius.
	Bomb			Ballistic or glide flight path. High temperatures and heating rates. Short exposure time.	Conventional gravity bombs or a powered missile.
Reconnaissance				Same as for bomber, but no bomb.	Same as for bomber.
Short-Range Interceptor		Air or ground launch. Boost throughout flight. High temperatures. Short exposure time.	Air or ground launch. Cruise flight path, possibly with rocket boost for combat. Ram-jets may have rocket boost after launch. Moderate temperatures. Moderate exposure time.	Ground launch. Boost and possibly powered combat. Glide return. High temperatures and heating rates. Short exposure time.	Ground launch. Cruise flight path with rocket boost for combat. Ram-jet may have rocket boost at launch. Moderate temperatures. Short exposure time.

Present-day turbojet engines are relatively limited in thrust, so that only moderate longitudinal accelerations are achieved. The use of solid-propellant rocket boosters does not materially change this situation, since their thrust duration is usually short. Optimistic extensions of present turbojet developments suggest that considerations should be limited to encompass velocities up to approximately Mach 3.0, and, since air-breathing engines are limited by air density and the requirement of flame propagation, maximum altitudes of approximately 65,000 feet. Hydrogen has the ability to sustain a flame under conditions of very low pressure, and its use as a fuel promises to raise altitude limitations of turbojet engines. For the present, 100,000 feet will be considered a maximum. These characteristics, together with its relatively high efficiency, make the turbojet engine most suitable for cruising-type paths at speeds less than Mach 3.0.

Since the ramjet has no rotating machinery, the speed limitation is raised arbitrarily to approximately Mach 5.0. At this speed, the ram air entering the engine would be at approximately 2000°F before the addition of combustion heat. Therefore, to achieve this speed, extensive development of existing materials or the addition of a complex cooling system to the engine would be necessary. Such a cooling system would require an expendable cooling material since ram-air cooling is, evidently, ineffective because of the air temperature. Because of the speed limitation, the ramjet requires a cruise-type flight path to achieve range, but rocket boost is also necessary to achieve the speed (approximately Mach 1.0 to 2.0) at which ram compression is sufficient to operate the engine; hence, the boost-cruise flight. Present rocket engine designs are such that large thrusts are obtainable with a small engine weight, so that large longitudinal accelerations of the rocket propelled vehicle are possible. Since this is not an air-breathing engine, it is unlimited in altitude and speed, but this characteristic is also its most serious restriction. The rocket-propelled vehicle must carry its own supply of oxidizer, of generally three times the weight of fuel. When this fact is combined with the low efficiency of the rocket, it is evident that such vehicles will carry large quantities of propellants. In a Mach 20 vehicle, for instance, 80% of the take-off weight may be propellants, while the corresponding value for a conventional subsonic fighter is approximately 20%.

For large rocket-propelled vehicles, therefore, it becomes important, in the achievement of minimum take-off weight, to minimize the energy used in accelerating propellant to the maximum speed. Evidently, all propellants should be exhausted at the instant that maximum speed is achieved, so that the propulsive effort should be used entirely in accelerating. The necessary range is then attained either by gliding or by following a ballistic trajectory in which aerodynamic forces are essentially absent. A "skip" path is also possible, in which the vehicle descends quite rapidly into the atmosphere and then climbs above the atmosphere again by exchanging some kinetic energy for altitude. This process is repeated several times with a net loss of velocity at each "skip." The energy loss appears as aerodynamic heating, but the airframe has an opportunity to dissipate this heat by radiation while the vehicle is in the ballistic part of the projection outside the atmosphere. This flight path is complex for a generalized analysis, and is neglected in the present work since it has already been shown (Reference 1) to give more severe aerodynamic heating problems than the glide path.

For reasons discussed previously, cruising flight at constant speed and altitude is usually inefficient with rocket propulsion, unless the range is less than approximately 100 nautical miles.

The following is a more detailed description of the vehicles summarized in Table II-1.

1.   Ballistic Missile

Since this vehicle follows a ballistic trajectory, it does not depend upon aerodynamic forces for lift. Some aerodynamic control may be provided during the initial stages of launch but, for most operations, the range will require a maximum altitude well beyond the earth's atmosphere where the only control is from rocket engines mounted on gimbals, auxiliary rockets, or vanes in the engine exhaust. Since only auxiliary rockets could be used after the boost phase, all guidance functions would generally be performed during boost.

Because this vehicle has no wings, it is not recoverable and, of course, is unmanned. The speeds and altitudes required to achieve range are generally such that only the rocket-type power plant is suitable. The principle of expending propellants as rapidly as possible is invariably followed in the design of these vehicles, with the rate of expulsion being a compromise between propulsion efficiency, power plant size and weight, and acceleration forces on the structure and equipment.

To minimize further the energy used in accelerating, these vehicles are usually constructed in stages; a small vehicle contains the payload, to which is attached one or more boosters. As propellant tanks are emptied, they can then be jettisoned, together with some of the engines, so that propellant is not used unnecessarily to accelerate structural weight.

Maximum velocities are generally achieved after the vehicle has left the atmosphere; at this point, aerodynamic heating does not exist. The booster stages are then subjected only to moderate heating for short times. The stage containing the payload, however, re-enters the atmosphere at maximum speed and thus experiences severe heating. The time of this heating is short because the vehicle traverses only the thickness of the atmosphere under heating conditions. For aerodynamic heating considerations, this flight path may be divided into the booster phase and re-entry phase.

2.   Long-Range Bombardment - Bomber - Boost Glide

As explained previously, the use of rocket propulsion for this vehicle requires a boostglide flight path, probably using a multistage vehicle from which propellant tanks and engines are jettisoned after the propellants are used. Considerable speed is required to achieve range by gliding, and this speed permits sufficient lift to be obtained at altitudes beyond the limits of airbreathing engines. Thus, the high altitude tends to reduce the severity of aerodynamic heating and to compensate somewhat for the high speed.

The manned boost-glide bomber must have range beyond the target so that it may return to friendly territory; this can only be achieved by speed. The manned boost-glide airplane will thus be larger and faster than the unmanned expendable missile having the same operational radius.

The final stage of this airplane must also have low-speed flight characteristics and a wing area sufficient for landing.

This type of flight path may be divided into a boost phase applying to booster and the final stage vehicle, and a glide phase in which only the final stage participates. Heating conditions are moderate and of short duration for the booster, and moderate to severe for the final stage, depending on speed. Heating will exist for an extended time on the final stage vehicles.

3.   Long-Range Bombardment - Bomber - Cruise

Air-breathing engines will be used in a long-range cruising vehicle so that speed and altitude will be limited as mentioned previously. Heating conditions will be moderate, but may persist for a number of hours, particularly with the manned bomber where extra range is required for return to base.

If turbojets are used with or without afterburners, the climb and acceleration is unlikely to be critical for thermal effects. If the ramjet is used, speeds may be greater, and the necessary rocket boost at launch may produce significant thermal transients. The flight path should thus be considered in two phases, boost and cruise.

4.   Long-Range Bombardment - Bomb

Manned recoverable bombers will continue to carry the warhead in a separate bomb which will be released into the airstream with the same velocity as the bomber. Aerodynamic heating must, therefore, be considered in the bomb design.

For bombers using air-breathing engines, the bomb may be of a conventional gravity-type or a powered and winged missile. Considering the speed and altitude limitations of air-breathing engines, it is unlikely that the gravity bombs will give critical heating conditions. The powered missile bombs should be treated as a bomber having the same power plant and flight path characteristics.

When launched from rocket-powered bombers, the bomb should have sufficient range potential from its launch velocity to make power or wings unnecessary. The bomb, therefore, enters a ballistic path. It descends into dense atmosphere at high speed so that heating conditions range from moderate to extreme, but exposure-time is short.

5.   Reconnaissance

The remarks made previously for the bomber apply equally to reconnaissance vehicles, except that such vehicles generally will be manned. Reconnaissance information transmitted by radio from an expendable missile is unlikely to equal the quality of photographic data, except by the use of equipment too costly to be expended after one flight. Since photographic results necessitate a recoverable vehicle, it will be advantageous to include a crew in order to minimize the unreliabilities and inaccuracies of presently developed automatic guidance, and to facilitate recovery.

6.   Short-Range Interceptor - Rocket-Propelled

The unmanned version of this vehicle may be air or ground launched, with air-launching being reserved for the smaller short-range missiles that can be carried to altitude rapidly in larger high-speed aircraft. Interception missiles of long-range (more than 20 nautical miles), or of high speed, will be large enough to make ground-launching more practical; the same is true of the manned interceptor.

The unmanned rocket-powered interceptor may be expected to use power for its complete flight to the attack, since most of this time will be consumed in climbing. Thus, the flight is entirely boost, with temperatures and heating rates reaching high values if the speed is sufficient, but the total flight time is of short duration.

The manned rocket-powered interceptor, while also using power up to and including combat so that speed and altitude can be achieved and maintained, will then glide back to base. The flight would be boost-glide with heating rates and temperatures reaching high values, depending on final speed, and with short exposure times but considerably longer than those of the unmanned interceptor.

The manned rocket-powered interceptor, while also using power up to and including combat so that speed and altitude can be achieved and maintained, will then glide back to base. The flight would be boost-glide with heating rates and temperatures reaching high values, depending on final speed, and with short exposure times but considerably longer than those of the unmanned interceptor.

7.   Short-Range Interceptor - Air-Breathing Engines

Whether manned or unmanned this vehicle may be expected to be rocket-boosted for rapid acceleration to altitude. The manned vehicle may also use boost for combat. The flight path is thus divided into boost and cruise phases. Heating rates and temperatures will be moderate because of the limited speed capability of the air-breathing engine. Times during which heating is present will be short for the unmanned vehicle, extending to moderate for the manned vehicle.

From the previous discussions it is evident that the flight paths of future high-speed vehicles can be divided generally into one or more of the following phases:

(1)   Boost, with increasing speed and either constant or increasing altitude

(2)   Cruise at constant speed and altitude

(3)   Glide at decreasing speed and/or altitude

(4)   Ballistic re-entry, decreasing speed and altitude

Each phase has peculiar characteristics with regard to the effects of aerodynamic heating, and these are discussed in more detail in the following paragraphs.

The boost phase of flight is characterized by increasing speed with or without increasing altitude, so that the aerodynamic conditions are changing. This leads to temperature gradients within the structure with corresponding thermal stresses or deformation. Except for the re-entry missile, the boost phase can be expected to be the primary cause of temperature gradients and thermal stresses in a structure. Other causes, such as change in angle of attack during maneuver or deceleration prior to landing, are comparatively unimportant.

Changes in angle of attack or altitude, for instance, result only in changing the heat transfer coefficient, and this produces a change in structural temperature only to the extent that radiation from the surface is present. Radiation will be significant if the speed is high so that surface temperature is high, or if the altitude is considerable so that convective heat transfer to the structure is small. In either case, it is clear that to achieve these conditions severe boosting will be required, so that boost will probably still give the critical thermal stresses. The deceleration prior to landing can presumably be performed so that, here too, the thermal stresses are no more critical than during boost; it should be noted that in this case the thermal stresses are of opposite sign from those produced by the boost. Since the rate of change of speed, without power, is dependent upon drag, it may be necessary to limit the rate of descent to lower altitude so that speed can be lost more gradually under conditions of low drag.

For the boost-glide or boost-cruise flight paths, the boost phase will have no significance other than to produce maximum transient conditions. The reason for this is that the maximum structural temperatures will almost always be reached shortly after completion of boost due to the thermal inertia of the structure.

For the unmanned rocket-powered interceptor, which is powered continuously to the target, the entire flight may be considered as boost so that this phase produces not only maximum thermal stresses, but also maximum temperatures. Similarly, the boost phase of the ballistic flight path will generally continue until the missile is beyond the atmosphere; aerodynamic heating is then reduced to zero and maximum temperatures will have been passed.

During boost flight, maximum acceleration for a manned vehicle will be determined by the physical limitations of the crew, while the rate of climb will probably be such that minimum energy is expended. For an unmanned vehicle, the maximum acceleration is obtained by optimizing between an increasing power plant weight and a reducing propellant weight (due to increased efficiency by more rapid usage), as the thrust is increased. Rates of change in altitude for an unmanned vehicle probably will be chosen for minimum energy expenditure if the vehicle is intended for cruise or glide, but for interception the criteria may well be minimum time to altitude.

In view of these complexities, the boost flight will be studied analytically by assuming a constant angle of climb and by expressing the results in terms of the ratios of thrust to take-off weight, and propellant weight to take-off weight. With these variables, any required combinations of velocity and altitude can be obtained for the start of a glide or cruise flight; the boost required to achieve the necessary range with a ballistic missile is included, since velocity and climb angle at the end of boost are related to the required range.

While the assumption of a constant angle of climb is only approximate, it permits the development of generalized relationships between speed, altitude, and time in terms of few parameters. Such generalization is the objective of this section of the report. In using the data to be presented, it is suggested that the flight path angle be selected to give the desired combination of speed and altitude at the end of boost.

This approach also permits a limiting acceleration to be used as a criterion for a manned vehicle, since maximum acceleration is related to thrust and propellant weight ratios. The altitude-time relationships will also show the climb angle to reach altitude in minimum time, and, if no other requirements are specified, this may be used as the criterion for the boost of an unmanned interceptor.

As previously explained, most cruising will be done with air-breathing engines and these have definite altitude and speed limitations. Relationships exist between speed, altitude, required range, lift-drag ratio, and fuel weight, but these are not simple. They depend closely upon vehicle configuration and engine performance (particularly as affected by intake duct shape).

Because of this difficulty, the thermal parameter values for cruise flight will be based upon arbitrary combinations of speed and altitude values which will be representative of both present and anticipated future practice.

The glide phase of a flight path is similar to the cruise phase except that speed and altitude change gradually, but usually not sufficiently fast to cause transient heating effects within the structure.

For glide flight, maximum values of speed are limited only by orbital speed, but attention during this study will be confined to values not exceeding Mach 20. Speed, altitude, and range, however, are related in a manner that depends primarily upon lift-drag ratio; consequently, when values of this ratio are assumed, a relationship exists between the temperatures or heating rates and the time of application.

The re-entry phase of a ballistic missile involves a decreasing altitude, and a decreasing speed due to drag. Re-entry speed and angle are related to the required range, but the deceleration after entry into the atmosphere depends upon body shape and is therefore arbitrary.

A ballistic missile is usually required to strike with the maximum possible velocity, but this must be accomplished with a body shape that has stability about the pitch and yaw axes, and also with heating rates that can be dealt with structurally. Many combinations of expendable dragbrake surfaces are possible to reduce speed at high altitudes where heating rates are low, but to establish the boundaries of the problem it will be assumed that these devices are undesirable and that a constant drag coefficient applies throughout re-entry. The drag coefficient can then be obtained from a specified striking velocity, and the resulting velocity-time-altitude relationships are readily calculated from the equations of motion. Maximum values of temperature, heating rate, and total heat will then follow.

Date: Nov 1956   Title: The Rocket Propelled Commercial Airliner

Brief Autobiography - Dr. Walter R. Dornberger

Dr. Walter T. Dornberger was Guided Missile Consultant for the Bell Aircraft Corporation of Buffalo, New York, from May 1950. Prior to that date he had, for three years, held a similar post at the Wright-Patterson Air Force Base at Dayton, Ohio.

He was born at Giessen, Germany, on September 6, 1895, and began his military career two days after the outbreak of World War I by volunteering for service in a German heavy Artillery Regiment. He was commissioned as an officer a few months later and saw active combat service until captured by the Allies in October, 1918.

After the Armistice had been signed, Dornberger continued in the small professional army Germany was allowed to maintain and attended a number of engineering and technical schools. Berlin Technical Institute awarded him three degrees — a B.A., a M.A., and an engineering Doctorate — during the period 1925 - 1930.

In 1930, when he held the rank of Captain in the German Board of Ordnance, he was given charge of the development of solid fuel rockets for use as modern weapons.

Dornberger was placed in command of the German Army's new experimental station at Kummersdorf in 1931. Pioneering work was begun there in the development of liquid fuel rockets in 1932.

From 1932 to 1945 he was division chief for rocket development, German. Board of Ordnance and commanding officer of the German Rocket Development Center at Peenemünde, where the A1 - A10, the V-2 and the Wasser Fall weapons were developed. From 1944 - 1945 he was also special commissioner of the German Reich for all V-weapons and anti-aircraft guided missiles. Dornberger held the rank of major general when he became a prisoner of war in 1945.

He was the author of the book "V-2" which has been published in several countries. He was an associate fellow of the Institute of Aeronautical Sciences, a member of the Buffalo Aero Club, and an honorary member of the German Society for Space Research.

THE ROCKET-PROPELLED COMMERCIAL AIRLINER
by
Dr. Walter R. Dornberger Guided Missile Consultant Bell Aircraft Corporation

Shortly before and during World War II, four new propulsion systems for flying vehicles were invented - the turbo-jet, the pulse-jet, the ram-jet, and the liquid fuel rocket. All four use the reaction principle to move forward.

Three of them utilize the oxygen of the air to sustain combustion in their engines, and are thereby bound to the atmosphere. Only one, the rocket power plant, by carrying along its own oxygen supply, is able to operate outside the atmospheric shell of our globe. Three of them saw operational use in the last war: turbo-jets in some hundred fighters, pulse-jets in some ten thousand V-1's, and liquid fuel rockets in almost 5000 V-2's. Hence the reaction power plant, this newcomer in aircraft engine techniques, has proved its reliability and its suitability for mass production.

At the end of the war, the state of the art of the turbo-jet, pulse-jet, and liquid fuel rocket was almost at the same level. But only one of these engines, the turbo-jet, very hesitatingly is presently used in commercial aircraft, despite the fact that all these reaction power plants furnish higher flying speed than any reciprocating engine with propellers. It is generally known that ram-jets will give us higher speeds than turbo-jets, and rockets even higher speeds than ram-jets.

What, then, is the reason why let's say commercial air traffic is so reluctant in applying these new drives? Are rocket-driven commercial airplanes technically feasible? Are they safe? And, above all, are they economical?

In spite of the fact that from a weight-thrust viewpoint rocket engines are the lightest, that their efficiency increases with altitude, and that they operate best with high flying speed, I think it is correct to assume that the rocket power plant, up to now, has not been regarded as a suitable drive for commercial aircraft for economic reasons. The specific fuel consumption of rocket power plants at maximum continuous thrust is so poor that no one could expect them to compete from an economic standpoint.

Now, if we assume for a moment that high flying speed is really something for which we should strive, then we have to prove that a rocket-propelled commercial airplane is technically feasible, and that economy will be attained by the proper selection of flight courses and flight paths.

Let's first have a look at the present trend in commercial airplane development. If the British hadn't developed that turbo-jet driven Comet, I am convinced we would not, even today, be considering turbo-jets as power plants for our commercial airliners. We would stick to our reciprocating engines and propeller-driven aircraft, increasing their flying speed every five years about 50 miles an hour. And we would be quite satisfied with that tremendous progress.

Now, however, we will have in three to four years the first turbojet or turbo-prop-driven airlines flying at 500 to 600 miles an hour. Maybe in 15 years from now commercial aircraft will approach the speed of sound and some 10 years later we will have the first supersonic commercial airliner. And we will be very proud of this achievement, also.

Then approaching the so-called "temperature barrier", we will encounter the "big stop" in our ever-climbing flying-speed curve. Here, by air-friction heat, the shell of our airplane becomes so hot that we will have to give thought to new approaches from the material, design and structural viewpoint. Having solved these problems we will slowly increase our flying speed up to the limit of turbo-jet performance. Finally, perhaps in 50 years from now, we may use boosted up, ram-jet-driven aircraft and will fly with a maximum cruising speed within the atmosphere of almost Mach No. 4.0 - 2600 miles an hour or almost four times the speed of sound - which presently seems the upper limit obtainable with air-breathing engines.

But we will achieve that goal only if we don't run into another imaginary "barrier" like the "sonic" and the "temperature barrier". (Prediction: Turbulence barrier)

Deliberately, I used the word "imaginary". Any aerodynamic engineer knows that approaching Mach No. 1 (speed of sound) within the transonic range, we get a sudden increase of the drag coefficient. In our airframe we get a lot of buffeting. We get shock waves attacking our structure, causing it to flutter and imposing a lot of additional load. But the same aerodynamic engineer also knows that after passing this zone of danger, in the supersonic region, the drag coefficient goes down again, that the shock waves now behave more reasonably, and that any aircraft flying at such speed regains a reasonable degree of maneuverability.

This fact has been proven ever since guided missiles entered the picture 10 years ago. They fly with speeds several times the speed of sound, and, wonder of wonders, they are guided and they maneuver. They don't mind the sonic barrier. The reason! They are from the very beginning designed and built to overcome this known trouble spot. Only a fool tries to maneuver his aircraft extensively in the transonic range, thus deliberately increasing the already tremendous load on airframe and structure.

Driving straight through this transonic range is the best course we can take. Furthermore, no missile starts burning up when entering the so-called "temperature barrier". Such missiles from the very beginning are designed, material and structure-wise, to withstand high temperatures created by air-friction heat.

I believe that what is valid for the guided missile - a vehicle originally designed according to aircraft principles - should be valid for the high-speed, passenger-carrying aircraft, too, when there is a need for it. All we have to do then is to use in the design of such an aircraft the experience gained with guided missiles. We have to build the future rocket-propelled, high speed aircraft with guided missile characteristics, and we have to fly it on a flight path similar to that of long-range guided missiles. Only this way it seems feasible to fly at the upper border of the atmosphere with manned vehicles.

But this flight at the upper border of the atmosphere obviously implies operation in an extremely low density atmosphere and requires extremely high flight velocities - at hypersonic Mach numbers - to obtain the dynamic pressures necessary to support the aircraft. These extreme conditions bring many new considerations compared with flight we normally encounter.

For flight at altitudes of 100,000 feet and higher, air-breathing engines are useless. Big and heavy liquid rocket boosters with rocket motors in cluster arrangement, which are released or dropped after their fuel is exhausted, will bring the hypersonic aircraft to its operational altitude.

The characteristics of the rocket engines and especially of the multi-stage principle are such that to obtain long-range performance a boost-glide flight path seems to be the best. That means that such a hypersonic airplane of the future attached to its boosters either in piggy-back fashion according to Professor Crocco or on top of the boosters according to conventional multi-stage rocket design, will take off vertically like a multi-stage rocket. Guided-Rocket characteristics will predominate from take-off during the reflection of the aircraft into a curvilinear trajectory up to burning cut-off of the last stage. Aerodynamic characteristics will predominate after the release of the boosters, the hypersonic aircraft then being in a horizontal position. When the propellants of the booster stage or stages are exhausted, they are separated, expended or flown back to the take-off base and landed. The final stage, the hypersonic aircraft, then glides along a maximum lift-drag ratio path to its destination over a range dependent on its initial velocity at burning cut-off. The hypersonic airplane flying at the upper border of the atmosphere will be a rocket-boosted glider.

At such hypersonic flight the stagnation temperature of the air is many thousands of degrees above ambient air temperature. This temperature may be closely approached by the air in the stagnation areas at the aircraft leading edges and in the boundary layer flow adjacent to all exposed surfaces. Heat transfer from these areas to the aircraft skin is very critical and is probably the dominant problem of hypersonic flight. It is presently felt that the heat transfer can be predicted with some confidence up to approximately Mach No. 10. Above this Mach number the inherent temperatures are so high that chemical changes in the air are probable, i.e., dissociation of the air molecules, which will produce a very different set of physical properties. The quantitative effect of dissociation on the boundary layer flow and the heat transfer process remains largely to be determined. At the present the overall effects are thought to be small. Because the heat transfer increases rapidly with local air density and angle-of-attack of the aircraft surfaces, it will be necessary to carefully control both the aircraft design and the flight path in this respect. Maneuvers such as turns or pull-ups will increase heat transfer and should be kept to a minimum. Low wing loading is desirable as it will allow flight at correspondingly higher altitudes (low densities) and minimize aerodynamic heating.

Flight in low air densities and at high Mach numbers produces several new gas dynamic phenomena. The usual concept of a very thin boundary layer about the aircraft no longer applies. The boundary layer thickens to appreciable size with respect to the aircraft dimensions. It interacts with the local stream flow and the shock waves enveloping the aircraft to produce increased surface pressures, skin friction and heat transfer. It might be visualized as effectively thickening the aircraft dimensions, particularly near the leading edges. The concept of air as a continuous fluid also begins to break down; the action of the individual air particles on the surfaces begins to become important. This brings us into the field of super-aerodynamics and the consideration of slip and free molecule flows.

The range that can be obtained is directly dependent on the aerodynamic lift/drag ratio of the design. Prediction of the lift/drag ratio is subject to several new gas dynamic problems as indicated previously but lift/drag ratios on the order of 4 to 5 seem possible at hypersonic speeds. These are small L/D's compared to many present-day aircraft (for which L/D is 10 or better), but it must be remembered that in flight about the earth, i.e., in a great circle about the earth's center, there is a large centrifugal force which opposes gravity and effectively increases the L/D. For example, at 20,000 fps (about 13,600 mph) this centrifugal force reduces the apparent gravity to approximately 40% of actual sea level gravity, and the effective L/D would be 10 for an aerodynamic L/D of 4.

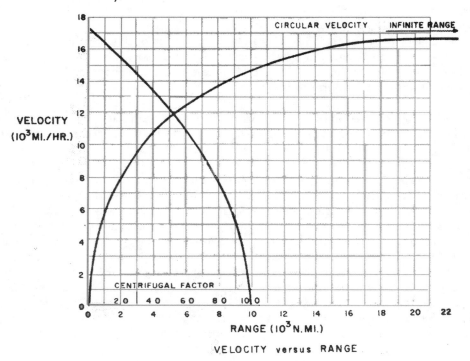

VELOCITY versus RANGE

At the very high speeds one must also pay very close attention to the mechanics of the flight path. Effects generally disregarded in present low speed practice produce significant accelerations that must be accounted for. As an example, because of the earth's rotation, flight to the east over the equator at 20,000 fps initial glide velocity will give about 2,000 miles greater range than flight to the west from the same initial velocity. Flight across the equator or the poles will require constant application of side force to the aircraft if it is to remain on the great circle between the launch and landing points because of the Coriolis effect.

Control of the boost stage configurations and the aircraft over the great range of flight conditions encountered in boost and glide will require a combination of reaction and aerodynamic control. During the boost phases control will be predominately through swiveling rocket engines; during the glide, aerodynamic stability and control, probably with the aid of artificial damping, appears feasible as long as the dynamic pressure is sufficient - say about 10 psf. The concern is for sufficient dynamic pressure. (The aerodynamic forces are not large as one might naturally think at these hypersonic speeds. At hypersonic speed at that altitude they may not be as large as if we fly with 300 knots at sea level.) Beyond this, i.e., when satellite speeds are approached and equilibrium altitudes are such that "q" is very small, some type of reaction control will again be necessary. Here it could be by gyroscopic or inertial means as well as rocket thrust.

Figure on opposite page shows a few of the aerodynamic limits with regard to altitude and speed.

Control: This curve corresponds to a 10 lbs/sq ft dynamic pressure. NACA has suggested this as the lower limit which will still permit use of aerodynamic control. Flying above these curve other types of control, for example jet control have to be used.

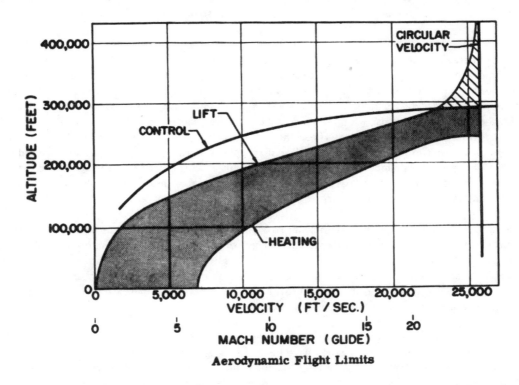

Aerodynamic Flight Limits

Lift: This curve corresponds to a low static wing loading of 10 lbs/sq ft. (It also includes reduction in effective gravity due to centrifugal force.) Note that approaching circular speed effective gravity becomes zero and the lift drag ratio loses its significance.

Heating: This curve; illustrates a typical flight path limitation with respect to aerodynamic heating. It corresponds to a constant 1600°F outer surface temperature at a point one foot from the leading edge of the wing. Flying below that curve the temperature of the wing becomes too high.

Considering these three limitations, the shaded area represents a useable region in which we will be able to fly.

As mentioned previously, an airplane traveling at hypersonic speeds will, in essence, be surrounded by a layer of kinetically heated atmosphere. The temperature of this layer rises as the square of the Mach number, and so at speeds of Mach 16 - 22, the temperature of this surrounding air will be about 20,000°F. This does not mean, however, that the temperature of the structure will attain these boundary layer temperatures since most of the heat will be re-radiated by the structure to the atmosphere. A typical temperature for most of the structures at the altitudes and speeds would be on the order of 16 - 1800°F. These temperatures are not prohibitive for the high strength alloys which have been developed recently. Certain portions of the structure, however, like the leading edges, will experience higher temperatures than those indicated, but there are means envisioned and under development and test which provide promising solutions.

However, we have not yet all the answers. I would like to explain some of the problems we have in connection with the development of such structure.

(1) We have always considered weight to be an extremely important factor in aircraft design and operation. With the use of a rocket power plant, the significance of weight, for instance, on the overall size or gross weight of the aircraft is exceedingly more impressive; for instance, one pound of added weight would add approximately 50 lbs to the gross weight of a three-stage vehicle. If weight is not controlled very carefully and means found for reducing structural weight, the proportions of such air-

craft would soon be beyond reason. This fact gives the structural designer greater impetus toward invention.

(2) A multi-stage rocket powered hypersonic aircraft, because of its power plant characteristics and for the purpose of economy of operation, is subject to relatively high longitudinal accelerations in take-off. This results in a high rate of change of the boundary layer temperature which in turn produces a high rate of heat flux to the structure. Since conventional aircraft structure is composed of stiffened thin shells, thermal gradients exist within the structure wherever there is a local thickening of the skin such as at stringers, stiffeners, or frames. These thermal gradients, because of the expansion of metals with temperature, result in thermal stresses sufficient to buckle, yield and distort the structure without the addition of externally applied loads. This fact alone requires the structure of the vehicle to be different from what has been used in the past and the change is a sizeable one, comparable to the change from wooden and fabric aircraft to metal construction.

(3) Suppose we were to ignore the thermal stress problem, consider the use of conventional aircraft construction for this vehicle and estimate the weight of the structure as a function of one to operate under ambient conditions, we would find at temperatures of 16 - 1800°F, (considering the reduction in allowable strength and including that due to creep), that the elevated temperature structure would weigh on the order of 150 - 200% more. Obviously, this is a sizeable increase in structural weight and some other type of structural design must be utilized.

For many years now, many factories have been studying the problem of structural design at hypersonic speeds in a systematic manner. Under study have been the design of thermal stress-free structures, the use of insulation, and the mechanical cooling of the structure. Each one of these methods does not lead to obvious designs; for example, it is not obvious how to make an almost completely thermal stress-free structure or how to cover the structure with insulation so that it can withstand the erosion of a hypersonic flight or be sufficiently small in thickness so as not to occupy valuable structural space. However, the use of insulation and mechanical cooling of the structure appears to be quite promising since a great portion of the airplane such as propellants, passengers, crew and equipment must be maintained at or near ambient temperature conditions anyway. This problem will be eased somewhat since the time of flight over any distance on this globe will be on the order of about one to two hours. The weight of insulation and mechanical cooling is time-dependent. The hypersonic structure can then be described as one which, in addition to forming and maintaining the shape of the aircraft and carrying the loads imposed upon it, also provides a tolerable environment for the pilot and the instruments.

When we pursue this course, I predict that rocket power will become commercially available to the public as a drive for a new kind of plane destined to fly with fantastic speed along intercontinental routes. Rocket power is here to stay, and not only as a means of modern warfare. Stretching along far-flung paths through the upper atmosphere at the border of space, carrying their passengers safely around the globe in a matter of hours, these ultra-planes will make rocket flight a profitable business.

Passenger carrying rocket planes will be no small stuff. They will never be used for such limited distances and in such great numbers as present-day propeller planes. Like the gigantic ocean liners, there will be only comparatively few, and used only for the longest routes on our planet. They will essentially serve inter-continental connections and may run on a few transcontinental routes.

In contrast with conventional propeller and jet planes, the rocket plane will not be driven continuously, but will be accelerated for the comparatively short period of a few minutes. During this time it will attain a speed which hurls it in free flight over distances of thousands of miles. From the great altitude reached at the end of the propulsion period, the plane begins its long glide path, which may last one hour or more, depending on its initial velocity and altitude. Despite its brevity, the propulsion period will not be uncomfortable for the passengers.

Much has been written about the frightening effects of the high acceleration to which the crew of a manned

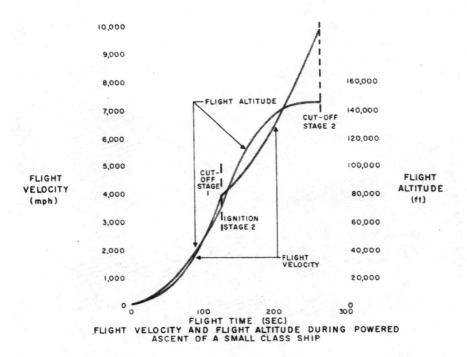

FLIGHT VELOCITY AND FLIGHT ALTITUDE DURING POWERED
ASCENT OF A SMALL CLASS SHIP

rocket vehicle would be subjected. This is not true, since large rocket-driven vehicles actually must take off at a very low acceleration, in order to avoid use of an excessive number of motors and too-heavy propulsion systems, and in order to avoid high air-friction heat because of excessive speed within the denser layers of the atmosphere. It is true that the efficiency of rocket drives goes down as the initial acceleration is reduced, but the increase in propellant consumption is always cheaper than the use of much more expensive hardware. It is more important, however, that in rocket planes, which glide for most of their path, the gliding weight should be as small as possible in order to obtain maximum range with the smallest wing area. Limitation of the wing area is desirable in order to keep the structural weight down as much as possible. You can assume that at the moment of take-off your weight would increase to something like 25 percent above your normal weight. At termination of burning, the point of cut-off, your weight

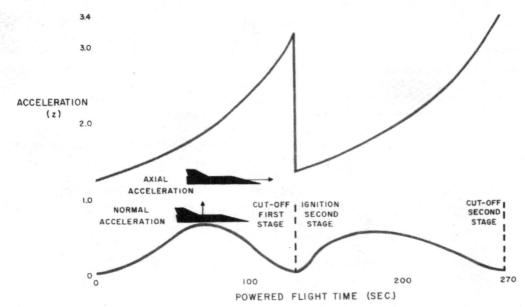

AXIAL AND NORMAL ACCELERATION IN SHIP OF THE SMALL CLASS

would then be about 3-1/2 times normal. Although not everyone, particularly persons with heart ailments, could stand this increase, it is certainly more tolerable than the famous 10 g's (10 times your weight) or more which seems standard in every science fiction movie, which gives the spectator a creepy sensation as he sees the distorted faces of nearly crushed human beings.

"10 g's"

In contrast with the conventional plane, the reliance of the rocket plane on air for flying depends on its flight speed. At moderate speeds - and even 3000 miles per hour is still moderate in this connection - air is needed to support the plane's weight during the glide. But as the speed increases, a new and significant effect makes itself felt. This is the centrifugal effect caused by the fact that the vehicle must follow a curved path due to the earth's gravitation. (Consequently, the same force which causes the passenger in a car to be pushed in a direction opposite to the direction of turning of the car in a curve is the very same force which pushes the rocket plane away from the direction of path curvature, that is, away from the earth into space.) As a result of this apparent support of the rocket plane's weight by the centrifugal effect, air is needed only to support a portion of the weight of the vehicle. The higher the speed attained by the rocket plane, the larger is the portion of its weight which, in effect, is supported by the centrifugal effect. To a lesser and lesser degree air is needed for the flight. Ultimately the ship is kept in its path exclusively by the centrifugal effect. That means we are in a satellite orbit. Air is no longer needed for lift and therefore is completely dispensable. Moreover, at this speed of about 17,400 mph, it would only impede the rocket plane's motion.

This shows that the rocket plane is the link between aeronautics and astronautics. The way to space flight logically leads via the rocket aircraft as a vital intermediate step. With the rocket plane, a generation of pilots will grow up and be educated to possess much more understanding, experience and training for space flight than we can imagine today. For them space flight will not be something unbelievably new and mysterious, but more a matter of course.

But before the first large rocket plane will take off, science and technology must close many gaps still existing in our knowledge and know-how of the art of flying at hypersonic speeds and at extreme altitudes.

What we must learn may be summarized in four points:

1. We must learn to develop and operate reliable large rocket power plants consisting not of one single huge motor, but of clusters of many motors.

2. We must learn to design and develop airplane configurations that operate with reasonable efficiency at hypersonic and supersonic, as well as subsonic speeds.

3. We must learn to tackle the problems connected with the simultaneous mechanical load due to air pressure and the thermal load due to the high temperature of the compressed air in the immediate vicinity of the plane, particularly its nose and the leading edge of its wings.

4. Last, but not least, we must learn to combine the principles that govern the flight of a guided missile with those applied in a piloted airplane. This is often not fully understood. As mentioned before, the rocket glider is the link between the continuously propelled airplane and the satellite. As to its mode of flight, the rocket glider stands between the airplane and the guided missile. Or, if you wish, the space ship. As I have already mentioned, even to the layman unfamiliar with airplane flight, it is obvious that acrobatic flight is possible only at relatively low speed. A jet fighter approaching the speed of sound is definitely limited in its freedom of motion. An intercontinental rocket glider at cut-off behaves more like a projectile than an airplane. This condition prevails from take-off, through power cut-off, into the glide phase, until the flight speed is reduced to about 2700 mph. During this time the pilot can only supervise, monitor and adjust the automatic navigation, stabilization and control equipment which is very much like that of a guided missile.

When the rocket plane's speed drops below 4 times the velocity of sound, airplane conditions are approached again and soon thereafter the pilot can take over in a more or less conventional manner. It would be a fallacy to believe that a plane traveling 12 times the speed of sound can be flown like an aircoach. The manned rocket glider can be operated only as a combination of airplane and guided missile, and the duties of the pilot will be comparable with the duties of the engineer in a railroad locomotive bound to a given track.

Years before the first commercial rocket plane will take off, smaller planes will have probed the high altitude atmosphere, and developed a pattern of environmental conditions which makes flight at the border of space safe and comfortable for persons of normal health, at least for the short time of its duration.

The coming decade will witness the development of rocket planes flying faster and higher than today's advanced rocket planes, such as — the X research series of Bell Aircraft-Corporation and the Douglas Skyrocket. The bigger planes will no longer be carried up by a "mother" plane or take off like a conventional Jet plane; they will take off vertically like the V-2's or other ballistic guided missiles. The flight path first will be straight up. Shortly after take-off they will follow a curved path which, at the end of the propulsion period, will deliver them in horizontal position for the subsequent glide. Such planes will be limited to a certain speed capacity which may be in the neighborhood of 4500 mph or 3850 knots. That however, is not sufficient even for a glide over the relatively modest distance from New York to San Francisco, about 2000 miles apart on the great circle route. For such distance an initial speed of about 7400 knots is necessary.

The additional speed is attained by attaching the plane to a so-called booster, that is, to a bigger rocket which carries the plane up to a certain speed and altitude and then is separated after burn-out.

The plane then adds its own velocity increment to that imparted to it by the booster.

In order to simplify and facilitate the handling of the combined set, the passenger-carrying rocket plane will not be mounted on top of the booster, but ride on it, piggy back fashion. If both are erected vertically, the final stage will be attached to the side of the booster. This limits the height of the ship and makes it possible to give both vehicles an aerodynamical shape, enabling them to fly as individual airplanes if the booster stage is also equipped with wings.

During take-off, the engines in both rocket planes burn simultaneously but the engines of the second stage are fed with propellants from the booster plane. The booster in turn saves weight by using the engines of the second stage. The booster plane has its own crew which, after separation, pilots the ship back to the airport from which the combination took off. Thus, the booster stage is not only salvaged, but brought right back to its take-off point where it can be readied for a new flight while the passenger-carrying second stage continues on to its destination.

Imagine such rocket planes operating commercially at the turn of our century. A thin, but far-reaching net of flight routes will lead across the globe. Assuming no major shifting of cultural and civilizational centers by that time, the rocket sky lanes will radiate from two major centers - the United States and Europe - to all continents. In the United States, the east and the west coasts will have one rocket airport each, say, New York and San Francisco. Both ports will be connected by a rocket plane. From San Francisco one line will lead nonstop to Tokyo with connections to Calcutta. Another line will lead to Sidney via Hawaii. In Sidney and Calcutta transfer to other rocket liners will be possible, leading from Calcutta to London in nonstop flight.

In an eastward direction, the San Francisco rocket port will offer a flash connection to New York, from where you can fly nonstop to Rio de Janeiro, or to London, Europe's huge rocket port. Air taxi service connects each rocket port with all airports on the continent. Thus, we will have the huge Pacific triangle, San Francisco - Hawaii - Sidney - Calcutta - Tokyo - and back to San Francisco, and the smaller Atlantic triangle connecting New York - Rio - and London. The route London - Calcutta connects Europe with Asia and the route London - Cape Town serves the traffic to Africa.

With this flight pattern two groups of distances exist. One group ranges from 2000 to 3600 nautical miles, the other from 4500 to 5200 miles. Not all of these distances can be covered by the same rocket plane, since the maximum flight speeds necessary to cover these distances vary too greatly.

However, it is not necessary to have a different sized rocket plane for each route. This would make production of and facilities for servicing the rocket planes much too expensive. It is sufficient to have two types corresponding to the main groups of distances. Variation of the necessary flight speed within each

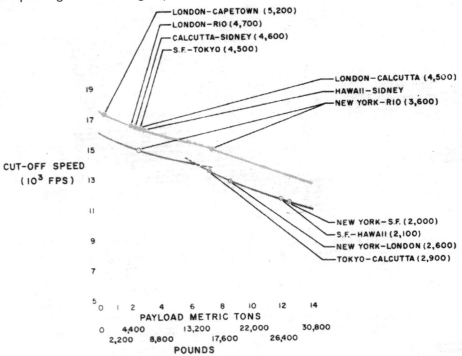

Cut-Off Speed versus Payload

group is accomplished by reducing or increasing the amount of payload carried. This performance of rocket vehicles is more sensitive to changes in payload weight than that of conventional airplanes. The maximum payload weight for a flight from New York to San Francisco can be more than four times as much for the same rocket plane as for a flight from New York to Rio de Janeiro. For a flight from Hawaii to Sidney, this rocket plane could not take any payload at all, and therefore a larger vehicle is required on the long routes.

For economical reasons we will have a smaller class of rocket planes with a take-off weight of 600,000 pounds and a larger class with a take-off weight of 770,000 pounds.

The reduction in flight time accomplished by these rocket planes is spectacular. New York is but a mere 1-1/2 hours away from San Francisco, and even the long flight from Hawaii to Sidney, Australia, takes only a little more than 1-1/2 hours.

This must be paid for, however, with a lot of propellants. The small rocket planes consume about 57,000 gallons of liquid oxygen and gasoline during each ascent. Those of the larger class do the job with 75,000 gallons, their flight range being about 5200 miles at the most, with 1000 pounds payload from London to Cape Town.

In comparison, the large Convair B-36 high altitude, intercontinental bomber, driven by four jet engines and six propellers, has a fuel capacity of 30,000 gallons gasoline for a range of 10,000 miles. If one subtracts the liquid oxygen from the 75,000 gallon capacity of the larger class of rocket planes, the

STAGE	VEHICLE	DATA SMALL 1	CLASS 2	LARGE 1	CLASS 2
WEIGHT PAYLOAD		110,000	8,800	139,400	8,800
DRY	(lb)	124,000	33,000	147,000	36,400
PROPELLANT	(lb)	360,000	66,000	477,000	90,000
AUXIL FLUIDS	(lb)	6,000	2,200	6,600	3,400
GROSS	(lb)	600,000	110,000	770,000	138,600
THRUST, TOTAL	(lb)	750,000		900,000	
STAGEWISE	(lb)	600,000	150,000	730,000	170,000
NO. OF MOTORS		5	3	6	3
PROPELLANT		LIQUID OXYGEN - GASOLINE			
SPECIFIC IMPULSE	(sec)	270	310	270	310
CHAMBER PRESSURE	(at)	70	50	70	50
EXIT PRESSURE	(at)	0.7	0.1	0.7	0.1
NOZZLE AREA-RATIO		11.8	35.6	11.8	35.6
BURNING TIME	(sec)	130	136	143	161
DEFLECTION PROGRAM		90°-10°	10°- 0°	90°-10°	10°- 0°
INITIAL		1.25	1.36	1.16	1.25

FLIGHT SCHEDULE

FROM	TO	DISTANCE ALONG GREAT CIRCLE (n.mi.)*	HIGHEST SPEED (kts)**	NON-STOP FLIGHT TIME RS hr min	JET*** hr min	PROP[4] hr min
NEW YORK	SAN FRANCISCO	2,000	7,400	1 15	4	6 42
NEW YORK	LONDON	2,600	7,340	1 15	5 12	8 40
NEW YORK	RIO de JANEIRO	3,600	8,880	1 20	7 12	12
SAN FRANCISCO	HAWAII	2,100	7,270	1 15	4 12	7
HAWAII	SIDNEY	4,400	9,760	1 38	8 48	14 40
SAN FRANCISCO	TOKYO	4,500	10,200	1 30	9	15
SAN FRANCISCO	SIDNEY	6,300	11,210	1 37	12 36	21
TOKYO	CALCUTTA	2,900	8,640	1 19	5 48	9 40
LONDON	NEW YORK	2,600	8,290	1 17	5 12	6 42
LONDON	RIO de JANEIRO	4,700	9,900	1 38	9 24	15 42
LONDON	CAPETOWN	5,200	10,200	1 30	10 24	17 18
LONDON	CALCUTTA	4,500	9,000	1 22	9	15
CALCUTTA	SIDNEY	4,600	9,240	1 24	9 12	15 18

THE DIFFERENCES IN GREATEST SPEED AT EQUAL OR SIMILAR DISTANCES ARE DUE TO
THE EARTH'S ROTATION. LESS SPEED IS REQUIRED FOR SAME DISTANCE WHEN
FLIGHT IS EASTWARD RATHER THAN WESTWARD.

*   n.mi. = NAUTICAL MILES
*** BASED ON 500 KNOTS

** kts = KNOTS = NAUTICAL MILES PER HR
4) PROPELLER, BASED ON 300 KNOTS

remaining quantity is exactly the same amount of gasoline as for the bomber, 30,000 gallons. However, the range is only half of that of the conventional plane, while the flight time on a route such as London-Cape Town is only about one-sixth of that of the jet plane.

In other words, for this speed the passenger would have to pay for twice the amount of gasoline at equal flight distance than in a jet propelled plane and, in addition, for the oxygen which is not expensive when ordered in large quantities. Under these conditions a ticket from New York to San Francisco in the small rocket plane carrying about 20 passengers would cost about twice as much as it costs now.

Let's have a look at how such a future rocket plane will take off. The rocket plane will be launched from a so-called "crater", a mammoth circular concrete pit connected by 60-foot deep trenches called canyons. The word "canyon" will become a professional term like Gate No. so-and-so of our present airports. These canyons run straight from the maintenance hangars to the crater, which is on the same level as the canyons.

The Rocket Plane

In huge hangars each rocket plane, consisting of two independently powered stages, is assembled in the "pit" from which the canyon originates. Standing on a specially-shaped movable splash plate, the empty rocket plane is pulled by a tractor along the canyon to the crater. Thus, the different sections of the ship are always easily accessible from gangways in the wall of the canyon, and the rocket plane with its large wings is protected from gusts while it takes off out of the crater.

Upon arrival in the crater, the ship is turned so that the edges of its wings are directed into the wind in order to avoid side movements which might damage the ship on the crater walls when pulling out into the

Boarding the Plane

open on stormy days. Since the rocket plane ascends very slowly at the beginning, the first few feet after take-off are the most critical ones. After having reached a safe distance from the ground, the rocket plane will be rotated, if necessary, into its correct direction of departure.

Arriving at the crater, the ship is filled with propellants from large tanks buried close to the rim. Not until the fueling is completed and the vehicle is ready for take-off do the passengers arrive by bus from the main building a few miles away. A small elevator carries them 20 feet underground near the rim. Through a short tunnel section, they arrive at a platform projecting from the vertical crater-wall like a balcony. They are now at the same level as the entrance hatch through which they can board the final stage of the rocket airliner over a gangway. The passenger space is in the rocket plane's midsection, close to its center of gravity to secure smooth riding.

This also allows putting the tanks so as to reduce changes in the rocket plane's flight stability when emptying the tanks during powered flight, and when passing through a wide range of flight velocities in the subsequent glide phase. In front of the passenger space is the fuel tank, pipelines from which run through a special conduit to the tail section which contains the liquid oxygen tank and the engines. The pilots compartment is in the nose section.

The rocket plane takes off vertically because it is far too heavy to attain sufficient speed for a normal take-off on a runway. But soon after having left the ground the rocket plane begins to tilt toward a horizontal flight position, a process which is not completed until the final stage has attained its maximum speed. Therefore the passenger seats are tiltable to counteract the rocket plane's motion.

The rocket plane will be impressive, black-colored, towering some 90 feet above the splash plate. At the side of the huge booster is the final stage, some 60 feet tall. This stage will carry the passengers to their destination, while the booster plane returns to its takeoff base. Nothing is expended except fuel.

Now, fasten your seat-belts, and let's make a trip with such a rocket plane, say, a trip from San Francisco over Hawaii to Sidney, Australia.

We board the rocket plane through the entrance hatch at the upper end of the passenger section. A light elevator runs down to seats in tiltable plastic bowls on the left and right, one below the other at the presently erect position of the rocket plane. The plastic bowls will rotate about horizontal axis during the flight, so that the passengers will always sit in the most comfortable position. The hostess helps us to get into our seats and to fasten our seat-belts. After we are all bundled up, our pretty hostess may welcome us on behalf of her company and may give us some interesting information. But, sorry! The stewardess will not accompany the ship on long routes. Limitation of the payloads forbids that, and no hot meals will be served because of the short flight time.

Let's assume the hostess will give us a welcome speech. It may run like this:

"In exactly five minutes we will take off. The motors of the booster plane as well as of our plane will operate, that is, burn, together. Our motors are being fed by the booster tanks until separation of the planes, so that our propellant supply will not be exhausted prematurely. The booster has five motors; our plane operates with three motors. Together, they deliver a thrust of some 760,000 pounds. You will not be bothered by excessive noise from the motor section, since the oxygen tank in between will absorb most of the sound and vibration.

Passenger Cabin

"At take-off you will be pressed into your seats with a 25 percent greater weight than you have now. During the first propulsion phase your weight will slowly increase up to three times your normal weight at the moment of separation, which occurs 130 seconds after take-off. Thereafter, you will become lighter temporarily. During the second burning phase of our motors alone, you again will gradually gain weight until, immediately before attaining our maximum speed and the cut-off of our engines, you will weigh neatly-3-1/2 times your normal weight.

"Then, comes the fastest reducing job you have ever experienced in your life. Your weight will drop to 75 percent of what you now weigh. Gradually, as the plane loses speed, your weight will become normal.

"Since we at no time follow a free-fall path like a projectile or an artificial satellite, we will never be weightless.

"At the beginning of our free glide we will be 145,000 feet or 29 miles high. You will notice that a dark glass covers the inside of your normal window. It serves to protect your eyes from the radiation of the sun at this altitude. You cannot shift this inner glass aside until its holding mechanism is electrically unlocked by the pilot. This will be done at a lower altitude where, your eyes are no longer endangered. You will be able to observe the beautiful panorama of space through the window. Our large triangular wings do not permit any observation of the earth underneath, but you will see the horizon.

"Finally, one important factor. The company is proud of its safety record and takes every conceivable precaution to further increase your safety during the flight. Our rockets have outstanding reliability. However, should trouble occur in the motor section during the burning phase, the pilot will jettison the tail section containing the oxygen tank and the motor section and will dump the gasoline out of the fuel tank. The rear portion of the wing, doweled into the forward section, will be separated with the tail section.

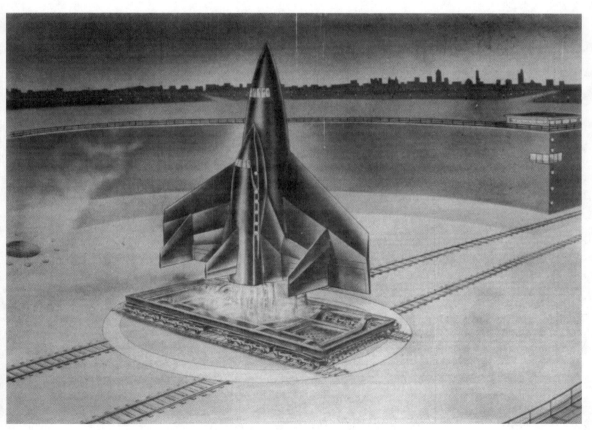

Take-off

Release

"With the front portion of our plane, we will glide to safety without your having to leave your seats.

"The burning of the motors last only 4-1/2 minutes. The remaining hour and 10 minutes of the flight will be very comfortable and quiet. I hope you will enjoy your trip very much. Thank you!"

Our entrance hatch is closed. Outside they retract the telescope-type gangway leading to the rocket plane, retreat from the platform into the tunnel and close heavy steel doors. The usual "No Smoking" and "Fasten Your Seat-Belts" signs appear; then a red sign flashes up: 3-2 -1 go!

Beneath us eight huge rocket motors ignite. The fire which furiously erupts from their nozzles illuminates our wings and the walls of the crater. It is early in the morning. The sky already has the blue transparency which heralds the imminent sunrise. Yet we, in the crater, are still in the deep shadows until the flames shed their weird light over us.

Then we are captured by a new strange sensation. A fantastic power tightens its grip on our vehicle and pushes this 600,000 pound tower upward, slowly, but irresistibly. We rise above the crater into the cool morning sky. The earth falls away beneath our blazing jets. As we gain speed and altitude, our rocket plane begins to tilt into horizontal position. But still we are climbing. Our seats tilt in the opposite direction so that we always have the most comfortable position, with our bodies pressed into the seats.

The rocket plane's acceleration is too small to be disturbing. At the moment of take-off a clock began to count the seconds. Continuously the big hand moves toward the 130 second mark. We are still climbing at an angle of 10 degrees with respect to the horizontal, 70,000 feet up and 30 miles west of our launching site.

The moment of separation has arrived. This is the most exciting event of the whole flight, not only for the passengers, but also for our skillful and extremely well-trained and selected pilots. The pilots in the passenger ship and in the booster stage exchange remarks via intercom. Our pilot actually is in control of both planes until the booster pilot takes over his plane after separation.

Our pilot tilts both ships, nose down, into a slightly negative angle-of-attack, in order to facilitate separation. The motors of the booster plane have ceased to burn. Then he releases the holds which kept our plane rigidly connected with the booster plane. Rapidly, our rocket plane glides along rails over the back of the booster plane. As soon as we are separated, our plane again assumes a positive angle-of-attack, turning its nose skyward, while the booster plane remains in its position. Drag and thrust quickly drive us apart to a safe distance.

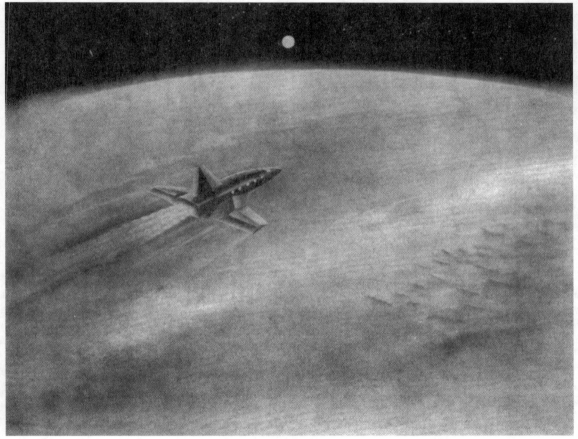

Passenger Ship

Then we lose sight of the booster plane which seems to fall back into the bottomless depth beneath us. We climb - 80,000-100,000-120,000 feet - as we pick up speed along our more than 150-mile-long runway into the sky. Finally, 136 seconds after separation, we are 140,000 feet high. The motors are cut off. At the fantastic speed of 7,300 nautical miles per hour we begin our far-flung glide path.

We are light again. The sudden silence is strange, maybe even somewhat boring. It is apparent that, in spite of all the sober routine action, this is more than an ordinary airplane flight. The earth is far beneath us, strangely remote it seems, but space is near. The early morning hour was selected as the time of departure for our western flight to keep the sun behind us as much as possible, in order to avoid excessive glare and still give the pilot in the nose section good visibility with respect to the earth.

In the darkness of space above and beside us, the moon and the stars are shining with a brightness

Take-Off Site

unknown to the dust bound human of the past. The outside air pressure is but one five-hundredth of what it is on the ground. The tremendous speed of our rocket plane still gives it a substantial lift. But otherwise we do not feel the speed. The earth seems to rotate only very slowly underneath us. We feel like being in space, and this is the most unforgettable experience of our trip.

Time passes quickly. Gradually the sky begins to brighten and the stars begin to fade. Under the irresistible pull of our planet we are losing altitude. The Hawaiian Islands appear on the forward horizon, green and yellow spots in the bluish infinity of the ocean. A few minutes later we touch the ground at the Honolulu rocket port, with 100 miles per hour landing speed.

There, we are informed that we have a half-hour for a light breakfast before we can board the bigger rocket plane which will carry us over the 4000 miles to Sidney in a little bit more than 1-1/2 hours. Larger than the rocket plane which brought us to Hawaii, our new vehicle flies even higher and faster, but for the passenger the trip will be little different. The main difference which we will feel is that we get a trifle heavier at the end of each burning time, and that the passenger-carrying capacity is notably smaller.

At 6 o'clock in the morning, Pacific Standard Time, we started from San Francisco. At 9:30 o'clock, Pacific Standard Time, we arrived at Sidney. If we did not have to add one full day when crossing the International Date Line from East to West, it would have been a real fast trip.

From Sidney, London is only 3-1/2 hours flight time away, not counting the layover time in Calcutta.

This generation, for the first time in the history of mankind, will obtain a practical conception of the earth as a small unit in space. From these commercial rocket airliners, the transition to the manned artificial

Landing at Honolulu

earth satellite, which everybody is talking about will be only a very small step.

 The first rocket ships bringing us up to a satellite orbit are only little more than twice as big as our rocket planes in which we made a trip from San Francisco to Sydney in 3 hours.

Then, we finally approach the turning point in the life of mankind, where the sky is no longer "the limit". That this holds true will become evident to every rocket plane passenger on our intercontinental routes, with planes flown by men who are space-conscious. Altitudes of 10,000 feet and speeds of 11,000 miles an hour are for them a matter of mere routine, just as 30,000 feet in altitude and 300 miles per hour were to their fathers 30 years before.

HEADQUARTERS
AIR RESEARCH AND DEVELOPMENT COMMAND
UNITED STATES AIR FORCE
Post Office Box 1395
Baltimore 3, Maryland

ADDRESS REPLY TO
COMMANDER, ARDC, ATTN:

RDZPR   24 October 1957

SUBJECT: Meeting on Dyna Soar I

TO: National Advisory Committee for Aeronautics
          Attention: Mr. Clotaire Wood
          1512 H Street, Northwest
          Washington 25, D. C.

        The inclosed Abbreviated Development Plan (Weapon System 464L)
has been approved by ARDC and sent to Headquarters USAF for a decision.
This copy is provided for your information and retention. Please
destroy the five copies of the rough draft edition we left with you
in August (ARDC control number, C7-115361). Our next step is to
determine courses of action which will be taken when the Air Force
decision is made. A meeting with Detachment No. 1 at Dayton is
scheduled for 7 November to discuss a preliminary work statement for
Dyna Soar I, management responsibilities, and selection of a contractor
for Dyna Soar I. Your participation in the discussion would be appreciated. I will call you about the 30th with detailed information
about the meeting.

        FOR THE COMMANDER:

1 Incl
  Abbr Dev Plan, WS 464L,   CHARLTON G. STRATHY
  (S)   Lt Colonel, USAF
                                   Chief, Research & Target Systems Division

                                   C7-115361

INTRODUCTION

This abbreviated plan is ARDC's recommended approach to investigating a new concept in weapon systems- boost glide. The plan specifically applies to GOR 92.

The boost-glide concept was conceived and partially tested by the Germans in the early 1940's. In 1948 Rand published studies on boost, glide and in 1951 industry began to show interest in its application to weapon systems. In 1954 ARDC and NACA decided that sufficient information was available to warrant assessment of its technical feasibility. ARDC and industry have spent $8 million in the last three years, in addition to the extensive NACA effort, to conclude that boost glide is a feasible weapon system concept.

This development plan consolidates the previously proposed Brass Bell and HYWARDS Systems and discusses the potentiality of a boost-glide strategic bombing system. It is a step by step program requiring relatively low funding during the first three years. ARDC recommends approval of pre-Phase I and Phase I of the initial program (designated Dyna Soar I, see chart 23).

A comprehensive test and operational analysis of boost glide and. its application to weapon systems is contained in Attachment 3 of this plan.

C7-115361

Weapon Sys 464L Abbreviated Development Plan

TABLE OF CONTENTS

ABBREVIATED SYSTEM DEVELOPMENT PLAN

A.   Abbreviated System Development Plan

R&D PROJECT CARD	TYPE OF REPORT Proposed Weapon System	REPORT CONTROL SYMBOL DD-R&D/A/119
1. PROJECT TITLE (Conf) Hypersonic Glide Rocket Weapon System (Uncl) Hypersonic Strategic Weapon System	2. SECURITY OF PROJECT Secret 4. INDEX NUMBER N/A	3. PROJECT NO. System 464L 5. REPORT DATE 23 August 57
6. BASIC FIELD OR SUBJECT Supporting Systems	7. SUB FIELD OR SUBJECT SUB GROUP SS	7A. TECH. OBJ. SS-1

8. COGNIZANT AGENCY Hq ARDC 9. DIRECTING AGENCY Hq ARDC, RDZP 10. REQUESTING AGENCY Hq USAF	12. CONTRACTOR AND/OR LABORATORY To be determined	CONTRACT/W. O. NO.

11. PARTICIPATION AND/OR COORDINATION

NACA (P)
AMC (P)
SAC (I)
U. S. NAVY (I)
U. S. ARMY (I)

13. RELATED PROJECTS Project 7990 Task 89774	17. EST. COMPLETION DATES	
	RES.	"Cont."
	DEV.	"Cont."
	TEST	"Cont."
	OP. EVAL.	

14. DATE APPROVED	18. FY	FISCAL ESTIMATES
	58	3,000
	59	5,000
15. PRIORTY 16. MAJOR CATEGORY	60	8,000
1A A		

19. REPLACED PROJECT CARD AND PROJECT STATUS This report supercedes
DD Form 613 dated 31 Dec 56, System 459L (Brass Bell); DD Form
613 dated 28 Dec 56 System 610A (HYWARDS): and Task No. 89774
(ROBO) of DD Form 613 dated 8 Jan 57, Project 7990.

20. Requirement and/or Justification

This system is proposed to satisfy System Requirement 131, dated 14 November
1956, title:(U) Hypersonic Weapons Research and Development Supporting Systems;
GOR No, 92 (TA-4e-1-59), dated 12 May 1955, title:(U) GOR For A Piloted Very
High Altitude Reconnaissance Weapon System; and SR No. 126, dated 12 June 1956,
title classified, Unclassified short titles ROBO.

The logical development of weapon systems utilizing the boost-glide concept
encompasses the above requirements in the order mentioned. In keeping with the
philosophy of "more Air Force per dollar" and the desirability of obtaining operational weapon systems of this type as early as practicable, it is essential that.
weapon systems using the boost-glide concept be developed under a completely
integrated program. This program consolidates the DD Form 613's previously
submitted on projects. HYWARDS and Brass Bell (see Item 21e); the ARDC Form 111's
dated 29 March 1957 and 10 May 1957 on Project 7990, Task 89774 (ROBO); and the
Evaluation Report of the Ad Hoc Committee for ARDC System Requirement No. 126
ROBO dated 1 August 1957. This program is identified as Weapon System: 464L and
given the nickname Dyna Soar, which stands for Dynamic Soarer. (Confidential)

21. OASD (R&D)	SN.	CN.	C.		X.	L.	C.

DD FORM 613
1 APR 55
REPLACES DD FORM 613,
1 JAN 56, WHICH MAY BE USED.

SECRET   C7-115361   PAGE 1 OF 10 PAGES

SECRET

SECURITY CLASSIFICATION

R/D PROJECT CARD
CONTINUATION SHEET

1. PROJECT TITLE	2. SECURITY OF PROJECT	3. PROJECT NUMBER
(Conf) Hypersonic Glide Rocket Weapon System	Secret	System 464L
(Uncl) Hypersonic Strategic Weapon System	4.	5. REPORT DATE
		23 August 57

21. Brief of Project and Objective

a. Brief and Military Characteristics

It is proposed that a development program for advanced weapon systems
utilizing the boost-glide concept be initiated to take advantage of the tremendous capability and potential this concept offers in the accomplishment of
Air Force missions of the 1967 through 1980 time period and beyond.Briefly,
this concept is characterized by a manned, winged vehicle, rocket-boosted to
hypersonic speeds and altitudes above 100,000 feet, whereupon the vehicle
operates in an unpowered gliding mode, trading kinetic and potential energy for
ranges on the order of 5,000 nautical miles to 22,000 nautical miles depending
upon the mission and time period.

The hypersonic boost-glide concept offers a major technological break-
through in performance and mission capability which should be exploited in.
future reconnaissance and bombardment aircraft weapon systems.The tremendous
improvements in speed, altitude, and range capabilities attainable simultaneously
with such a system, warrant immediate initiation of a concerted development
program to provide solutions to the technical problems involved in this concept.

Knowledge of the flight characteristics and equipment operation problems
for the boost-glide flight regime is very limited.Ground based facilities
such as wind tunnels, shock tubes, ballistic ranges and small scale flight
models can provide an important portion of the required information.However,
the inability of these ground facilities to simulate simultaneously the parameters encountered in actual flight requires design and development decisions
to be made in many areas on the basis of unsubstantiated theory, inference,
and extrapolation, unless actual flight data are available.The final critical
evaluation of systems and components can only be accomplished with a full scale
vehicle flying at the proper altitudes and speeds,

The development program, described herein, is designed to introduce
boost-glide weapon systems into the USAF inventory in an appropriate time period
compatible with weapon systems existing at that time and as a follow-on to the
weapon systems becoming obsolete at that time. Considerable care has been
taken in laying out the development schedule to avoid the necessity of a crash
program in the development of boost-glide weapon systems.

(1) Dyna Soar I
     (a) General

     The first flight article proposed under the program is a

SECRET
SECURITY CLASSIFICATION

R&D PROJECT CARD CONTINUATION SHEET

1. PROJECT TITLE	2. SECURITY OF PROJECT	3. PROJECT NUMBER
(Conf) Hypersonic Glide Rocket Weapon System	Secret	System 464L
(Uncl) Hypersonic Strategic Weapon System	4.	5. REPORT DATE 23 August 57

conceptual test vehicle (see Attachment 1 for definition) with performance characteristics and capability significantly beyond those of the X-15 research aircraft system, which is designed to investigate speeds of 6,600 fps and altitudes up to 250,000 feet). The mission of this manned, hypersonic glide rocket test vehicle is to supply information and data required for the development of boost-glide weapon systems. It will provide development information on aerodynamic, structural, human factor, and component problems involved in glide flight for lengthy periods at high altitudes and speeds.In addition, this vehicle will provide for full scale testing of breadboard models of components and subsystems of boost-glide weapon systems. This test vehicle will provide a capability of hypersonic speeds and altitude commensurate with a thorough investigation of the glide rocket concept and its flight regime.

The attainment of such speeds is essential to support future weapon systems development because, in the region of Mach 10 and above, flow phenomena peculiar to the boost-glide flight regime are predicted and boundary layer temperatures reach magnitudes which cause chemical changes in the air at the surface of the airframe. At Mach 7, the top speed of the X-15, the flight phenomena confronting hypersonic intercontinental weapon systems would not be encountered. No system currently under development is capable of providing this urgently needed data.

As a flight test bed supporting development of future hypersonic, manned or unmanned, weapon systems, this test vehicle will provide a significant contribution to the USAF effort to maintain technological superiority.

(b)Brief and Military Characteristics (Dyna Soar I)
The Hypersonic Conceptual Test Vehicle, Dyna Soar I, will provide research and development information on aerodynamic, structural, human factors, and component problems and will serve as a test bed for equipment and subsystem development for weapon systems based on the hypersonic glide rocket concept. A speed of approximately 18,000 feet per second and altitudes of about 350,000 feet will be the initial design objectives.By incorporating a high degree of design flexibility at the start of the program, the performance capabilities of this system may be extended to higher speeds with a minimum of modification and by employment of additional booster units.

Details of the airplane configuration and associated components have not as yet been investigated.The preliminary design specification will be prepared after about one year of laboratory testing and analysis of certain important technical problem areas, principally those involving high heat flux.

R&D PROJECT CARD
CONTINUATION SHEET

SECRET
SECURITY CLASSIFICATION

1. PROJECT TITLE	2. SECURITY OF PROJECT	3. PROJECT NUMBER
(Conf) Hypersonic Glide Rocket Weapon System	Secret	System 464L
(Uncl) Hypersonic Strategic Weapon System	4.	5. REPORT DATE
		23 August 57

(c) Technical Approach

Data being generated and components and subsystems being developed by other programs will be used to the maximum extent in the development of this conceptual test vehicle. Current USAF and national effort in hypersonic ballistic missiles and satellites, plus current development research aircraft such as the X-15, will materially assist in achieving the state-of-the-art advances required for the hypersonic glide vehicle.

Large thrust, high impulse liquid rocket engines, which would produce the necessary propulsive force, are currently being built and acceptance tested. It is estimated that a liquid rocket engine delivering about 60,000 lbs thrust will be required for Dyna Soar I. The power plant selection will be made during the Pre-Phase I Investigations Program (see Attachment 1 for definition). Several engines appear to be suitable for this vehicle. A fluorine-ammonia or fluorine-hydrozine engine is attractive primarily because of its high specific impulse. Component development of an experimental fluorine-ammonia 35,000 lb thrust chamber is under contract at Bell Aircraft Corporation. Two of these engines would be required for this research aircraft. The feasibility of pumping fluorine has already been successfully demonstrated. The major problem involved in pump development lies in the design of the shaft seals. If it appears that the fluorine engine cannot be made available in time for Dyna Soar I, at least two other possibilities remain. Modification of the NAA 60,000 lb Lox/JP-4 liquid rocket sustainer engine for WS 107A, which is already under development, is one. The other is the RMI 50,000 lb liquid rocket engine being developed for the X-15.

Advancements in the fields of heat resistant materials and fabrication methods will permit high airframe surface temperatures when combined with accepted and proven techniques of insulation and cooling. The primary structural problem is one of providing adequate heat protection and structural integrity simultaneously. Although a tremendous amount of work remains to be done, several promising solutions to the problem have been proposed, and materials research and development may result in timely availability of new materials with the required high temperature properties for use in Dyna Soar I. However, it is likely that the structural design will utilize both conventional fabrication methods and new techniques such as double wall construction and transpiration cooling.

R&D PROJECT CARD
CONTINUATION SHEET

SECRET
SECURITY CLASSIFICATION

1. PROJECT TITLE	2. SECURITY OF PROJECT	3. PROJECT NUMBER
(Conf) Hypersonic Glide Rocket Weapon System	Secret	System 464L
(Uncl) Hypersonic Strategic Weapon System	**4.**	**5. REPORT DATE** 23 August 57

Component developments in the areas of navigation and control have progressed to the point where actual flight experience in the expected operational environment is the logical next step in the subsystem development. Reaction controls such as will be incorporated on the X-15 will most likely be required in addition to conventional aerodynamic controls.

Initial contractual effort will be directed toward wind tunnel model testing and other technical investigations required to verify the theoretical work accomplished to date and to provide design data for the Dyna Soar I vehicle. It is estimated that this Pre-Phase I Investigations Program will take from 12 to 18 months. Contingent upon successful progress in this initial phase, a Phase I development program for this conceptual test system could begin about 12 - 18 months after initiation of the program.

It is planned that initial flight testing will be accomplished using the air drop technique similar to that to be used for the X-15. Subsequent flights will be made using rocket boosters for launching from a ground base.

(2) Dyna Soar II (previously Brass Bell)

(a) General

The first weapon system proposed under this part of the program is a Hypersonic High Altitude Reconnaissance Weapon System (Dyna Soar II).

(b) Brief and Military Characteristics (Dyna Soar II)

This system is proposed as a manned two-stage rocket-powered glide aircraft. The rocket engines will boost it to an altitude of about 170,000 feet and a speed of about 18,000 feet per second, after which it will be able to glide over a range of over 5,000 nautical miles. The vehicle will be controllable both during the rocket-powered and glide portions of flight. The crew will monitor the automatic system operation and make corrections as required, operate the reconnaissance equipment, observe activities and areas of interest over which the aircraft flies, land the aircraft and perform other duties as required. The aircraft will be capable of providing high quality photographic, radar and ferret intelligence data. It may be able to perform an additional function of strategic bombing with a high degree of accuracy at speeds and altitudes sufficient to insure a large measure of invulnerability. The desirability and potential complications involved in providing this additional capability will be determined at a later date.

R&D PROJECT CARD
CONTINUATION SHEET

SECRET
SECURITY CLASSIFICATION

1. PROJECT TITLE	2. SECURITY OF PROJECT	3. PROJECT NUMBER
(Conf) Hypersonic Glide Rocket Weapon System	Secret	System 464L
(Uncl) Hypersonic Strategic Weapon System	4.	5. REPORT DATE
		23 August 57

(c) Technical Approach

Considerable effort will necessarily be directed toward the solution of the aerodynamic heating problem. The relatively high heat fluxes appear to be tolerable through the use of a cooling liquid and a double-wall construction technique. More work to determine the mechanical and chemical behavior of the boundary layer of air surrounding the airframe is required to more accurately determine probable heat transfer rates and pressure loads. Further testing of the materials will be required to determine which are best suited to cope with the high temperature conditions that will prevail.

Liquid rocket engines now appear to be best suited for propulsion of this vehicle. "Man-rated" WS 107A engines appear to be usable in the boosters. A "man-rated" version of the WS 107A sustainer engine could be utilized in the airplane; however, in the Pre-Phase I Investigations study, the desirability of initiating development of a fluorine-hydrozine engine for the airplane will be investigated. The high specific impulse of this fuel combination permits a considerable saving in size and weight of the aircraft and booster. Possible use of solid propellants will also be further investigated.

Automatic control and guidance systems presently under development could be employed. Additional work is required to determine the probable effectiveness of the aerodynamic control surfaces and to determine the extent of stability augmentation required. The reconnaissance sensors will require additional investigation to determine the effect on resolution and intelligence quality of the heated boundary layer which surrounds the aircraft. The possible attenuation and distortion resulting from the shock wave resulting from the high speeds will also warrant investigation.

(3) Dyna Soar III (previously ROBO)

(a) General

The second weapon system proposed under this program is a hypersonic, global, strategic bombardment/reconnaissance weapon system (Dyna Soar III). This system is proposed as a manned multi-staged rocket-powered glide aircraft.

(b) Brief and Military Characteristics (Dyna Soar III)

The rocket engines would boost the glider stage to approximately 300,000 feet and a speed of about 25,000 feet per second, after which the

SECRET
SECURITY CLASSIFICATION

R&D PROJECT CARD
CONTINUATION SHEET

1. PROJECT TITLE	2. SECURITY OF PROJECT	3. PROJECT NUMBER
(Conf) Hypersonic Glide Rocket Weapon System	Secret	System 464L
(Uncl) Hypersonic Strategic Weapon System	4.	5. REPORT DATE
		23 August 57

vehicle would be capable of circumnavigation of the globe. The system would be
designed to accomplish a bombardment or reconnaissance mission or a combination
thereof. Capabilities characteristic of those described for Dyna Soar II might
apply in this vehicle also, although designed for operation at the higher
altitudes and speeds of Dyna Soar III.

    (c) Technical Approach

        Considerable effort will be directed toward the solution of
the problems peculiar to this weapon system. Dyna Soar III will suffer from
aerodynamic heating in a manner different from its predecessor. The temperature
problem is not expected to be any more severe than in Dyna Soar II; however,
there will be an increased cooling requirement over a longer period of time for
Dyna Soar III. Stability and control will also be a more severe problem since
the vehicle operates during the time it is at satellite speed essentially out of
the atmosphere. Particular emphasis will be placed on accurate delivery of a
weapon to the target since the number of weapons required, hard target capability,
contamination of the atmosphere, and potential unnecessary population destruction
are intimately tied to CEP. The preliminary estimates of attaining a 3,000
foot CEP for a manned global vehicle are considered feasible in the time period
being discussed, providing a reasonable continuing effort on the advanced
technical programs is evidenced.

    b. Approach

    (1) Development Approach

        The development approach of the Dyna Soar program has been designed
to provide a logical development of the proposed weapon systems allowing timely
entry into the USAF operational inventory. Every aspect of the development has
been correlated to provide a low-risk program at minimum cost, thus avoiding
a crash program. The development schedule is arranged such that both weapon
systems proposed are an integral part of the over-all development cycle. Success
or failure in the conceptual test vehicle phase will greatly affect the development of the weapon system, which is as it should be. Check points have been
carefully placed at timely intervals in the development schedule to prepare for
this possibility. Approval of this program does not mean that once it has
started a committment has been made for a full development program. The program
could be cancelled if it was ascertained that the boost-glide concept was not
practicable. Any time saved in the development and testing of the conceptual
test vehicle, however, will result in earlier operational dates for the weapon
systems. The important thing to consider is that time cannot be economically
bought and if we are to have boost glide weapon systems in time to be of some
use, we must start now.

DD FORM 613-1   PREVIOUS EDITIONS OF THIS FORM MAY BE USED.   SECURITY CLASSIFICATION   C7-115361

R&D PROJECT CARD
CONTINUATION SHEET

SECRET
SECURITY CLASSIFICATION

1. PROJECT TITLE	2. SECURITY OF PROJECT	3. PROJECT NUMBER
(Conf) Hypersonic Glide Rocket Weapon System	Secret	System 464L
(Uncl) Hypersonic Strategic Weapon System	4.	5. REPORT DATE 23 August 57

(2) Administrative Approach

It is planned that management responsibility for this program will be assigned to the Directorate of Systems Management, Headquarters ARDC. A supporting team of technical personnel from WADC Laboratories and other appropriate ARDC Centers will be established to provide effective utilization of personnel and facilities available within ARDC. Support by the NACA in the form of over-all technical guidance and participation in the testing phases of this program is considered essential to the successful development of the proposed systems. Arrangements for NACA support and participation already initiated in this area will be expanded and finalized. A single contractor will be selected to accomplish the exploratory research part of the Pre-Phase I investigation program giving due consideration to all capable contractors in the field. This work must be non-proprietary in nature and must be accomplished under paid Air Force contract. The second part of the Pre-Phase I program, that of design studies and systems analysis, will be carried out on a competitive basis under paid and/or voluntary studies and the data generated will be proprietary. Two or more contractors will be selected for competition under this part of the Pre-Phase I program. It will be made clear at the outset that all of the contractors involved in the Pre-Phase I program will have an equal opportunity to compete for the Phase I of the conceptual test vehicle. This philosophy is applied throughout the development schedule for all Dyna Soar vehicles. Proprietary rights of contractors competing in the design study and system analysis phase will be protected as necessary to maintain their competitive position. Technical data from WS 107A has been and will be utilized to the maximum during this program. In addition, studies of the designs and concepts of the Dyna Soar program will show cognizance of the facilities, support equipments, concepts, etc. being developed for the WS 107A.

A more thorough treatment of the Dyna Soar program, including justification for the weapon systems proposed and the test vehicle required, the development philosophy, and a thorough discussion of the development schedule is contained in Attachment 3.

c. Background History

The boost-glide concept is not new. In the early Forties, Dr. Sanger, German rocket scientist, proposed a skip-glide vehicle to bomb New York from a launch site in Germany. Serious consideration was given this proposal by the Germans. A program known to the Germans as the A9/A10 development was designed to use a winged V-2 rocket as the second stage of a two-stage system. This vehicle was under development and test by the Germans when the war ended. At the close of the war Dr. Walter Dornberger, ex-German general and head of the Peenemunde Rocket Research Institute in Germany went to work for Bell Aircraft

DD FORM 613-1 PREVIOUS EDITIONS OF THIS FORM MAY BE USED.
SECURITY CLASSIFICATION **SECRET** C7-115361

PAGE 8 OF 10 PAGES

R&D PROJECT CARD
CONTINUATION SHEET   **SECRET**
                                          SECURITY CLASSIFICATION

1. PROJECT TITLE	2. SECURITY OF PROJECT	3. PROJECT NUMBER
(Conf) Hypersonic Glide Rocket Weapon System	Secret	System 464L
(Uncl) Hypersonic Strategic Weapon System	4.	5. REPORT DATE
		23 August 57

Corporation in this country. It is not surprising then that Bell approached
the USAF in 1952 with an unsolicited proposal for a Manned, Hypersonic Boost-
Glide Bomber/Reconnaissance Weapon System. Rand conducted investigations of
this concept in 1948 and the NACA published work on the subject in 1954. Since
1954, the ARDC has sponsored a considerable amount of work in the boost-glide
field. The following table summarizes this effort.

Contractors	Program	Effort	Year	P-600 Funds	Contractor Funds
Bell	BOMI	Feasibility	Apr 54-Apr 55	$246,000	$200,000
Bell	BOMI	Design Study	Sep 55-Dec 55	174,000	400,000
Bell	Brass Bell	" "	May 56-Aug 57	1,736,000	
Bell	ROBO	Feasibility & Design	1956 - Jun 57	None	Total for ROBO volun-
Boeing	ROBO	" "	Jan 56-Jun 57	None	tary studies by all
Convair	ROBO	" "	Jun 56-Dec 56 Jan 57-Present	245,000 None	contractors $3,200,000
Douglas	ROBO	" "	Jun 56-Jan 57 Jan 57-Present	374,000 None	
Martin	ROBO	" "	Jan 57-Present	None	
North American	ROBO	" "	Jun 56-Dec 56 Jan 57-Present	240,000 None	
Republic	ROBO	" "	Jun 56-Jun 57	None	
Lockheed	ROBO	" "	Jul 57-Present	None	

                                   TOTALS: $3,015,000 $3,800,000

      d.   Future Plans

          Upon receipt of Headquarters USAF Development Directive together with
necessary funds and other essential resources, ARDC and AMC will select a contractor for the exploratory research part and two or more contractors to

R&D PROJECT CARD
CONTINUATION SHEET

SECRET

SECURITY CLASSIFICATION

1. PROJECT TITLE	2. SECURITY OF PROJECT	3. PROJECT NUMBER
(Conf) Hypersonic Glide Rocket Weapon System	Secret	System 464L
(Uncl) Hypersonic Strategic Weapon System	4.	5. REPORT DATE
		23 August 57

accomplish the Design Study and system analysis of this program. In accordance with 80-4 a complete System Development Plan will be prepared and submitted to Headquarters USAF for approval.

    c. References

        (1) Requirements

            (a) GOR 92 (TA-4e-1-59) dated 12 May 1955, title: GOR for a Piloted Very High Altitude Reconnaissance Weapon System.

            (b) System Requirement 126 (Study), dated 12 June 1956, title: (C) Rocket Bomber Study.

            (c) Headquarters USAF letter, dated 29 June 1956, subject: (C) Research System 455L, Hypersonic Glide Rocket Research System.

            (d) System Requirement 131, dated 14 November 1956, title: ... (S) Hypersonic Glide Rocket Research System.

        (2) Development Plans

            (a) HYWARDS - title: (C) Hypersonic Glide Rocket Research System No. 455L, dated 28 December 1956.

            (b) Brass Bell - title: (U) Hi Fi Recce System No. 459L, dated 31 December 1956.

        (3) Summary Reports (see Attachment 2).

3 Attachments
    1. Definitions - Project Dyna Soar
    2. Summary Reports
    3. Presentation Material

DD FORM 613-1   PREVIOUS EDITIONS OF THIS FORM MAY BE USED.   SECURITY CLASSIFICATION C7-115361

SECRET

DEVELOPMENT SCHEDULE - PROJECT DYNA SOAR

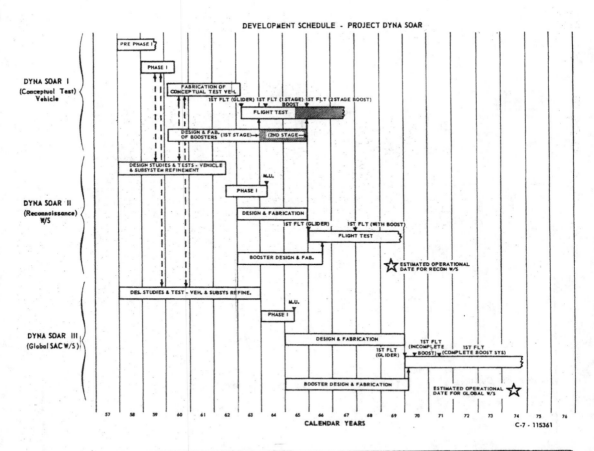

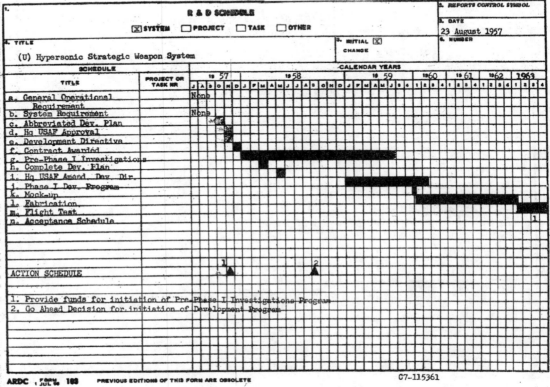

R & D SCHEDULE

[X] SYSTEM [ ] PROJECT [ ] TASK [ ] OTHER

2. REPORTS CONTROL SYMBOL

3. DATE 26 August 1957

4. TITLE
(U) Hypersonic Strategic Weapon System   Part II

5. INITIAL [X]
CHANGE

6. NUMBER

SCHEDULE TITLE	PROJECT OR TASK NR	1957	1958	1959	1960	1961	1962	1963
ACCEPTANCE SCHEDULE								1
Airframe						X	O	
Propulsion						X	O	
Personnel & Equipment						X	O	
Environmental Protection								
Communications (airborne)						X	O	
Navigation (airborne)						X	O	
Flight Control						X	O	
Communications (ground)						X	O	
Navigation (ground)						X	O	
Data Processing						X	O	
Test Instrumentation						X	O	
Air-gas-liquid generating Handling & Storage						X	O	
Maintenance, Handling and servicing							X	O
Ground and Special purpose vehicles						X	O	
Support Aircraft								
Personnel operations						X	O	
Personnel & Material Safety Protection & survival						X	O	
Training equipment						X	O	
ACTION SCHEDULE								
1-Determination of additional flight vehicles								

X - Subsystem Completed
O - Complete Subsystem integrated into system

RDC FORM 103 PREVIOUS EDITIONS OF THIS FORM ARE OBSOLETE   G7-115361
1 JUL 55

R & D COST ESTIMATE RECAPITULATION

[X] SYSTEM [ ] PROJECT [ ] TASK [ ] OTHER

2. REPORTS CONTROL SYMBOL

PAGE 1 OF 1 PAGES

3. DATE 23 August 1957

4. UNCLASSIFIED TITLE
Hypersonic Strategic Weapon System

5. INITIAL [X]
CHANGE

6. NUMBER

	ITEM	A. PREVIOUS YEARS		B. 1st FISCAL YEAR		C. 2d FISCAL YEAR		D. 3d FISCAL YEAR		E. TO COMPLETE	
		000	OTHER	000	OTHER	000	OTHER	000	OTHER	000	OTHER
7. CONTRACT	A. TOTAL			3,000,000	—	5,000,000	—	8,000,000	—		—
	B. AVAILABLE			0		—		—	—		—
	C. NEW REQ			3,000,000	—	5,000,000	—	8,000,000	—		—
8. MATERIEL	A. TOTAL										
	B. AVAILABLE										
	C. NEW REQ			Column E Estimates will be available 2 years after plan is approved.							
9. FACILITIES											
10. MANPOWER											
11. TRAINING											
12. TEST ITEMS											
13. TEST SUPPORT AIRCRAFT											
14. SUBTOTAL											
15. TOTAL											

ARDC FORM 116 PREVIOUS EDITIONS OF THIS FORM ARE OBSOLETE.   G7-115361
1 JUL 55

B. Attachment I - Definitions
PROJECT DYNA SOAR DEFINITIONS
RESEARCH SYSTEM
I. A vehicle designed to obtain basic or fundamental flight data in an environment which has not been sufficiently well defined.
2. Not developed to directly support 2. any given weapon system development.
3. Designed to research area not test solutions to problems.
4. Resulting data non-proprietary.
5. Example : X-15

CONCEPTUAL TEST VEHICLE
I. A vehicle designed to obtain flight data in an environment which has not been sufficiently well defined.
2. Intended for use in development of weapon systems utilizing a specific concept - developed to provide maximum data on the flight regime associated with a specific weapon system concept.
3. Designed to research area and test solutions to problems - serves as test bed for breadboard models of subsystems.
4. Resulting data non-proprietary.
5. Examples: Dyna Soar Conceptual Test Vehicle for Boost Glide Flight Regime of Brass Bell and ROBO family.

PROTOTYPE VEHICLE
I. A vehicle designed to obtain flight data in an environment which has not been sufficiently well defined.
2. Intended for use in development of a specific weapon system and is part of development program of that weapon system - flies in same mode of operation as actual. weapon system.
3. Designed to test solutions to problems, not research areas used in development of operational subsystems.
4. Resulting data proprietary.
5. Example: X-10 and XSM-64 prototype vehicles.

DYNA SOAR PRE PHASE I PROGRAM (PRACTICABILITY INVESTIGATION)

I.   Exploratory Research and Testing
        A. Objectives:
                I. Studies and tests designed to validate assumptions, theory.. and data generated under previous boost glide study programs.
                2. Provide sound basis for development of new theory and techniques required.
                3. Provide design data.
                4. Establish capability of boost glide vehicles.
                5. Determine optimum flight profile of conceptual test vehicle.

        B. Requirements
                I. One contractor.
                2. Data non-proprietary.
                3. Dissemination of all data to all Dyna Soar contractors.
                4. Paid program.

II.   Design Studies and System Analysis
        A. Objectives:
                I. Refine conceptual test vehicle design.
                2. Establish capability of test vehicle.
                3. Define subsystems and research instrumentation required.
                4. Define vehicle as operational system.

B. Requirements:
1. Two or more contractors.
2. Data proprietary.
3. Dissemination to military only.
4. Paid and/or voluntary program.

C. Attachment 2 - Summary of Technical Reports

SUMMARY REPORTS

CONTRACT REPORT NO.	TITLE OF REPORT	REPORT NO. ARDC	ASTIA NO.
Bell Acft Co.			
D143-941-001	Criteria and Loads	TR 56-43	AD 113081
D143-941-002	Liquid Metal Cooling for Hot Spots	TR-56-44	AD 113082
D143-941-003	Secondary Structure	TR-56-45	AD 113083
D143-941-004	Outer-Wall-Panels & Insulants for Primary Structure	TR-56-48	AD 113086
D143-941-005	Aux Power from Aerodynamic Heating	TR-56-47	AD 113085
D143-941-006	Water Cooling System	TR-56-48	AD 113086
D143-941-007	Primary Structure	TR-56-49	AD 113087
D143-941-008	Gusts and Wind Shears	TR-56-50	AD 113088
D143-941-009	Acoustic Loadings	TR-56-51	AD 113089
D143-941-010	Double-Wall Construction	TR-56-35	AD 113073
D143-941-011	Cooling System-Primary Structure	TR-56-36	AD 113074
D143-941-012	Internal Airflow in Double-Wall	TR-56-37	AD 113075
D143-941-013	Test Program Plans-Structures	TR-56-38	AD 113076
D143-941-014	Test Results and Program-Materials	TR-56-39	AD 113077
D143-941-015	Liquid Metal Cooling System	TR-56-40	AD 113078
D143-945-018	Advanced Strategic Weapon System, dated 29 April 1955		
D143-945-029	Reconnaissance Aircraft Weapon System, dated 1 December 1955		
D143-945-031	Ames Double-Wall Tests	TR-56-52	AD 113090
D143-945-033	Aerodynamics Design Vol. 1	TR-56-23	AD 113061
D143-945-033	Aerodynamics Design Vol. 2	TR-56-24	AD 113062
D143-945-034	Aerodynamics Research	TR-56-25	AD 113063
D143-945-035	Flight Control	TR-56-26	AD 113064
D143-945-036	Radar	TR-56-27	AD 113065
D143-945-037	Structures	TR-56-28	AD 113066
D143-945-038	System Design	TR-56-29	AD 113067
D143-945-039	Summary Report	TR-56-34	AD 113072
D143-945-032	Program Plan		
D143-945-041	Electrical System	TR 56-42	AD 113080
D143-945-044	Request for NACA/Langley Gas Dynamics Branch Experimental Panel Flutter Tests		
D143-945-045	Request for NACA/Langley Transonic Dynamics Branch Experimental Panel Flutter Tests		
D143-945-046	Request for Hot Spot Tests of Double-Wall Construction in NACA/PARD Ethylene Jet.		
D143-945-102	SR-126 Briefing, dated 16 January 1057		
D143-945-103	System Design	TR-57-51	AD 117971
D143-945-104	Aerodynamics	TR-57-52	AD 117972
D143-945-105	Structures	TR-57-53	AD 117973
D143-945-106	Avionics	TR-57-54	AD 117974
D143-945-107	Summary Report	TR-57-55	AD 117975
D143-945-108	SR-126 Briefing, dated 16 June 1957		
D143-978-001	Flat Plate	TR-56-68	AD 116210
D143-978-002	Delta Wing	TR-56-69	AD 116601

D143-978-003	Sweep Angle	TR-56-70	AD 116602
D143-978-004	Trailing Edge Flap	TR-56-71	AD 116003
D143-978-005	Flat Plate	TR-56-41	AD 113079
D143-984-001	Work Statement for Materials	TR-56-53	AD 113091
GE-R56A0161	Navigation	TR-56-30	AD 113068
CIA-1725	Photographic Analysis	TR-56-31	AD 113069
AIL-3887-1	Ferret	TR-56-32	AD 113070
PRA-56-16	Human Factors	TR-56-33	AD 113071

Boeing Airplane Co.

D2-1062	(C) Propulsion System for Project Robo: Preliminary Calculations Based Upon Fluorine-Hydrazine Rocket Technology.		
D2-1571	SR-126 Progress Report (no dtd)		
D2-1706	(C) ROBO:Hypersonic Testing Program	TR-57-81	
D2-3053	(C) ROBO Rocket Bomber Studies Vol. 1 Final	TR-57-79	
D2-3053	(C) ROBO Rocket Bomber Studies Vol. 2 Final	TR-57-80	
D2-3094	ROBO Atran/Inertial-Guidance	TR-57-91	AD 131480
D-17385	(C) Hypersonic, Rocket Booster, Strategic Vehicles		
D-17744	Statement of Work - ARDC Task #89774, Feasibility Study		
D-18440	Preliminary Work Outline Project ROBO		
D-18470	Derivation of the Equations of Motion of a Hypersonic Quasi-Satellite Vehicle		

Convair

FZM-449	ROBO - Program Planning Report		
FZM-672	ROBO - Progress Review dated 26 November 1956		
FZM-941	ROBO SR-126 Feasibility Summary, Vol I	TR-57-58	AD 117976
FZM-942	ROBO SR-126 Primary Technical Studies Vol. II, Part I	TR-57-59	
FZM-942	ROBO SR-126 Primary Technical Studies Vol. II, Part II	TR-57-60	AD 117978
FZM-942	ROBO SR-126 Primary-Technical Studies Vol. II, Part III	TR-57-61	AD 117979
FZM-942	ROBO SR-126 Primary Technical Studies Vol. II, Part IV	TR-57-62	
FZM-943	ROBO SR-126 Weapon Development Planning, Vol. III	TR-57-63	AD 117980
FZM-944	ROBO SR-126 Design Developments Vol. IV	TR-57-64	
FZM-945	ROBO SR-126 Supplementary Propulsion system Studies Vol. V	TR-57-65	AD 115601
FZM-946	ROBO SR-126 Supplementary Technical Studies, Vol. VI	TR-57-66	AD 115602

Douglas Aircraft Co.

LB-25300	Weapon System Feasibility Summary	TN-56-15	AD 120035
LB-25301	Weapon System Design	TN-56-15	AD 120036
LB-25302	Warhead Effectiveness Analysis	TN-56-15	
LB-25303	Electronics System Feasibility Summary	TN-56-15	AD 120038.
LB-25303 A	Consultant Study of Integrated Electronics System		AD 120039
LB-25303-B	Consultant Study of Inertial Navigators and Control System	TN-56-15	AD 120040
LB-25303-C	Consultant Study of Communications System	TN-56-15	AD 120041
LB-25303 D	Consultant Study of Radar System	TN-56-15	AD 120042
LB-25304	Human Factors	TN-56-15	AD 120043
LB-25305	Performance	TN-56-15	AD 120044
LB-25306	Propulsion & Auxiliary Power Systems	TN-56-15	AD 120045
LB-25306A	Consultants Comments on Auxiliary Power Systems	TN-56-15	AD 120046
LB-25307	Applicability of Nuclear Propulsion	TN-56-15	

LB-25308	Aerodynamic Heating	TN-56-15	AD 120048
LB-25309	Structural Design Considerations	TN-56-15	AD 120049
LB-25310	Stability & Control	TN-56-15	AD 120050
LB-25311	Environmental Control: System	TN-56-15	AD 120051
LB-25312	Estimated Program Costs and Schedule	TN-57-3	
SM-27148	Weapon Delivery Feasibility.	TN-56-15	AD 120034
LB-25121	Preliminary Technical Data		
LB-25072	Project ROBO Engineering Report		
LB-25072	Engineering Report - Appendix I		
LB-25185	Project Review Charts		
LB-25371	Feasibility Briefing		AD 120033

LB-25371 "Preliminary Spec. On the Feasibility Study of the Integrated Electronic System for Douglas Model 1,377", prepared by: Radio Corp. of America, Minneapolis-Honeywell Regulator Company, Sylvania Electric Products, Ind

RMI-136-F Reaction Motors, Inc. Report on "Feasibility Study of AD 120053, Propulsion System for SR-126."

AGC Ser. No. AeroJet-General Corp. Report on Technical Information　　　　AD 120055
3705 Pertinent to Propulsion Data Requirements for Douglas Model 1377.

SB-579-X-001 Radio Corp, of America Specification No. SB-579-X-001, "Preliminary Spec. on the Feasibility Study of the Integrated Electronic System for Douglas Model 1377."

(S)Grand Central Rocket Co. report for Douglas Aircraft Co. on the　　AD 120054
Feasibility of a Solid Propellant Rocket Propulsion System for the DAC
Model 1977 Vehicle.

General Electric Company Report for DAC Model 1377　　　　　AD 720057
"Feasibility Study".

R-413P Rocketdyne division, North American Aviation, Inc.　　　　　　AD 120052
Report Do. R-413F

LB 25494 DAC Engineering Report, Model 1377 Alternate Development Program Concepts

AGC Ser. No. 3751 Aerojet-General Corp, Report on "Technical Discussion　　AD 120056
on the Propulsion System for the Propulsion System for the Douglas
Model 1377".

LB-25401	Comparison of Economic Effectiveness of Model 1377(SR-226) with an Alternate System.	TR-57-68	
LB-25403	Technical Study of an Alternate System for Accomplishing the SR-126 Mission	TR-57-69	AD 116610
LB-25378	Electromagnetic Wave Propagation in Ionized Media, Special Application to a Hypersonic Vehicle.	TR-5770	AD 115604
LB-25456	Vehicle Defense Considerations	TD-57-71	
LB-25457	Aerodynamic Heating Studies	TR-57-72	AD 115605
LB-25458	(S) Multiple Missile Capability	TR-57-73	
LB-25459	(S) Effects of Gravity Anomalies & Earth's Oblateness on Performance	TR-57-74	AD 215614
LB-25462	(S) Preliminary Study of an Air- Breathing Booster Stage	TR-57-77	AD 115619

McDonnell Acft Corp.

4590 (C) Proposal for a Feasibility Study of a Hypersonic
Bombardment/Reconnaissance Weapon System dtd 27 Feb 56.

North American Aviation

AL-2530	ROBO Technical & Operations Report	TR-57-56
AL-2540	ROBO Report on Development Plan & Feasibility Testing Program	TE-57-57
AL-2484	(C) ROBO Hypersonic Bomber System	
AL-2260	(C) Proposal for a Hypersonic Bombardment Reconnaissance Weapon System	
NAD-50-105	Briefing for Proposed Hypersonic Weapon System Feasibility Study Task No. 89774	
AL-2412	Development Testing Facilities for Advanced Ballistic Missiles	
PC-413P	Fluorine as a Rocket Propellant	

Republic Aviation Corp.

ED-AP 82-901	(C) Appendix "A" Preliminary Design of a Hypersonic Orbiting Weapon
GMD-109-950	Preliminary Design Study of a Strategic Bombardment W/S under ARDC SR No. 126 (SLIM)
AP82-901	(C) Preliminary Design Study of a Hypersonic Orbital Weapon
D2-3101	ROBO Presentation (BAC Report)

D. Attachment 3— Technical Analysis

Analysis of Boost Glide Concept

INTRODUCTION

The analysis consists of six parts: introduction, boost-glide capabilities, advantages of boost glide, a development concept, a development schedule, and ARDC's conclusions and recommendations. The subject of this analysis is the application of the boost-glide concept to long range Air Force missions.

We'll begin by discussing three approaches to the SAC mission. The first is the old Sustained-Flight method using Air Breathing Engines. The second one is the Boost-Glide Concept, and the third, the Ballistic Missile. For the past 50 years our aircraft have performed their mission by taking off and climbing to some altitude, maintaining a constant speed cruise to and from the target, and finally descending and landing.

Figure 1 is a plot of altitude in thousands of feet versus speed in Mach number defining the sustained-flight regime for long range vehicles. Any long range vehicle which sustains itself in level flight through aerodynamic lift will be confined to the shaded region on the chart in accomplishing a long range mission. It is possible to propel a winged vehicle into the area above the sustained flight regime, but this constitutes a fish-out-of-water type flight. For a given speed, maintaining level flight above this line imposes an increasingly severe penalty in range on the vehicle as the altitude increases. This range penalty is the result of a reduction in aerodynamic lift to drag ratio, as the lift coefficient necessary for level flight increases above the lift coefficient corresponding to L/D max.

The upper limit shown is defined by a constant wing loading line of 20 pounds per square foot. As the wing loading increases, a practical limit is reached where-in the length of the runways required becomes excessive, take-off and landing speeds are too high, and from a structural standpoint, the dynamic pressure and aerodynamic heating encountered at the lower altitudes becomes prohibitive. Two hundred pounds per square foot has been arbitrarily chosen as the lower limit of the sustained-flight regime, representing about the largest wing loading which might be considered practical.

It is of major significance to the hypersonic boost-glide concept that the curve of both the upper and lower limits of the sustained flight regime reverse themselves in the neighborhood of Mach 12 and become asymptotic to satellite speed. The reason for this is that in level flight, as the speed increases the vehicle enjoys a larger and larger contribution from centrifugal lift. The entire lift process becomes increasingly efficient until upon achieving satellite speed, no aerodynamic lift at all is required to maintain level flight - the vehicle has become a satellite. Thus, this chart defines the flight regime of long range vehicles designed for operation at or near L/D max in level flight. Above 100,000 feet, as shown by the curve, the minimum Mach number required for efficient sustained flight increases rapidly for a small gain in cruising altitude.

At speeds of about Mach 4, winged aircraft will be characterized by tail-less or canard configurations, low aspect ratio, highly swept wings and large fineness ratio fuselages. As a result, for a given speed above Mach 4, it does not appear (Figure 2) that the magnitude of the optimum lift coefficient associated with L/D max. will vary appreciably between one long range vehicle and the next. Consequently, the upper limits of the sustained-flight regime can be fairly clearly defined for long range vehicles. At speeds below Mach 4, however, a large spread in optimum lift coefficient is possible through a variation in configuration. The difference between the F-104 and the F-102A configurations is an example of the latitudes possible. Therefore, the upper limits of the sustained-flight regime cannot be as clearly defined at subsonic or low supersonic speeds and are left vague on the chart.

Having defined the sustained-flight approach and its flight regime, we will now describe (Figure 3) a boost-glide vehicle and its approach to mission accomplishment. The vehicle part of the boost glide weapon system will be composed of an aircraft stage and one or more booster stages. Rocket boosters will take the multistage vehicle off vertically and boost it to altitudes between 100,000 and 300,000 feet and to hypersonic speed. When all the fuel has been expended in the boosters, they will drop off and the aircraft will start gliding flight trading kinetic and potential energy for range.

Although defined for level flight conditions, this flight regime (Figure 4) also reflects the glide path of a boost-glide vehicle. The aircraft stage of a boost-glide weapon system is generally characterized by a low wing loading, on the order of 25 to 35 pounds per square foot, which remains almost constant for the remainder of the flight. The glide portion of a boost-glide system can be shown on this chart lying along the upper limits, of the sustained-flight regime.

Thin concept is not new. During the early 1940's a German scientist, Dr. Sänger, developed a plan to bomb New York, using a skip-glide vehicle launched from Germany. This plan was seriously considered by the Germans. The Peenemünde group independently conceived the A9/A10 boost glide program. This system utilized a winged V-2 missile as the second stage of a two-stage boosted unmanned vehicle and was under development and test when World War II ended. In 1948, Rand began studies of the boost glide concept; the USAF expressed official interest in June 1954.

After World War II, Dr. Walter Dornberger, ex-German general and head of rocket development at Peenemünde Rocket Research Institute in Germany, came to this country and went to work for the Bell Aircraft Corporation. Early in 1951 Bell began investigating the boost-glide concept. In 1952 Bell made an unsolicited proposal to Headquarters USAF for a global manned boost-glide weapon system. It was not until June 1954, however, that a contract was awarded to Bell to investigate the feasibility, problem areas, and offer solutions to the problems of manned, hypersonic, boost glide flight. A follow on contract was awarded Bell in 1955 to investigate the application of the boost-glide concept to a manned, long range, high altitude reconnaissance weapon system. Bell gave this program the nickname BOMI, which stood for bomber missile. The work on the reconnaissance system is still underway with Bell under Air Force contract; however, it is now known as Brass Bell.

Early in 1956, many other contractors were brought into this type of program under Project ROBO, which stands for rocket bomber. The contractors engaged in this program are Bell, Boeing, Convair, Douglas, Martin, North American, Republic, and Lockheed. All of these contractors, with the exception of Martin, and Lockheed who only recently entered the program, presented the results of their feasibility studies and investigations to the Air Force in June 57. Without exception, these contractors came to the conclusion that the boost-glide concept is feasible; that it offers a tremendous capability in performance and mission accomplishment with unlimited growth potential for the future. The NACA agrees with these conclusions, and is supporting this type of program through investigations of its own.

Turning to capabilities of a boost-glide weapon system, let us first consider a reconnaissance weapon system. A weapon system of the type envisioned by Bell Aircraft, under the Brass Bell program is a manned, rocket-boosted, hypersonic, boost-glide vehicle, which would be boosted to an initial Mach number of about 18 at 170,000 feet, and would be capable of glide flight for about 5,000 to 6,000 nautical miles. It would be designed for subsystem capability of photo, radar, ferret, infrared, search and detail, and be capable of detecting and identifying hard targets. One major advantage of such a system would be the simultaneous attainment of high speed, range, and altitude. It will be recoverable and one of the major features of such a system is that it would not require prior information of the exact location of the target area in order to do its job.

Some of the difficulties peculiar to this particular weapon system would result from the problem of looking through a radiating and disassociated boundary layer. The reconnaissance subsystems would be required to provide adequate target resolution. One of the subsystem development requirements would be a coherent Doppler side-looking radar, which could effectively operate from the extreme high altitude and speeds that would be characteristic of such a vehicle. It now appears possible to obtain such a weapon system about 1969 through a normal development program. The details of this particular system are contained in the reports published under the Brass Bell program, a list of references for which is contained in System 464L Development Plan.

A typical global weapon system envisioned by the contractors under the ROBO program would be a manned, hypersonic, multi-boosted weapon system, with the initial Mach number of 22-25, at an altitude of about 300,000 feet, and would have a global capability. Weapons delivery would be accomplished with a bomb or an ASM, with sufficient accuracy for destruction of hard targets. This system would also have a recall capability. The possibility of a multi-target capability also exists. Prior knowledge of the exact location of the target is not required. Aside from the advantages mentioned previously for the reconnaissance systems, this system would have a global capability, multiple target attack, and the distinct advantage of multiple approaches to the target.

One-of the problems peculiar to this system would be weapon delivery. The methods of weapon delivery proposed by the contractors have been varied and a decision at this time as to the exact mode of operation is considered premature. Guidance and control will be a problem, since at satellite speed, it is operating essentially out of the atmosphere. A severe cooling problem will be experienced by the global type weapon system because it experiences a glide re-entry problem. This problem differs from the re-entry problem of a ballistic missile since, although a ballistic missile will experience temperatures of an order of magnitude greater than those experienced during the re-entry of a glide vehicle, the total heat flux to the vehicle is much greater for the glide situation. The penalty for glide re-entry is the weight of coolant required to withstand aerodynamic heating over a longer period of time.

From anticipated state-of-art progress, it is reasonable to expect that an operational global system can be obtained about 1975. Further detailed information concerning these global weapon systems can be obtained from references listed in Attachment 2 of the Development Plan.

We will now discuss problems which will be common to any boost-glide vehicle. From an aerodynamic standpoint, the physical state of the air is different at the high Mach numbers of the boost-glide flight regime. Problems of viscosity, diffusion, conductivity, dissociation, and ionization will be experienced. The boundary layer characteristics are relatively unknown, and one of the vital problems which will require investigation very early in the program is the problem of transition Reynolds number and the characteristics of turbulent and laminar flow. From a structural standpoint, designing the vehicle for structural integrity will require knowledge of surface effects, the problem of melting, evaporation, sublimation, oxidation, sputtering, emission, ion reflection, etc.

One of the more difficult design problems that will be encountered is the use of materials in the construction of the vehicles. It appears possible, through structural design innovations, that existing materials can be utilized for the fabrication of boost-glide vehicles, however, the development of new high temperature materials would certainly be an advantage. The problem is one of designing heat protected structures, utilizing techniques such as double-wall construction and leading edge cooling systems, the problem here being one of developing light, heat protected structures. From a subsystems standpoint, one of the problems, is providing satisfactory environment for the subsystems and crew.

The development of what will be called a conceptual test vehicle to investigate the boost-glide flight regime and concept is a prerequisite to the successful development of a boost-glide weapon system. Although boost-glide weapon systems offer great promise, if there are so many problems to be solved, and we have to resort to a conceptual test vehicle before an operational weapon system can be safely developed, why should we undertake such a program? Is there a real need for boost-glide weapon systems, and if there is, are they worth all of the trouble and expense which is apparently going to be required in their development? This question can be treated through the consideration of three factors trends, characteristics, and mission capability.

Let us first look at the trends. Figure 5 is a plot of altitude in thousands of feet versus years, which shows the altitude trends of U. S. operational aircraft from about 1925 through the present. The top line shows the official altitude records of U. S. aircraft for the last 30 years. The next line represents the service ceilings of the best of any type of fighter or bomber, and the bottom line, the service ceilings for the best heavy bombers. Notice that the trend in every case is a gradual decrease in slope, indicating a leveling off trend of altitude capability for air breathing engine aircraft.

Of course, this trend is to be expected, since it is obvious that air breathing engine vehicles cannot have an infinite altitude capability, however, it is interesting to note that the official altitude record line and the "best any type" line meet at about 60,000 feet in our time period, indicating that the last drop of altitude capability is being wrung out of contemporary air breathing engines. It is apparent that some breakthrough is required in order to break this trend and fly at altitudes on the order of 100,000 feet with air breathing engines between 1960 and 1970.

If we now look on Figure 6 at the speed trends for air breathing engine aircraft, we find a little brighter picture. It appears that regardless of the type of aircraft - fighter or bomber - there is a continual increase in slope of the speed curves with no indication of leveling off. If we break this down into types of air breathing engines, such as propeller driven and turbojet, we find a definite leveling off in slope for the various propeller driven engines through the years, and also a similar picture for turbojets. However, in general, it does not appear that speed-wise air breathing engines have reached their limits.

Investigation of the range trends (Figure 7) through the years tells an interesting story. Prior to World War II, not too much emphasis was placed on range, and therefore the range capabilities of heavy bombers were quite low. During World War II, a tremendous emphasis was placed on range, and therefore, we find a sharp increase in the slope of the curve. At first glance a gradual decrease in the slope of this curve at the close of World War II might indicate to some that the emphasis on range was again on the decline. However, we know that this is not the case. The reason for the decrease in slope is actually indicative of the difficulty of attaining range. One of the reasons for this is the fact that we are now flying supersonic aircraft.

Since range (Figure 8) is directly proportional to the L/D ratio, we can see from this chart that supersonic flight imposes a severe penalty in range, since L/D's on the order of 22, such as is attainable in the B-52 drop to about 5 as we exceed Mach 1. What must we pay then in order to accomplish ranges in excess of 6,000 nautical miles with air breathing engine aircraft? One thing we pay is size. The evolution of USAF aircraft from 1907 to 1957 has shown a definite trend towards extremely large vehicles. One indication of the trend for the future of air breathing engine aircraft can be seen from the vehicles proposed by North American and Boeing under the Weapon System 110 program. One approach to this mission (which desired a range of 11,000 nautical miles) was the use of an air breathing engine aircraft, utilizing hydrogen fuel. The Boeing design for this requirement was well over the length of a football field.

Another factor of interest is the weight trend. Figure 9 plots take-off weight in millions of pounds versus range in thousands of nautical miles. Plotted are a manned and an unmanned version of a vehicle in cruising flight, powered by ramjet propulsion at a speed of Mach 8. In both cases, the take-off weight is becoming intolerable, considerably short of the 10,000 nautical mile range. A practical limit on the ratio of gross weight to empty weight of 5 is shown where it would fall on the chart. Now this is just a trend of one particular type of air breathing engine vehicle. However, the trend is similar for supersonic air breathing engine aircraft in general. Note that the characteristic trend for a manned orbital glider is considerably different than that of the air breathing engine trend, in that the slope of the curve is decreasing as we approach global range. The trends tell one story.

Suppose we now take a closer look at the characteristics of the sustained flight regime which we defined earlier. We have shown on Figure 10 the upper and lower limits dictated by aerodynamic consideration on the altitude versus Mach number chart. Let us first treat the aerodynamic heating picture (Figure 11). Aerodynamic heating is a complex problem and cannot be completely defined on a chart of altitude versus Mach number. However, a good insight to the problem can be obtained by plotting an overlay of constant temperature lines on the sustained flight regime. The solid lines are lines of constant equilibrium wall temperatures one foot behind the leading edge in turbulent air at an angle of attack of 5° and a conservative emissivity of .6. This is not a complete picture, and no limit is defined; however, this plot allows a discussion of the characteristics of aerodynamic heating.

The dates associated with these temperatures are predictions by the WADC Materials Lab of the time periods wherein it might be reasonably expected that operational weapon systems could be flying and experiencing such temperatures. In estimating these dates, the assumption was made that accommodating temperatures by the dates shown requires research and development in materials fabrication, and design, and the state-of-the art in materials and structures will have advanced to the point where the temperatures and dates are compatible. Note that the slope of the temperature lines decrease with increasing altitude and Mach number.

A direct attack on the aerodynamic heating problem by adding mass, insulation, cooling, etc., increases weight and complicates the technical development of the weapons systems. However, it is apparent that one means of avoiding the aerodynamic heating problem is flying as near as possible to the upper limit of the sustained flight regime. This immediately imposes the requirement for low wing loading. For example a temperature of 1800° F. is experienced at an altitude of 60,000 feet, and a speed of Mach 6 at the lower limit of the sustained flight regime while an altitude of 130,000 feet and about Mach 10 can be reached without exceeding this temperature by flying at the upper limit of the sustained flight regime. Turbulent flow conditions are a reasonable assumption for the lower Mach number and altitude region of the sustained flight regime. However, at the higher Mach numbers and altitudes, the Reynolds numbers encountered are of the right order of magnitude for the realization of laminar flow. The dashed curves are equilibrium wall temperatures one foot behind the leading edge in laminar flow at an angle at attack of 5o at an improved emissivity of about .9. This condition produces a considerable shift in the constant temperatures lines for turbulent and laminar flow, which offers interesting possibilities with respect to boost-glide vehicles which would normally initiate their gliding phase at these extreme altitudes and Mach numbers.

Let us consider the future, with respect to propulsion systems. The air breathing engine picture (Figure 12) for the present concerns turbojets and ramjets, hydrocarbon fuel, with a capability of speeds up to about Mach 2.75 and altitudes to about 70,000 feet. We can see from the temperature lines that only moderate aerodynamic heating is to be expected in this time period. Best estimates (Figure 13) of our potential capability in 1964 show that pre-cooled turbojet or vapor cycle engines utilizing hydrogen fuel should have a capability of Mach numbers from 3 to 4 and altitudes up to 100,000 feet, which is a considerable improvement. However, at this time, none of these engines are under development; their existence in 1964 is predicated on their development being initiated now. Also, hydrogen fuel is not an Air Force inventory item at the moment.

If we began work now, (Figure 14) it is quite possible that a conceptual test vehicle investigating the boost-glide flight regime out to a speed of about Mach 18 in an altitude of about 160,000 feet should be realized in 1964. This is a considerable improvement, well over four times the capability promised by air breathing engines. If we took (Figure 15) to the year 1970, one contractor in the air breathing field (Marquardt) has proposed a hypersonic ramjet for a speed of about Mach 8 to an altitude of 140,000 feet. If such an engine were available by 1970, which would be admittedly a considerable development program, it would represent 100% improvement, speed-wise, over what has been predicted for the 1964 time period. However, in 1970, (Figure 16) if the conceptual test vehicle program had met with reasonable success, it is quite possible that hypersonic boost-glide manned bombardment reconnaissance weapon systems could be operating in the entire boost-glide flight regime up to satellite speeds.

At this point, we might summarize the advantages of boost-glide. Hypersonic boost-glide vehicles offer a simultaneous break-through in speed, range, and altitude. The low flight wing loadings desired are possible because fuel and engines are not carried in the glider. The aerodynamic heating problem is reduced by flight at the upper limits of the sustained flight regime. Because it uses rocket propulsion, the boost-glide propulsion system is free of air breathing engine limitations, and these propulsion systems are already under development for the Air Force. Most important, the large contribution to lift that results from centrifugal force significantly improves the efficiency of the entire lift process.

One important feature of boost-glide vehicles is that they can be recoverable or non recoverable, and either manned or unmanned, and they represent a major step towards manned space flight. Another factor concerned with our question "why boost-glide" is mission capability. Since the development of the B-17, two primary reasons for replacing weapon systems have been to increase range and to decrease vulnerability. As the time of operational missiles approaches and as systems become more and more complex, two other factors become worthy of consideration. They are: (1) the yield-accuracy combination of the weapon system, and (2) the total system cost to perform the mission as defined.

In considering the strategic advantages of boost-glide systems, the following assumptions will be made: (1) The B-52 will remain in the inventory until at least 1970. (2) The B-58 will be in the inventory in 1962. (3) SNARK should phase-in beginning in 1959. (4) ICBM will be operationally available in 1962. (5 Weapon System 110A could phase in beginning in 1965. (6) Weapon System 125A will not be operationally available until 1970 or later, if at all.

The first question which must be answered is. Can the entire strategic mission in the post-1970 period be entrusted to ICBM? And the question is further complicated by the possibility of improved or advanced ICBM's. If ICBM's with CEP's of 500 feet can be operational by 1970, there may be no requirement, for a boost-glide bomber, but we cannot at this time safely assume that the 1970's will find us in a position to destroy all necessary targets with an ICBM. Also there still will be a requirement for a high performance reconnaissance system. Much time can be spent in arguing guidance state-of-the-art in 1975, but the problems of unlocated targets, hardened targets, expected CEP's warhead yields, reliability and combinations of these things seem to indicate that a 100% missile Air Force cannot perform the required mission in 1970 to 1975. At least, we cannot plan for missiles alone because, if it is determined later that another manned vehicle is required, it might be too late to begin development.

Let us assume that the entire strategic mission cannot be entrusted to ICBM. The next question which arises is if some targets require manned vehicles for destruction, why won't the manned vehicles in being and under development perform the mission? Our previous discussion of trends and characteristics of the classic sustained-flight approach to mission accomplishment answered this question in part. However, from a mission standpoint, the answer to this question is quite complicated. A recent SAC-Rand study investigated the capabilities of the B-52 to cope with the Russian defensive environment in 1960. In general, the conclusions were that the B-52, provided with ECM, air-to-ground launched decoys, ARM's and ASM's can do the required job in 1960. As enemy SAM's, supersonic fighters, radar, and IR detectors improve in capability, the B-52 will encounter even more difficulty in the ten year period following 1960.

Current studies are investigating the capabilities of combinations of B-52's, B-58's, and Weapon System

110A's to perform the required mission in 1965. Although the results of there studies are not available in final approved form, preliminary indications are that without decoys and other penetration aids, even a Mach 2 to 9 bomber may experience considerable difficulty in depositing bombs on some Russian targets in 1965.

In 1972, the B-52 would have been in the strategic inventory for approximately 18 years, the B-58 and Weapon System 110A may have been in the inventory seven to ten years. It is possible that all of these bombers will have difficulty in penetrating enemy defenses and destroying the required strategic targets. It is also possible that missiles will not be effective against hardened, dispersed and/or inaccurately mapped ICBM sites and other important strategic targets. A manned hypersonic boost-glide vehicle is a logical and attractive weapon system for this time period. In addition to requiring no refueling, it promises global range, multiple attack trajectories, a 3000' CEP, the capability of recall and detection warning time of three minutes as compared to fifteen minutes for ICBM.

It is felt that a SAC reconnaissance capability is extremely vital. In general what is required is a reconnaissance capability of high order photo, ferret, radar and infrared, with a capability of detection and identification of hard targets. It is also desirable that such a system have an immediate action capability - that is, a capability of providing the required reconnaissance data almost immediately after the initiation of hostilities. Some of these requirements may be met by the ARS system, however, the objectives of a boost glide reconnaissance system and those of ARS are quite different. The Advanced Reconnaissance System is intended to collect intelligence through routine surveillance, accepting whatever level of information detail it is possible to attain.

Boost glide reconnaissance vehicles, on the other hand, could collect detailed technical intelligence. The information could be obtained when it is desired due to the short flight times over areas as desired because it is controllable, at comparably low altitudes (25 to 50 miles as versus 300 miles). hence obtaining more detailed information. In addition, the vehicle is recoverable whereas the Advanced Reconnaissance system is not. It is appropriate to note at this time that these two systems are complementary, not competitive. Each has unique capabilities not attainable by the other.

Our analysis of the mission requirements raises a question as to the ability of air breathing engine aircraft to accomplish a SAC mission beyond 1965; and there are definite indications of the superiority of boost glide weapon systems in this time period. It is quite apparent that air breathing engine aircraft have a very limited growth potential for the future, whereas the growth potential for boost glide appears to be practically unlimited.

DEVELOPMENT CONCEPT

If it is decided that this is what we want, how do we go about the development of boost glide weapon systems? At this point then, we introduce a specific program. "Dyna Soar." The name stands for the combination of words Dynamic Soarer. The program consists of a development of three separate vehicles. Dyna-Soar I would be a conceptual test vehicle, designed, to obtain vital information is the boost-glide flight regime for both weapon systems envisioned, and is the test vehicle previously mentioned as a requirement for boost-glide development, Dyna-Soar II is a long range strategic reconnaissance weapon system and Dyna-Soar III, a global range strategic bombardment-reconnaissance weapon system.

In order to understand the meaning of "conceptual test vehicle," it is necessary to recognize the difference between conceptual test vehicle, research vehicle and prototype vehicle. All of these have one thing in common - each is a vehicle designed to obtain basic or fundamental flight data in an environment which has not been sufficiently well defined; but at this point, the similarity ends. A research vehicle normally is not developed to directly support any given weapon system development, although it may contribute to it. A prototype vehicle, on the other hand, is intended for use in development of a specific weapon system, and is part of the development program of that system. In addition, it flies in the same mode of operation as the actual weapon system. A conceptual test vehicle differs from these two in that it is intended for use in development of weapon systems utilizing a specific concept - developed to provide maximum data on the flight regime associated with the specific weapon system concept.

A research vehicle is designed to research areas, not solutions to problems. A prototype vehicle is designed to test solutions to the problems, not research areas, and is used in the development of operational subsystems. The conceptual test vehicle again differing from the other two types, is designed to research areas, test solutions to problems, and serve as a flight test bed for breadboard models and experimental prototypes of subsystems for the weapon system to follow.

As an example to each of these types, the X-15 exhibits the characteristics described under the research vehicle, the X-10 and the XSM-64 prototype vehicle for the Navaho, exhibit the characteristics described under the prototype vehicle, and the Dyna-Soar Conceptual Test Vehicle for Boost-Glide Flight-Regime of Brass Bell and ROBO type family of boost-glide vehicles will exhibit the characteristics described for the conceptual test vehicle.

Having defined the conceptual test vehicle, we will resort again to an altitude versus speed plot of boost-glide flight regime with the constant temperature lines imposed (Figure 18). Also superimposed on this chart are the flight areas investigated by the X-2 and the X-15 research vehicle and the initial glide points of Dyna-Soar II and Dyna-Soar III. The objective of Dyna-Soar I is to conduct flight testing in the boost-glide flight regime and obtain information of particular interest in the development of the two weapon systems, Dyna-Soar I will be designed to operate in the environment Dyna-Soar II is in at the initiation of glide.

Dyna-Soar I would be boosted to a speed of about 18,000 feet per second to an altitude on the order of 300,000 feet, which is considerably above the operational altitude of Dyna-Soar II of about 170,000 feet. As you can see from Figures 19 - 22, boosting Dyna-Soar I to an altitude of about 300,000 feet places it out of the most severe aerodynamic heating condition, so we may gradually enter the more severe aerodynamic heating area and ultimately arrive at the most severe condition of Dyna-Soar II at an altitude of about 170,000 feet. This approach should also be of considerable value in investigating the problems of stability and control of vehicles operating essentially outside the earth's atmosphere.

If this mode of operation is found to be successful, there is every reason to believe that it would be possible, with minimum modification of Dyna-Soar I, to give additional boost to the vehicle so that it could be boosted to the complete initial conditions of Dyna-Soar III. (See Figure 22) Using the same technique, it could work its way down the boost-glide flight regime with the possibility of completing the investigation. Boosting it to the conditions of Dyna-Soar III as an initial design goal is not considered advisable due to the complicated design and development of multiple boost systems, the more severe coolant requirements which would be imposed, and the sizable delays in initiation of a flight article which would be encountered through the necessity of designing Dyna-Soar I for Dyna-Soar III initial conditions.

Sufficient information is not available today to allow the initiation of a Phase I of any of the Dyna-Soar vehicles. Preliminary investigations, designed to provide data sufficient to enable the initiation of a Phase I of Dyna-Soar I must first be accomplished. These preliminary investigations will be done in a two part Pre-Phase I program. Objectives of part I are: (1) validation of assumptions, theory, and data gathered on the previous boost-glide study programs; (2) to provide sound basis for development of new theory and techniques required; (3) to provide design data; (4) to establish the capability of boost-glide vehicles; and (5) to determine the optimum flight profile for the conceptual test vehicle.

The requirements of part I of the Pre-Phase I program area that there be one contractor - that the data be non-proprietary - that there be dissemination of all data to all Dyna-Soar contractors participating in voluntary studies - and that it be a paid program.

The second part, that of design studies and system analyses, has as its objectives: (1) to refine the conceptual test vehicle design, (2) to establish the capability of the test vehicle, (3) to define subsystems and research instrumentation required, and (4) to define the vehicle as an operational system. The requirements here are that there be two or more contractors - that the work be done on a competitive basis, and therefore, the data be proprietary - the dissemination of the information will be to military only - and that it be a paid and/or unpaid program.

DEVELOPMENT SCHEDULE

The Pre-Phase I program (Figure 23) for Dyna-Soar I could last anywhere from 12 to 18 months before enough information was generated to justify the initiation of a Phase I program. With a reasonable amount of emphasis placed on the Pre-Phase I effort, Phase I could begin about one year from the start of the Pre-Phase I program. At that time, all of the contractors engaged in the Dyna Soar program would be invited to submit a design proposal for Dyna-Soar I. One contractor would be selected for the development of the conceptual test vehicle.

At the same time the Pre-Phase I program is started, design studies and tests of vehicles and subsystems for Dyna-Soar II and Dyna-Soar III will be initiated. The work conducted under Dyna-Soar II and Dyna-Soar III during this time period will be under the voluntary study program. Since the exploratory research part of the Pre-Phase I program is nonproprietary, all of the contractors involved in Dyna-Soar II and Dyna-Soar III studies will have been receiving valuable inputs from the conceptual test vehicle program. Once a contractor has been selected for the Phase I of the Dyna-Soar I program, all competition ceases with respect to Dyna-Soar I and all of the information generated under that program will be dissemination to all of the other contractors in the field participating in the weapon systems studies. It is planned that all of the contractors involved in the entire Dyna-Soar program will be brought together right in the beginning, before the initiation of the Pre-Phase I program, and briefed on this development philosophy.

The contractors participating under Dyna-Soar II and Dyna-Soar III will be expected to participate in the over-all development of these weapon systems in the following way. Although their design study and test work under these programs will be proprietary until a contractor is selected for each one, this approach will be used to provide a mutual exchange of data and information throughout the development cycle. Since the early realization of weapon systems under Dyna-Soar II and Dyna Soar III will depend upon how well Dyna-Soar I accomplishes its task of investigating the boost-glide flight regime, it is extremely important that the conceptual test vehicle be designed to provide data on components and subsystems of the type which will be contained in the weapon systems operating in the system environment.

Therefore, the contractors under Dyna-Soar II and III will be asked to submit recommendations as to the type of equipment and instrumentation which should be carried by Dyna-Soar I in its flight investigations. This will not compromise the contractors' position from a competitive and proprietary standpoint, since they will be asked to submit this information to the Weapon System Project Office managing the program. These suggestions and proposals will then be evaluated by appropriate government agencies, such as WADC and NACA, etc., and ultimately correlated midway through the Phase I program of Dyna-Soar I.

Final recommendations will be made by the Weapon System Project Office for the specific items to be designed or included in the conceptual test vehicle. They will include such things as the specifications for the installation and design of radar antennas, communication equipment, research instrumentation, and proposals for the solution to such problems as looking through a radiating and disassociated boundary layer. Using this approach, it is expected that the contractors may freely contribute their ideas and requirements and provide an input to the design of Dyna-Soar I without adversely affecting their relative position in the competition for the weapon systems. At the same time, all of the contractors will benefit from the results of the work conducted under Dyna-Soar I.

This approach is expected to accomplish the following.- (1) the realization of the truly non-proprietary conceptual test vehicle program, (2) maximum utilization of the conceptual test vehicle, (3) minimum development time and development risk for the weapon systems, (4) greatly reduced necessity for "production fixes," (5) maximum participation of all of the capable contractors in the field, (6) minimized development cost, (7) establishing of a common denominator for all of the contractors so as to base the selection of the contractor for each weapon system on the fairest and soundest basis, (8) encourage active participation by the contractors in the unpaid design studies and test program, at least until the contractor is selected for Dyna-Soar III, (9) provide for sufficient capability among the major aircraft producers in the country, so that if a multiple source development program is found to be necessary, the delays incurred in educating the second source can be avoided.

Phase I for Dyna-Soar II is shown on the development schedule as being initiated about the middle of 1962. Although the mock-up for Dyna-Soar II is shown in the middle of 1964, it will probably be available for inspection about a year earlier. Final approval of the mock-up is scheduled to occur after the Dyna-Soar I vehicle is tested at Mach 18. Phase I of the Dyna-Soar III vehicle is shown as being initiated early in 1964. If, after a year and one half of flight testing Dyna-Soar I, it is found that the flight technique previously described has been successful, minor modification and additional boosters will be accomplished on Dyna-Soar I, and flight testing near satellite speeds will be initiated early in 1966. The design information gained from this program will provide an early input in the design and fabrication stage of the development of the Dyna-Soar III and should greatly aid its development and fabrication. The estimated operational date for Dyna-Soar II is 1969, Dyna-Soar III, 1974.

Figure 24 shows the predicted funding trends for the Dyna-Soar program representing the total cost. No attempt has been made to assign a dollar value to the development of any of the Dyna-Soar systems. This chart is merely intended to indicate the trend of the costs over the years. It is expected that the peak of the funding of Dyna-Soar I will occur about 1962, the peak for Dyna-Soar II about 1965, and the peak for Dyna-Soar III about 1969.

Figure 25 shows the estimate of the P-600 funds required for the initiation of the development of Dyna-Soar program through the year 1961 and is compared on the chart with the funding schedule for the X-15. It is seen that the P-600 funding of the X-15 intersects the Dyna-Soar funding at approximately 1960, as the X-15 funding is phasing out. The modest funding of the Dyna-Soar program will not interfere with the funding of the X-15 program. The first flight date of Dyna-Soar I could be accomplished earlier if additional funds are made available during the fabrication phase.

So far as the Dyna-Soar III program is concerned, the development cycle is such that a 1974 operational system would be initiated by a GOR dated 1960. The recommended course of action, as already discussed, will allow a thorough examination of identifiable problem areas during the next two years, thus providing two significant advantages. (1) allow a lower total and rate of expenditure by making good use of the available time; (2) provide time to examine risks and consolidate the opinions of other agencies. Continued work might easily result in either a recommendation against a GOR or a valuable input into any GOR which is forthcoming.

CONCLUSIONS AND RECOMMENDATIONS

In conclusion, it is believed that the boost-glide concept is feasible, and that the classic sustained flight air breathing engine approach to strategic mission accomplishment is very limited and has very little growth potential. With respect to the reconnaissance boost-glide weapon system, it is believed that there is a very great need to provide "eyes" for the ballistic missile and boost-glide weapon systems, and therefore, a Dyna-Soar II type vehicle appears to be a very realistic design goal for the first weapon system.

Realizing that the development procurement, and operation of any future weapon system will have a basic effect on our economy we must look beyond technical feasibility in determining if we should embark on a new weapon system. We are planning to obtain cost estimates during the next two years, using data provided by RAND, AMC, SAC and the contractors.

It is, therefore, recommended that the first part of the Dyna-Soar program be initiated now at a low funding level of $3 million for FY 58. By substituting time now for a potential crash program if we hold up program initiation until several years later, we can determine while the monetary investment is still low whether or not such a system will meet our present expectations.

THE SUSTAINED FLIGHT REGIME

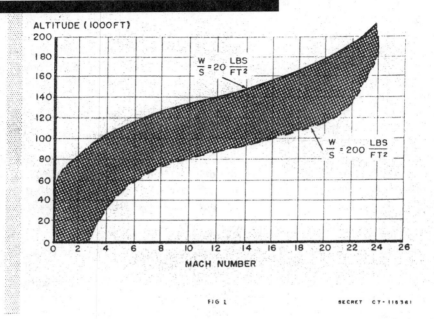

FIG 1   SECRET   C7- 118361

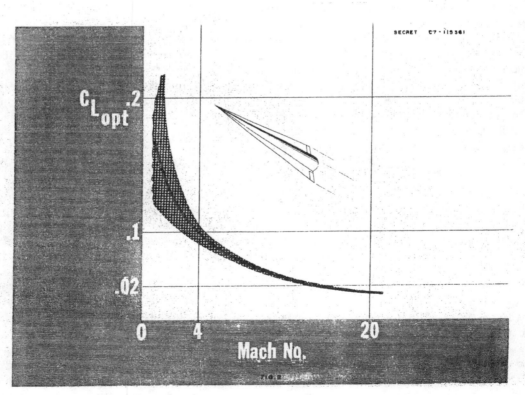

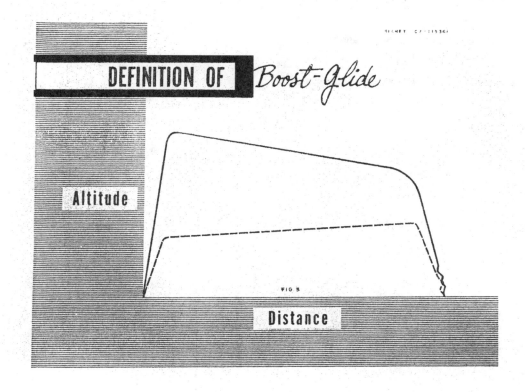

DEFINITION OF *Boost-Glide*

Altitude

FIG 3

Distance

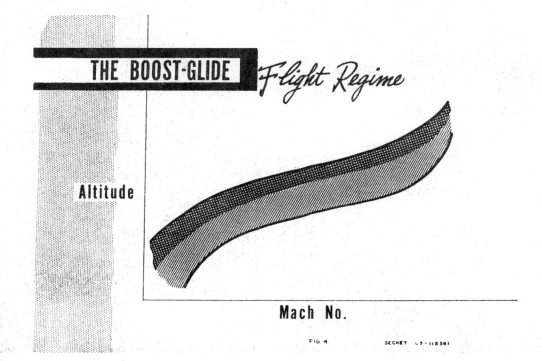

THE BOOST-GLIDE *Flight Regime*

Altitude

Mach No.

FIG 4

ALTITUDE TRENDS

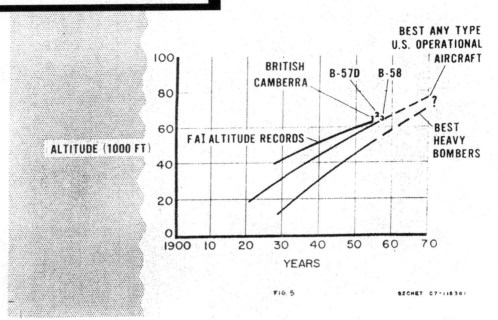

FIG. 5   SECRET C7-118361

SPEED TRENDS

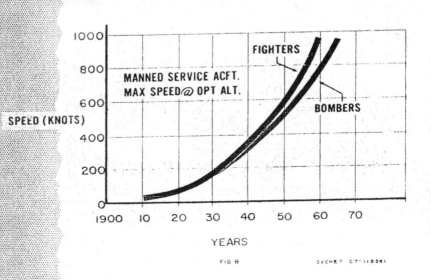

FIG 6   SECRET C7-118361

RANGE TRENDS

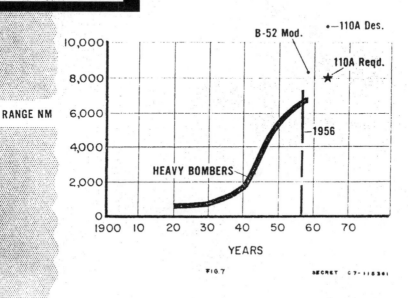

FIG.7   SECRET C 7-118361

L/D TREND

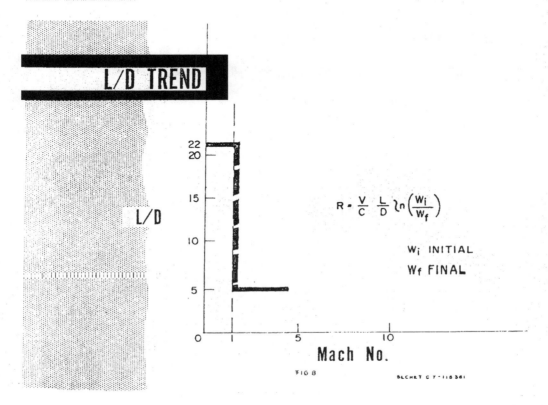

$$R = \frac{V}{C} \frac{L}{D} \ln\left(\frac{W_i}{W_f}\right)$$

W_i INITIAL

W_f FINAL

FIG 8   SECRET C 7-118361

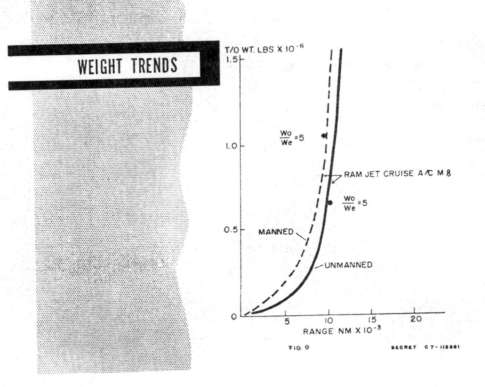

WEIGHT TRENDS

FIG 9

Upper and Lower limits dictated by aerodynamic consideration

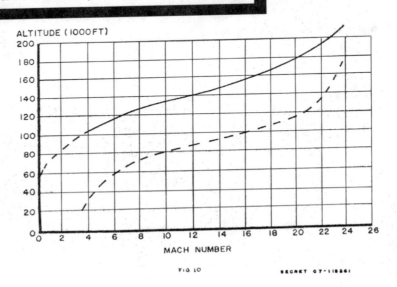

FIG 10

Aerodynamic Heating Predictions

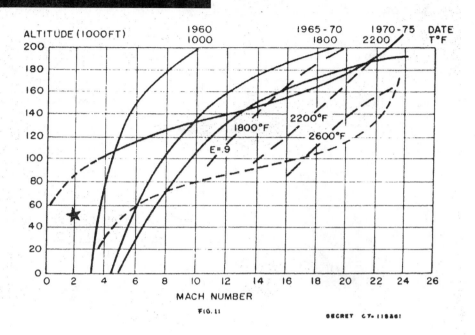

FIG. 11

SECRET CT-118261

Current air-breathing comparison

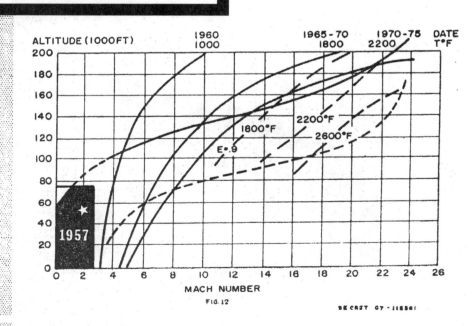

FIG. 12

SECRET CT-118261

Estimated potential capability for 1964

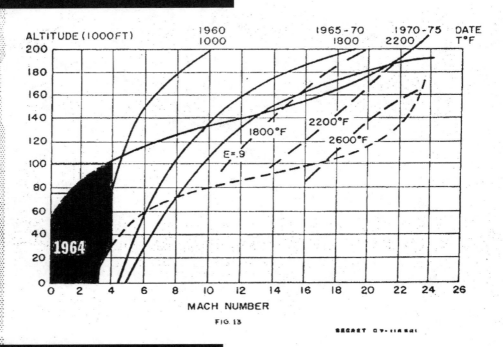

FIG. 13

Projections based on starting today

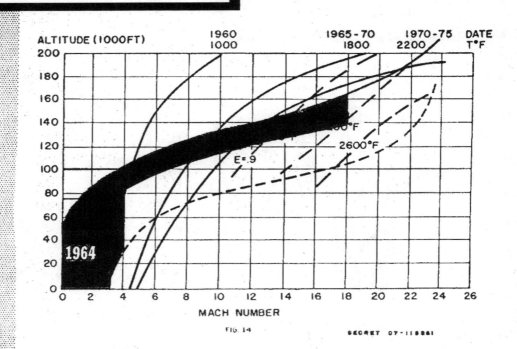

FIG. 14

1970 Prediction for Marquandt Hypersonic Ramjet

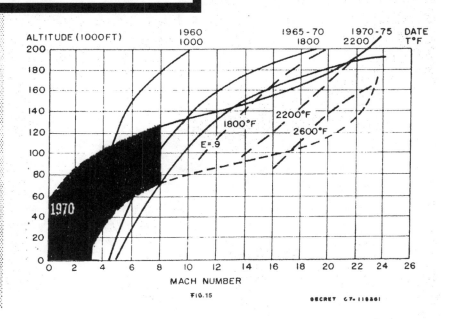

FIG.15

SECRET C7- 118361

Successful Conceptual test vehicle program

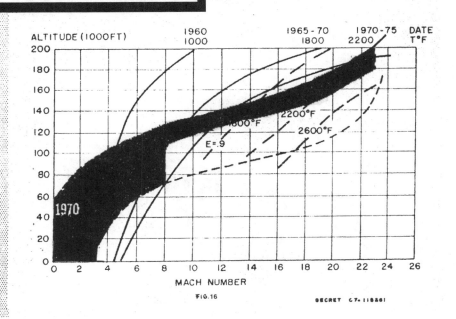

FIG.16

SECRET C7- 118361

DYNA-SOAR I TEST APPROACH

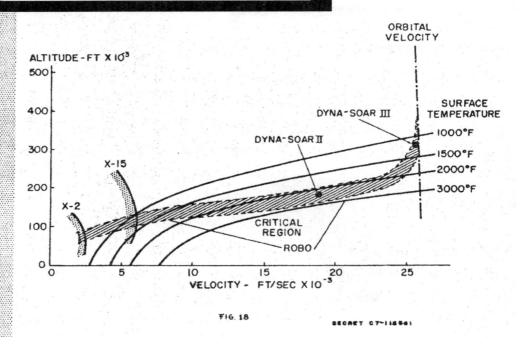

FIG. 18

CHARACTERISTIC FLIGHT OF DYNA-SOAR II

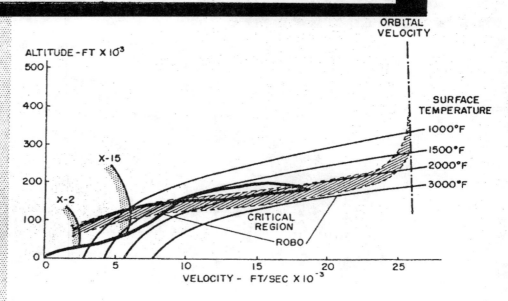

FIG. 19

CHARACTERISTIC FLIGHT OF DYNA-SOAR III

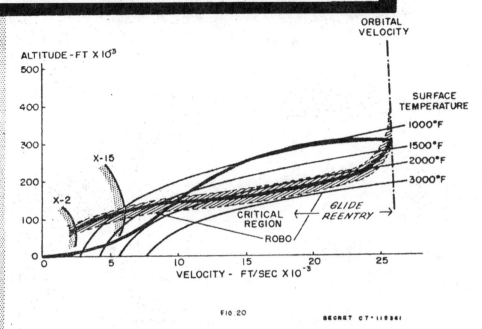

FIG. 20

SECRET C7-115341

DYNA-SOAR I

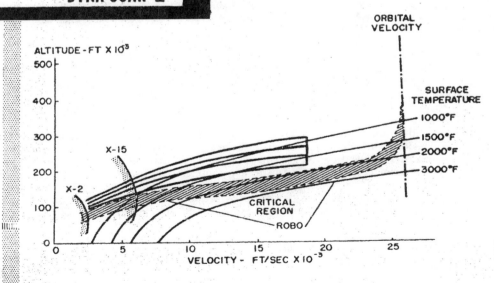

FIG 21

SECRET C7-115331

Date: Dec 1957   Title: Development Requirement System 464L Dyna-Soar

TAB A

DEVELOPMENT REQUIREMENT
HYPERSONIC CONCEPTUAL TEST VEHICLE
SYSTEM 464L PROJECT DYNA-SOAR

RDZSBA
20 December 1957

GROUP 4
Downgraded at 3 year intervals; declassified after 12 years

Revised: 29 January 1958

1.   OBJECTIVE:

1.1   The objective of the Dyna Soar I program is to confirm the feasibility of boost-glide weapon systems and gather design data for their most rapid development. Boost-glide (as defined for the Dyna Soar Program) is characterized by the boost to hypersonic speeds and high altitudes of winged vehicles which use aerodynamic lift to extend range, provide terminal flight path control, and, for recoverable vehicles, the capability of landing.

Dyna Soar II – A family of weapon systems developed at the earliest possible date, designed for a range of (5,000 - 6,000 N.M.). These may be useful in one or more of the following forms: (a) a manned or unmanned recoverable system with bombardment / reconnaissance system, (b) an unmanned one way bombardment / reconnaissance system.

Dyna Soar III – The ultimate objective is to have a manned or unmanned global recoverable bombardment / reconnaissance system available at the earliest possible date.

1.2   The Dyna Soar program will consist of study, hardware construction and testing.

1.2.1   A program of study and ground testing to determine the best flight paths and configuration for the Conceptual Test Vehicle (CTV) in order to obtain data in support of Dyna Soar II and Dyna Soar III type systems (proposed by the weapon system contractor).

1.2.2   A cost study to determine the over-all funding required for developing and operating Dyna Soar I, Dyna Soar II, and Dyna Soar III type systems as major elements of the national defense system.

1.2.3   The design construction and operation of a conceptual test vehicle to verify conclusions or further define problems indicated in the study and test program. The vehicle is expected to test not only aerodynamics, structures, propulsion, flight control, but also military mission problems such as operation of weapon system "breadboard components," store separation, photography through heated boundary layers, radio transmission through ionized layers, the value of manned operation. If practicable, growth potential of the Dyna Soar I should be considered so that with minimum modification, future development and test capabilities could be realized for such items as: (a) future boost-glide systems, including TAC, AD, logistics, and follow-on bombing and recon system, (b) manned controllable re-entry flights from extremely high ballistic trajectories to aerodynamically controlled flight, (c) sustained (such as orbital) test operations outside the atmosphere, and (d) test flight of the first flyable experimental nuclear rocket engine and other ultra performance propulsion systems. These will not, however, be allowed to compromise its prime purpose of most rapid accumulation of data for boost-glide weapon system development.

2.   SYSTEM CAPABILITY AND REQUIREMENTS:

2.1   Performance

2.1.1   The specific performance will be determined by the contractor.

2.1.2   A ground launched booster system capable of propelling the CTV to a speed of at least 18,000 ft/sec will be acceptable as an initial objective.

2.1.3   Performance estimates shall be based on use of the "ARDC Model Atmosphere 1956" and a rotating earth.

2.2   Operational Factors

2.2.1   The earliest possible flight date of Dyna Soar I vehicle, consistent with attaining the objectives of the complete Dyna Soar development programs, is required.

2.2.2   Careful consideration must be given to selection of launching and landing sites since the rage of the vehicle will very likely vary between less than 1,000 N.M., and, over 3,000 N.M. during the flight test program. This factor shall be taken into consideration in the design of the vehicle.

2.2.3   It is expected that drop tests (both unpowered and powered) from a suitable carrier aircraft, followed by ground launch with a booster system will characterize the Dyna Soar I flight test program. In any event. the size and weight of the vehicle must be compatible with all changes in mode of operation with a minimum of modification required to the vehicle system. The vehicle must be capable of landing at a predetermined landing site.

2.2.4   Economy of operation of the Dyna Soar I shall be an important design consideration within limits compatible with the overall objectives of the system, allowing for some compromises in performance, mode of operation, etc.

3.   VEHICLE DESIGN:

3.1   Stabilization and Control

3.1.1   A control system is required which will provide the necessary control and stabilization during all phases of flight, including flight at angles of attack corresponding to G_L maximum.

3.1.2   Controls shall be provided to permit changing airplane attitude and flight path in the absence of sufficient aerodynamic forces.

3.1.3   Reliability of the integrated automatic flight control systems shall be a prime design objective.

3.1.4   Adequate flight control shall be provided to ensure that allowable thermal, structural and human limits are not exceeded in flight during exit and re-entry from maximum speeds and altitudes.

3.2 Structure

3.2.1   The CTV shall be capable of flight investigation of all environmental conditions encountered by the Dyna Soar II and Dyna Soar III weapon systems and have design limit load factors of 8g axial and 7.33g normal.

3.3 Propulsion

3.31   The propulsion system shall be designed for highly automatic operation throughout the boost portion of flight

3.3.2   Reliability of booster units shall be a prime design requirement in order to provide adequate crew safety.

3.4 Crew Provisions

3.4.1.   Provisions for at least one (I) crew member pilot) required.

3.4.2   Provisions for cockpit pressurization and air conditioning and other environmental protection shall be adequate for all flight conditions.

3.4.3   Practicable escape provisions shall be provided for the crew. Consideration will be given to the use of an escape capsule and to use of the CTV and final boost stage for escape.

3.4.4   All instruments, equipment and controls and associated equipment necessary for the proper performance of the airplane shall be provided for the crew to enable satisfactory accomplishment of the mission.

3.4.5   Ground support activities (handling, control and maintenance), will not compromise ground crew environment, work load or safety.

3.5   Communication and Landing Aids

3.5.1   Air-to-air and air-ground-air communication capability is required to provide short range line of sight communication for launch, recovery, and vehicle/chase plane control and reporting. The 225 to 400 Mc band should be considered for voice and data transmission to and from the vehicle. Reliability is essential for this function as it may be the only tie to tie ground during testing.

3.5.2   Tracking Aids: Tracking aids are desired to enhance the ground radar tracking range capability and to provide identification to the ground Air Defense Electronic Environment.

3.5.3   Landing Aids: Suitable aids to safe recovery or landing consistent with practical vehicle weight and space considerations are required.

3.5.4   Rescue: Electronic aids to location and recovery of the crew and/or vehicle are desired.

3.6   Ground Handling and Test Equipment

Ground handling and support and test equipment is required concurrent with the delivery of the CTV and carrier aircraft.

3.7   Training Equipment

Suitable training aids are required concurrent with delivery of the CTV.

3.8   Navigation

3.8.1   An automatic system is required for navigating the CTV from its launch point to the initial point of approach control.

4   RESEARCH AND WEAPON SYSTEM COMPONENT TESTING:

4.1   The system is required to carry research instrumentation and component equipment to accomplish purposes as outlined below. It is not expected that all of these tests will be accomplished at the same time or that only the tests listed will be accomplished. The extent of capability of the CTV to meet all of the requirements will be determined by an evaluation of the trades and penalties involved and the needs of subsequent boost-glide weapon systems to be developed.

4.1.1   The vehicle is required to have the capability of exploring fundamental problems in areas such as geophysics, aerodynamics, materials structures, human factors, flight control, etc.

4.1.2   The system shall have the capability to stow and eject stores.

4.1.3   The system shall incorporate provisions for evaluating materials to be used as radomes and windows to transmit electromagnetic radiation, infrared and light waves. This installation should be sufficiently flexible to allow testing of various materials in the hypersonic environment to determine optical properties and physical characteristics. Reconnaissance sensors may be used as instrumentation tools. The system shall also be capable of evaluating electromagnetic, infra-red and photographic equipment. For this application, radiation from the window shall be minimized.

4.1.4   Long Distance Communication: Voice and/or automatic processed data transmission is desired for long range flights beyond line of sight. Determination of optimum frequencies consistent with vehicle performance, propagation and attenuation problems, and a practical ground electronic environment will be made jointly by the Air Force and the contractor.

<u>TAB B</u>

<u>PHASE I STATEMENT OF WORK</u>
<u>SYSTEM 464L (PROJECT DYNA SOAR)</u>

RDZSBA
20 December 1957
Rev. 21 January 1958

I. GENERAL:

1.1 The Contractor shall conduct a Phase I program, as outlined herein, leading towards the development of a Hypersonic Glide Rocket Conceptual Test Vehicle (Dyna Soar I) as defined by Development Requirements, Hypersonic Conceptual Test Vehicle, System 464L, dated 20 December 1957, incorporated herein by reference. This program is to consist of tests and investigations directly applicable to Dyna Soar I as required to provide the technical data and other information necessary to the successful development of this test vehicle. The contractor's work will be considered non-proprietary and will be disseminated in accordance with the requirements established by the USAF.

1.2 The objectives of the Phase I program are to:

1.2.1 Accomplish preliminary design and mockup of the conceptual test vehicle system.

1.2.2 Establish the capability of the test vehicle.

1.2.3 Define subsystems and flight test instrumentation required in the test vehicle.

1.2.4 Accomplish necessary investigations and tests designed to validate assumptions, theory data currently in use to the extent necessary to assure a reasonable success of the conceptual test vehicle.

1.2.5 Establish the practical limits of capability of Boost-Glide vehicles as applicable to DS II and DS III.

2. DESIGN:

2.1 The contractor shall accomplish a Phase I engineering design to meet the requirements of "Development Requirements, Hypersonic Conceptual Test Vehicle, System 464L dated 20 December 1957." In accomplishing this design, the contractor shall:

2.1.1 Conduct analyses and studies to determine the flight profiles for the Dyna Soar I vehicle, as defined in the development requirement, which will yield the optimum in useful information and data concerning the Boost-Glide flight regime.

2.1.2 Accomplish studies to determine optimum space internal allocations required to accomplish as much of the testing as can be accomplished consistent with economy and efficient system design of the type outlined in paragraph 4 of the development requirements. These studies shall include consideration of the effect on vehicle design of providing various volumes to house the necessary test equipment.

2.2 The Contractor shall construct a full scale mockup of the CTV and associated booster systems which reflect the results of the Phase I design. Specification covering these mockups shall be submitted for approval prior to initiating manufacture.

2.3 The Contractor shall conduct general studies to determine performance trade data and to determine limits on performance and physical parameters associated with the Dyna Soar I system to include, but not be limited to:

2.3.1 Investigation of the effect of designing the vehicle for various aerodynamic heating conditions which would be encountered in glide flight, including designs up to L/D max.

2.3.2 Compare performance trade data of Dyna Soar I designed to operate at different maximum speeds from 18,000 ft/sec to 25,000 ft/sec holding the range constant as a parameter. Determine the effect this would have on system as a whole, economy of operation and on the objectives and requirements of Dyna Soar I as specified in the development requirements.

2.3.3 Investigation of the effect on the objectives and requirements and performance of Dyna Soar I as defined in the Development Requirement, and in addition, on the timing and difficulty of development of the vehicle if it were designed for initial operation at satellite speeds and altitudes on the order of 250,000 to 400,000 feet.

2.3.4 Preliminary design studies of Boost-Glide Weapon Systems of the type defined in Paragraph 1.1 of the Dyna Soar I Development Requirement. These studies are for the purpose of defining the detailed design requirements for the Dyna Soar I System. The effort shall include cost trade-off studies for developing, producing and operating the various Dyna Soar II and Dyna Soar III type weapon systems as elements of the national defense systems.

2.3.5 Investigation of the practicability of providing a low thrust propulsion system to the Dyna Soar I vehicle to allow a period of constant altitude flight at various end of boost conditions for the vehicle defined in the development requirement.

2.3.6 Investigation of the use of a landing engine in the conceptual test vehicle.

2.3.7 Investigation of the early use of the DSI as an unmanned system to include effect on development schedule, cost, and comparison with equivalent manned system with regard to data provided, performance, and economy of operation.

2.4 The following are general procedures to be applied in accomplishing the engineering design called for in Item 2.1. Items 2.4.1 through 2.4.3 pertain to the design and Items 2.4. through 2.4.6 pertain to supporting investigations.

2.4.1 The contractor shall investigate the applicability of components and subsystems already in development or planned under other programs in the design of Dyna Soar I.

2.4.2 The contractor shall incorporate into the vehicle design as many of the suitable and currently obtainable materials, components and/or subsystems investigated as are compatible with the technical objectives and development schedule of Dyna Soar I. Careful consideration shall be given to the extent of modification required to the subsystem or component before any selection is made. The contractor shall substantiate the selection of these components or combination thereof with suitable trade data including economy of operation to indicate the penalty, if any, in utilizing these equipment rather than developing new equipment specifically designed to meet the requirements of Dyna Soar I.

2.4.3 The contractor shall define the requirements for all subsystems and components in the design of Dyna Soar I and provide preliminary performance specification for them, where no applicable military specification currently exists.

2.4.4 Existing data and information including that generated by other programs will be utilized to the maximum extent. The contractor shall correlate these data in accomplishment of the objectives of this program.

2.4.5 The contractor shall cooperate with and give assistance to other agencies, as designated by the USAF, in accomplishing the objectives of this program. This cooperation and assistance will include planning and establishment of study and test programs, construction of models and other test items, conducting tests, reducing data and preparing reports.

2.4.6 The contractor shall conduct, as designated by the USAF, studies, analyses and tests supplementary to the efforts required by Items 2.4.4 and 2.4.5 necessary to complete program objectives.

2.5 In accomplishing the engineering design called for in item 2.1 the contractor's work shall include but not be limited to the areas listed below:

2.5.1 The contractor shall investigate and, as approved by USAF, (Ref. 2.4.4 through 2.4.6), conduct tests to determine the effects of Geophysical factors on design and operation of

Dyna Soar I. At least the following factors shall be included:

2.5.1.1   Atmospheric Composition

2.5.1.2   Ambient Temperature, Pressure and Density

2.5.1.3   Condensation and Ionization Trails

2.5.1.4   Wind, Wind Shear and Gusts

2.5.1.5   Ionospheric and Auroral Characteristics

2.5.1.6   Solar Radiation, including the far Ultra-violet and X-Rays

2.5.1.7   Reflection, absorption and re-radiation of solar energy by earth and lower atmospheric layers

2.5.1.8   Sputtering

2.5.1.9   Ambient Electric and Magnetic Fields

2.5.1.10 Meteors

2.5.1.11 Weather Effects on Radar

2.5.2   The contractor shall conduct with respect to Dyna Soar I investigations and, as approved by USAF, (Ref. 2.4.4 through 2.4.6) tests in the fields of Aerophysics, Aerothermodynamics and Aerodynamics and Dynamics to include, but not limited to:

2.5.2.1   Investigations of the following:

2.5.2.1.1 Configuration studies

2.5.2.1.2 Optimum staging and separation of booster stages

2.5.2.1.3 Recoverable boosters

2.5.2.1.4 Control Systems (aerodynamic reaction type etc.)

2.5.2.1.5 Launching Problems (aircraft drop and multi-stage boost)

2.5.2.1.6 Aeroelasticity (Flutter)

2.5.2.1.7 Dynamic Loads (Gust penetration, landing impact abrupt control disp. Boost and launch phases)

2.5.2.2   Detailed analysis and tests of the following as they effect Dyna-Soar I throughout its environments

2.5.2.2.1 Real gas effects

2.5.2.2.2 Disassociation and ionization

2.5.2.2.3 Boundary layer characteristics

2.5.2.2.4 Turbulent and laminar flow characteristics

2.5.2.2.5 Interference Effects

2.5.2.2.6 Aerodynamic stability

2.5.2.2.7 Aerodynamic heating

2.5.2.2.8 Pressure distribution

2.5.2.2.9 Stagnation properties

2.5.2.2.10 Vibration and Noise

2.5.3 The contractor shall conduct with respect to Dyna-Soar I investigations and tests approved by USAF (Ref. 2.4.4. through 2.4.6) in the field of structures to include:

2.5.3.1 Detailed analysis and tests to define the materials requirements and production methods for DS I. For new or improved materials include requirement dates and substantiation of their suitability. Particular attention should be paid to the producibility and fabrication-ability of these new materials.

2.5.3.2 With respect to the design of Dyna-Soar I, the contractor shall conduct analyses and tests to define the requirements for:

2.5.3.3.1 Cooling System

2.5.3.3.2 Heat protected structures (cooled and uncooled).

2.5.3.3.3 Sandwich construction

2.5.3.3.4 New manufacturing and fabrication processes

2.5.3.4   The contractor shall prepare a structural design criteria specification for Dyna Soar I.

2.5.4   The contractor shall conduct studies and tests as approved by USAF (Ref. 2.4.1 through 2.4.6) to determine the requirements for and the availability of the Dyna Soar I propulsion system to include:

2.5.4.1   Engines:

2.5.4.1.1 Liquid and solid propellant rocket

2.5.4.1.1.1 Performance Characteristics (Thrust Magnitude)

2.5.4.1.1.2 Physical characteristics

2.5.4.1.1.3 Controls

2.5.4.1.1.4 Throttleability

2.5.4.1.1.5 Reliability

2.5.4.1.1.6 Safety features

2.5.4.1.1.7 High energy propellants

2.5.4.1.1.8 Exhaust products control

2.5.4.1.2 Air Breathing Booster Engines

2.5.4.1.2.1 Turbo jet

2.5.4.1.2.2 Ram jet

2.5.4.1.2. Dual cycle

2.5.4.1.3 Landing engine

2.5.4.1.3.1 Performance Characteristics

2.5.4.1.3.2 Environmental criteria

2.5.5   The contractor shall conduct investigations and tests as approved by USAF (Ref. 2.4.1 through 2.4.6) to establish the subsystem requirements for the Dyna Soar I vehicle.

2.5.6   With respect to Dyna Soar I, the contractor shall conduct investigations and tests as approved by USAF (Ref. 2.4.1 through 2.4.6) to establish requirements in the field of human factors far both ground and flight crew. This shall include investigation of providing positive escape for crew and trade data associated with providing less than this capability.

2.5.7   The contractor shall conduct investigations to establish the requirements for flight testing the Dyna Soar I vehicle.

3.   DATA REQUIREMENTS:

3.1   The contractor shall prepare and submit all reports, drawings, specifications and other documents in accordance with Table I to:

Commander
Air Materiel Command
ATTN: MCXH
Wright- Patterson Air Force Base, Ohio

3.2   The quantities and the time schedule for the submittal of the data shall be in accordance with Table I. These data shall cover results of work accomplished under paragraph 2 of the Work Statement, as specifically referenced in the column of Table I designated "Work Paragraph Reference". The front cover and edge binding, where practical, shall contain the title, in accordance with Table I, and the company report number. It is desired that the report cover be light in color to facilitate marking.

TAB C

GENERAL MANAGEMENT PROPOSAL FOR SYSTEM 464L

A General Management Proposal in 25 copies will be submitted by 24 March 1958. Only a single proposal will be accepted from each contractor. The General Management Proposal shall include a management section and a technical approach section.

A. General Management Section. This section is not to exceed 25 pages in length. The required content is as follows:

(1) The contractor's statement of his general experience applicable to this type of system in areas of R&D, development engineering, test and production.

(2) Contractor shall indicate the organization he proposes for the accomplishment of System 464L and shall also indicate the quantities and grades of personnel to be assigned to the project and name all key personnel. For key personnel, full or part time assignment or utilization as consultants shall be specified. A chart shall be prepared to show the numbers of personnel to be allocated by months, including subcontractor personnel, and to distinguish various classes of personnel, such as engineering, technicians, supervisors, draftsman, etc.

(3) Contractor shall indicate the approximate completion dates of the various phases of the project as he would plan such phases. These phases are to be outlined. This will include dates for production release as required, and date of delivery production articles and flight test.

(4) Contractor shall indicate his plans for subcontracting parts of the project and to whom, by name(s), he would consider awarding such subcontracts.

(5) Cost estimate information will be supplied by the contractor based on the same phases as given above in (3) above. This information should be broken down into subtotals showing the apportionment of costs to engineering, facilities, fabrication, testing and subcontracts. Funding requirements for FY 58/59 also will be indicated.

(6) Contractor shall indicate in general terms his own company physical resources and those Government facilities presently under his or the subcontractors control that are contemplated for use in the project.

(7) Contractor shall indicate type of contract desired for R&D, production phases and profit provisions anticipated.

TABLE I

Item	Title	Work Paragraph Reference	Preliminary Ref No Later than 9 Mos. From Date of LTR Contract	Reproducible Copy	Data Required at End of Phase I or Periodic as Shown
1	Program Planning	N/A		Yes	3 mos and every 3 mos thereafter
2	Progress Reports	N/A			Every 3 mos.
3	Test Specifications	2.5			
4	Detail Model Specifications	2.1 2.4 2.5	X	Yes	X
5	Weight and Balance Report	2.1	X		X
6	General Arrangement Drawing	2.1	X		X
7	Combined Inboard Profile and Fixed Equipment Drawing	2.1	X		X
8	Standard Aircraft Characteristics Charts	2.1	X		X
9	New Materials Requirements	2.1 2.5	X		X
10	Structural Design Criteria Specification	2.5.4	X		X
11	Propulsion System Specification	2.1 2.4 2.5	X	Yes	X
12	Subsystem Performance Specifications (Nav, Com, Flt Control, Res Ins, Test Bed Prov.)	2.1	X	Yes	X
13	Summary Report	N/A	X	Yes	X
14	Systems Operational Concept Report	2.1	X		X
15	System Design Report	2.1	X		X
16	Geophysics Report	2.4 2.5	X	Yes	X
17	Aerodynamics Report	2.1 2.4 2.5	X	Yes	X
18	Flight Control Report	2.1 2.4 2.5	X	Yes	X
19	Structures Report	2.1 2.4 2.5	X	Yes	X
20	Propulsion Report	2.1 2.4 2.5	X		X
21	Human Factors Report	2.1 2.4 2.5	X		X
22	Subsystems Report	2.1 2.4 2.5	X	Yes	X
23	Flight Test Support Requirements Report	2.1 2.5	X	Yes	X
24	Flight Test Program Report	2.1 2.5	X		X
25	Test Reports	2.5		Yes	45 days following completion of test
26	Special Studies	2.3	X	Yes	Every 3 mos.
27	Other Data, Reports and Drawings	N/A	X		X
28	Motion Pictures Documentation	N/A			X

(8)   Contractor shall indicate in general terms what additional facilities he will expect the Government to supply for his own and his subcontractor's R&D, testing, and potential follow-on production.

B.   Technical Approach Section. This section shall not exceed 100 pages and shall include at least the following information:

(1)   A statement of the contractors concept for the overall technical approach to be utilized in the development of DS Weapon Systems.

(2)   The contractors proposed design approach for DS I.

(3)   Capability of contractor's proposed design in providing basis for development of future boost-glide weapon systems.

(4)   Major problems areas foreseen by the contractor, with contractor's ideas and technical approaches for solution.

(5)   Test programs planned in the development and flight test of the test vehicle.

Date: Jan 1958   Title: Letter to Hugh Dryden from USAF

Department of the Air Force
Headquarters United States Air Force
Washington 25, D.C.

31 January 1958

Dr. Hugh L. Dryden Director
National Advisory Committee for Aeronautics
1512 H Street, N.W
Washington 25, D. C.

In the last few months the dimensions of the contest for superiority in aircraft and missile technology have suddenly and drastically expanded.

This letter is addressed to a particularly important event in this contest - the matter of a research vehicle program to explore and solve the problems of manned space flight. Specifically, the Air Force is convinced that we must undertake at once a research vehicle program having as its objective the earliest possible manned orbital flight which will contribute substantially and essentially to, follow on scientific and military space systems.

The Air Force has set up a design competition for a hypersonic boost glide vehicle nicknamed Dyna-Soar 1. The objectives of this program closely conform to the recommendations of the NACA report of last summer. It appears probable that this vehicle will be able to orbit as a satellite since the aerodynamic heating problems of reentry appear less severe than those of the Dyna Soar 1 flight profile. However, it may be feasible to demonstrate an orbital flight appreciably earlier with a vehicle designed only for the satellite mission than would be possible with a vehicle capable of the boost-glide mission as well. It is necessary, therefore, to determine whether a research aircraft designed only as a satellite will give us an orbital flight of technical significance enough sooner than a vehicle designed for the glide mission to warrant a separate development.

Both the NACA and the Air Force are well along in investigations seeking the best approach to the design of a manned earth orbiting research vehicle. We earnestly believe that these efforts should be joined at once and brought promptly to a conclusion. Accordingly, the NACA is invited to collaborate with the Air Research and Development Command in this important task. Because of the advanced. stages to which the individual NACA and ARDC investigations have already progressed and because of the urgency of getting on with the job, we believe that the evaluation should be confined to existing and planned projects, appropriate available proposals, and competitive approaches already under study. We visualize that any program growing out of this joint evaluation will best be presented, managed and funded along the lines of the X-15 effort, with the Navy being brought into the picture as soon as possible without delaying further evaluation.

To provide further insight into Air Force thinking on this matter, the concluding paragraphs of the letter directing ARDC to make this evaluation are quoted:

"4.... It is desired that the evaluation consider separately the following approaches;

"a. What is the best design concept, the minimum time to first orbital flight and the dollar cost of demonstrating a manned one-orbit flight in a vehicle capable only of a satellite orbit? Time is a primary consideration, but to qualify, an approach must offer prospects of tangible contributions to the over-all Astronautics program.

"b. What is the minimum time to first orbital flight and dollar cost of demonstrating a manned one-orbit flight with a vehicle designed to utilize the boost-glide concept? In this approach it is not necessary that the first orbit flight be made within the atmosphere under typical boost-glide conditions — it could be made outside the atmosphere if an "outside" orbit offered the possibility of an earlier successful flight.

"5. The following additional guidance is provided:

"a. The program to meet the stated objective should be the minimum consistent with a high degree of confidence that the objective will be met. Maximum practical use must be made of existing components and technology and of the momentum of existing programs.

"b. The hazard at launch and during flight will not be greater than that dictated by good engineering and flight safety practice. If feasible, in order to save time and money, pilot safety may be provided by emergency escape systems rather than insisting on standards of component reliability normally required for routine repetitive flights of weapon systems. This statement is particularly pointed at the problem of qualifying boosters for initial orbital flights.

"6. It is requested that this Headquarters be furnished the results of your evaluation of each of the approaches specified in paragraph 4. Finally, your over-all conclusions and recommendations for accomplishing the objective stated in paragraph 1 are desired.

"7. The requested information should be forwarded at the earliest practicable date, but in no event later than 15 March 1958."

It is hoped that the Air Force-NACA team relationship which has proven so effective in earlier programs of the X-airplane series can be continued in the conception and conduct of this and other research vehicle programs directed to the extension of our knowledge and capability in upper atmosphere and space operations.

We look forward to receiving your comments and suggestions to this proposed course of action.

Sincerely

D.L. Putt
Lieutenant General
USAF Deputy Chief of Staff,
Development

Date: Feb 1958   **Title:** Proposal for Memo of Understanding between NACA & USAF

Suggested Memorandum of Understanding Between Air Force and NACA for Conduct of Dyna Soar I Project

A. The Dyna Soar Conceptual Test Vehicle shall be a joint project between the Air Force and NACA to explore the boost-glide concept of long-range hypersonic flight and gather design data for rapid development of boost-glide weapons systems.

B. Over-all direction of the project shall be the responsibility of (some Headquarters Air Force General) acting with advice and assistance of the Director, NACA.

C. The NACA will be responsible for the applied research to be conducted with Dyna Soar I. The data procuring system for the project will also be the technical responsibility of the NACA.

D. The Air Force will be responsible for the weapons systems development to be conducted with Dyna Soar I.

E. The Air Force will finance the project and administer the design and construction phases of the project.

F. Upon acceptance, the vehicle and its related equipment will be turned over to a joint operations committee of the AFFTC and NACA HSFS which shall conduct the flight tests. The NACA will process the data and supply the designated Air Force organizations with copies.

G.   The Director, NACA, will be responsible for disseminating technical information from the applied research accomplished with Dyna Soar I. The Air Force will use all the information for periodic evaluations of the advantages and disadvantages of boost-glide weapon systems and of the advisability of proceeding with Dyna Soar II and/or Dyna Soar III.

The conceptual test vehicle legally has to be related to a weapon system to expend weapon system funds. It would be desirable that entanglement be kept to a minimum with the realization that the work should be aimed toward a weapon system. Weapon systems can be better handled by a different group who could monitor, rather than direct, to see that their needs be given first priority. This group would take data, integrate it into their considerations, and make decisions as to desirability of the use of a boost-glide for weapon system and, if it is concluded desirable, the time at which to start weapon system hardware. This division of the D.S. project would simplify future management problems although at a later time might bring about conflicts for priorities.

It would be extremely desirable for NACA to be made an active partner in D.S. I project because the Air Force organization cannot adequately handle such a project. The alternate is an equivalent to BMD who could hire someone like R-W to play the NACA role.

If NACA is to be a partner, a memo of understanding will be needed because of the changing Air Force personnel on a project.

Whether or not NACA participates as a partner, the government – the NACA and the Air Force, or the Air Force alone – should conduct the major portion of the flight studies to be made with the vehicle. The Air Force cannot otherwise be in a position to appreciate the significance of findings relating to military mission. The cost and time of operations will be so high as to preclude independent investigations by the contractor and the Air Force.

Langley Field, Va.
February 13, 1958

Date: Feb 1958   **Title:** NACA memo for clarification of NACA role

Washington, D. C.
February 25, 1958

MEMORANDUM For The Director

Subject: Need for Clarification of Role of NACA in Development of Dyna Soar I

1. On February 6, 1958, Mr. Soulé met with General Haugen and Mr. Arthur Boykin (ARDC Detachment #1 Technical Director for Aircraft) in an effort to clarify the responsibilities of NACA in the development of Dyna Soar I, the "conceptual test vehicle" for boost-glide. ARDC would like to have extensive participation by NACA in the development of the Dyna Soar I. At the same time, in keeping with the only instructions (the Directive for the Dyna Soar Program), which have reached ARDC, NACA would have to participate as a consultant, just as in the case of any Weapons Systems.

2. On February 10, 1958, Mr. Soulé and I met with Mr. William Lamar (Assistant Chief, Bombardment Aircraft Division), to discuss NACA participation in evaluation of the design proposals which are to be submitted by the prospective contractors by March 24th. We found that we were unable to reach any conclusion until the NACA's role in the Dyna Soar I program is clarified. As things now stand, the NACA would participate in the evaluation in an advisory capacity. If their advice were not taken the NACA representatives would not be obligated to press their case in the areas of NACA competence (e.g., aerodynamics, structures, propulsion, flight operations.)

3. As NACA has previously advised ARDC, we believe that the best way to develop the conceptual test vehicle is to establish a joint Air Force-NACA program along the lines of the X-15 project. Because the Dyna Soar I must be more closely related to the Weapons Systems which will follow it, than were the X-15 and the preceding research airplanes, it appears that technical direction of the project should be the responsibility of the Air Force. As in the case of the X-15, it would be desirable to have final responsibility lie with an Air Force-NACA Committee. In this case, perhaps the Air Force member of the committee should be Chairman. There should be in writing an understanding that this is a joint project and that the NACA has certain responsibility in the program.

4. It appears to us that we can go no further in arranging details of NACA participation in the Dyna Soar program until the NACA's role in the program are clarified.

5. There is another point worth noting. The companies have been asked to propose the flight test program in its entirety for the conceptual test vehicle, with the understanding that the contractor would conduct the flight test program in its entirety through the Project Office. We view this possibility with concern. In addition to removing NACA pilots and research scientists from their usual role in obtaining research data, it would deprive Air Force pilots of operating experience upon which to evaluate the boost-glide concept. This is in direct contrast to the Air Force-NACA team approach which has been evolved to make full use of the X-15 research airplane.

Clotaire Wood

CloW:dlf

Date: Mar 1958   **Title:** Memorandum of understanding between NACA & USAF

19 March 1958

DRAFT MEMORANDUM OF UNDERSTANDING

Principles for Participation of NACA in Development and Testing the "Air Force System 464L Hypersonic Boost Glide Conceptual Test Vehicle (Dyna Soar I)".

1. System 464L is being developed to:
    (a) Determine the military potential of hypersonic boost glide type weapon system and provide a basis for such developments
    (b) Research characteristics and problems of flight in the boost glide flight regime up to and including orbital flight outside of the earth's atmosphere.
2. The following principles will be applied in conduct of the project:
    (a) The project will be conducted as a joint Air Force-NACA project.
    (b) Overall technical control of the project will rest with the Air Force, acting with the advice and assistance of the NACA. The two partners will jointly participate in the technical development to maximize the vehicle's capabilities from both the military weapon system development and aeronautical-astronautical research viewpoints.
    (c) Financing of the design, construction, and Air Force test operation of the vehicles will be borne by the Air Force.
    (d) Management of the project will be conducted by an Air Force project office within the Directorate of Systems Management, Hq ARDC. The NACA will provide liaison representation in the project office and provide the chairman of the technical teams responsible for data transmission and research instrumentation.
    (e) Design and construction of the system will be conducted through a negotiated contract with a prime contractor selected by the USAF on the basis of the recommendations of the ARDC-AMC-SAC-NACA Source Selection Board.
    (f) Flight test of the vehicle and related equipment will be Accomplished by the NACA, the USAF, and the prime contractor in a combined test program under the overall control of a joint NACA-USAF Committee, chaired by the Air Force.

FINAL MEMORANDUM of UNDERSTANDING

Subject: Principles for Participation of NACA in Development and Testing of the "Air Force System 464L Hypersonic Boost Glide Vehicle (Dyna Soar I)".

1. System 464L is being developed to:
    a. Determine the military potential of hypersonic boost glide type weapon systems and provide a basis for such developments.
    b. Research characteristics and problems of flight in the boost glide flight regime up to and including orbital flight outside of the earth's atmosphere.

2. The following principles will be applied in conduct of the project:
    a. The project will be conducted as a joint Air Force-NACA project.
    b. Overall technical control of the project will rest with the Air Force, acting with the advice and assistance of the NACA. The two partners will jointly participate in the technical development to maximize the vehicle's capabilities from both the military weapon system development and aeronautical-astronautical research viewpoints.
    c. Financing of the design, construction, and Air Force test operation of the vehicles will be borne by the Air Force.
    d. Management of the project will be conducted by an Air Force project office within the Directorate of Systems Management, Hq ARDC. The NACA will provide liaison representation in the project office and provide the chairman of the technical team responsible for data transmission and research instrumentation.
    e. Design and construction of the system will be conducted through a negotiated contract with a prime contractor selected by the USAF on the basis of the recommendations of the ARDC-AMC-SAC Source Selection Board, acting with the consultation of the NACA.
    f. Flight test of the vehicle and related equipment will be accomplished by the NACA, the USAF, and the prime contractor in a combined test program under the overall control of a joint NACA-USAF Committee, chaired by the Air Force.

General Thomas D. White Chief of Staff, USAF 13 May 1958
Hugh L. Dryden Director, NACA 20 May 1958

Date: Apr 1958   Title: Problems of manned orbital vehicles

NATIONAL ADVISORY COMMITTEE FOR AERONAUTICS RESEARCH MEMORANDUM
OPERATIONAL PROBLEMS OF MANNED ORBITAL VEHICLES

By Hubert M. Drake, Donald R. Bellman, and Joseph A. Walker

SUMMARY

Manned orbital vehicles, because of their extreme performance and relative inflexibility of operation, introduce many problems in the fields of escape, piloting, orbit selection, flight termination, and range requirements. The effects of some of these problems, including some effects of configuration, are discussed.

No insurmountable operational problems were found regardless of configuration, but it is indicated that the problems of the various vehicle types materially affect operations and must be considered early in design. Safety and survival requirements may force appreciable deviation from optimum procedures. The presence of the pilot may simplify design and increase reliability. The type of vehicle has a considerable effect on range requirements, the more simple vehicles generally requiring increased range and recovery complexity.

INTRODUCTION

Manned vehicles of orbital performance potential introduce many operational problems as a result of their extreme performance and the relative inflexibility of their operations. The present paper discusses a few of the major problems and indicates their possible effects on flight research operations.

The vehicles being considered for possible use as manned satellites fall into the three general categories shown in figure 1. Briefly, the first vehicle is the ballistic-type, characterized by the use of drag alone for entry deceleration and heat-load reduction. The second category, the semi-ballistic vehicle, employs lift to reduce the peak decelerations and to provide some degree of aerodynamic flight-path control. The final category consists of what might be termed winged vehicles; that is, vehicles capable of aerodynamically efficient flight. It should be noted that this category may also be considered of the semi-ballistic type because lift-drag ratios as low as zero can be obtained by operating at high angles of attack. In general, only the ballistic and winged types are discussed in detail, inasmuch as the capabilities of the semi-ballistic-type fall between these extremes.

Although both vertical rocket-boost take-off and air-launch might be considered for orbital flight, the major operational problems for the two types of launch differ only during the initial phases. Since the vertical take-off presents the more stringent problems, it is the type considered herein. The general problem areas of escape, piloting, orbit selection, entry and flight termination, range requirements, and flight test program are discussed briefly.

SYMBOLS

H	altitude, miles
L/D	lift-drag ratio
q	dynamic pressure, lb./sq. ft
V	velocity, ft/sec
V_v	vertical velocity, ft/sec
V_{orb}	orbital velocity, ft/sec
Y	lateral distance, miles
γ	flight-path angle, deg

DISCUSSION Escape and Survival

The presence of the human in the orbital vehicle requires that malfunctions be either nondestructive, or such that an escape system can provide survival. The provision of a means of escape from all reasonable emergency conditions requires an escape system with all the characteristics of the final vehicle. It therefore appears that the final stage should be designed to serve as the major element of the escape system.

A primary goal should be the design of the final stage and the tailoring of the entire flight operation to provide the greatest possible survival potential for this stage. In addition, a positive means of pilot separation from the final stage survival vehicle, such as a high-performance ejection seat, is required to permit the pilot to use his personal parachute for low-speed survival.

The presence of propellants, the take-off operation, stage separation and ignition, and high dynamic pressure combine to make the launch operation the most critical escape region (fig. 2). Significant survival regions are indicated generally by the lettered areas on the launch trajectory of figure 2 and are further described in table I. The boundaries of these various escape areas vary, of course, with the configuration and its design characteristics. In some designs or operations a given region might not exist, being absorbed by an adjacent one. An example of this is the case of air launching, where region A is in general absorbed by region B.

In the present discussion only two of the indicated regions, A and C, are discussed in any detail.

Escape at lift-off, region A, is difficult in that the use of the ejection seat would require its reorientation, and the normal final stage power plant, in general, possesses insufficient acceleration to permit satisfactorily rapid separation from a malfunctioning first stage. A possible escape technique consists of providing high-thrust, jettisonable, solid rocket units attached to the final stage survival vehicle. Rockets sufficient to remove this vehicle from the launching pad to an altitude of 1,000 feet and a speed of 300 knots within 3 to 4 seconds would probably be adequate for escape from all take-off accidents not involving an actual detonation. This end condition permits airplane final stages to be airborne and, should this stage have an engine, allows sufficient time for an attempted engine start. Should the engine fail to start, a gliding landing can be made if possible, or the ejection seat may be used. Ballistic or semi-ballistic vehicles can make a normal parachute landing. These auxiliary rockets and any necessary stabilizing surfaces should be retained to the altitude at which a normal separation and recovery can be made. During an investigation of an example of such a system, it was found that carrying the system to an altitude of about 20,000 feet reduced the first stage burnout velocity by only about 50 feet per second. It might be mentioned that, although the ejection seat is listed only for region A, it is assumed to be available in all cases for use at low speed as needed.

Another critical area for escape and survival is that indicated as region C in figure 2 and table I where a malfunction may subject the final stage to conditions which it, or its passenger, cannot survive. The criticality and extent of this region are greatly influenced by such design and operational factors as the type of vehicle, use of final stage power,. launch trajectory, lift-drag ratio, structural design, and lift or drag loading. With the ballistic vehicle in this region there is a danger that the man will be subjected to excessive decelerations in case of booster malfunction as shown in figure 3. In this figure the solid line indicates the decelerations encountered in the event of a malfunction during the normal gravity-turn launch of a ballistic vehicle. The decelerations in this case reach values of about 22g at a launch malfunction velocity near 15,000 feet per second. Substantially higher values are possible for other ballistic configurations. It might be well to note that the final satellite vehicle would also have a peak deceleration near 22g at malfunction speeds near 2,000 feet per second if it were separated from the boosters at this point. This results from the high dynamic pressure at this point. The lower decelerations shown for these low speeds result from retaining with the vehicle the final boost stage, unfired, to increase its sectional density during the coast to high altitude following malfunction.

Separation of the final vehicle at the peak of the coasting period will then result in decelerations near 2g.

Possible means of reducing the decelerations resulting from malfunctions at the higher speeds during launch of course include the use of lift (as with the semi-ballistic vehicle), provision of thrust to reduce flight-path angle, and variable drag geometry. Another possibility is the use of a launching trajectory which has been modified in such a manner that the vehicle will, in case of booster failure, always enter the atmosphere at a sufficiently flat angle to keep the decelerations to a tolerable level. A first approximation to such a trajectory has been calculated for the example ballistic vehicle and the resulting decelerations are shown as the line labeled "safety" trajectory (fig. 3). In this case the peak decelerations have been reduced one-half, with a peak value of about 11 g.

Figure 4 shows that this "safety" trajectory is considerably flatter than the optimum gravity-turn launching path, thus subjecting the boosters to higher aerodynamic and control loads and to increased heating. These factors may result in a further performance penalty above that incurred by the use of the non-optimum trajectory. This performance penalty must be judged against the costs of other means of insuring survival in this region. It might be noted that figure 4 does not show the entire launch operation for the "safety" trajectory. The conditions at the burnout point shown yield an elliptical orbit having an apogee at 150 miles, and an additional (small) speed increment must be applied at this point to obtain the desired circular orbit.

A similar condition exists for the winged vehicle in region C (table I). In this case there is a possibility, following booster failure, that the vehicle will be forced to perform a skipping entry under conditions that will expose it to excessive heating. A similar trajectory modification can be made to avoid this region. Here, again, possible use of final-stage thrust can greatly alleviate the problem. The investigation of "safety" trajectories has been a neglected field of research which must be explored for manned satellite operations.

It is difficult to envision a reason for evacuating the vehicle in orbit, region D; however, a malfunction in orbit may make the entry operation hazardous. Examples of such malfunctions are the failure or explosion of auxiliary power units and their fuel tanks, and the failure of stabilization systems. Adequate reliability, isolation, and duplication of such critical systems are the best safety and survival provisions. The prevention of such accidents should be a primary design goal.

In some designs if adequate reliability cannot be attained it may even be necessary to incorporate a special simple escape capsule of the drag-entry type for orbital or entry escape.

It might be well to emphasize a point that has been implied throughout the foregoing; that is, that the final stage should be designed with the most reliable power plant and auxiliary power system possible.

The final stage power plant can, by reliable stop and restart capabilities, greatly alleviate many otherwise dangerous emergencies.

Piloting

Although complete automatic stabilization and control of the manned satellite throughout its flight is feasible, it would be desirable to take advantage of the abilities of the pilot to simplify the system and thus increase the reliability and safety of the operation. An exploratory analog simulator investigation has been made to determine the accuracy with which a pilot could fly a three-stage vehicle to a desired orbit at an altitude of 100 miles. The guidance used consisted of a presentation of error between programmed and actual pitch angle, and indications of altitude, rate of climb, velocity, and angle of attack. Figure 5 shows some of the results of this investigation. The left side of the figure indicates the accuracy in angle and velocity required to maintain the orbit perigee above 75 miles and indicates the manner in which an error in angle can be compensated for by an increase in velocity. The right side of the figure shows the piloting accuracy for various conditions. The basic condition, using a rate-of-climb instrument of 25-feet per second indication for final guidance into orbit, gave a piloting accuracy of ±0.1°. Using sensitive or insensitive inertial altimeters increased the errors as shown. Reducing the damping augmentation to zero caused the vehicle to become difficult to control with sufficient accuracy to approach the desired orbital conditions consistently. It appears that the damping system must be extremely reliable. Although not shown, loss of the static stabilization system, resulting in an extreme value of aerodynamic instability during the first 200,000 feet of the trajectory, had little effect on the pilot's ability to establish a satisfactory orbit.

The effects of a malfunctioning climb program on the pilot's ability to place the vehicle at the desired orbital conditions were also investigated. The malfunctions simulated ranged from inaccurate and erratic indications to complete failure as early as 20 seconds after lift-off. The effect of these malfunctions was generally to increase the error in the final orbital altitude from the normal, ±2,000 feet, to about ±8,000 feet, which is still thought to be reasonable. The maximum deviation from the programmed altitude during the boost period was 20,000 feet, which might be critical for some vehicles.   In all probability a complete presentation failure would cause the pilot to abort the flight; however, there are certain regions of the launch, as discussed in the previous section, in which it would be safer to attempt to establish the orbit. It appears the pilot can do this with satisfactory accuracy by using several altitude and speed check points during the climb.

Although the simulation used in this investigation was by no means optimum or even desirable, the results indicate that pilot guidance of a launching vehicle with adequate accuracy was feasible.

It appears that proper design of presentation and proper use of the pilot may considerably reduce the complexity of the vehicle and increase its overall reliability, particularly in case of malfunctions.

<u>Orbit Selection, Entry, and Landing</u>

Although the orbital factors of eccentricity, altitude, and inclination each have a bearing on manned satellite operations, eccentricity is of very minor importance, provided it is reasonably small. The altitude of the orbit is of somewhat greater importance and will be determined primarily by the desired lifetime, ranging from near 100 miles for short duration vehicles to altitudes above 300 miles for semi-permanent installations.

The inclination of the orbit may determine, or be determined by, the factors of use, survival, and operational ease. With regard to use, military satellites will require, and geophysical satellites will probably desire, orbits as steep as 900. Satellites for vehicular research have less stringent requirements, while permanent, high-altitude, space terminals will undoubtedly have equatorial orbits, inasmuch as this orbit has the greatest stability and passes over the same points on the earth on each rotation. Thus the observation, supply, and rendezvous problems are considerably simplified.

The vehicle survival potential of an orbit is primarily associated with the problems of entry, landing, and rescue following landing and therefore differs for the various configurations. Considering first the ballistic vehicle, malfunction during the launching operation could possibly cause the vehicle to land anywhere around the world, approximately on the first orbital path. In actuality, malfunctions over 90 percent of the boost period would cause impact in the first 6,000 miles and proper use of the retro rockets could, in any case, limit this distance to about 12,000 miles. Intentional landings on later orbits can occur, in general, anywhere between the extreme latitudes obtained by the orbit. The only azimuth control, in this case, is the crude one of choice of orbit on which to enter. To offset this lack of azimuth control, and thus minimize the area to be searched for rescue, the equatorial orbit is an obvious choice, if the mission permits.

The passenger does, of course, have complete freedom of choice in range, since he is able to fire his recovery rockets at any point in his orbit and thus land wherever he desires.

The prediction of the impact point is least sensitive to errors if the retro rockets are fired at the apogee of an elliptical orbit, in which case the landing is made near perigee. The determination of the apogee point by the passenger will be relatively easy by use of a radio altimeter and clock. If everything progresses satisfactorily, a rather unlikely event, the point of landing can be predicted before launch within a circle perhaps 60 miles in diameter.

In the more probable case in which the flight and recovery are not executed with the desired precision, the area to be searched may be considerably greater. Consideration of the launch malfunction problem mentioned before indicates a possible landing area of 12,000 miles by, perhaps, 100 miles. The difficulties of search and rescue in an area this large composed of open sea or jungle cannot be overemphasized. This problem may give the purely ballistic vehicle an inherently lower survival potential than the other two types.

The winged vehicle places fewer requirements on the inclination of the orbit because the pilot is able to modify the entry path both in range and azimuth and thus may navigate to land at pre-selected areas which may be considerably off the projected flight path. In a launch emergency this would require that only about six or seven emergency landing areas be available in the first 12,000 miles of the first orbital path. A similar condition exists for the intentional landing from satellite orbit. Figure 6 shows the lateral deviation available to the semi-ballistic or winged vehicles as a function of lift-drag ratio for entry from a 100-mile orbit. Even the lowest lift-drag ratios result in making a large area available for landing. Although only a portion of the area for the L/D = 4 condition is shown (area extends to Y $\approx$ 8,500 miles at 20,000-mile range), this area is so large it is obvious that lift-drag ratios above 4 are probably not necessary for those satellite entry vehicles which can perform the complete entry at high lift-drag ratios. Figure 6 also shows a curve for the winged vehicle of a design such that aerodynamic heating requires the initial entry be made at a lift-drag ratio of unity down to a velocity of 16,000 feet per second, and a lift-drag ratio of 4 be available for the remaining distance. An indication of what this lateral maneuverability means to the pilot is given in figure 7, which depicts this latter case superimposed on a map. The large elliptical region indicates an area which, if intersected by the projected orbital track, will permit a landing anywhere within the smaller enclosed area. The orbits on which a possible landing could be made for the orbital conditions indicated are listed in the figure. In this case a landing could be made on any of the first four orbits and then on the tenth to the eighteenth. This gives the pilot greatly increased flexibility of operation both in normal operations and in case of emergency. An interesting point indicated here is that only slightly greater maneuverability would be required to enable landing in the continental United States from an equatorial orbit.

Such maneuverability can also be attained by use of thrust in space. However, the lateral deviation shown in figure 7 would require a mass ratio greater than 4 at a specific impulse of 250.

Consideration of the actual landing maneuver indicates a high probability of a water landing for the ballistic vehicle and a possibility of a similar landing for the other types. The vehicles should be designed, therefore, with water landing capability. The landing can be made by conventional landing gear or parachute with the winged vehicle, or by parachute with the ballistic vehicle.

The effects of orbit inclination on operational ease are considered only briefly. It is thought that the maximum operational ease will probably be attained with a winged-type vehicle launched and recovered within the continental limits of the United States, which restricts the inclination to greater than 20°. The equatorial orbit has the operational problems of shipboard or island launch of extremely large vehicles, logistics, the establishment of a sea-borne range, and, in the case of the ballistic vehicle, search and rescue in a large area roughly 22 percent jungle and 78 percent water.

Range Requirements

Undoubtedly, manned operations will require exact position and trajectory data, monitor and command data link, communications, long range GCA, and homing (for the winged vehicle) in certain parts of the orbit. The coverage desired of these facilities is again affected by the type of vehicle, while the number of installations is determined by this desired coverage and by the limits of line-of-sight radio propagation. This effectively limits any one installation to a radius of about 850 miles for an orbit altitude of 100 miles, and less for lower altitudes. These distances can be increased by about 1,200 miles for UHF communications by employing a repeater station in an aircraft at high altitude.

The range requirements of the ballistic and winged vehicles differ considerably, with the ballistic vehicle substituting complexity of ground installations for vehicle complexity. The winged vehicle, of necessity, has greater complexity in guidance and control than the purely ballistic type. The minimum coverage probably required for a ballistic vehicle in an equatorial orbit is shown in figure 8. This vehicle requires complete coverage, as shown, for a distance of about half way around the world for rescue in case of launch malfunction. The darker region shown in the Pacific Ocean is the primary launch and data-taking region while the other areas are, as indicated, only for location and communication. An additional station at 180° from the launch site is desirable for orbit verification and for aiding in entry initiation. It should be emphasized that this is the minimum coverage; for non-equatorial orbits it should be considerably increased, particularly when the possibility of an unscheduled flight termination is considered.

At the other extreme the minimum coverage required for the winged vehicle in a non-equatorial orbit is shown in figure 9. This coverage consists of the first 3,000 miles following take-off, the intermediate stations shown, and the last 2,000 miles before landing. The first region is used to monitor the take-off and initial portion of the orbit to make an initial determination of the orbit and check the pilot's instrument indications. The intermediate points are selected for orbit verification, to insure communication with the pilot at least once per revolution, and to assist the pilot in initiation of entry. The final coverage in the Pacific Ocean is in the nature of long-range GCA for the final approach. The intermediate points may well be chosen to provide coverage for the selected emergency landing areas.

It would probably be desirable during the launch operation and the first orbits to have the maximum communications coverage possible. The stations shown for the ballistic vehicle, of course, provide complete coverage for the first 12,000 miles, leaving about a 40-minute gap; whereas, those for the winged configuration leave gaps of as much as 20 minutes in which communications are lacking. As mentioned previously, this coverage can be improved easily and quickly on a temporary basis by use of airborne repeaters. The pilots have a natural desire for continuous worldwide communications; however, it appears improbable that such coverage can be achieved with a reasonable number of earthbound stations, particularly for non-equatorial orbits. A promising solution in this case is the provision of communication satellites in the "stationary" (22,000-mile altitude) orbit. The ability to establish such facilities may well precede the capability of establishing any but exploratory manned satellites.

Flight Testing

The flight test program should be included in any discussion of manned satellite operations. Any flight test program should utilize the usual procedure of a rational buildup of performance on successive flights in order to explore, with reasonable safety, successively higher performance ranges. This procedure would have the desirable effect of providing the longest period possible for the improvement and demonstration of booster reliability. The provision of boosters of sufficient reliability is, of course, one of the greatest obstacles to the accomplishment of manned orbital flight.

CONCLUDING REMARKS

Although this cursory survey has not indicated any insurmountable operational problems to manned orbital flight regardless of configuration type, it is indicated that the problems of the various satellite configurations do materially affect operations and must be considered early in the design. Safety and survival requirements must be taken into consideration in manned operations and may force appreciable deviation from optimum procedures. Although the presence of the human in the vehicle requires increased emphasis on reliability and safety, proper use of his abilities can greatly simplify design and increase reliability. The type of vehicle has a considerable effect on range requirements, the more simple vehicles requiring increased ground complexity for other than very special conditions.

High-Speed Flight Station,
National Advisory Committee for Aeronautics, Edwards, Calif., April 12, 1958.

TABLE I

LAUNCH ESCAPE REGIONS

	APPROXIMATE CONDITIONS			PROBLEM	ESCAPE PROVISIONS
	H, MI	V, FT/SEC	q, LB/SQ FT		
A	0 TO 4	0 TO 1,000	0 TO 600	ESCAPE FROM LAUNCH AREA	1. BOOSTED FINAL STAGE 2. EJECTION SEAT
B	4 TO 20	1,000 TO 5,000	300 TO ≈1,500	AS FOR NORMAL OPERATION	1. FINAL STAGE
C	20 TO 90	5,000 TO 24,000	300 TO ≈0	POSSIBILITY OF EXCESSIVE g OR HEATING ON ENTRY	1. PROPER TRAJECTORY 2. FINAL STAGE (POWER ON, IF POSSIBLE)
D	>90	24,000 TO ORB.		AS FOR ENTRY FROM ORBIT	1. FINAL STAGE

EXAMPLES OF MANNED ORBITAL VEHICLES

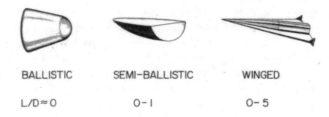

BALLISTIC   SEMI-BALLISTIC   WINGED

L/D≈0   0-1   0-5

Figure 1

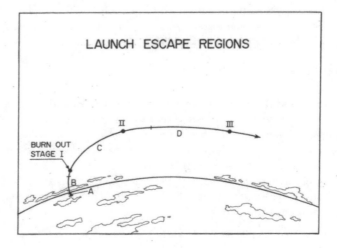

Figure 2

DECELERATIONS FROM ABORTED LAUNCH

BALLISTIC VEHICLE, 150 MILE ORBIT

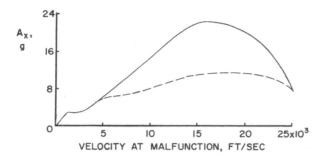

Figure 3

BALLISTIC VEHICLE LAUNCH TRAJECTORIES

ORBIT ALTITUDE, 150 MILES

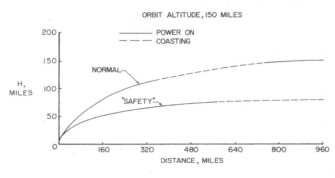

Figure 4

PILOTING ANGULAR ACCURACY AT BURNOUT

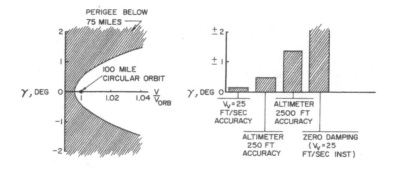

Figure 5

MANEUVERABILITY DURING ENTRY

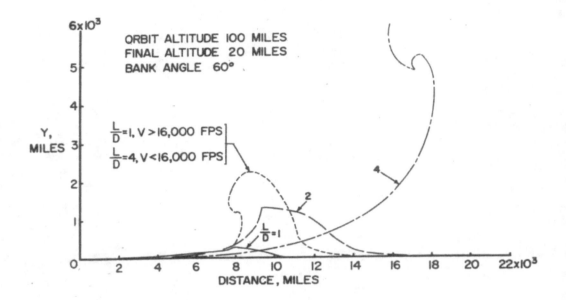

Figure 6

LANDING ORBITS

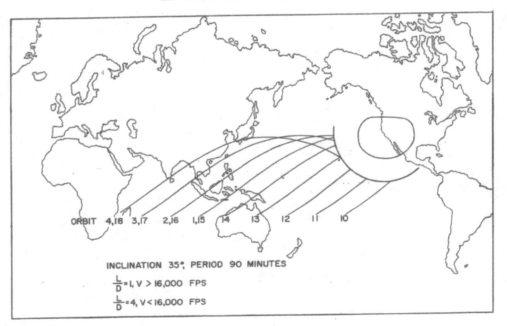

Figure 7

RANGE REQUIREMENTS
BALLISTIC VEHICLE

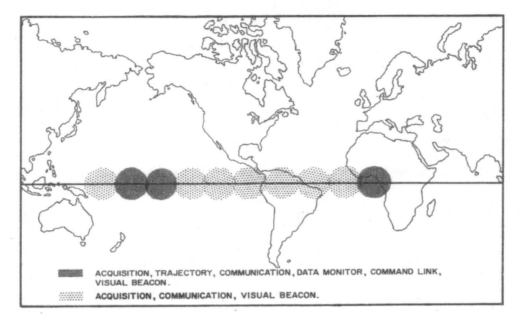

Figure 8

RANGE REQUIREMENTS
WINGED VEHICLE

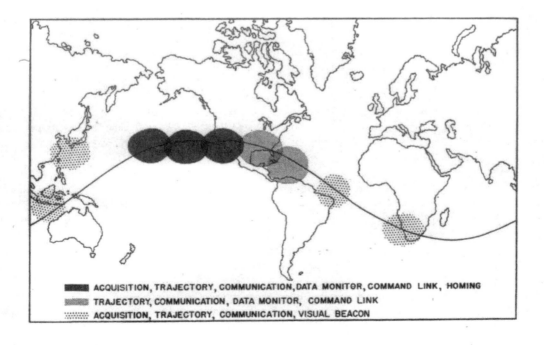

Figure 9

Date: Late 1958?   **Title:** Revised Memorandum of Understanding

MEMORANDUM OF UNDERSTANDING

Subject: Principles for NASA participation in Dyna-Soar Program System 620A. (Supersedes Memorandum of Understanding dated November 14, 1958)

1.   The Dyna-Soar glider is a manned reentry vehicle with the unique capabilities of high maneuverability at hypersonic speeds, with conventional tangential landing. One of the objectives of the Dyna-Soar program is to explore the problems and potentialities of this type of highly maneuverable manned reentry vehicle.

2.   NASA endorses this objective as necessary to the national aero-space program.

3.   The following principles of conduct will be applied in the remaining phases of the project:

   (a)   Dyna-Soar is an Air Force program.
   (b   Technical support (consulting and ground-facilities will continue to be provided by NASA Centers as initially agreed upon with the Air Force.
   (c)   Instrumentation and flight-test support will be provided by NASA Flight Research Center to an extent to be mutually agreed upon with the Air Force.

Secretary of the Air Force
Administrator, NASA

Date: Nov 1958   **Title:** Air Force Staff Summary Sheet

DEPARTMENT OF THE AIR FORCE AIR STAFF SUMMARY SHEET

NOV 28 1958

SUBJECT
(U) Support of Dyna-Soar Program

SUMMARY

1.   On 19 October 1958, the Assistant Secretary of Defense (Comptroller), acting on instructions from the Bureau of the Budget, withheld apportionment of $12.5 million ($10.0 million R&D support, $2. 5 million industrial facilities) previously released to the Dyna-Soar program. This action was taken "pending further justification."

2.   In August 1958 the Air Force has programmed an additional $14.5 million of R&D support funds for the Dyna-Soar program. The attached memorandum was prepared at Mr. Horner's suggestion for the signature of the Secretary of the Air Force to the Secretary of Defense. It asks the Secretary of Defense to urge the Bureau, of the Budget to release the $10.0 million of R&D support funds now withheld by the Bureau of the Budget and also asks for CSD approval of an additional $14.5 million of FY 59 R&D support funds previously programmed by the Air Force.

3.   A presentation to support this request has been given to the Deputy Chief of Staff, Development, the Vice Chief of Staff and to the Assistant Secretary of the Air Force (R&D). This presentation is available upon request.

4.   A proposed memorandum for the Secretary of Defense requesting release of these funds was

processed to Mr. Horner by Air Staff Summary Sheet, dated 14 Nov 58. Mr. Horner returned this memorandum with instructions that it be re-written to delete emphasis on the weapons system connotation of Dyna-Soar and to omit reference to any subsystem development other than the vehicle itself. The present memorandum is identical with the original except for deletions which have been made to comply with Mr. Horner's instructions. The original coordination is still valid for the re-written memorandum.

5. That the Secretary of the Air Force sign the attached memorandum to the Secretary of Defense.

I Incl:
Proposed Memo for Signature

JOHN L. MARTIN, JR.
Colonel, USAF
Deputy Director of Advanced Technology
DCS/Development

Date: Dec 1958   **Title:** Memorandum for Secretary of Defense

DEC 4 1958

MEMORANDUM FOR SECRETARY OF DEFENSE

SUBJECT: (U) Support of Dyna-Soar Program

1. In November 1957 the Air Force established the Dyna Soar program, basing it upon more than four years of study of the boost-glide flight regime. In June 1958 after a careful review of proposals from nine of the major aircraft companies of the country, two contractors, the Martin Company and the Boeing Airplane Company, were chosen to engage in a competitive development of approximately twelve months.

2. In September 1958 the Department of Defense, acting on instructions from the Bureau of the Budget, denied apportionment of $10.0 million of the R&D support funds that the Air Force had programmed for Dyna-Soar, leaving a total of $8.0 million on the program. The reasons given for this action were the "questionable necessity" and the extremely high total cost of this project. It was further stated that the remaining $8.0 million "will carry the studies currently underway at least through December 1958 by which time a decision can be made whether the project should be continued or not." At the time this action was taken the Air Force had already programmed an additional $14.5 million for the Dyna-Soar program and was preparing to ask the Department of Defense for approval. This addition would have permitted ending the present competition in April 1958 rather than July 1958 and all of the additional funds would have been spent by the single, winning contractor to advance the actual design of the experimental prototype aircraft for Dyna-Soar.

3. Because major technological advances are contemplated, it has been realized from the outset that an intermediate experimental prototype airplane was required prior to development of a Weapon System. One purpose of the experimental prototype is to prove out solutions to the aerodynamic heating, structure and stability and control problems of controlled atmospheric flight from near-orbital speeds down to a safe and normal landing on an airfield. In solving this problem the experimental prototype will become the basic building block for any weapon systems of the future. It appears certain that the Air Force mission in the future will require manned operations in the orbit flight mode and Dyna-Soar is an essential step to that future. Maneuvering flight within the atmosphere and a routine, uneventful landing as provided by the experimental prototype will be required. before such operations become practical. In opening up this flight regime the experimental prototype will make possible many new methods and concepts of operations and another purpose of the prototype is to

enlarge and improve our understanding of these methods and concepts. The military data produced will be fed into a weapon system concept study which will run concurrently with the experimental prototype hardware program. This study program will compare and select the best of the methods to exploit the newly created flight stability.

4. Since the sub-orbital Dyna-Soar experimental prototype will fill the gap in hypersonic aerodynamics between the performance of the X-15 and satellite speeds, there is a natural community of interest in the program between NASA and the Air Force. A NACA-Air Force Memorandum of Understanding was concluded on 20 May 1958 and establishes Dyna-Soar as a Joint NACA-AF program to be managed and funded by the Air Force. Absorption of the NACA by the NASA has recently reaffirmed it. NASA has willingly given unstinting technical support to the program and has furnished a representative to the WSPO. There is likewise a natural community of interest in the program between the ARPA and the Air Force since the Dyna-Soar program includes studies of possible future systems which could operate in space. Extensions of these studies will be managed and partially funded by ARPA as a separate effort from the development and test of the experimental prototype. The Dyna-Soar program has been discussed with Dr. Hugh L. Dryden of the NASA and Mr. Roy Johnson of ARPA since the BOB action was taken to withhold funds. The NASA position is that the Dyna-Soar experimental prototype should be developed as rapidly as possible and that it does not conflict with the NASA Man-in-Space project. The ARPA concurs in the importance of the project and confirms that it should proceed under Air Force management.

5. In view of the great technological and military advantages it will produce it is strongly recommended that the Dyna-Soar program be allowed to proceed. It is requested that the Secretary of Defense urge the Bureau of the Budget to release the $10.0 million of FY 59 R&D support funds now being withheld. CSD approval of the $14.5 million of FY 59 funds now programmed by the Air Force is also requested.

Date: Late 1959   **Title:** Dyna-Soar Memo of Understanding

DYNA SOAR – MEMO of UNDERSTANDING

PARTIES:
>   USAF & NASA

ASSIGNMENTS:
>   TECHNICAL DEVELOPMENT
>>   USAF & NASA JOINTLY PARTICIPATE
>   FINANCING & ADMINISTRATION
>>   USAF
>   FLIGHT TESTING
>>   USAF - NASA COMMITTEE
>   INSTRUMENTATION & DATA
>>   NASA - USAF COMMITTEE

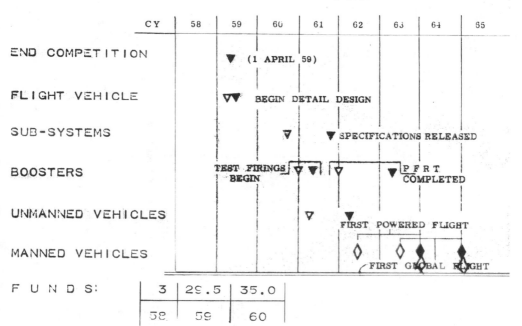

REVISED DS SCHEDULE

STATUS OF FUNDS

	AMOUNT (MILLIONS)	STATUS	CONTRACTORS	USE
FY 58	3.0	EXPENDED 31 DEC 58	2	PHASE I DESIGN STUDIES
FY 59	5.0	EXPENDED 31 DEC 58	2	W. S. STUDIES
FY 59	10.0	RELEASED JAN 59	2	COMPETITION
FY 59	14.5	PROGRAMMED BY AF RELEASE REQUESTED	1	PHASE II DETAIL DESIGN FOR DS 1
FY 60	35.0	PROGRAMMED	1	

REQUIRED FEATURES OF MILITARY TEST SYSTEM – DS I

MANNED
MANEUVERABLE IN ATMOSPHERE
PRECISE LANDING
SIGNIFICANT PAYLOAD
FLEXIBLE FLIGHT PROFILE
VERSATILE TEST OPERATIONS

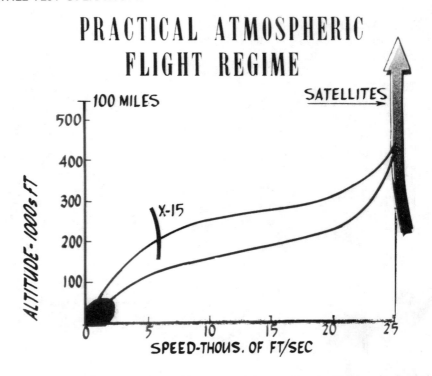

PRACTICAL ATMOSPHERIC FLIGHT REGIME

FLIGHT PATH COMPARISON

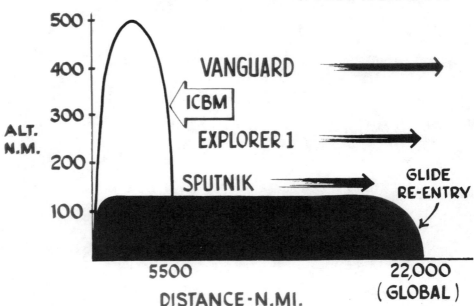

DS - PRESENT PROGRAM

DEVELOP MILITARY TEST SYSTEM - DS-1

      SOLVE MILITARY & TECHNICAL PROBLEMS OF HYPERSONIC, ATMOSPHERIC FLIGHT UP TO NEAR ORBITAL VELOCITY

      AF – NASA

ARPA   WEAPON SYSTEM STUDIES
        DEFINE & EVALUATE CONCEPTS
            STRATEGIC RECON
            STRATEGIC BOMBING
            SPACE INTERCEPTION & INSPECTION
            AF – APRA

BACKGROUND

1954-1957 - FEASIBILITY & DESIGN STUDIES

NACA	(IN HOUSE)
AF	$3,015,000
CONTRACTORS	3,800,000
Bell, Boeing, Convair	$6,815,000
Douglas, Martin	
No. American, Republic	
Lockheed	

Nov '57 - DYNA SOAR DEVELOPMENT DIRECTED
JUNE '58 - MARTIN & BOEING SELECTED
1 APR '59 - COMPETITION ENDS

OBJECTIVES

IMMEDIATE OBJECTIVE IS DYNA SOAR I, A MANNED MILITARY TEST SYSTEM.

OVERALL OBJECTIVE IS THE CAPABILITY FOR FLEXIBLE, EFFECTIVE MILITARY OPERATIONS TO EXPLOIT GLOBAL RANGE & HYPERSONIC, HIGH ALTITUDE FLIGHT UP TO ORBITAL.

GROUP 4
Downgraded at 3 year intervals;
declassified after 12 years

Date: Jan 1960   **Title:** Dyna-soar Step I Flight test Program

DYNA-SOAR STEP I FLIGHT TEST PROGRAM

By Lt. Col. Harold G. Russell, USAF, and B. Lyle Schofield Air Force Flight Test Center
and Thomas F. Baker Flight Research Center

INTRODUCTION

The objectives of the Dyna-Soar project have been stated to be the development of a piloted, maneuverable, hypersonic glider capable of a controlled landing following reentry from orbital flight. The Step I flight-test objectives of Dyna-Soar, as shown in figure 1, are twofold: exploration of the flight regime of the glider and development of satisfactory subsystems and vehicle.

Development and verification of the operational concepts and requirements for a Dyna-Soar type vehicle are significantly important from military, astronautical, and possibly commercial standpoints. Verification of the vehicle and subsystems design and modification and development of the hardware as problems arise is the historic role of flight testing and constitutes a primary objective of the Dyna-Soar flight-test program.

The cost, effort, and complexity of conducting ground-launched flights of the Dyna-Soar will be nearly an order of magnitude higher than those on previous airplanes, including the North American X-15. As a consequence, the number of flights that can be expended in developing a satisfactory vehicle and in exploring the flight regimes must be held to the absolute minimum.

FLIGHT REGIME

The general configuration contemplated for Dyna-Soar was described in a previous paper by R. L. Rotelli as being a winged glider with a hypersonic lift-drag ratio on the order of 2. The flight envelope of the Dyna-Soar glider is shown in figure 2 in terms of altitude and velocity.

The equilibrium glide corridor is the primary regime to be explored during the flight tests, although some semi-ballistic flights above the corridor will be performed. The booster currently planned for Step I of the Dyna-Soar program is a modified Titan ICBM which will limit the maximum Step I velocity to about 19,000 feet per second. For comparison, the design flight envelope of the X-15 is shown at the left in figure 2, and the nominal reentry trajectory of the Project Mercury capsule is indicated by the heavy line.

Within its relatively limited envelope, the X-15 will provide very valuable information on aerodynamic heating, flight control at high altitude, atmospheric reentry, piloting techniques, and terminal guidance.

Project Mercury experience in utilizing an ICBM for boosting a manned vehicle, developing man's capabilities in a space environment, operating a global range, and developing recovery techniques will, likewise, be of much value.

As may be seen, however, the flight regime of the Dyna-Soar is a tremendous extension of the X-15 envelope and there is a basic conceptual difference between the lifting-vehicle Dyna-Soar glider and the ballistic-vehicle Project Mercury capsule. Additionally, as the flight regime extends to higher velocities, the capabilities of wind tunnels and rocket models to support the design and development of the vehicle become substantially reduced. To reiterate, exploration of the hypersonic-glide corridor is the primary objective of the Dyna-Soar flight test.

DATA OBJECTIVES

The general flight-test areas of interest and the data objectives are shown in figure 3. In each area, onboard instrumentation will provide data by which the conduct of the flight and operation of the systems may be monitored to either confirm the design or provide information to correct deficiencies.

The aerodynamics area is perhaps the most important in that it encompasses aerodynamic heating, flow characteristics, performance, and stability and control. Adequate aerodynamic-heating information for progressive conduct of the flight test can be obtained from a knowledge of the temperatures that exist throughout the skin and airframe during the flight, the flight conditions, and the structural properties (fig. A large number of temperature sensors will be located to provide for determination of experimental heat-transfer characteristics and verification of the structural design.

Detailed analysis of experimental heat-transfer data requires a knowledge of free-stream and local-flow conditions and local gas properties. Use of the non-dimensional heat-transfer coefficient, Stanton number S_T, is convenient in comparing experimental results and theory and is given in the following expression;

$$S_T = \frac{h}{p C_p V} = f(R_e, P_r, T_s, T_w, M_\infty, \alpha ...).$$

where

h	heat-transfer coefficient
p	density
C_p	specific heat of fluid
V	free-stream velocity
R_e	Reynolds number
P_r	Prandtl number
T_s	stagnation temperature
T_w	wall temperature
M_∞	Mach number
α	angle of attack

Some measurements of both free-stream local-flow characteristics pertinent to heat-transfer analysis are planned for a few specific locations. As shown in figure 5, the required free-stream data consist of total pressure P_T, total temperature T_T, angle of attack α, and angle of sideslip β.

Local-flow conditions will be determined primarily from surface pressure measurements, together with such measurements of surface and boundary-layer temperatures, boundary-layer pressures, dissociation, and gas composition as are possible. The extent of the flow characteristics measurements obtained during Dyna-Soar flight tests and the quality of information that can be attained depend to a large extent on successful development of both transducers and flight-measuring techniques.

Acquisition of accurate performance data is essential to the conduct of the Dyna-Soar flight program and can only be obtained during flight of the full-scale glider. Performance measurements during gliding flight require vehicle velocities, accelerations and attitudes, and a measure of the atmospheric environment. Ground-tracking trajectory information will be utilized as backup for onboard data.

Aerodynamic stability and control considerations are virtually inseparable from the vehicle's flight-control and guidance systems. These areas are considered under the general heading "Flight Controls" in figure 6. The flight-controls test objectives are determination of stability derivatives and control effectiveness parameters throughout the flight corridor. Such information is essential for the flight program buildup discussed subsequently and also is of general research interest. Also, development of an adequate flight-control system is mandatory, and full-scale flight testing is required for final development and evaluation. A description of the flight-control system envisioned for the Dyna-Soar was presented in a previous paper by

Alan H. Lee and Leroy J. Mason. The automatic and redundant features of the primary flight-control system, the guidance and navigation system, and cockpit-display equipment will require development in the course of the flight-test program.

One of the basic concepts of the Dyna-Soar flight-control system is to provide for maximum pilot utilization. Also of considerable interest is the determination of desirable handling qualities of hypersonic vehicles. The information gained from the Dyna-Soar flight-test program will be directly applicable to the verification of man's role and capabilities in piloting space and reentry vehicles and in the establishment of design guidelines for hypersonic handling qualities.

The data requirements in the area of flight controls for the Dyna-Soar flight-test program will be much like those of the X-15.

Basically, it is necessary to establish the flight conditions, determine the control motions, and measure the vehicle response. The analysis procedure to be used for data evaluation will take various forms. Where possible, as with the trim evaluation, analysis will be made directly from the flight records. For maneuvering or dynamic analyses, where changes in flight conditions are appreciable or glider response is altered by spurious control inputs, a data-matching procedure utilizing analog computer synthesis methods will be used.   It is anticipated that the X-15 flight-research program will develop new techniques and methods in this area of stability analysis which can be utilized in the Dyna-Soar program.

In the areas of dynamics, loads, structures, and materials, the objective, and subsequent contribution, of the Dyna-Soar flight-test program is primarily one of demonstration. The usual accelerations, noise measurements, and strains required to verify the integrity of the vehicle will be obtained. Additionally, some measure of the distortion of the external shape of the glider will be made. The structural and aerodynamic measurements, when analyzed together, will provide useful design information on aerodynamic and heating loads.

The human-factor aspects of reentry from orbital and near-orbital speeds and altitudes will continue to be of importance. Reentry flight times during Step I testing will require up to 30 minutes, wherein longitudinal decelerations of from 0.39 to 2.0g will be experienced. Physiological effects of decelerations, time, and cockpit environment on pilot operation of the glider during reentry will be studied.

Development of reliable and efficient subsystems, such as environmental control and secondary power, and demonstration of their operation in the Dyna-Soar flight environment is no less an objective than exploration of the flight corridor. Adequate monitoring sensors will be included in the instrumentation package to assure acquisition of significant subsystem operating data.

The areas of military applications and geophysical research are additional flight-test objectives. The suitability of the Dyna-Soar type vehicle for military applications will be determined during the course of exploring its flight corridor, as will its suitability as a platform for conducting geophysical experiments.

FLIGHT-TEST PROCEDURE

The flight-test procedure to be utilized in developing the Dyna-Soar I and in exploring the hypersonic flight regime has been developed to stay within cost limitations but, at the same time, maintain a high degree of confidence in extending the flight envelope. The resulting flight-test program (fig. 7) consists of manned air-launch flights covering the subsonic and supersonic flight regimes, unmanned ground launch flights for investigation of conditions from launch to hypersonic speeds, and the main test-program objective - manned exploration of the hypersonic flight corridor.

The air-launch phase of the test program, utilizing the Boeing B-52 for air drop, will provide the first opportunity to evaluate the test article under actual flight conditions. There are several important objectives (fig. 8) which must be accomplished during this phase before the test program can proceed to the manned ground-launch tests.

The first of these objectives is systems checkout and demonstration. Some systems development, including data-acquisition systems, will be most easily accomplished during the air-launch flights. Aerodynamic and structural verification, including investigation of stability and control characteristics, will be accomplished throughout the attainable speed and lift-coefficient range. Another objective is pilot familiarization with the low-speed flight and landing characteristics of the Dyna-Soar, together with the development of optimum approach and landing techniques.

The maximum velocity that can be achieved during the air-launch phase utilizing a rocket-boosted glider is uncertain. Attainment of a supersonic Mach number of about 7 is highly desirable, but, because of technical and economic factors, the maximum feasible velocity for the air-launch phase may be a Mach number of about 2.

The unmanned ground-launch test phase will be conducted on the Atlantic Missile Range, with launch from Cape Canaveral. Although there will have been approximately 40 Titan firings prior to this time, modifications for Dyna-Soar - such as the addition of first-stage stabilizing fins, structural beef-up, and any booster-subsystems changes - will require flight testing. The prime requirements of unmanned tests are demonstration of the booster-glider combination and glider separation from the booster. Some assessment of the reliability of both the first and second stages of the booster is necessary before the manned portion of the test program can the initiated, and each of the unmanned test flights will be carried through ignition and separation of the second stage. Escape system tests will also be accomplished during this phase.

The third and major phase of the flight-test program consists of a manned systematic expansion of the Dyna-Soar flight envelope, with launch at Cape Canaveral down the Atlantic Missile Range utilizing down-range islands as intermediate landing sites. Improved landing strips, 8 to 10,000 feet in length, have been specified as landing-site runway requirements. The locations of the landing sites must be compatible with the glider range and maneuverability, desired burnout velocities, and test objectives. A summary of the results of the landing-site study is shown in figure 9. The limits of injection velocity for each landing site are determined from the lift coefficient (or lift-drag ratio), angle of bank, and the permissible launch azimuths of Cape Canaveral.

The first manned ground-launch flight has been planned for an injection velocity of approximately 9,000 feet per second. Selection of this speed was dictated by economic and geographical considerations as well as the knowledge that flight environment up to speeds of approximately 7,000 feet per second will have already been investigated during the X-15 program. The landing sites which would permit the most comprehensive coverage of injection velocities from 9,000 feet per second to approximately 19,000 feet per second are Mayaguana, Santa Lucia, and Fortaleza, Brazil.

The test-flight tracks down the Atlantic Missile Range for maximum and minimum burnout velocities with landings at Mayaguana, Santa Lucia, and Fortaleza are indicated in figure 10.

It should be noted that all the possible landing sites lie at approximately 1300 azimuth from Cape Canaveral, but, since a maximum launch azimuth of 1100 must be observed during boost, turning flight must be performed to arrive at the high key point over each of the landing sites.

The step-by-step expansion of the hypersonic flight regime during the manned test phase will provide a reasonable degree of confidence in exploring the unknown flight regime and attainment of test results.

This test phase will commence with a glider injection at an optimum lift coefficient and a velocity of approximately 9,000 feet per second. Each successive flight in the speed range above 9,000 feet per second is a moderate extension in both speed and lift coefficient over the previous flights.

Data obtained from each flight will be analyzed sufficiently to reveal possible danger areas so that they may be cautiously approached or avoided during follow-on test missions.

A typical flight for the systematic expansion of the Dyna-Soar flight envelope is presented in figure 11. The clear area denotes that portion of the flight envelope which has previously been explored, and the grey area denotes the unexplored regions of flight. The crosshatched areas indicate the portion of the flight envelope which will be expanded by this particular test mission. It should be noted that the vehicle injection takes place at a mid lift coefficient and that the lift coefficient is increased as velocity decreases. A similar technique will be utilized for investigation of the lower lift coefficients. Testing at any particular lift coefficient and velocity combination
will be a moderate extension in speed or lift coefficient, or both, over a previous test mission.

Once the flight envelope has been extended, the remainder of the flight will be devoted to both data fill-in and energy-management requirements for arrival over the landing site.

The test flight outlined in figure 11 was simulated on an analog computer, and a time history of pertinent trajectory parameters is presented in figure 12. Although the simulation was limited, in that it was only a three-degree-of-freedom point mass simulation, it does provide an insight into the times which will be available for flight testing.

As can be noted, something less than 5 minutes is available for testing during the flight-envelope expansion, while the remainder of the flight will provide for data fill-in and energy management.

Extensive use of six-degree-of-freedom flight simulation is a prerequisite to all Dyna-Soar flight planning. It is also necessary that the pilot fly each proposed mission on the simulator, including the higher probability boost-abort situations.

All flights will be planned to allow maximum assurance of a successful landing of the glider in case of a boost abort.

CONCLUDING REMARKS

In summary, the Dyna-Soar flight-test program will consist of three phases: (1) air-launched tests using a powered glider to checkout and demonstrate the operating characteristics of the vehicle; (2) unmanned ground-launched tests to demonstrate the integrity of the booster glider combination; and (3) the major phase, ground-launched manned exploration of the hypersonic flight regime.

Development and verification of the Dyna-Soar design and its operational concepts and requirements will be accomplished during the flight test program which will be conducted as a joint operation by an Air Force - NASA - contractor team. This program will provide aerodynamic data from the reentry flight corridor, will verify the design requirements of reentry vehicles, and define their operational capabilities in this flight regime. These results, which must be timely for proper military exploitation of the aerospace medium, will provide valuable contributions to other astronautical ventures.

It is readily admitted that there are many unknowns to be discovered in this program. One thing is certain, however, this will be one of the greatest testing efforts that the free world has ever known.

DYNA-SOAR I STEP I
FLIGHT TEST
OBJECTIVES

**EXPLORE FLIGHT REGIME
TO M ≈ 20
DEVELOP VEHICLE AND
SUB-SYSTEMS**

- ◆ **definition and solution of problems**
- ◆ **development of operational concepts**
- ◆ **verification of design**

Figure 1

FLIGHT ENVELOPES

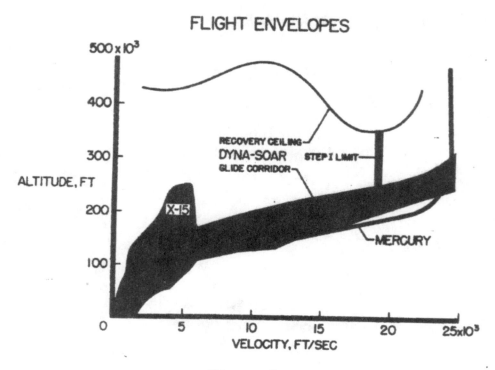

Figure 2

FLIGHT TEST DATA OBJECTIVES

aerodynamics
control and guidance
dynamics and loads
structures and materials
human factors
vehicle subsystems

— — — — — — — — —

military applications
geophysics

MONITOR
CONFIRM
DEVELOP
RESEARCH

Figure 3

AERODYNAMIC HEATING

FLIGHT TEST:

$$h \Longleftarrow \begin{cases} \text{skin temperatures} \\ \text{flight conditions} \\ \text{structural properties} \end{cases}$$

ANALYSIS:

$$S_T = \frac{h}{\rho C_P V} = f\left(R_e, P_r, T_s, T_w, M_\infty, \alpha \cdots \right)$$

$$= f \begin{cases} \text{free-stream conditions} \\ \text{local flow conditions} \\ \text{local gas properties} \end{cases}$$

Figure 4

FLOW CHARACTERISTICS

- FREE-STREAM CONDITIONS, P_T, T_T, α, β.

- LOCAL FLOW CONDITIONS,
  surface pressures
  surface temperatures
  boundary layer profile
  disassociation detection
  gas composition
  skin friction

Figure 5

FLIGHT CONTROLS

FLIGHT TEST OBJECTIVES:
- determine stability derivatives
- develop flight control and energy management systems
- verify pilot's role and desirable handling qualities

DATA REQUIREMENTS:
- flight conditions
- control motions
- vehicle responses

Figure 6

DYNA-SOAR I FLIGHT TEST

TEST PHASE	FLIGHT REGIME COVERAGE	PRIME TEST INTERESTS
air launch manned	landing ⟶ supersonic	glider
ground launch unmanned	launch ⟶ hypersonic	booster-glider
ground launch manned	supersonic ⟶ hypersonic	pilot-booster-glider

Figure 7

AIR-LAUNCH FLIGHT TEST

▲ sub systems checkout and demonstration

▲ aerodynamics and structural verification

▲ pilot checkout and training

▲ landing characteristics and requirements

Figure 8

LANDING SITE REQUIREMENT

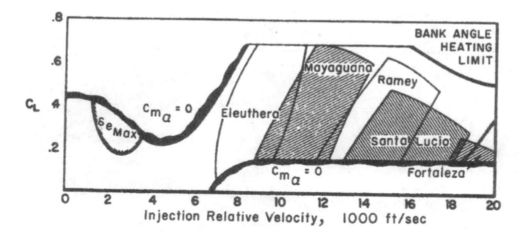

Figure 9

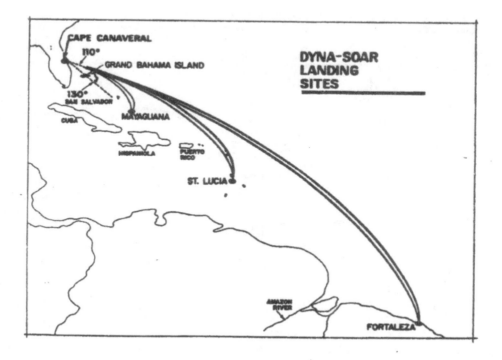

Figure 10

FLIGHT ENVELOPE EXPANSION

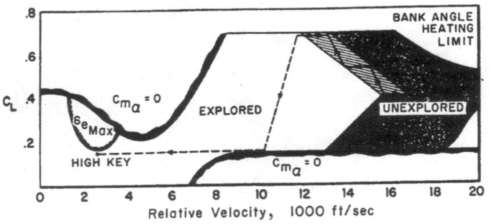

Figure 11

TYPICAL MANNED GROUND LAUNCH FLIGHT

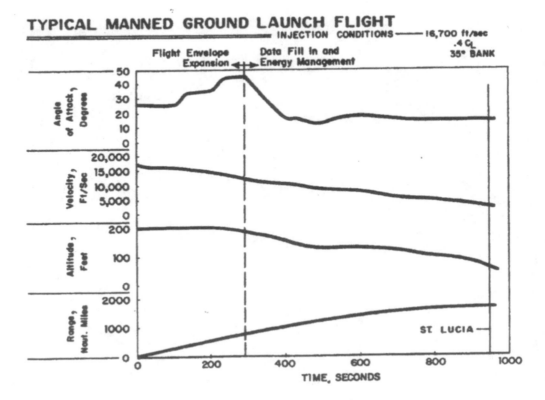

Figure 12

Date: May 1960   **Title:** Information for Dr. Charyk

DEPARTMENT OF THE AIR FORCE
HEADQUARTERS UNITED STATES AIR FORCE
WASHINGTON 25, D.C.

5 May 1960

REPLY TO ATTN OF: AFDAT

SUBJECT: Information for Dr. Charyk – Dyna Soar

TO: SAFUS (Col. Butcher)

1.   Reference telecon between Colonel Butcher, SAFUS, and Colonel Munroe, AFDAT, this date.

2.   Attached is the information requested pertaining to Dyna Soar for Dr. Charyk's use in the Committee hearings.

> 7 Atch
>> 1. Dyna Soar Chronology
>> 2. Press Release – 27 Apr 60
>> 3. System Development Tests
>> 4. DS Step I Schedule (Master)
>> 5. Ships & Landing Sites
>> 6 . Instrumentation Coverage
>> 7. Dyna Soar Funding – 19 Apr 60

DYNA SOAR CHRONOLOGY

Prior to 1945	Boost-glide in Germany brought to point of first vehicle design.
1954-1957	Intensive study by AF, NACA, industry. Funds, AF-30M, NACA in-house, industry 3.8M
15 Jan 57	HYWARDS to BRASS BELL development submitted.
30 Apr 57	HYWARDS to BRASS BELL not approved. ARDC directed to combine.
Jun-Aug 57	ARDC, NACA, RAND evaluation: boost-glide feasible.
24 Aug 57	Dyna Soar Development Plan submitted.
10 Oct 57	Dyna Soar Development Plan submitted.
1 Nov 57	General White approved Dyna Soar.
25 Nov 57	$3.0M of FY 58 released.
25 Nov 57	Dyna Soar Development Plan approved. Development. Directive No. 94 was issued.
31 Jan 58	Request for proposals to Industry.
24 Mar 58	Proposals received from Martin-Bell, Boeing, Convair, Douglas, Lockheed, McDonnell, Republic.

1 May 58	Source Selection Board recommendations to Hq USAF.
9 Jun 58	Recommendations approved.
16 Jun 58	Selection of Boeing and Martin announced.
17 Sep 58	Feasibility of ending competition 1 April 59 established.
19 Sep 58	Release of 10.0M FY 59 funds withdrawn.
30 Sep 58	ARDC directed to make 8.0M released last until 1 Jan 59.
Week of 2 Oct 58	Meetings, AFDAT, General Swofford, Mr. Horner on Dyna Soar funds.
15 Oct 58	Memo, SAF to C/S: ARPA's interest is sustained orbit – Dyna Soar is sub-orbital.
Week of 21 Oct 58	Presentations on Dyna Soar prepared.
7 Nov 58	ARPA Memo to Secretary of Defense supported Dyna Soar. Announced intention to take over weapon system studies.
2 Dec 58	General White signed a new AF-NASA Memo of Understanding on Dyna Soar signed by Dr. Glennan 14 Nov 58.
4 Dec 58	Memo, SAF to Secretary of Defense requested release of $10 and $14.5.
7 Jan 59	Deputy Secretary of Defense released $10.0M.
6 Apr 59	Proposals received from Martin and Boeing.
13 Apr 59	DDR&E approves 14.5M and states objectives are: Primary – 22,000 ft/sec, manned, maneuverable controlled landing vehicle; Secondary – military payload orbit capability.
25 May 59	Source Selection Board recommendations to Hq USAF.
16 Jun 59	Asst Secys Charyk and Taylor briefing on Source Selection Board recommendations.
7 Jul 59	Memo, Secy MacIntyre to C/S says: 1. Reconsider booster and 2. Vehicle size. 3. Consider total program cost with view to reducing, 4. Restudy management plan, and 5. sub-contract structure.
28 Jul 59	Memo from DDR&E to SAF and DIR, ARPA, says consider a joint development for Dyna Soar booster and Saturn second stage; report to DDR&E before making firm commitments.
19 Aug 59	Drs. York & Wagner briefed on Dyna Soar at AFBMD. Mr. Donovan, STL, presents data on alternate booster uses. Dr. York requires more cost data.
27 Aug 59	Booster cost data for Dyna Soar, and cost, configuration and performance data on common development for Dyna Soar and for space booster and 160" Saturn second stage given to Dr. Charyk.
21 Oct 59	OSD budget "mark-up" sheet dated 10/11/59 received. FY 61 funds reduced from $25.0 to 0. Explanation: Program is challenged within AF; $25.0 only token funding;

questionable value for future military use. Statement made that a policy decision to pursue program needed. Elimination of Dyna Soar contemplated."

21 Oct 59	General Daushey reports OSD FY 61 action to Dr. Charyk who says Dyna Soar will not be eliminated.
6 Nov 59	Dyna Soar step program presented to SAF Douglas. Secy Douglas approves with reservation that there is no commitment on FY 61 fund amount .
9 Nov 59	Selection of Boeing and Titan A Booster announced.
17 Nov 59	C/S approves: Step I for implementation, Step II for planning, selection of Boeing & Martin, Dyna Soar management plan.
30 Nov 59	ARDC directed to prepare work statements for Step I, but to include a Phase Alpha which will examine technical approach to Dyna Soar development.
11 Dec 59	Contract signed with Boeing. Work statement covers all of Step I; actual work authorized limited to Phase Alpha.
28 Dec 59	Dr. Charyk is given Phase Alpha statements and agrees to release additional funds for Phase Alpha.
30 Dec 59	Because of Titan review, Cmdrs AMC and ARDC decide to delay contract with Martin for booster.
4 Feb 60	Phase Alpha contract with Martin signed.
25 Feb 60	Review of Boeing Phase Alpha results to date at WPAFB.
26 Feb 60	ARPA Order No. 84 cancelled by Order 84-60, Amend #1 – ". . . ARPA has no further responsibility for the Dyna Soar project."
28, 29 & 30 Mar 60	Results of Phase Alpha reviewed by Aero Space Vehicles Panel at Boeing.
5,6 & 7 Apr 60	Air Staff reviewed Phase Alpha.
8 Apr 60	Under-Secy Air Force reviewed Phase Alpha.
22 Apr 60	DDR&E reviewed Phase Alpha results and approved Dyna Soar program.
22 Apr 60	D&F signed and transmitted to AMC 13.5M FY 59 and 16.2M FY 60 funds approved. Revised Development Plan requested of ARDC.

Date: July 1960   Title: Memo: Dyna-Soar low velocity booster study

NASA FRC
July 29 1960

Memorandum for Director, NASA Flight Research Center

Subject: Dyna Soar low velocity booster study

1. Messrs. Baker, Bellman, Jordan, Weil, and Armstrong discussed the subject study on this date and determined a definition of the problem and established the approach to be taken in the solution of the problem.

2. Glider problem: Determine practical injection conditions for the glider at velocities above 10,000 feet per second. Booster problem: determine required boost trajectory to place glider at the practical injection conditions and determine the required modifications to the Titan ICBM to so accomplish.

3. The general approach taken was to separate the study into two projects, one concerning the glider and one concerning the booster. Both projects to be run in parallel. It should be noted and understood by all concerned that this study must be completed by September 1, 1960, and has the highest priority at the Flight Research Center. Overall direction of the study will be provided by the above Branch Heads and Mr. Armstrong with Mr. Baker having overall responsibility. The project engineer for the glider will be Mr. Alan Brown, the project engineer for the booster will be Mr. Joseph Washko. It was agreed that no definitive schedule for the study could be established at this time, other than to emphasize the September 1 completion date.

The broad approaches to be taken in the glider injection project and in the booster trajectory and modification project are attached hereto.

Thomas F. Baker
Head, Dyna Soar Branch

Attachment 1

Glider Injection Project Approach

1. Determine general criteria for injection, considering such things as glider stabilization after separation, flight safety, mission success, injection inaccuracies, glider system malfunctions, and proximity to placard questionable flight regions.

2. Compile glider flight characteristics and placards.

3. Use injection velocities of 10,000, 12,000, and 15,000 and 19,000 feet per second in this order and priority. Run analog (see item 5) and/or IBM studies of glider flight from separation to stabilization at a mid C_L within the glide corridor, but don't throw out runs that do not stabilize at a mid C_L since stabilization at a high and low C_L will be required later.

4. Conclude the practical injection conditions for these four velocities and mid C_L stabilization and conclude the criteria used to determine the injection conditions.

5. Use three-degree-of-freedom analog for initial studies and review other analog requirements including the usefulness of the AFFTC analog. Present the analog requirements and the project group's

recommendations to the Branch Heads as soon as possible.

6. The following personnel will comprise the glider injection project team: A. D. Brown, Project Engineer, E. C. Holleman, W. Stillwell, L. Thomas, H. Washington, R. Banner, D. Kordes, J. Gibbons, W. Dana, G. Waltman, J.Perry

Attachment 2

Booster Trajectory and Modification Project

Approach

1. Ascertain rational possible modifications of the Titan ICBM that will permit injection of the glider at 10,000 feet per second and above at conditions described in item 4 below.

2. Using two-degree-of-freedom analog or IBM computers as required, determine the effect of and the amount of modifications on the boost trajectories. Run the boost trajectories to second stage burnout or to flight conditions at least below the glider corridor.

3. Conclude the general obtainable burnout conditions resulting from the above modifications and trajectory studies and review the conclusions with the branch heads, for velocities of 10,000, 12,000, 15,000 and 19,000 feet per second in that order and priority.

4. The family of trajectories should be bounded by the nominal BAC 19,000 foot per second trajectory and a trajectory that ends at the velocities of paragraph 3 in the middle of the corridor at $\gamma = 0$ flight path angle

5. Flight path angle at boost termination should vary from $\gamma = 0$ to whatever maximum may be desired from the booster standpoint.

6. The following personnel are assigned to the booster trajectory and modification project team: J. Washko,
Project Engineer, J. W. Smith, E. Holleman, H. Washington, R. Banner, W. Stillwell, D. Kordes, J. Love.

Date: Nov 1960   **Title:** Dyna-Soar Guidance Systems

ROUGH DRAFT

November 29, 1960

DS GUIDANCE SYSTEMS

A review has been made of possible guidance systems for the Dyna-Soar vehicle. This review was made with the objective of examining all techniques to study the trade-off between simplicity and the ability to meet the guidance requirements. These requirements are:

I. Boost trajectory guidance
    (1) Step I
    (2) Step II

II. Glide trajectory guidance
    (1) Step I
    (2) Step II
    (3) Aborted-boost Corridor exploration In addition to these requirements it is believed the system should make maximum utilization of the pilot, and that some form of redundancy or backup guidance is desirable.

I. Boost

A. Open loop, programmed attitude
    1. Least complex system
    2. Autonetics studies have indicated that
    boost injection conditions cannot be accurately controlled by open loop techniques.

B. Radio
    1. Basic system presently in use for Titan.
    2. Present guidance computer mechanization requires extensive modification from ICBM
    or distance-velocity to DS altitude-velocity type of programming.
    3. One additional downrange station will be required for Step I boost (350 nm) .
    4. Two additional stations may be required for Step II boost trajectories (900 nm for Titan-Centaur, 1,500 nm for Saturn).
    5. Technique for switching guidance to different stations must be developed.
    6. Most complex system.

C. Inertial, ICBM type
    1. Proven components available.
    2. Same modifications to guidance equation mechanization as required for radio guidance.
    3. Neither platform nor computer are adequate for glide phase guidance.
    4. Adequate for Steps I or II.

D. Inertial, New DS type
    1. One system for boost and glide phase guidance.
    2. Adequate for Steps I or II.
    3. New platform or modification to existing platform required.
    4. New computer or modification to existing computer required.

II. Glide

A. No active guidance

    1. Least complex system

    2. Marginal accuracy for flights to St. Lucia.

    3. Inadequate accuracy for Steps I, II or aborted boost flight conditions.

    4. Pilot has no indication of V, H, α, β.

    5. Corridor exploration difficult.

B. Radio - DR

    1. Existing tracking stations available for flights to St. Lucia.

    2. Little development work required.

    3. Accuracy marginal for flights to Fortaleza because no ground stations are available for last 1,000 nm.

    4. Inadequate accuracy for Step II missions.

    5. Corridor exploration difficult because of lack of information showing approach to temperature or g limits.

    6. System inoperative if communications blackout occurs.

    7. Pilot has no indication of V, H, α, or β.

C. Radio - Trajectory computation

    1. Identical computer mechanization required at each ground station as is required for airborne system.

    2. Accuracy for Step I, Step II and abort boost flights depends upon an adequate number of ground tracking stations.

    3. Corridor exploration possible.

    4. System inoperative if communication blackout occurs.

    5. Pilot has no indication of V, α, β.

D. Inertial - DR

    1. Inertial platform and computer somewhat more advanced than X-15 type will have to be developed.

    2. Marginal accuracy for Step II and abort boost flight conditions.

    3. Corridor exploration difficult because of lack of information showing approach to temperature or g limits.

E. Inertial Trajectory computations

    1. Inertial platform requirements same as for II-D.

    2. More complex computer and display required than for II-D.

    3. System and accuracy adequate for Steps I, II, or abort boost.

    4. Provides pilot with sufficient information to establish margins.

    5. Makes maximum utilization of pilot.

F. Inertial-Automatic Navigation

    1. Frees pilot from any active piloting task.

    2. Most complex system.

    3. Closed loop guidance and control mechanization must be developed.

    4. Automatic navigation technique must be developed.

    5. Does not give flexibility for corridor exploration.

BOOST

The DS boost guidance system will command the vehicle along a velocity-altitude profile. This guidance technique differs considerably from the Titan velocity-distance guidance method, and, consequently, a new system must be developed for the DS.

To insure that the vehicle follows the prescribed path, some form of redundancy or guidance monitoring is required during the development flights of a new guidance system. The redundancy is required not merely to insure attaining the desired end conditions, but also to insure that the dynamic pressure limits for the critical high-q and staging flight regions are not exceeded. Should the boost guidance system be in error, then these limits, which are structural limits, can be easily exceeded. The problem is made more critical by the present escape system concept which makes no provision for escape in event of vehicle structural failure.

The desired redundancy can be attained by either monitoring the primary system or the trajectory and either switch to another system or command abort should any disagreement occur. It is believed that a switchover to another system is more desirable during the Step I test program to avoid aborting flights for minor guidance system failures. Subsequent to the Step I guidance demonstration program, it should be practical to only monitor the boost trajectory and command abort if any limits are exceeded.

The radio guidance technique offers many advantages in the way of reliability for the Step I mission. However, for Step II a radio guidance system would be very complex and require much development effort because of the required number of ground guidance stations. A satisfactory technique, therefore appears to be the development of an inertial system for step II during the Step I program using a radio guidance system during Step I as a backup. Such an inertial system should also be capable of providing guidance during the glide phase.

GLIDE

The glide phase guidance system must provide commands for accurate control of the vehicle for abort-boost, corridor exploration, and Step I and II flight conditions. For maximum utilization of the pilot for these conditions, it is believed that it will be necessary to provide the pilot with an indication of velocity and altitude. These quantities can only be obtained from inertial platform measurements together with an airborne computer of the digital type. Therefore, the glide phase guidance method will include an inertial platform system to provide the basic velocity-distance information.

Aborted-Boost: A technique somewhat similar to that in use with the X-15, wherein simulator studies are used to define the correct action to take for any abort during boost, should be adequate for the DS. The boost phase can be zoned into velocity increments for which the pilot knows the correct recovery procedure and the commands to follow to attain specified landing sites. Backup guidance for Step I can be provided for the ground tracking radar stations. No backup will be available for all Step II abort-boost conditions.

Corridor Exploration: Some form of trajectory calculation is required for accurate control of the glide path for corridor exploration after injection at or above the top of the corridor. This calculation can be made from inertial platform measurements or from tracking radar measurements. It may be possible to use a simple technique such as flying a pre-programmed vertical velocity versus velocity program; however, simulator studies are required to further explore such simple methods. Backup guidance can be obtained through the use of a pre-programmed alpha versus velocity command or through some form of temperature monitoring. An automatic navigation method would be extremely complex for corridor exploration guidance.

Further studies, including simulator runs, are required to more accurately assess the various guidance techniques for corridor exploration.

Step I Mission: The airborne DR navigation technique appears to be the simplest form of guidance which will provide adequate information for the pilot. Backup for most of the Step I missions can be obtained through ground tracking radar. Little advantage can be seen in the use of an automatic navigation method.

Step II Mission: It is questionable whether DR navigation techniques will be adequate for the Step II mission. Trajectory computation techniques would provide adequate accuracy and somewhat greater flexibility for flights requiring large heading changes than DR techniques. However, it is believed that the adequacy or advantages of either technique can only be determined through extensive simulator evaluations.

Date: Early 1961   Title: Project Streamline

PROJECT STREAMLINE

STUDY No. 3

INTRODUCTION

The proposed program represents a maximum effort plan to advance as much as possible the date of first piloted orbital flight, consistent with minimum increase in risk and over-all program cost.

SUMMARY

The key to the plan is the elimination of Step 1 and the temporary substitution of available hardware for schedule-pacing subsystems in the glider, in order to make the glider available at the earliest practical date. When high performance subsystems are available, they will be incorporated and full system performance attained. Experience gained from the earlier flights will provide greater assurance of successful performance of the final configuration.

Use of an existing booster capable of orbital boost will eliminate the need for re-orientation and additional effort associated with a new and different booster part way through the program. Multi-orbit missions would require only an increase in the expendables in the booster-glider transition section and a method for updating the glider guidance system.

PROJECT STREAMLINE

Dyna-Soar Objectives & Approach
Glider Design & Growth
Booster requirements
Booster Availability
Alternate program
Costs

PLAN

Program Go-ahead 1 June 1961
Simplify, centralize, strengthen Program MGT
Use existing booster
Go Orbital to Edwards AFB first flight
Use Glider Model 2050 (Current Model)
  Substitute available equipment for schedule pacing subsystems on early flight
Increase funding & effort in critical development areas
Plan orderly & timely block change to incorporate full-performance subsystems

CONFIGURATION VERSUS FLIGHT

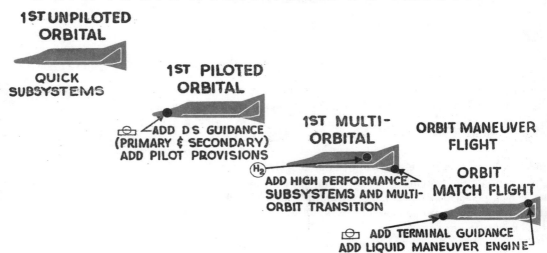

1ST UNPILOTED ORBITAL

QUICK SUBSYSTEMS

1ST PILOTED ORBITAL

ADD D S GUIDANCE (PRIMARY & SECONDARY) ADD PILOT PROVISIONS

1ST MULTI-ORBITAL

ADD HIGH PERFORMANCE SUBSYSTEMS AND MULTI-ORBIT TRANSITION

H_2

ORBIT MANEUVER FLIGHT

ORBIT MATCH FLIGHT

ADD TERMINAL GUIDANCE
ADD LIQUID MANEUVER ENGINE

D2-80253 7.

GLIDER CONFIGURATION
(MODEL 844 -2050)

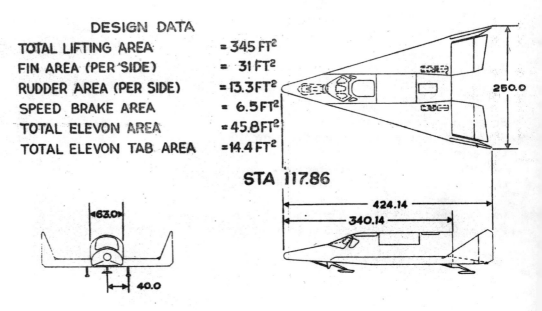

DESIGN DATA

TOTAL LIFTING AREA	$= 345\ FT^2$
FIN AREA (PER SIDE)	$= 31\ FT^2$
RUDDER AREA (PER SIDE)	$= 13.3\ FT^2$
SPEED BRAKE AREA	$= 6.5\ FT^2$
TOTAL ELEVON AREA	$= 45.8\ FT^2$
TOTAL ELEVON TAB AREA	$= 14.4\ FT^2$

STA 117.86

250.0

424.14

340.14

63.0

40.0

D2-80253 - 9

SINGLE ORBIT GLIDER-TRANSITION CONFIGURATION
(HIGH PERFORMANCE SUBSYSTEMS)

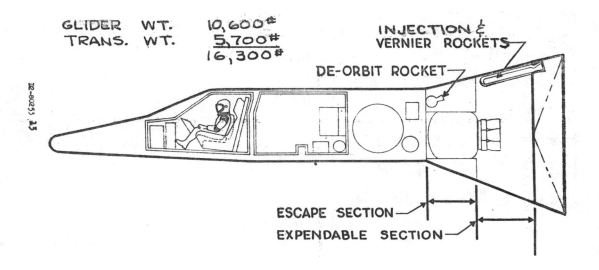

GLIDER WT. 10,600#
TRANS. WT. 5,700#
16,300#

INJECTION & VERNIER ROCKETS

DE-ORBIT ROCKET

ESCAPE SECTION

EXPENDABLE SECTION

EARLY GLIDER-TRANSITION
QUICK SUBSYSTEMS

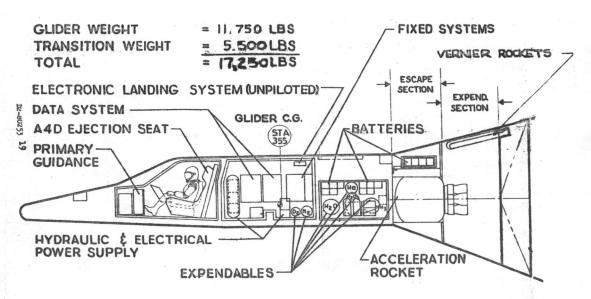

GLIDER WEIGHT = 11,750 LBS
TRANSITION WEIGHT = 5,500 LBS
TOTAL = 17,250 LBS

FIXED SYSTEMS

VERNIER ROCKETS

ESCAPE SECTION

EXPEND. SECTION

ELECTRONIC LANDING SYSTEM (UNPILOTED)

DATA SYSTEM

A4D EJECTION SEAT

PRIMARY GUIDANCE

GLIDER C.G.
STA 355

BATTERIES

HYDRAULIC & ELECTRICAL POWER SUPPLY

EXPENDABLES

ACCELERATION ROCKET

FIRST FLIGHT GLIDER SCHEDULE

PROGRAM GO-AHEAD

ENGINEERING

FABRICATION

SUB, MAJOR & FINAL ASSY.

SYS. INTEGRATION LAB.

GRD. SYS. TESTING

PRELAUNCH OPER.

1ST UNPILOTED ORBITAL

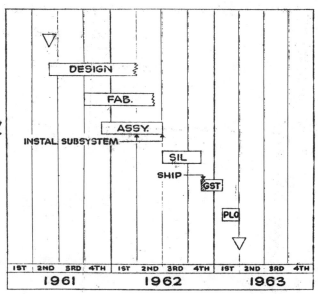

ORBIT MATCH GLIDER-TRANSITION CONFIGURATION
(High Performance Subsystems)

GLIDER   WEIGHT   10,600 #

TRANSITION WEIGHT   11,400 #
                            ―――――――
                            22,000 #

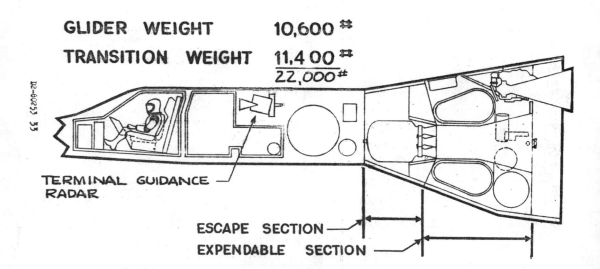

TERMINAL GUIDANCE RADAR

ESCAPE SECTION

EXPENDABLE SECTION

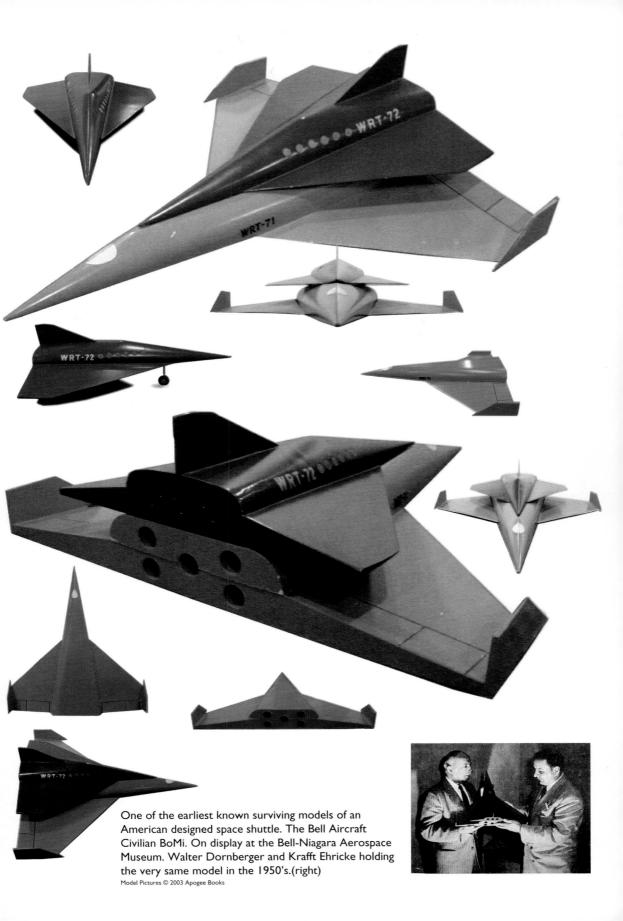

One of the earliest known surviving models of an American designed space shuttle. The Bell Aircraft Civilian BoMi. On display at the Bell-Niagara Aerospace Museum. Walter Dornberger and Krafft Ehricke holding the very same model in the 1950's.(right)

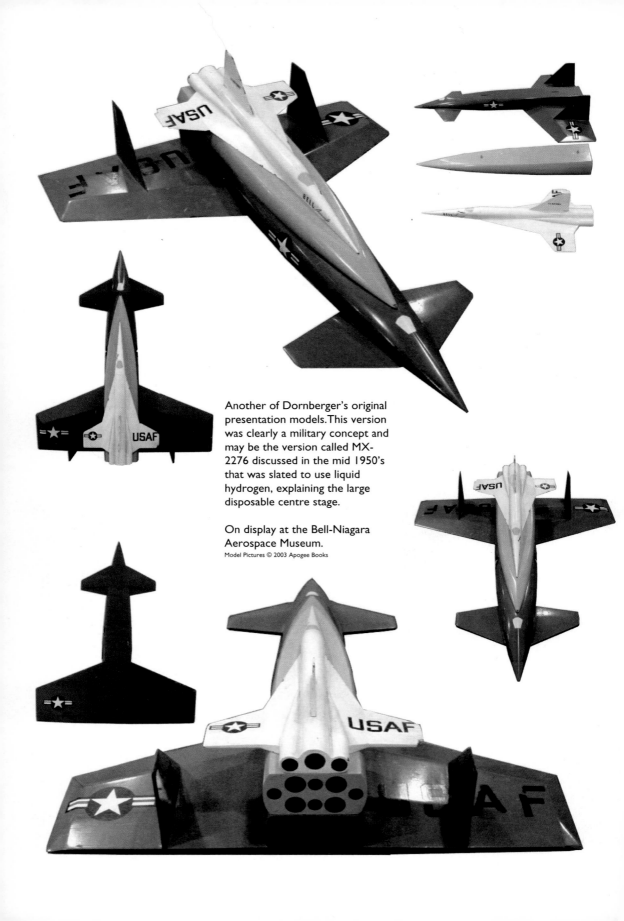

Another of Dornberger's original presentation models. This version was clearly a military concept and may be the version called MX-2276 discussed in the mid 1950's that was slated to use liquid hydrogen, explaining the large disposable centre stage.

On display at the Bell-Niagara Aerospace Museum.

Model Pictures © 2003 Apogee Books

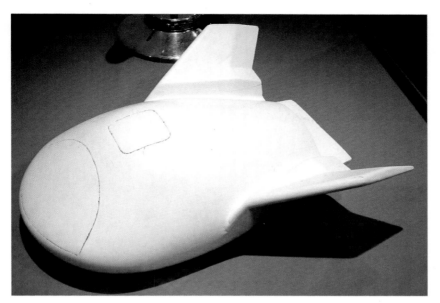

Another model on display at the Bell-Niagara Aerospace Museum. The provenance for this model is harder to determine.

It is said to have originated from Dornberger's department at Bell. It clearly illustrates a different concept for the re-entry shape. This version more towards the disc shape which is still being investigated today.

Model Pictures © 2003 Apogee Books

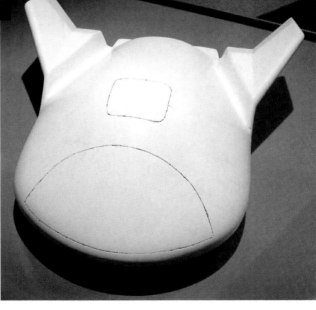

Screen capture from an Air Force documentary of an early Boeing model with its launch structure the design for the glider is probably model 1047 circa November 1958 (left)

A variety of models from the Boeing archives. Top left is a late model X-20 on a finned Titan Centaur. Top right shows a similar glider with the second stage and transtage attached. Bottom left are two possible designs with what appears to be models 1047 and 1010 on unidentified liquid boosters, probably 2-stage.
Bottom right is a NAA-McDonnell Douglas design for a Dyna-Soar I recoverable booster.

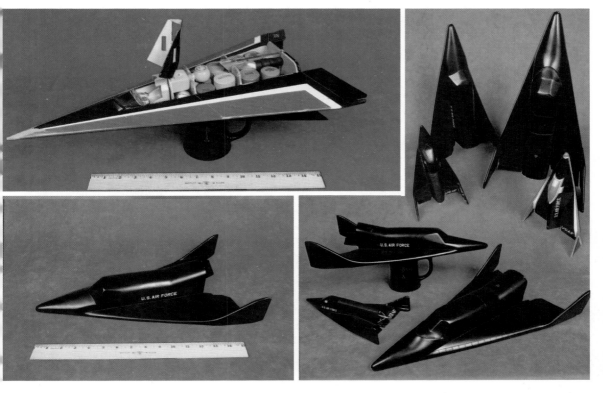

An assortment of models (above) from the Boeing archives. The red stripes represent early models. The one at top left shows the early delta shape. Later models are all black and show a variety of slight modifications.

More Boeing models. Top left shows the landing skids. Top right are two early proposals, 814-1047 is at left. Bottom left is a very early Boeing-Vought dorsal finned glider (possibly 814-1010) with an enormous flyback booster. Bottom right shows Dyna-Soar 814-2005 attached to the Centaur rocket.

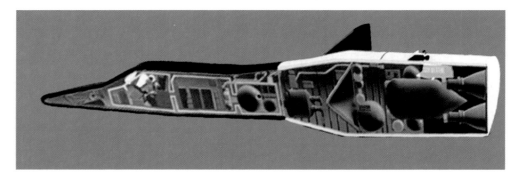

Cutaways of two late designs for Dyna-Soar probably sometime in 1960. The image shows the transtage innards. Above is a standard version while below is the advanced version (later dubbed X-20X) showing a manned mid-deck. (below)

An Air Force model showing the Dyna-Soar in place atop a Martin Titan 3C booster.

An early Boeing photograph showing a markedly different Dyna-Soar (possibly model 1047) dropping from a B-52 model. This may have been the version with an ejectable crew cabin (see DVD) (above)

The Martin Dyna-Soar on display at the New York World's Fair in 1964. (below) The shape bears some resemblance to the Martin X-23A/SV-5D reentry model and appears to be an interim design falling between the X-20 and later lifting bodies such as the X-24, the M2-F2 and the HL-10. This life-size model shown docking with a space station was animated. It ended up at the Oklahoma Omniplex and may be the nearest thing left in existence to a Dyna-Soar.

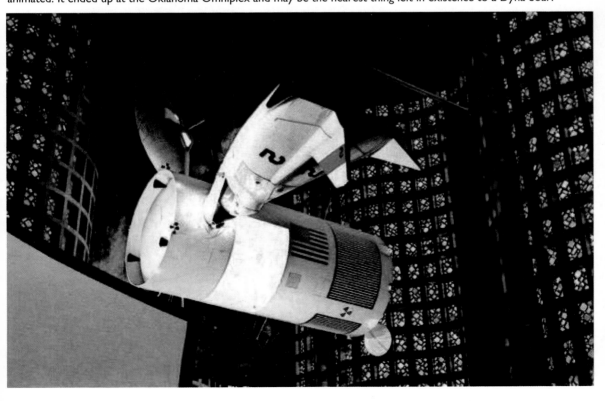

A collage illustrating the flight sequence proposed for the final manifestation of Dyna-Soar. The orbiter is launched on Martin's Titan 3C heavy lift booster. After discarding the United Technology solid rockets the subsequent stage fires to inject the manned vehicle into orbit. After discarding the transtage, orbital operations are conducted until reentry begins. Once the vehicle has cleared the plasma and heating stage the pilot ejects the window cover to allow a clear view of the landing site. Touchdown was to be back at the point of origin using wirebrush skids made from the experimental alloy René 41.

By April 1960 Dyna-Soar began to take shape and the first launcher assigned to the program was the Martin Titan I missile. It was 98 feet long and had a diameter which shrunk from 10 feet on the first stage to 8 feet on the second. It used perishable fuels (i.e. kerosene and liquid oxygen) and weighed in at 220,000lbs (without Dyna-Soar). The total thrust of the two stages was 380,000 lbs. Note the addition of stabilising fins on the booster to compensate for the awkward flight characteristics of the glider. The pitch stabilisers were 17.5 feet long while the yaw were 9.25 feet long.

In January 1961 the proposed launcher for Dyna-Soar was upgraded to a Martin Titan II (above). It was five feet longer than the Titan I and the upper stage had been widened to make the missile a consistent 10 feet in diameter. The main improvement was the use of storable fuels, a combination of 50% Hydrazine and UDMH with Nitrogen Tetroxide as the oxidizer. Overall thrust was increased to 530,000 lbs.

In February 1962 the launcher for Dyna-Soar was upgraded once more for fully operational orbital flight by using a Titan III with its large strap-on solid rockets. The boosters generated 1.3 million pounds of thrust each, creating a combined lifting thrust of 3.1 million pounds. Note that in this concept painting the solid rockets have stabilising fins attached. The top inset is an Aerojet painting showing the same arrangement. The center inset is a Feb 1962 Air Force rendering showing a distinctly different shape to the solids which originally were slated to include a dousing system. The lower inset is Boeing's final artist concept. The Spaceworld magazine cover is from Feb 1964, two months after the programs' cancellation and it still shows the finned Titan III.

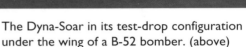

The Dyna-Soar in its test-drop configuration under the wing of a B-52 bomber. (above)

Early USAF animations showed a somewhat truncated version of the Titan (above) The stages were to separate at the lower gold line.

Editor's rendering showing the Titan 3C launch vehicle at staging. The peel-away staging concept for the solid rockets came from the British Bloodhound anti-aircraft missile. (above)

Early Dyna-Soar atop a finned NOVA launch vehicle (artistic liberty by the Editor from a period sketch and painting, left) NOVA was dismissed because it was considered too complex and expensive.

USAF model from the late 1950's before the addition of the sloped transition (below)

As with any experimental vehicle the cockpit went through rigorous testing and subsequent modifications. These screen captures from a USAF film show some of the instrument panel arrangements.

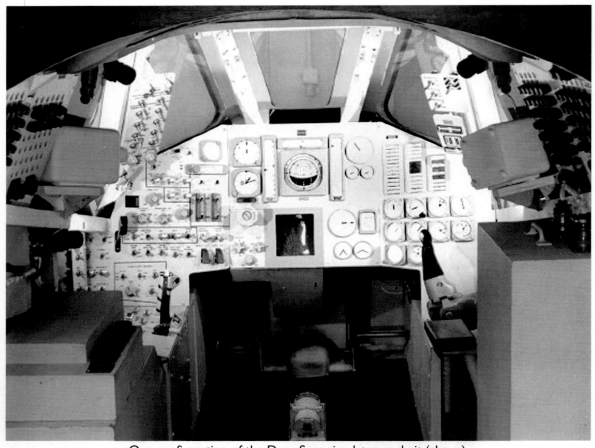

One configuration of the Dyna-Soar simulator cockpit (above)

A rare color photograph of the Boeing Mock-up sporting a bright yellow nose. (below)
Notice the window cover is in place.

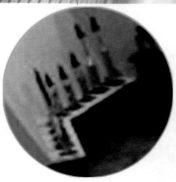

A full size mock-up of Dyna-Soar at the Boeing plant atop a General Dynamics/Convair Atlas-Centaur A or B booster. The AC-B had a LOX/kerosene 1st stage with a LOX/LH2 2nd stage. Capable of lifting 4670 kg. The AC-B was Boeing's preference but the DOD opted for the Titan. The surrounding display looks like it may been some sort of exhibit, perhaps for visiting Air Force dignitaries. Notice the array of possible launch vehicle models highlighted at left.

The great lost opportunity, Dyna-Soar in orbit. A Boeing artist's conception. (right) The essentials of the vehicle are still being reviewed today as America looks to the future of manned space flight.

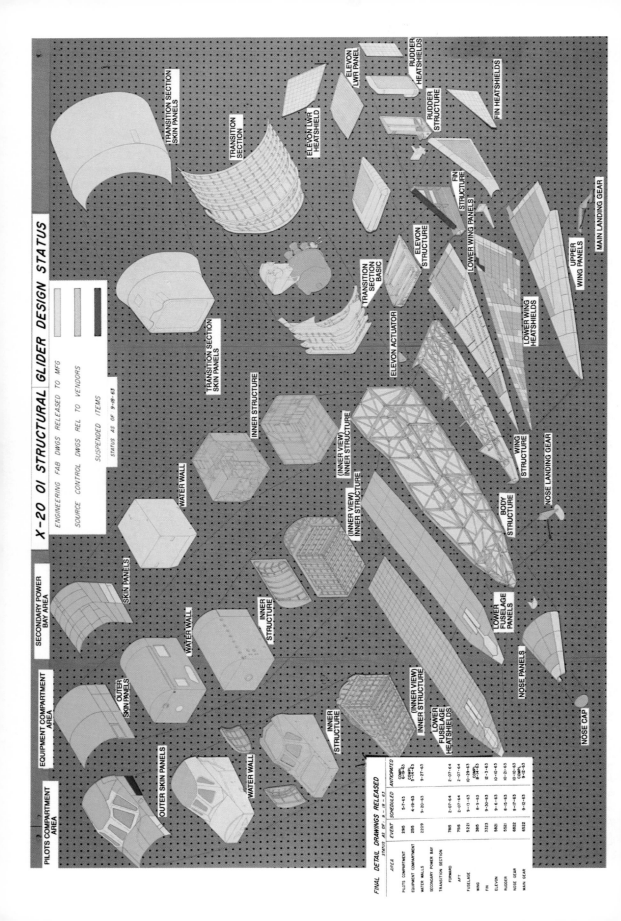

X-20 01 STRUCTURAL GLIDER DESIGN STATUS

PILOTS COMPARTMENT AREA   EQUIPMENT COMPARTMENT AREA   SECONDARY POWER BAY AREA

ENGINEERING FAB DWGS RELEASED TO MFG

SOURCE CONTROL DWGS REL TO VENDORS

SUSPENDED ITEMS

STATUS AS OF 9-18-63

OUTER SKIN PANELS
WATER WALL
INNER STRUCTURE
(INNER VIEW) INNER STRUCTURE
LOWER FUSELAGE HEATSHIELDS
NOSE PANELS
NOSE CAP

OUTER SKIN PANELS
WATER WALL
INNER STRUCTURE
(INNER VIEW) INNER STRUCTURE
LOWER FUSELAGE PANELS
NOSE LANDING GEAR

SKIN PANELS
WATER WALL
INNER STRUCTURE
(INNER VIEW) INNER STRUCTURE
BODY STRUCTURE
WING STRUCTURE

TRANSITION SECTION SKIN PANELS
TRANSITION SECTION
TRANSITION SECTION SKIN PANELS
TRANSITION SECTION BASIC
ELEVON ACTUATOR
ELEVON STRUCTURE
LOWER WING PANELS
LOWER WING HEATSHIELDS

ELEVON LWR PANEL
ELEVON LWR HEATSHIELD
RUDDER HEATSHIELDS
RUDDER STRUCTURE
FIN HEATSHIELDS
FIN STRUCTURE
UPPER WING PANELS
MAIN LANDING GEAR

FINAL DETAIL DRAWINGS RELEASED

STATUS AS OF : 9-18-63

AREA	EVENT	SCHEDULED	ANTICIPATED
PILOTS COMPARTMENT	395	5-7-63	COMP. 6-21-63
EQUIPMENT COMPARTMENT	355	4-19-63	COMP. 5-14-63
WATER WALLS	2209	9-20-63	9-27-63
SECONDARY POWER BAY			
TRANSITION SECTION			
FORWARD	785	2-07-64	2-07-64
AFT	795	2-07-64	2-07-64
FUSELAGE	5221	9-13-63	10-29-63
WING	385	8-9-63	8-14-63
FIN	3323	9-30-63	12-3-63
ELEVON	5821	9-6-63	10-10-63
RUDDER	5531	8-15-63	10-21-63
NOSE GEAR	6822	9-17-63	10-10-63
MAIN GEAR	6522	9-12-63	COMP. 9-12-63

MANNED HIGH W/S GLIDER

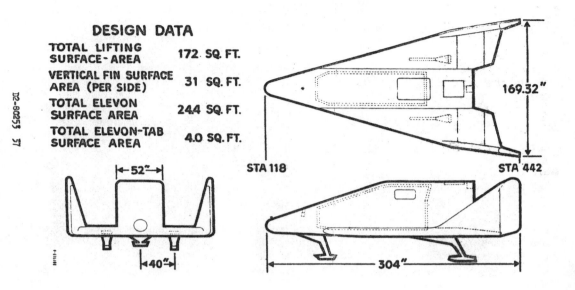

DESIGN DATA

TOTAL LIFTING SURFACE-AREA	172 SQ. FT.
VERTICAL FIN SURFACE AREA (PER SIDE)	31 SQ. FT.
TOTAL ELEVON SURFACE AREA	24.4 SQ. FT.
TOTAL ELEVON-TAB SURFACE AREA	4.0 SQ. FT.

169.32"

STA 118   STA 442

52"

40"

304"

D2-80253   57

PROGRAM BLOCK CHANGES

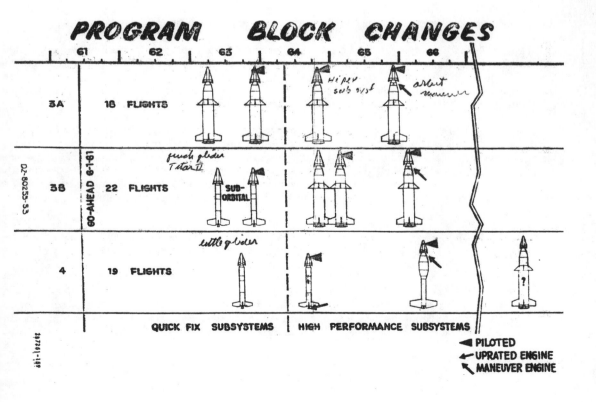

D2-80253-53

		61	62	63	64	65	66
3A	18 FLIGHTS						
3B	GO-AHEAD 6-1-61 22 FLIGHTS			SUB-ORBITAL			
4	19 FLIGHTS						

QUICK FIX SUBSYSTEMS   HIGH PERFORMANCE SUBSYSTEMS

◄ PILOTED
← UPRATED ENGINE
↖ MANEUVER ENGINE

Date: Mar 1961   Title: Primary Vehicle Guidance Subsystem Specs.

21 March 1961

WWZRG/Mr. Chandler/36196/B2O/R219

Dyna-Soar Primary Vehicle Guidance Subsystem Specification

National Aeronautics and Space Administration Flight Research Center
Attn: Mr. Paul F. Bikle Box 273, Edwards, California

1. Reference is made to NASA Flight Research Center letter, subject: Dyna-Soar Primary Vehicle Guidance Subsystem Specification., dated 2 November 1960, which forwarded comments on Boeing Aircraft Company Document D2-7408.

2. The subject letter has been reviewed by the SPO and the following comments are offered:

a. Paragraph 2. The concept which NASA proposes is reflected in the Primary Guidance Subsystem Associate Contractor's Work Statement. It is agreed that maximum utilization of the pilot's ability and judgment should be made by providing him with sufficient data for navigation and control of the vehicle from launch to landing. Display information will be such that the pilot can make decisions based on the margins available and can subsequently monitor the progress of the flight in relation to margins.

b. Paragraph 3. It is essential for manual piloted control of the vehicle that suitable handling qualities be provided with appropriate stability augmentation. Automatic attitude hold in the manual control mode is considered to require more complexity for the piloted tasks than is desirable considering the added cost, weight, reliability and other factors. No specific minimum "hand off" attitude rates have been specified in the Flight Control Electronics Subsystem (FCES) specification to date. It is planned that the next revision of the FCES specification will include such requirements. These requirements will be such that the pilot may easily control attitude since the drift rate in each axis will be sufficiently low to require little attention in maintaining the desired attitude when stability augmentation is "on". In the design of the FCES the pilot will have the capability to "fly" the glider or the glider-vehicle. This is being done in the glider with stability augmentation utilizing an adaptive technique where in angular rates and/or coordinated turns are commanded by the pilot's controls. It should be recognized that secondary attitude reference is available in event of inertial guidance failure.

c. Paragraph 4. Before judgment can be given on the method of approach to the Energy Management scheme, it will be necessary for both the Primary Guidance Subsystem Associate Contractor and the System Contractor to complete their portions of the Energy Management Study. The Associate Contractor for primary guidance will make maximum utilization of the work done by Chance-Vought and has contracted with them to do their Energy Management Analysis.

d. Paragraph 5. The Primary Guidance Subsystem Associate Contractor's work statement reflects the need and the requirement for close coordination with the System Contractor and other Associate Contractors to insure a completely integrated system. In reference to sideslip control being specified as a function of both FCES and the guidance subsystem, it is desirable that the FCES provide the aerodynamic turn coordination and that the guidance subsystem provide for alignment of the vehicle heading axis with the velocity vector (non-aerodynamic sideslip) prior to initiating a re-entry maneuver.

e. Paragraph 7. The Energy Management studies of the System Contractor and the Primary Guidance Subsystem Associate Contractor will either confirm or deny the necessity for an alpha (angle of attack) display. It can be approximated by theta (pitch angle).

f. Paragraph 9. The use of temperature and/or temperature rate information is a good idea for use as cross-check or as an error limitation function. This area will be evaluated. The use of self adaptive techniques will be studied; however, it is not felt that sensor instrumentations have sufficient accuracy at this time to warrant the preference of that data over pre-flight computed data.

GEORGE B. JOHNSON
Chief, Guidance and Communications Branch Dyna-Soar SPO
Directorate of Systems Management

Date: Apr 1961   **Title:** Dyna-Soar stability and control characteristics

MEMORANDUM for Dyna-Soar Files   April 5th 1961

Subject: Dyna-Soar stability and control characteristics

1. Recently, many discussions have been held concerning the longitudinal stability characteristics of the DS during abort. Since this condition produces an unstable vehicle, major concern has been expressed on this subject to the extent that it appears there is little concern for other problems. It seems appropriate, therefore, to outline other regions which may also contribute to DS stability and control problems .

2. When one considers the complete flight envelope of the DS, one can appreciate the magnitude of the task for designers to satisfy the stability and control criterion. The flight regions to be considered are not only the normal hypersonic $\alpha = 15$ to $55°$ region and the low speed approach and landing conditions, but also the injection, abort, and recovery from abort conditions. Control during abort is receiving some attention; however, neither injection nor recovery from abort have been investigated. A summary of what is known about these flight regions is summarized below. When discussing characteristics , it is important to point out which DS configuration is under consideration.

(a) Hypersonic flight - The hypersonic flight region has been studied by both Ames and FRC. Briefly, these studies have shown regions of positive stability where the vehicle was in fact, uncontrollable without stability augmentation. The FRC studies showed that over a large part of the hypersonic flight envelope the DS is uncontrollable without using special control techniques which, although successful on a simulator, have not been proven successful during normal or emergency flight conditions. It should also be remembered that the display for this evaluation consisted of an oscilloscope display of a considerably different type from a normal pilot display. These factors point out the lack of an absolute indication of controllable regions for the DS, since studies to date indicate that it may not be possible to obtain satisfactory control characteristics (unaugmented) even with a stable vehicle.

(b) Landing - Studies to date indicate no stability and control problems.

(c) Injection - The latest data available (2035 configuration) shows the DS to be longitudinally unstable at the low alpha injection conditions. The q at injection is low and this instability should not present a difficult control problem; however, since the pilot will be presented with a control problem abruptly at injection, some investigation of this region seems in order. No data are available upon which to base a "best estimate" of this condition.

(d) Abort - The longitudinal stability during abort has been thoroughly investigated by Crane of Ames with no unique solution in sight. A condition resulting from abort, and not evaluated by Crane, is the injection from abort. Since some of the abort injection conditions involve unstable low alpha regions (2035), this factor combined with the high q of an abort may produce as difficult a control problem as the instabilities discussed by Crane. (I think they may be more difficult.)

(e) Recovery from Abort - The recovery from an aborted boost requires a high alpha - high q maneuver similar to an X-15 reentry from high altitude. Although the DS is stable and controllable for this condition, the combination of α-q produces high forces and a difficult control problem to prevent the vehicle structural limits from being exceeded. One FRC investigation showed that it is difficult to remain below structural limits at q> 300psf. This study concerned longitudinal control and lateral control was not involved. The recovery from abort appears to be a serious problem (unaugmented), however, a constant gain damper should prove very effective for this condition.

3. The above considerations result in great emphasis being placed upon the satisfactory operation of the adaptive control system. The approach taken by Boeing is to provide dual or triple redundancy in the various channels of the adaptive control system. The undersigned has recommended that the work statement be changed to state that the adaptive control system should be designed so that no single failure can result in a significant degradation in performance, since dual or triple redundancy - per se - does not guarantee satisfactory control in event of a component failure. In addition it has been recommended that it should be demonstrated that a constant gain damper mode provides satisfactory control, before this mode is accepted as the backup for the adaptive control system.

Wendell H. Stillwell Aerospace Technologist

Date: Apr 1961   Title: Memo: Dyna-Soar Guidance System

April 6th 1961

NASA-FRC

Memorandum for Chief, Research Division
Subject: Dyna-Soar Guidance System discussions

1. A visit was made to the Boeing Airplane Company on March 30 and 31 1961, to discuss the Dyna-Soar Guidance System. Mr J. Dodgen of Langley Research Center participated in the meeting March 30 and M. Thompson of FRC and Major J. Woods of AFFTC participated part time on the same day. Discussions were held with Messrs. R. Curren, B. Muzzey, A. Nagel and F. Reynolds of BAC.

2. Mr Muzzey described the complete BAC guidance concept as of this date. Basically, this concept is the same as that presented in BAC document D2-8088; however. A number of studies are underway at Boeing which could alter the guidance system. These studies are concerned with the secondary or backup guidance modes, and the information required by the pilot for manual control. Considerable effort is being devoted to the development of a system called the Flight Integrator which will present the pilot with the information required to operate the vehicle within the acceleration and temperature limits.

3. The division of responsibility for the complete guidance system design and development was subject to considerable discussion without significant clarification. Since Minneapolis-Honeywell, Autonetics, Chance Vought, and General Electric are concerned with parts of the guidance system as associate or subcontractors under WADD, the responsibilities of BAC have not been too clear. The usual response to the question of BAC responsibility was, that since BAC had accomplished most of the work to date, it was expected that the associate contractors would rely upon this effort. It is believed that a reasonably well coordinated program may result from this relationship; however, the usual tendency of contractors to disregard other efforts could create many problems in regard to the integration of all components into one system and is felt to be of major concern.

4. The BAC concept of manual control seems to be that it is a design objective and will be provided for; however, the basic mode of operation will be automatic control. Manual control, therefore, is considered by some to be the backup or secondary mode of operation and attempts are being made to provide the pilot with information from different sources than the automatic guidance system. This has generated a great deal of interest in determining velocity, altitude, and angle of attack from pseudo-air date systems. Several studies are underway, but it is too early to determine if a practical system can be developed for Dyna-Soar. BAC plans to provide the pilot with a map display, in combination with the flight integrator display, should provide the pilot with all information required for guidance and energy management for manual control.

5. Several techniques are being studied for application to the flight integrator display. These studies are concerned with the various methods of combining velocity, altitude, angle of attack, lift, and vertical velocity into a composite display which will permit the pilot to monitor the progress of the flight in relation to temperature and acceleration limits. Such a display should be of considerable aid during the recovery from an aborted boost or for corridor exploration. The same display component will be used for the flight integrator and the mapping display, and pilot may select one of the displays at any time.

6. Efforts to obtain information on the development and test program for the guidance system were unsuccessful inasmuch as Mr. R. Hare, who is in charge of the program, was visiting the FRC on these dates. It is planned to contact Mr. Hare at an early date to discuss this subject.

Wendell H. Stillwell
Aeronautical Engineer

Date: Apr 1961   **Title:** Minutes: NASA Dyna-Soar Committee Meeting

FOR INTERNAL NASA USE ONLY
MINUTES: NASA DYNA SOAR COORDINATING COMMITTEE (4th Meeting)
DATE: April 14, 1961 (last meeting 12/20/60)
TIME: 9:30 a.m.
PLACE: Room T-910, 1512 H Street, N. W., Washington, D. C.

PRESENT: Chairman: H. A. Soulé
Members: J. V. Becker P. F. Bikle R. M. Crane
Secretary: M. G. Rosche
NASA Dyna Soar Office: P. F. Korycinski
Also: M. B. Ames
                      H. H. Kurzweg NASA HQ
                      R. W. May   (Part-time)
                      C. Wood
                      T. F. Baker - FRC
                      W. H. Stillwell - FRC
                      D. Fetterman - LRC

SUMMARY OF MINUTES:
The main items of discussion at the meeting were:
    (a) System Development Test Plan.

    Plans were made for the preparation and transmittal to the Air Force   of NASA comments on the
    Boeing Document D2-5697-14, "System Development Test Plan."

    (b) Stability and control problems of DS-1.

The discussion revealed the number of stability and control problem areas of DS-1. The Air Force is being
informed of NASA's activity and concern regarding the most critical of these problems.

REVIEW OF ACTION ITEMS FROM THE LAST MEETING

Mr. Soulé reported that a letter had been sent to ARDC from NASA Headquarters dated January 13, 1961,
outlining NASA's position regarding the number of unmanned flights on the DS-1 glider in the flight test
program. The Air Force reply to this letter dated March 21, 1961 expressed agreement with the NASA
views.

Mr. Becker reported that he had drafted a letter which had been sent to the System Project Office from
NASA Headquarters dated February 20, 1961 outlining the NASA proposal for flight tests utilizing the
RVX-2A. He explained that the SPO has allocated $3,000,000 to be used for two RVX-24/Titan II shots.
Mr. Korycinski reported that the first installment will be transferred to NASA in the very near future.
NASA will have primary responsibility for the tests to be conducted during the first shot in July 1962 and a
secondary role in the tests in the second shot in September 1962. It was agreed that a meeting between
NASA and the Air Force should be held soon to discuss these experiments. Mr. Ames reported the
approval by Headquarters for Langley to spend up to $100,000 for payload design for these experiments.
He asked Langley to prepare and submit to Headquarters a formal request for this work.

Mr. Soulé referred to the monthly report which had been requested by the Project Office surveying the
NASA manpower being devoted to the Dyna Soar project. He noted that the members are supplying this
information and instructed them to continue to do so.

Mr. Baker discussed the need to decrease the number of instrumentation channels in the DS-1 glider. He explained that at first it appeared that weight would be the factor which controls the number of measurements possible. However, the controlling factor now appears to be the availability of access space. He reported upon recent attempts to decrease the number of instrumentation channels. He explained that this topic will be the subject of a meeting to be held at Wright-Patterson Air Force Base during the middle of May.

Mr. Baker reported that because of lack of manpower he has not been able as yet to define instrumentation requirements for supplying the pilot with air data.

Mr. Soulé asked for suggestions as to whether the NASA Technical Advisors should attend the Development Engineering Inspection in June or the Mock-up in September. It was agreed that the Mock-up would be the preferred time. Mr. Korycinski reported that the recommendations by BSD regarding the booster for Dyna Soar Step II will not be complete until August of this year.

DYNA SOAR SYSTEM DEVELOPMENT TEST PLAN

Mr. Soulé referred to Boeing Document D2-5697-14, "System Development Test Plan," which had been sent to NASA for comment. He questioned whether NASA should submit detailed comments on this test plan outlined in the document. Mr. Bikle explained that FRC had reviewed and compiled detailed comments on the document. He emphasized that because of the size of the document, and the short time available for its review, these comments do not cover in detail every aspect of the test plan. However, he felt that they would be of use to the Air Force.

Mr. Bikle offered a number of general comments on the test plan. He criticized it as being extremely optimistic from the standpoint of money and time. He said that it is based upon obtaining complete success, leaving little margin for failures, partial successes and unanticipated problems. The committee agreed that experience based upon past flight tests shows such optimism to be unrealistic.

Another question was raised regarding the wisdom of detailed advanced planning of the flight tests. Mr. Bikle explained that even at present such planning cannot be successful for X-15 flights. Mr. Soulé reviewed experiences of flight tests on other research airplanes which also revealed that such detailed advanced planning is meaningless. Based upon this experience, he suggested that the most useful planning which could be done at this time for the Dyna Soar flight tests involves establishing clearly defined objectives.

Discussion of the over-all objectives and philosophy of the Dyna Soar flight tests revealed considerable differences of opinion between the Dyna Soar System Program Office and NASA. Dissatisfaction was expressed regarding NASA's role in the Dyna Soar project, particularly in the flight test operations aspects of the project. A synopsis of these points was prepared by Mr. Baker for possible future reference.

It was agreed that Mr. Bikle would prepare a draft of a letter for Mr. Abbott's signature in which NASA comments would be sent to the Air Force. This draft will be sent to Mr. Soulé and Mr. Ames by April 18, and the final letter will be sent to the Air Force by May 1.

STABILITY AND CONTROL PROBLEMS OF DS-1

Mr. Soulé expressed concern regarding the stability and control characteristics of DS-1. He asked Mr. Crane to report on the results of recent studies underway at Ames.

Mr. Crane first discussed the stability of the glider alone. He reported that estimates made at Ames indicate adequate directional stability throughout the flight corridor; however, longitudinal stability appears marginal in the M = 3 to 4.5 regime at high angles of attack. He indicated that instability could also be expected at extreme angles of attack. However, he was not concerned about this since these extreme angles of attack are not anticipated.

Mr. Crane expressed the opinion that the most severe stability problem appears to be associated with air launch. He said that considerable negative margins occur following drop from the B-52. He said that some changes in geometry must be made unless one wished to rely upon augmentation. He emphasized that Ames does not advocate reliance on stability augmentation systems.

Mr. Crane expressed the opinion that the stability during an abort at max q may be an even more important problem than during an air drop. He felt that a "fix" could be found for the air launch condition, however, he knew of no way to provide adequate stability for the abort-from-pad condition.

Mr. Crane concluded that the present glider + abort-rocket configuration is uncontrollable without stability augmentation. He stated he has informed Mr. Klepinger, Dyna Soar SPO, and Boeing of these conclusions pointing out that reliance on augmentation is contrary to Air Force philosophy as stated in the DS-1 work statement.

Mr. Soulé strongly questioned the wisdom of basing the success of Dyna Soar on an unproven and undeveloped adaptive control system. Mr. Crane was inclined-to rely on such a system provided it is not required during the early flight tests. This would permit experience to be gained about the system during the air launch flights. Mr. Stillwell felt that effort should be placed upon improved adaptive control systems. Mr. Becker suggested that NASA investigate possible alternatives to the use of such systems. It was agreed that Mr. Crane would investigate such alternatives and that Mr. Korycinski would so inform the SPO.

Mr. Crane mentioned that there appears to be some lack of roll control during launch; however, he estimated that this would not constitute a serious problem. It was agreed that Mr. Korycinski would inform the SPO of NASA's concern regarding the stability of DS-1 during the Configuration Control Committee meeting on April 27.

M.G. Rosche Acting Secretary NASA Dyna Soar Coordinating Committee

Date: Apr 1961   Title: Boeing Contract Changes

24 April 1961

<u>Daily Activity Report From Dyna-Soar SPO</u>

Dyna-Soar System 620A

The following contract changes were recently issued against the Boeing Dyna-Soar Contract:

a. Secondary Attitude Reference subsystem to be supplied as GFAE, associate contractor, in lieu of contractor furnished (credit for this contract period ending 30 September 1961 – estimated at $145,000).

b. Re-issuance of CCN on Titan II change, FY 62 reduced funding and stretchout of FY 61 contract from 30 June 1961 to 30 September 1961 (previous estimate for CCN of $10M reduced to $8M based on firm cost proposal submitted by Boeing).

A visit was made to Langley Research Center to obtain a wind shear and gust loads analysis to be applied to Dyna-Soar air vehicle boost. This analysis can be easily programmed for digital or analog computation. At present LRC is using this analysis on the Mercury-Atlas and Saturn. Other significant problems in the dynamics area were discussed such as glider instrumentation for flutter and vibration and booster fin flutter model tests. The booster fin flutter model tests will begin next week in the 26' transonic blow down wind tunnel.

A study of the Dyna-Soar communication development problems by RCA, the Dyna-Soar Communications Subsystem Associate Contractor, reveals that the communications range will probably be reduced slightly from the estimated range used for planning purposes by the SPO and AFMTC in selecting the sites along the Atlantic Missile Range. The change in range is from a slant range of 300 miles radius to 270 miles. This reduction should have little effect on the location of the instrumentation sites and ships along the Atlantic Missile Range, but it will be taken into account as one factor as the SPO examines the requirements for voice communications, tracking, and real-time display panels at the instrumentation sites, on the ships, and at the landing sites in an effort to determine optimum site location and equipment.

In keeping with the philosophy that the Dyna-Soar glider is being developed as a manned system and not as an unmanned system in which a pilot is inserted, the Dyna-Soar SPO has determined that the Secondary Altitude Reference System should not be installed in the gliders for the unmanned ground launches. To be of value in the unmanned gliders, the SARS would have required either the addition of an automatic switching system or of a command control signal to switch from the primary guidance system and probably the addition of other inputs for automatic control purposes. This additional complication would have increased the weight, complexity, and cost of the SARS and would result in a less reliable manned system.

John B. Trenholm

Date: May 1961   Title: Memo: Synopsis of objectives

MEMORANDUM   May 17 1961

To FLIGHT - Attn: Mr. W. H. Stillwell
From Mr. A. J. Evans
Subject: Forwarding of Synopsis for Dyna Soar Project for Information and Reference

1. Attached is a Synopsis of objectives, Philosophy, Management and Procedures for Dyna Soar Project, which was prepared by Mr. T. F. Baker as a result of discussion at the April 14, 1961 meeting of the Dyna Soar Coordinating Committee, mentioned on top of page 4 of the Minutes of the meeting.

2. This Synopsis is not attached to the Minutes, but a copy is forwarded herewith for your information and possible future reference.

Encl. AJE:hac

SYNOPSIS

OBJECTIVES, PHILOSOPHY, MANAGEMENT AND PROCEDURES FOR DYNA SOAR PROJECT

(Discussed at Meeting of Dyna Soar Coordinating Committee - April 14, 1961)

1. The NASA requires the adoption of a realistic approach in the flight testing of Dyna Soar. Involved are the objectives of both the Dyna Soar project and of the flight test program; the design and test philosophy; the management of both the program and the test activities; and the procedures and techniques to be utilized in the flight testing of the Dyna Soar air vehicle and glider.

2. The objective of the Dyna Soar program is to research the characteristics and problems of flight in the hypersonic, boost-glide flight regime. In accomplishing this research, the military potential of hypersonic boost glide weapons systems can be determined. The glider development objectives must be compatible with the program objectives and should be aimed at producing a basically "good" flying machine without undue reliance on "automatic" or "electronically-tailored" features. It is one of the objectives of the test program to develop and optimize the systems that are required and that are helpful to flight in the hypersonic and reentry flight regime. The NASA recognizes that an additional test objective is the utiliza-tion of the glider as a "test bed" for military systems development once the basic problems of flight are assessed and adequately overcome.

3. The design philosophy to be utilized in developing the air vehicle and glider should be documented by the contractor and reviewed by the NASA. Cardinal principles of the design philosophy must be the development of a vehicle to specifically accomplish the required research experiments (of which research on the glider in its flight environment is paramount); the development of a glider for operation by the pilot exclusively; the development of as simple a glider (and air vehicle) as possible with every effort being made to minimize the complexity of the system.

4. In order to properly discharge its responsibilities to the Dyna Soar project, the NASA must participate in the management of the program, not only with regard to performance, but also as regards budget and schedule. One means of increasing the NASA participation in the program management (and also of assuring that a proper glider is developed) is to provide deputy (or associate deputy) directors of Development and of Test to the Dyna Soar Air Force Systems Project Office. The NASA persons would work directly with the Air Force Deputies (Lamar and Svimenoff). It should be recognized by NASA that the Air Force Deputy for Tests has very limited prerogatives relative to the latitude given corresponding

NASA positions. Accordingly, it is proposed that an over-all flight test executive committee (similar in stature to the D. S. Management Council) be established, above the level of Air Force Dyna Soar Deputy for Test, to set guidelines, establish policy, etc. The NASA should regard such an executive committee as constituting the "flight test team" of the memo of understanding.

5. It is time that NASA laid its cards on the table in regard to its role in the test operations of Dyna Soar. NASA has more experience and know-how than anybody else in the testing of research aircraft and manned ballistic vehicles. The NASA expects to be responsible for and to provide the majority of the personnel staff required for the flight test engineering, data analysis aspects of the Dyna Soar flight test operations.

6. The NASA regards the detailed procedures and techniques for testing Dyna Soar as items to be worked out by the joint NASA-AF-Contractor flight test team under the general direction of an over-all flight test executive committee. It regards as imperative, however, several aspects of the Dyna Soar flight test procedures; namely, the glider should be completely demonstrated and developed during the relatively inexpensive air launch program - a Mach number of 2 appears adequate for the air launch program; a realistic schedule and budget for both air launch and ground launch tests must be established. It must be recognized and admitted that Dyna Soar is an R&D program and that "plan for success" is asinine; some unmanned tests of the air vehicle, i.e., the booster-glider combination, are required. Such tests must however be clearly recognized as demonstrations that the design is adequate—verification or establishment of booster-glider reliability cannot be and need not be accomplished; exploration of the hypersonic flight regime must be accomplished in steps of both velocity and proximity to structural placards.

Date: June 1961   Title: Memo: NASA Technical Advisors List

NASA Technical Advisors for Dyna-Soar Program

Technical Area	Cognizant Research Center	Technical Advisor
Booster Aerodynamics	Ames	Edward W. Perkins
Booster Dynamics	Langley	A. Gerald Rainey
Propulsion (Booster)	Lewis	Carl. F Schueller
Glider Aerodynamics	Langley	David E. Fetterman Jr.
Structures & materials	Langley	Eldon E. Mathauser
Landing Gear	Langley	Upshur T. Joyner
Simulation (Dynamic Stability & Control)	Ames	Steven E. Belsley
Instruments	Langley	Francis B. Smith
	Langley	William D. Mace
	Flight	George F. Schwartz
Flight Operations	Flight	Thomas F. Baker
Cockpit & Escape System	Flight	Neil A. Armstrong
Guidance & Controls	Flight	Joseph Weil
Reliability & Safety	Flight	Milton O. Thompson
Range	Flight	Kenneth C. Sanderson
Communications	Flight	Kenneth C. Sanderson
Aux. Power inc. Glider Engine	Flight	William P. Albrecht
(Low Speed Aerodynamics)	Ames	George G. Edwards
(Booster Ground Wind Loads)	Ames	Donald A. Buell
Life Sciences	Headquarters	G. Dale Smith, Dr.

NASA –FRC   June 7, 1961

MEMORANDUM for Chief, Research Division
Subject: Dyna-Soar guidance and control discussions at Wright-Patterson Air Force Base

1. Mr. Wendell Stillwell visited the ASD, Aeronautical Systems Division, WPAFB on June 1 and 2, 1961, to attend a Dyna-Soar Guidance System Meeting and for other discussions pertinent to DS guidance and control. Mr. J. Dodgem of Langley Research Center also attended the guidance system meeting. The purpose of the guidance meeting was to review, discuss, and redirect the Honeywell energy-management work, and to review boost guidance, the guidance malfunction detection subsystem, and inertial guidance system reliability. Representatives from Honeywell, Boeing, Martin, NASA and the SPO attended the meeting.

2. Mr. R. Schiedenhelm of Honeywell discussed their energy management system, which actually was conceived by Chance Vought under a subcontract from Honeywell. This system is similar to the Boeing system during automatic guidance; however, for manual control, the Boeing system will have a display which shows the maximum capabilities of the vehicle, whereas Honeywell will use the automatic guidance equations, for displaying vehicle capabilities. Boeing proposes to use a vertical situation display which will provide the pilot with information pertinent to operation within the temperature-load limits. Honeywell does not propose to have such a display for the early flights, and a rather complicated mechanization would be required for displaying all energy-management information on one instrument, and also provides steering command information in a more direct manner than the Boeing system.

3. Subsequent to the review, a meeting was held between SPO and NASA representatives to reach a decision as to which energy-management technique should be used with Dyna-Soar. The undersigned pointed out possible limitations of the Honeywell system and the desirability of additional studies before a firm decision is reached. However, Mr. Levine of the SPO emphasized that funds were not available for further work and a decision had to be made at this time. It was generally agreed that, at this time, there were more factors favorable to the Boeing system than the Honeywell system, and, therefore , the Boeing energy-management system was selected for implementation for Dyna-Soar.

4. The items discussed on boost guidance, guidance malfunction detection system, and inertial guidance system reliability can be reviewed from the copies of figures used by the Honeywell representatives which are available. Of significance to the FRC is the fact that the functions and importance of the malfunction detection system seems to be growing rather rapidly to the point where the pilot's role during boost may be that of a monitor of the MDS indicators and comparators. In regard to this, the SPO still has not taken a position as the role of the pilot during boost.

5. Discussions on the 3-axis controllers were held with Lt. Gregory and Lt. Stone of the Flight Control Laboratory. They are installing a Lear 3-axis force-grip type of controller on one of their flight simulators and plan to commence controller tests in about two weeks. The controller has a good "feel"; however, coordinated control inputs could not be evaluated at this time. FRC representatives were invited to evaluate the controller after the installation is completed. Lt. Gregory stated that the Minneapolis-Honeywell 3-axis controller will be installed and flight tested in the MH F101 adaptive-controlled airplane during late summer or fall of this year. He also mentioned that G.E. is installing a 2-axis controller in an F4H for test at Patuxent River.

6. Dr. Vale of the Bioastronautics Branch was contacted relative to the DS centrifuge program. This program will be conducted in the Johnsville centrifuge in August, and it now appears that 2-axis and 3-axis controllers and rudder pedals will be used, and that a center stick may not be available for evaluation. Dr. Vale will be in charge of the program and emphasized his interest in making it a combined SPO-AFFTC-NADC-Boeing-NASA effort.

Wendell H. Stillwell
Aerospace Engineer

Date: Jun 1961   **Title:** Memo: Guidance and Control Aspects

NASA-FRC   June 19, 1961

MEMORANDUM : for Dyna-Soar Project Manager

Subject: Comments on Boeing Airplane Company Document D2-6909-2,

Guidance and Control Aspects of Dyna-Soar System Description

1. I concur with the Dyna-Soar design in the area of my responsibilities of guidance (and controls) with the exception of the following items:

(a) The flight control system should incorporate attitude and/or angle-of-attack hold modes. Hold modes are presently a function of the automatic guidance system, and, therefore can only be used when automatic guidance is selected. In addition, the hold modes in the automatic guidance system are disconnected for any stability augmentation or primary guidance system failure. Manual control and manual energy management would be more precise and relieve the pilot of the constant control task with incorporation of hold modes in the flight control system. This capability is being provided by Minneapolis-Honeywell in a similar adaptive-control system in the X-15.

(b) Acceleration limiting is specified as a function of the flight control system. Such limits have to be a function of Y, H, and α, and any limiting technique must operate as a function of these variables. The flight control system does not have the capability to control acceleration for these variables, and it appears that this limiting can only be accomplished through the energy-management display.

(c) The design of the flight control electronic system should be such that no single failure can result in a significant degradation in the performance of the adaptive control system. If single failures cause the system to revert to a constant gain damper system, it should be demonstrated that this mode of operation is satisfactory for all flight conditions. Also, it is desirable to provide the pilot with the capability to adjust the fixed gains.

Wendell H. Stillwell Aerospace Technologist

STANDARDIZED LAUNCH VEHICLES for SPACE APPLICATION
HEADQUARTERS SPACE SYSTEMS DIVISION AIR FORCE SYSTEMS COMMAND SSR 61-2

1. Possible Choices for SLV-4 (Titan II)

The next launch vehicle configuration should be picked to permit boosting Dyna-soar into orbit, since at the moment this is the most clearly foreseen requirement for military space missions in this weight class. Furthermore, this represents about a four-to-one jump in payload capability over SLV-3 (Atlas-Agena-D), and this is a proper interval. In order to permit the Dyna-Soar glider some growth in capability, the booster should be capable of injecting at least 18, 000 lbs. into an 80-mile circular orbit, which is equivalent to something like 16,000 lbs. in a 300 mile orbit, or perhaps the booster should do a little better than this.

At this point advantage can be taken of a recent study aimed at selecting a booster configuration for Dyna-Soar Step II. The most attractive launching systems considered are indicated in Figure 10. It can be seen that there was a wide variety of configurations proposed. The first vehicle on the left is a two-stage oxygen-hydrogen configuration; the next uses the Titan first stage as its first stage and an oxygen-hydrogen second stage; the next is the Saturn C-1; the next a solid boosted Titan II; and finally, a new space launching system using a single oxygen-hydrogen stage as a second stage with solids as the first stage. This last configuration will be referred to as Space Launching System in the discussion which follows.

After considerable study, the evaluation of these and other alternative configurations led to an elimination of all but two. It was felt that the preferred first stage for a new space launching vehicle should be large segmented solids, for reasons of cost, flexibility, stage reliability and development time. For this reason the Convair Astro IV proposal, and the Martin Plan C proposal were eliminated. The Saturn C-1 did not appear from a previous evaluation to be more attractive than these first two configurations, due to cost and complexity. The all-solid possibilities were eliminated because the cost effectiveness and the flexibility for other missions gets poorer as upper stages become solid. The Titan II with high energy upper stage was eliminated because of inadequate payload for a growth version of Dyna-Soar. This led finally to the solid boosted Titan II and the new Space Launch System as being the two best choices.

2. Description of Titan II Plus Solids

This configuration is shown in Figure 11. The lift-off gross weight of 856, 000 lbs. includes the 508, 000 lb. weight of the segmented solids, the Titan II itself, and the glider, assumed here to weigh 17,700 lbs. This configuration, however, can boost more than this — more than 20,000 lbs., if needed — by putting on somewhat larger solids. Fins are supplied for stability. The inherent reliability of this system is estimated to be 82% after development bugs have been worked out. The drawing on the right part of the chart simply shows the way the solid boosters would be staged off the vehicle at burn-out. The schedule is shown as Figure 12. Assuming program approval in July of this year, the vehicle confirmation firings for the modified Titan II could begin in April of 1962 and would go on through the summer of 1963. The solid motor PFRT (Preliminary Flight Ratings Test) would start in June of 1963 and end in October of 1963. This is a key date in pacing the program. The AMR (Atlantic Missile Range Cape Canaveral) launch complex modifications installation and checkout and availability of AGE also tend to pace the program and would be accomplished in June or July of 1963. This leads to a first flight date of the solid boosted Titan II of October or November 1963. If the primary purpose of this configuration is to launch a manned Dyna-Soar glider, a schedule of three booster tests, followed by four tests of the manned glider, could lead to an earliest flight test with the manned system in the fall of 1964, or a little later.

As a standard space launch vehicle it could be considered operational as shown in July 1964. Figure 13 shows the estimated development program costs for this booster. The total is $229,600,000, of which

$110,000,000 is the actual modification of the Titan II itself. Note that the captive and flight test program of the Titan II and of the solid motors comes to more than $70,000,000; in addition the costs of the PFRT of the solid motors is a significant item. incidentally, the costs shown do not include the R &D for the solid motors, about which more later.

3. Description of Space Launching System

Figure 14 shows the physical configuration and some of the characteristics of the Space Launching System, which is the alternative to the Titan II. In this system there is a single liquid stage using oxygen-hydrogen and using a single engine, the J-2 engine. The engine thrust at sea level for this engine is 130,000 lbs. and vacuum 200,000 lbs. The first stage is again segmented solids, slightly smaller than in the previous configuration with Titan; the lift-off gross weight is slightly less: the inherent reliability is slightly greater— estimated to be about 87%, due to the greater simplicity in the liquid stage. In this particular drawing the boosted weight is shown as 18,700 lbs. This type of configuration could, however, boost something over 20,000 lbs. to low orbit. Again, the drawing on the right shows the way in which the solid motors would be staged at burn-out.

A few words are in order at this point about the design philosophy of the Space Launching System. Studies have indicated the desirability of segmented solids for a first stage to achieve low cost, high reliability and flexibility of basic booster size by adding or subtracting segments. Studies further show that oxygen-hydrogen with its very high specific impulse is a preferred choice for upper stages; where weight is more important. And again this choice results in minimum systems cost. By stressing the concept of a single liquid stage and a single engine, it is felt that a high reliability for the over-all system can be achieved. Furthermore, by starting out from the very beginning with the concept that this is to be a standardized vehicle for a wide variety of space missions, it is felt that a basically good design can be achieved which will be useful for at least five, and perhaps ten years as a work-horse booster. Figure 15 shows the estimated schedule for this launching system to be developed, assuming program approval in July of this year. The liquid stage engine PFRT can be accomplished in early 1963.

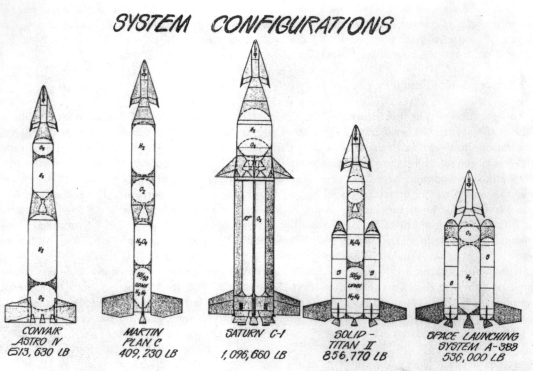

SYSTEM CONFIGURATIONS

| CONVAIR ASTRO IV 613,630 LB | MARTIN PLAN C 409,230 LB | SATURN C-1 1,096,660 LB | SOLID-TITAN II 856,770 LB | SPACE LAUNCHING SYSTEM A-388 536,000 LB |

FIGURE 10

BOOSTER CHARACTERISTICS
SOLID BOOSTER - TITAN II

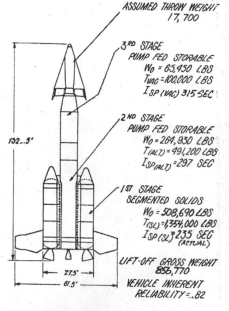

ASSUMED THROW WEIGHT
17, 700

3RD STAGE
PUMP FED STORABLE
W_0 = 65,450 LBS
T_{VAC} = 100,000 LBS
$I_{SP (VAC)}$ 315 SEC

2ND STAGE
PUMP FED STORABLE
W_0 = 264,930 LBS
$T_{(ALT)}$ = 491,200 LBS
$I_{SP(ALT)}$ = 297 SEC

1ST STAGE
SEGMENTED SOLIDS
W_0 = 508,690 LBS
$T_{(SL)}$ = 1,354,000 LBS
$I_{SP(SL)}$ = 235 SEC
(ACTUAL)

LIFT-OFF GROSS WEIGHT
858,770

VEHICLE INHERENT
RELIABILITY = .82

152.5'

27.5'

61.5'

FIGURE 11

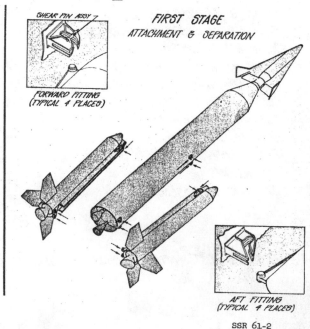

SHEAR PIN ASSY

FIRST STAGE
ATTACHMENT & SEPARATION

FORWARD FITTING
(TYPICAL 4 PLACES)

AFT FITTING
(TYPICAL 4 PLACES)

SSR 61-2

SOLID BOOSTED TITAN II DEVELOPMENT PROG SCHED

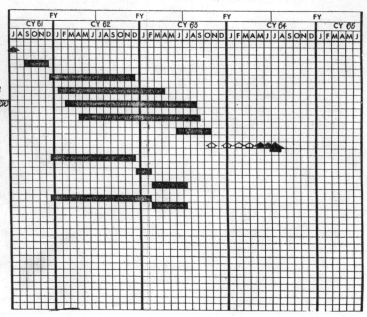

PROGRAM APPROVAL
CONFIGURATION REFINEMENT
BASIC DESIGN, BOOSTER
FABRICATION & DEVELOPMENT TESTS
VEHICLE COMPATIBILITY & DYNAMIC TESTS
VEHICLE CONFIRMATION FIRING
SOLID MOTOR PFRT
FLIGHT TESTS
AGE BASIC DESIGN
FUNCTIONAL TEST AGE AVAILABLE
AMR LAUNCH AGE AVAILABLE
LAUNCH COMPLEX MODIFICATIONS
INSTALLATION & CHECKOUT

⬠ BOOSTER FLIGHT TEST
⬟ HIGH RISK PAYLOAD LAUNCH
⬟ FIRST OPERATIONAL LAUNCH

FIGURE 12

SSR 61-2

LAUNCH VEHICLE DEVELOPMENT COSTS
SOLID BOOSTED TITAN II

ITEM	TOTAL COSTS
TITAN II MODIFICATIONS	87.0
PROPULSION SYSTEM MODIFICATIONS	23.0
SOLID MOTOR PFRT	27.4
CENTAUR MODIFICATION (FOR HIGH ALTITUDE MISSIONS)	1.0
TITAN II HARDWARE (CAPTIVE AND FLIGHT)	31.0
SOLID MOTORS (FLIGHT)	12.9
CENTAUR HARDWARE	7.8
GUIDANCE	2.8
PROPELLANTS	5.9
OPERATIONS	21.9
SYSTEMS MANAGEMENT	8.9
TOTAL	229.6

NOTE:

FACILITY COSTS AND SOLID MOTOR DESIGN AND DEVELOPMENT COSTS
NOT INCLUDED.

FIGURE 13   SSR 61-2

BOOSTER CHARACTERISTICS
SPACE LAUNCHING SYSTEM VEHICLE A 388

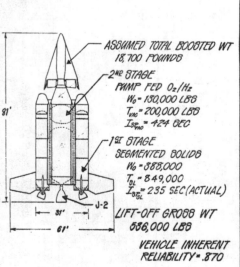

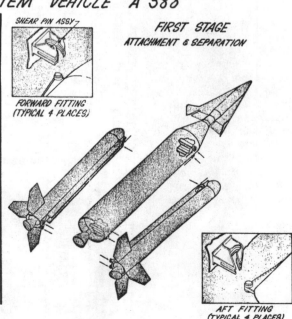

FIRST STAGE
ATTACHMENT & SEPARATION

ASSUMED TOTAL BOOSTED WT
18,700 POUNDS

2ND STAGE
PUMP FED O_2/H_2
W_0 = 130,000 LBS
T_{VAC} = 200,000 LBS
$I_{SP_{VAC}}$ = 424 SEC

1ST STAGE
SEGMENTED SOLIDS
W_0 = 388,000
T_{SL} = 849,000
$I_{SP_{SL}}$ = 235 SEC (ACTUAL)

LIFT-OFF GROSS WT
556,000 LBS

VEHICLE INHERENT
RELIABILITY = .870

SHEAR PIN ASSY
FORWARD FITTING
(TYPICAL 4 PLACES)

AFT FITTING
(TYPICAL 4 PLACES)

FIGURE 14   SSR 61-2

PROGRAM SCHED- SPACE LAUNCHING SYS VEH A388

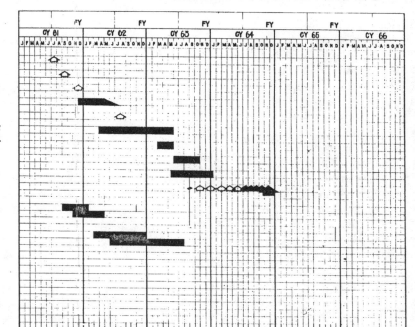

FIGURE 15

SSR·61-2

Date: Late 1961   Title: Structure Description Report

Air Force Systems Command Wright-Patterson Air Force Base, Ohio

Attn: ASZR, Production Group (8cys)
NASA, 1520 H. St. N. W.
Washington 25, D. C.

Attn: A. J. Evans (RAA) NASA,
Langley Research Center Langley Field, Virginia

Attn: P. Korycinski NASA,
Ames Research Center Moffett Field, California

Attn: Robert Crane
NASA, Flight Research Center
P.O. Box 273 Edwards, California

Attn: Paul Bikle
AF Flight Test Center,
Edwards AFB, California Attn: FTGS

Space Systems Division AF Unit Post Office
Los Angeles 45, California
Attn: SSVS (3cys) AFPR/RWRSB

TABLE OF CONTENTS	LIST OF ILLUSTRATIONS

INTRODUCTION

This document is initiated in compliance with paragraph B(1.1.1.1.2)7, Statement of Work, System 620A, Dyna-Soar (Step 1), Exhibit 620A-60-14, dated 6 August 1960, revised 1 May 1961. The purpose of this document is to provide description of the Dyna-Soar structure for the information of the contracting agency. This document will be expanded by constant revision as the design progresses and details become available.

SECTION 1. CONFIGURATION

1.0 CONFIGURATION

1.1 The configuration of the glider is described in Boeing Document D2-6909-2. Figures 1.1 and 1.2 are reproduced herein to provide continuity. Should differences exist between the configuration data shown in this section and that shown in D2-6909-2, the latter shall be considered as correct.

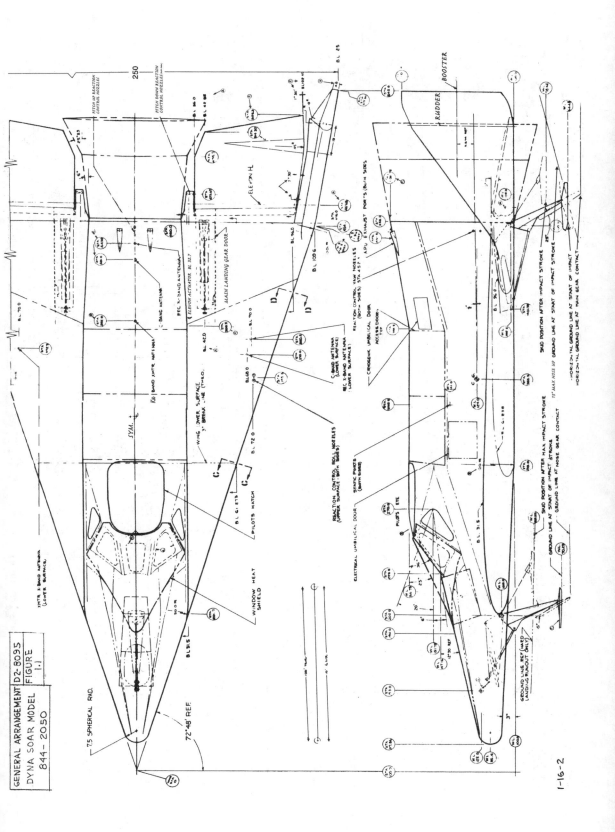

GENERAL ARRANGEMENT
DYNA SOAR MODEL
844- 2050

D2-8095
FIGURE
1.1

1-16-2

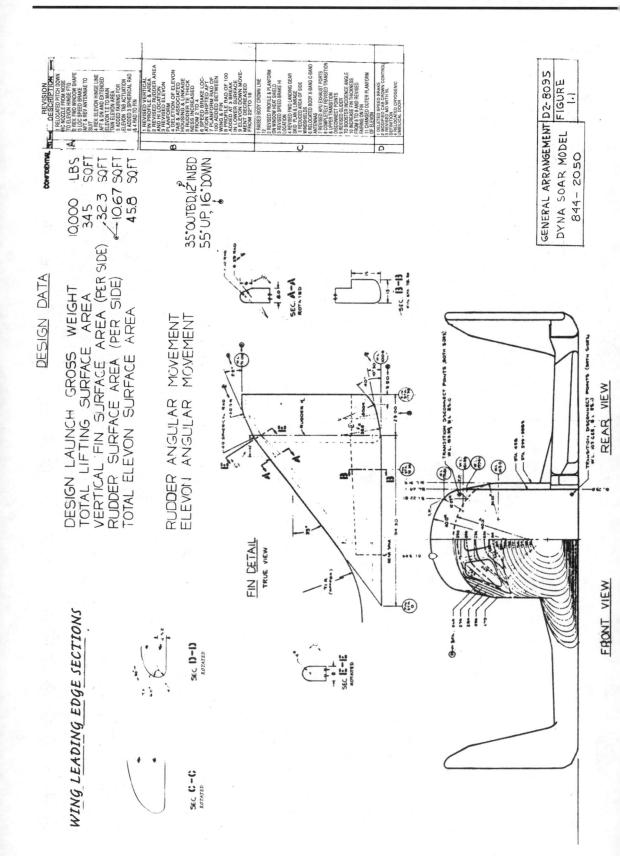

DESIGN DATA

DESIGN LAUNCH GROSS WEIGHT	10,000	LBS
TOTAL LIFTING SURFACE AREA	345	SQ FT
VERTICAL FIN SURFACE AREA (PER SIDE)	32.3	SQ FT
RUDDER SURFACE AREA (PER SIDE)	10.67	SQ FT
TOTAL ELEVON SURFACE AREA	45.8	SQ FT

RUDDER ANGULAR MOVEMENT   35° OUTBD, 12° INBD
ELEVON ANGULAR MOVEMENT   55° UP, 16° DOWN

WING LEADING EDGE SECTIONS

SEC C-C
ROTATED

SEC D-D
ROTATED

FIN DETAIL
TRUE VIEW

SEC A-A
ROTATED

SEC B-B

SEC E-E
ROTATED

REAR VIEW

FRONT VIEW

GENERAL ARRANGEMENT   D2-8095
DYNA SOAR MODEL   FIGURE 1.1
844-2050

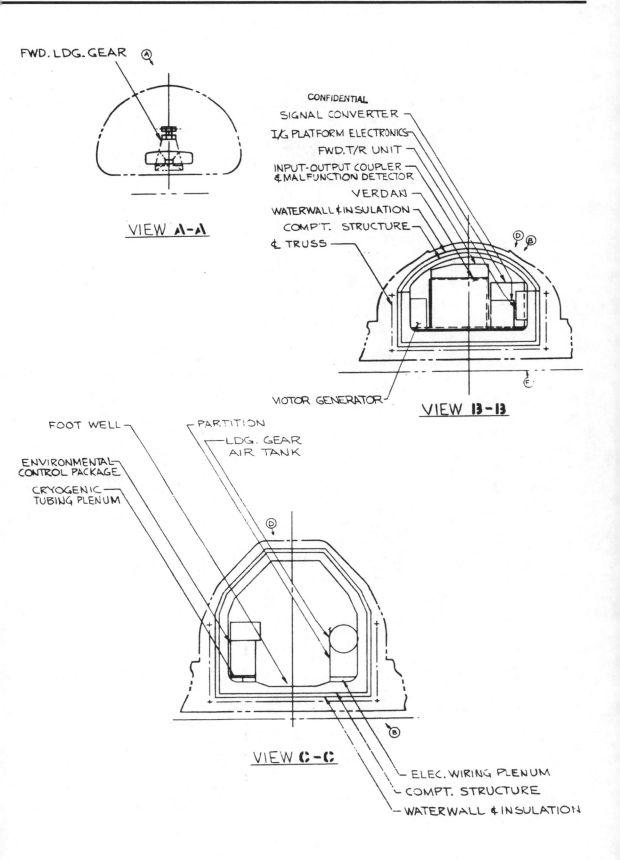

FWD. LDG. GEAR Ⓐ

VIEW A-A

CONFIDENTIAL
SIGNAL CONVERTER
I/G PLATFORM ELECTRONICS
FWD. T/R UNIT
INPUT-OUTPUT COUPLER
& MALFUNCTION DETECTOR
VERDAN
WATERWALL & INSULATION
COMP'T. STRUCTURE
& TRUSS

MOTOR GENERATOR

VIEW B-B

FOOT WELL
ENVIRONMENTAL
CONTROL PACKAGE
CRYOGENIC
TUBING PLENUM

PARTITION
LDG. GEAR
AIR TANK

VIEW C-C

ELEC. WIRING PLENUM
COMPT. STRUCTURE
WATERWALL & INSULATION

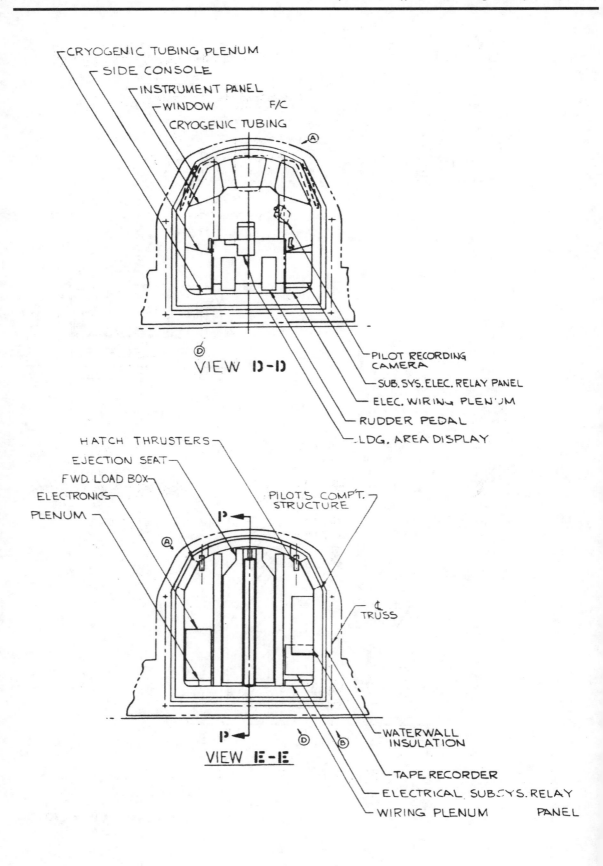

CRYOGENIC TUBING PLENUM
SIDE CONSOLE
INSTRUMENT PANEL
WINDOW   F/C
CRYOGENIC TUBING

VIEW D-D

PILOT RECORDING CAMERA
SUB. SYS. ELEC. RELAY PANEL
ELEC. WIRING PLENUM
RUDDER PEDAL
LDG. AREA DISPLAY

HATCH THRUSTERS
EJECTION SEAT
FWD. LOAD BOX
ELECTRONICS
PLENUM

PILOTS COMP'T. STRUCTURE

TRUSS

VIEW E-E

WATERWALL INSULATION

TAPE RECORDER
ELECTRICAL SUBSYS. RELAY
WIRING PLENUM   PANEL

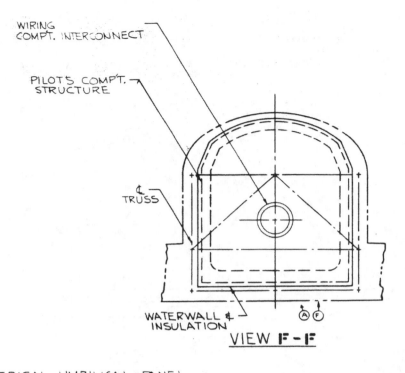

WIRING
COMP'T. INTERCONNECT

PILOT'S COMP'T.
STRUCTURE

℄
TRUSS

WATERWALL &
INSULATION

Ⓐ Ⓕ

VIEW F-F

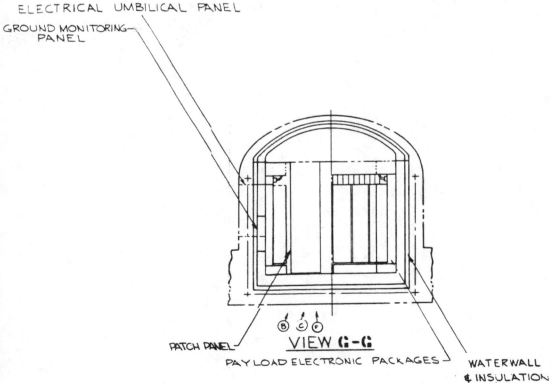

ELECTRICAL UMBILICAL PANEL

GROUND MONITORING
PANEL

PATCH PANEL

Ⓑ Ⓒ Ⓕ

VIEW G-G

PAYLOAD ELECTRONIC PACKAGES

WATERWALL
& INSULATION

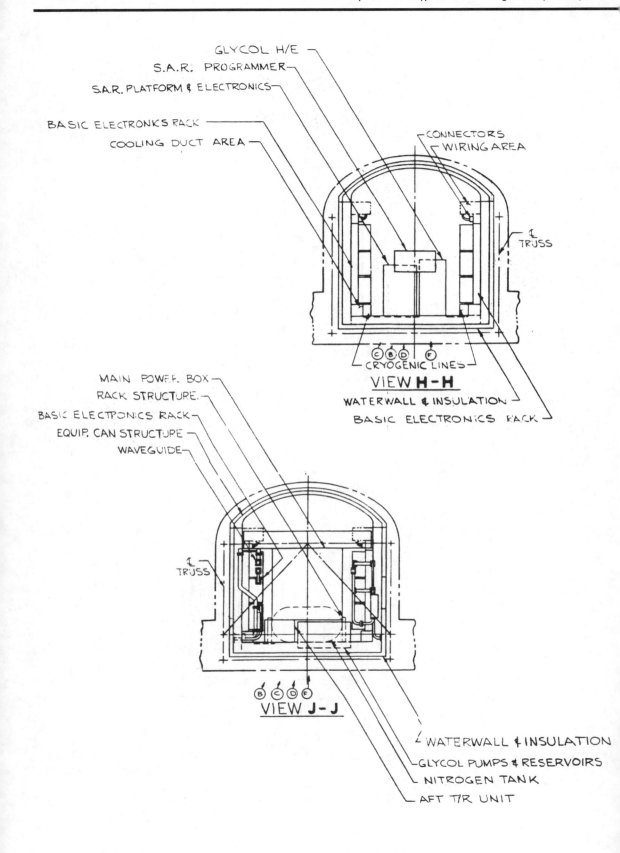

GLYCOL H/E
S.A.R. PROGRAMMER
S.A.R. PLATFORM & ELECTRONICS

BASIC ELECTRONICS RACK
COOLING DUCT AREA

CONNECTORS
WIRING AREA

TRUSS

C B D F
CRYOGENIC LINES

VIEW H-H

WATERWALL & INSULATION
BASIC ELECTRONICS RACK

MAIN POWER BOX
RACK STRUCTURE
BASIC ELECTRONICS RACK
EQUIP. CAN STRUCTURE
WAVEGUIDE

TRUSS

B C D F

VIEW J-J

WATERWALL & INSULATION
GLYCOL PUMPS & RESERVOIRS
NITROGEN TANK
AFT T/R UNIT

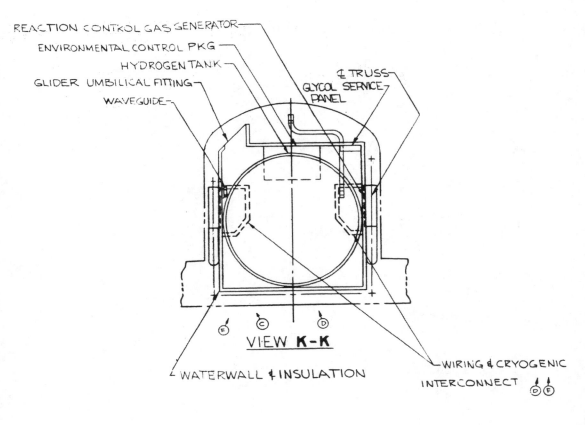

REACTION CONTROL GAS GENERATOR
ENVIRONMENTAL CONTROL PKG
HYDROGEN TANK
GLIDER UMBILICAL FITTING
WAVEGUIDE

₵ TRUSS
GLYCOL SERVICE PANEL

VIEW **K-K**

WATERWALL & INSULATION

WIRING & CRYOGENIC INTERCONNECT ⒟Ⓕ

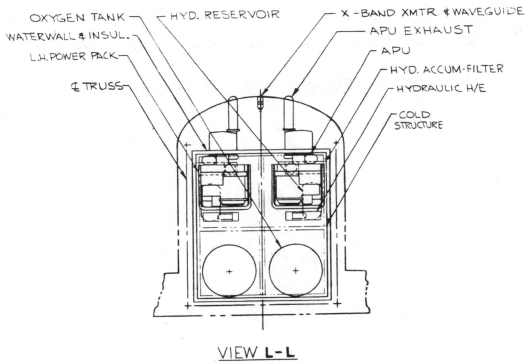

OXYGEN TANK
WATERWALL & INSUL.
L.H. POWER PACK
₵ TRUSS

HYD. RESERVOIR

X-BAND XMTR & WAVEGUIDE
APU EXHAUST
APU
HYD. ACCUM-FILTER
HYDRAULIC H/E
COLD STRUCTURE

VIEW **L-L**

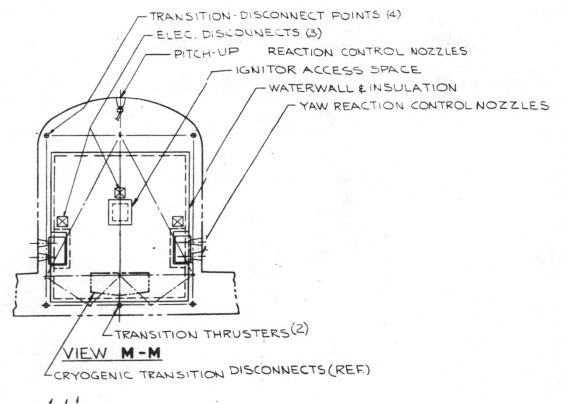

VIEW **M-M**

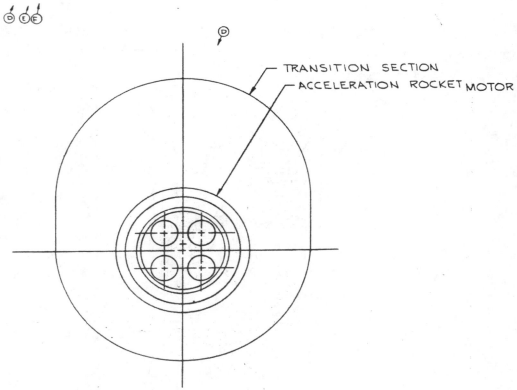

VIEW **N-N**

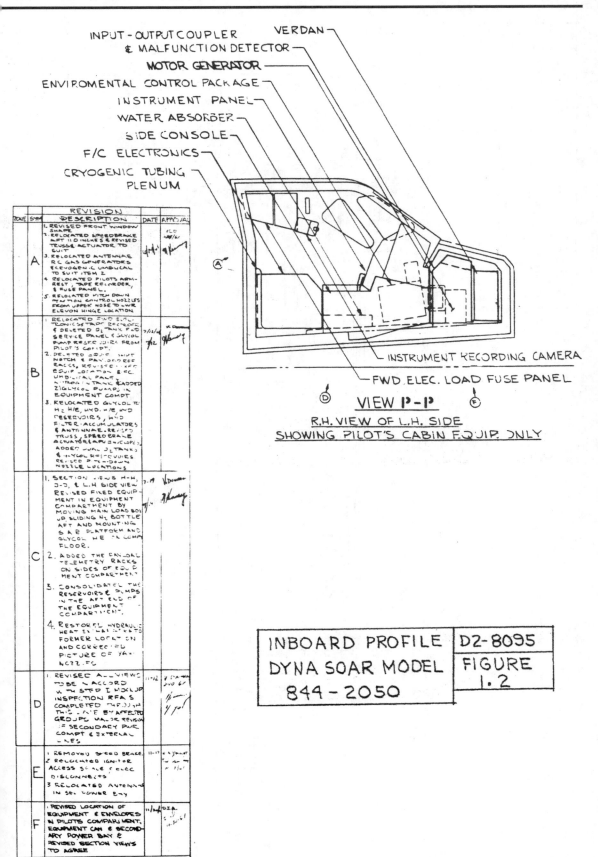

INPUT-OUTPUT COUPLER
& MALFUNCTION DETECTOR
MOTOR GENERATOR
ENVIROMENTAL CONTROL PACKAGE
INSTRUMENT PANEL
WATER ABSORBER
SIDE CONSOLE
F/C ELECTRONICS
CRYOGENIC TUBING
PLENUM

VERDAN

INSTRUMENT RECORDING CAMERA

FWD. ELEC. LOAD FUSE PANEL

VIEW P-P

R.H. VIEW OF L.H. SIDE
SHOWING PILOT'S CABIN EQUIP. ONLY

INBOARD PROFILE | D2-8095
DYNA SOAR MODEL | FIGURE
844 - 2050 | 1.2

ZONE	SYM	REVISION		
		DESCRIPTION	DATE	APP'D/AU.
	A	1. REVISED FRONT WINDOW SHAPE. 2. RELOCATED SPEED BRAKE AFT 110 INCHES & REVISED TRUSS ACTUATOR TO SUIT 3. RELOCATED ANTENNAE, RC GAS GENERATORS & CRYOGENIC UMBILICAL TO SUIT ITEM 2. 4. RELOCATED PILOTS ARM-REST, TAPE RECORDER, & FUSE PANEL. 5. RELOCATED PITCH DOWN REACTION CONTROL NOZZLES FROM UPPER NOSE TO LWR ELEVON HINGE LOCATION.		
	B	1. RELOCATED FWD ELEC-TRONICS, STRAP RECEIVER & DELETED O2 TANK FWD SERVICE PANEL & GLYCOL PUMP RESERVOIRS FROM PILOT'S COMPT. 2. DELETED EQUIP JOINT MATCH & PANDORA EQF RACKS, REVISED LWR EQUIP LOCATION ELEC UMBILICAL PANEL NITROGEN TANK ADDED 2 GLYCOL PUMPS IN EQUIPMENT COMPT 3. RELOCATED GLYCOL H2 H/E, H/E, HYD. H/E AND RESERVOIRS, HYD FILTER ACCUMULATORS & ANTENNAE, REVISED TRUSS, SPEED BRAKE ACTUATOR(APU ENCLOSED) ADDED DUAL O2 TANKS & HYDOL RESERVOIRS REVISED PITCH-DOWN NOZZLE LOCATIONS.		
	C	1. SECTION VIEWS H-H, J-J, & L.H SIDE VIEW REVISED FIXED EQUIP-MENT IN EQUIPMENT COMPARTMENT BY MOVING MAIN LOAD BOX UP, SLIDING N2 BOTTLE AFT AND MOUNTING S.A.R. PLATFORM AND GLYCOL H/E ON COMPT FLOOR. 2. ADDED THE PANDORA TELEMETRY RACKS ON SIDES OF EQUIP-MENT COMPARTMENT 3. CONSOLIDATED THE RESERVOIRS & PUMPS IN THE AFT END OF THE EQUIPMENT COMPARTMENT. 4. RESTORED HYDRAULIC HEAT EXCHANGER TO FORMER LOCATION AND CORRECTED PICTURE OF YAW NOZZLES		
	D	1. REVISED ALL VIEWS TO BE IN ACCORD WITH STEP 1 MOCKUP INSPECTION RFAS COMPLETED THROUGH THIS LINE BY AFFECTED GROUPS MAJOR REVISION IS SECONDARY PWR COMPT & EXTERNAL LINES		
	E	1. REMOVED SPEED BRAKE. 2. RELOCATED IGNITOR ACCESS SCALE & ELEC DISCONNECTS 3. RELOCATED ANTENNAE IN SEC LOWER BAY		
	F	1. REVISED LOCATION OF EQUIPMENT & ENVELOPES IN PILOTS COMPARTMENT, EQUIPMENT CAN & SECOND-ARY POWER BAY & REVISED SECTION VIEWS TO AGREE		

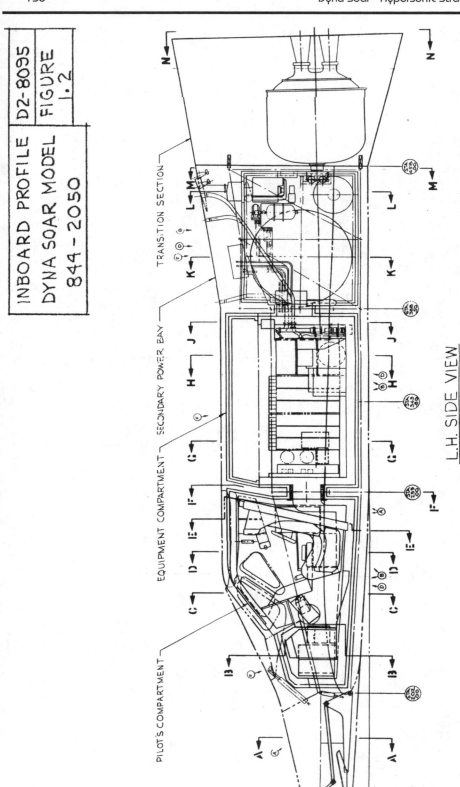

INBOARD PROFILE D2-8095 FIGURE 1.2

DYNA SOAR MODEL 844-2050

TRANSITION SECTION

SECONDARY POWER BAY

EQUIPMENT COMPARTMENT

PILOT'S COMPARTMENT

L.H. SIDE VIEW

SECTION 2 AIRFRAME DESIGN - GENERAL

2.0 AIRFRAME DESIGN

2.1 The structural section breakdown is shown in Figure 2.1. This breakdown is used to identify major components and subassemblies.

2.2 Figures 2.2 through 2.4 are structural clearance envelopes at the nose section through the pilot's compartment and through the equipment compartment. The data is limited to envelopes at the present time. As design details evolves these envelopes will be refined to reflect the final part shape.

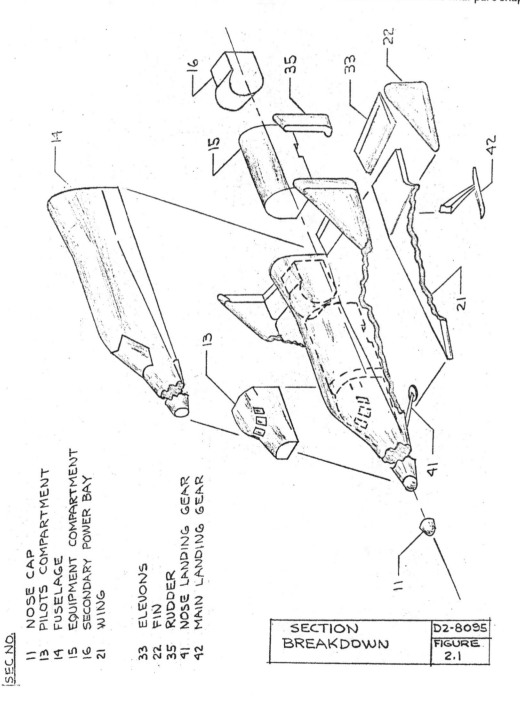

SEC. NO.	
11	NOSE CAP
13	PILOTS COMPARTMENT
14	FUSELAGE
15	EQUIPMENT COMPARTMENT
16	SECONDARY POWER BAY
21	WING
33	ELEVONS
22	FIN
35	RUDDER
41	NOSE LANDING GEAR
42	MAIN LANDING GEAR

SECTION BREAKDOWN	D2-8095
	FIGURE 2.1

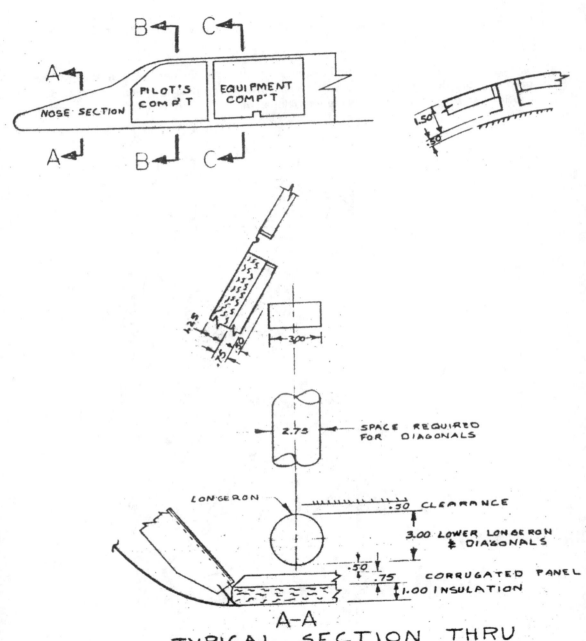

A-A

TYPICAL SECTION THRU
NOSE SECTION FWD OF
PILOT'S COMPARTMENT

STRUCTURAL	D2-8095
CLEARANCE	FIGURE
REQUIREMENTS	2.2

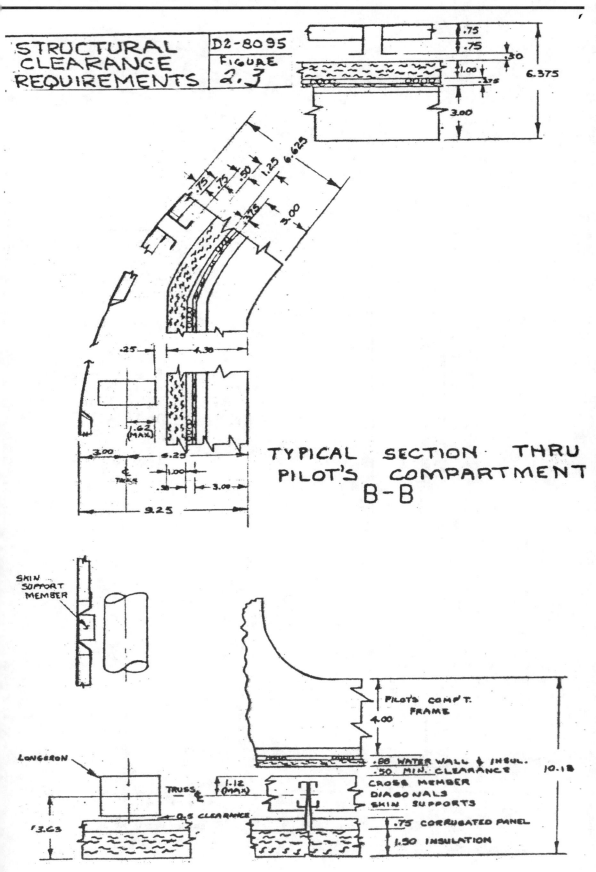

STRUCTURAL CLEARANCE REQUIREMENTS

D2-8095 FIGURE 2.3

TYPICAL SECTION THRU PILOT'S COMPARTMENT B-B

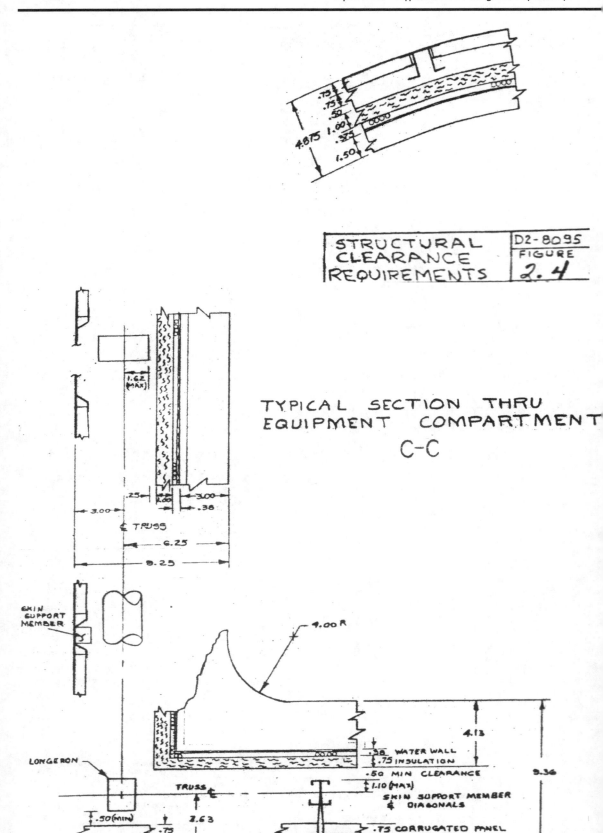

STRUCTURAL CLEARANCE REQUIREMENTS

D2-8095

FIGURE 2.4

TYPICAL SECTION THRU
EQUIPMENT COMPARTMENT
C-C

SECTION 3. FUSELAGE

.0 FUSELAGE - Description

3.1   The major components of the fuselage are: The nose cap assembly, the primary structure, the skin panels, the pilot's compartment, the equipment compartment and the secondary power bay.

3.1.1 The nose cap assembly, as shown in Fig. 3.1 consists of the nose cap, instrumentation, heat shield and support ring. The nose cap is a graphite shell that is further protected in the high temperature stagnation area by a zirconia face, instrumented to record temperatures and pressures during flight. The heat shield is a thermal barrier located at the aft face of the nose cap to maintain the temperature in the area of the support structure within acceptable limits. The nose cap is flexibly mounted to a support ring which in turn is supported at three locations by a linkage arrangement attached to the glider truss.

3.1.2 The fuselage primary structure is to carry the primary loads of the glider by means of a René' 41 super-alloy truss structure. The arrangement of this truss structure is shown in Figure 3.2. A main forward and aft beam is provided on each side of the fuselage and cross frames are provided at six stations. In the aft bay a superstructure of truss members spans between the main beam upper chords to react the skin support beam loads and to provide a load path for torsional loads induced in flight. Wherever possible, axially loaded truss members are round, swaged-end tubes. Sheet metal tabs or machined fittings are fusion welded to the swaged ends. Swaging the tube ends results in efficient and compact joints, especially where many members (up to eleven) intersect in a common joint.

Members subjected to loads other than pure axial (bending or combined bending and axial) are built-up, formed sheet metal sections of rectangular tubes using either fusion or resistance welding for fabrication. Rectangular tubes are preferred over "free-flanged" sheet metal section because of the warping of the free edges due to thermal gradients. Sheet metal tabs, doublers, or machined fittings are either fusion or resistance welded to the member ends. The space frame, in general, provides a structure capable of changing shape due to thermal gradients without introducing major stresses. This is accomplished by providing rotational capability at the ends of the individual members or joints. Rotation may be accomplished by attaching the members to one another or the joint fitting by pins or bolts, oriented so that the members may rotate in the plane of the maximum thermal growth. Another method of providing rotation, especially where small movements are expected, is through use of thin flexible tabs at the member ends. The tabs are also oriented to allow rotation in the plane of maximum thermal growth. Wherever possible the pins and member end-tabs are arranged so that the pins are loaded in double shear, thus resulting in smaller diameter pins. Hollow pins or bolts are used where it is necessary to develop bearing strength in thin members without the requirement for high pin shear loads. Where thermal gradients are so small that no appreciable rotation occurs between the members, fixed or partially fixed joints are used which results in a lighter weight design over pinned joints.

3.1.3 The fuselage leading edges are segments constructed of refractory alloy attached to a René' 41 beam. A typical cross-section through a fuselage leading edge is shown . in Figure 3.4.

3.1.4 The fuselage skin panel arrangement is shown in Figure 3.51 Upper panels are René' 41 resistance-welded corrugations with a covering of thin gage René' 41. Lower panels are René' 41 corrugations with a thermal protection of insulation and an exterior refractory alloy shield.

3.1.5 The pilot's compartment is a semi-monocoque aluminum structure pressurized to a nominal design pressure of 7.35 psig. Air leakage is minimized by utilizing machined integral stiffened skin panels welded together. The structural arrangement of the pilot's compartment is shown in Figures 3.6 and 3.7. The compartment is supported from the primary truss in such a manner to minimize stresses due to thermal expansion. The compartment structure is protected against high temperature by a blanket of insulation and a passive cooling system. The pilot's windows and windshields consist of outer panes of high temperature glass and an inner laminated pane of low temperature glass.

3.1.6 The equipment compartment is a D-shaped, semi-monocoque aluminum structure pressurized to 10.0 psig. (nominal design pressure). By utilizing integrally machined stiffened skin panels welded together, air leakage is held to a minimum. The structural arrangement of the equipment compartment is shown in Figure 3.8. Both the pilot's and equipment compartment are constructed of aluminum alloy and are protected from the aerodynamic heat by a passive cooling system consisting of insulation and wick-held water heat sink. The cooling system is described in Boeing Document D2-6909-2. The equipment compartment is supported from the primary truss in a manner such that thermal stresses are minimized and loads in the truss are not transferred to the equipment compartment from the truss. Access to the compartment interior is provided through a hinged lid which forms the upper portion of the compartment. Electrical and cryogenic lines penetrate the compartment structure in numerous locations.

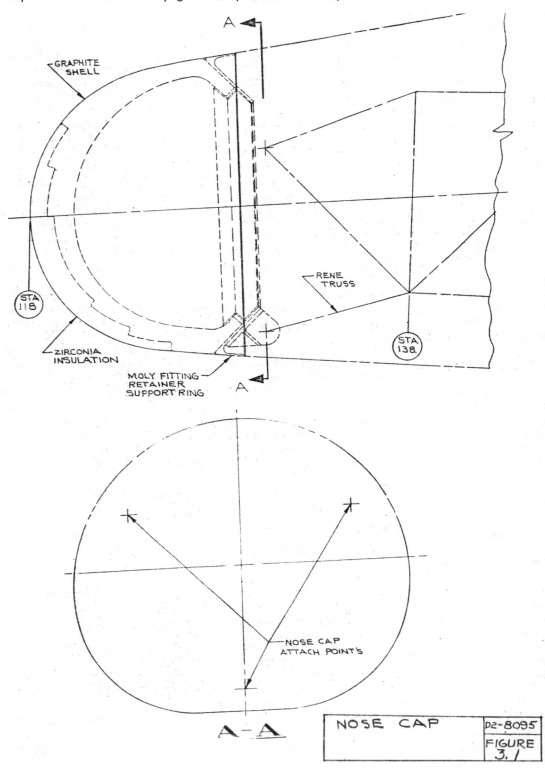

A-A

NOSE CAP

D2-8095

FIGURE
3./

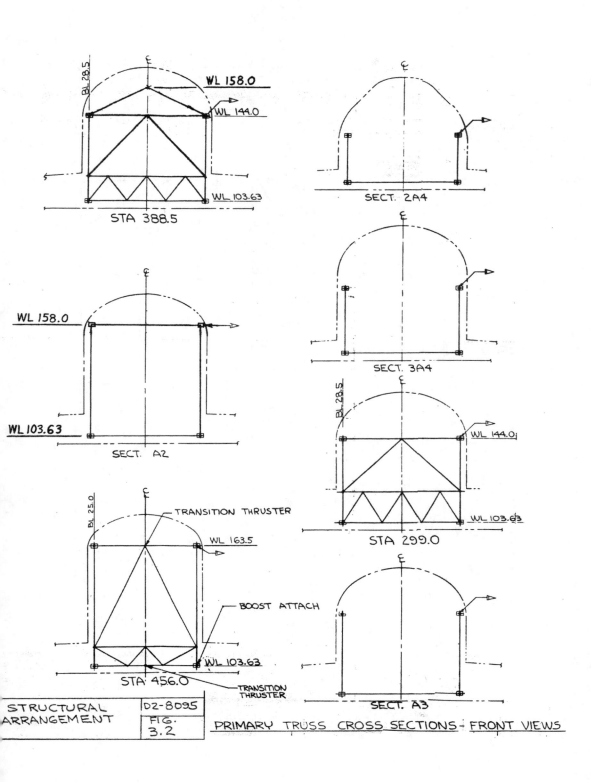

WL 158.0
WL 144.0
WL 103.63
STA 388.5
BL 28.5

SECT. 2A4

WL 158.0
WL 103.63
SECT. A2

SECT. 3A4

BL 28.5
WL 144.0
WL 103.63
STA 299.0

BL 25.0
TRANSITION THRUSTER
WL 163.5
BOOST ATTACH
WL 103.63
STA 456.0
TRANSITION THRUSTER

SECT. A3

| STRUCTURAL ARRANGEMENT | D2-8095 |
| | FIG. 3.2 |

PRIMARY TRUSS CROSS SECTIONS - FRONT VIEWS

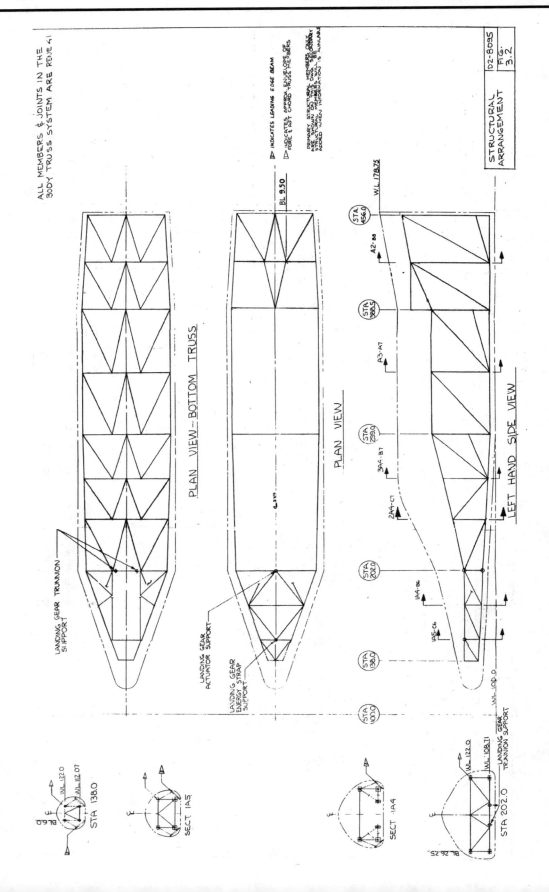

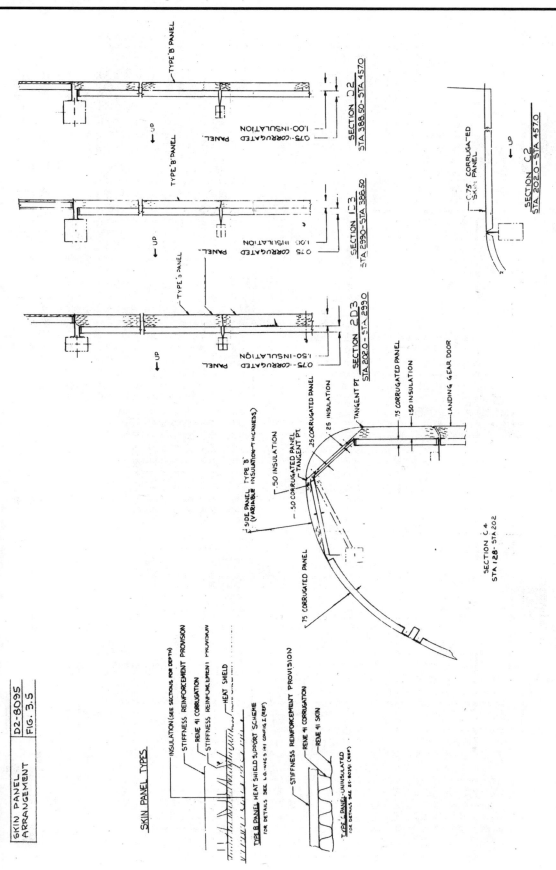

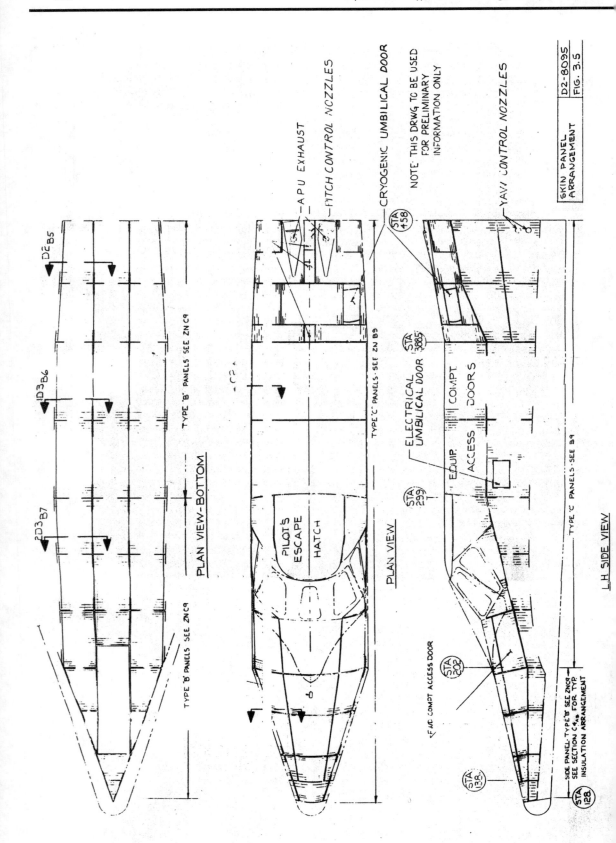

PLAN VIEW-BOTTOM

D2B5

D3B6

D3B7

TYPE 'B' PANELS SEE ZN C9

TYPE 'B' PANELS SEE ZN C9

APU EXHAUST

PITCH CONTROL NOZZLES

CRYOGENIC UMBILICAL DOOR

NOTE: THIS DRWG TO BE USED
FOR PRELIMINARY
INFORMATION ONLY

YAW CONTROL NOZZLES

STA 458

PILOT'S ESCAPE HATCH

TYPE 'C' PANELS · SEE ZN B9

PLAN VIEW

ELECTRICAL UMBILICAL DOOR

EQUIP. COMPT. ACCESS DOORS

STA 385

STA 299

STA 202

TYPE 'C' PANELS · SEE B9

LH SIDE VIEW

EQUIP COMPT ACCESS DOOR

STA 202

STA 138

STA 128

SIDE PANEL TYPE 'B' SEE ZNC9
SEE SECTION C4.13 FOR TYP
INSULATION ARRANGEMENT

| SKIN PANEL ARRANGEMENT | D2-8095 |
| | FIG. 3.5 |

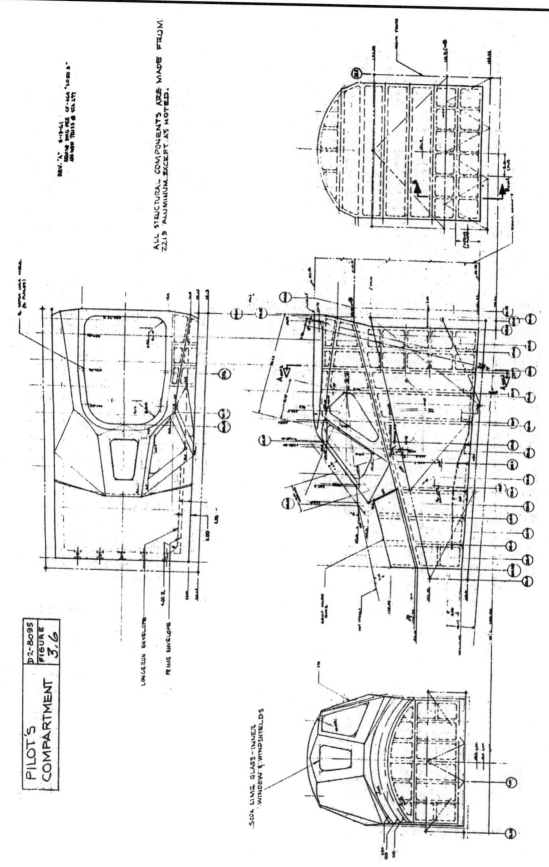

PILOT'S COMPARTMENT

D2-8095

FIGURE 3.6

ALL STRUCTURAL COMPONENTS ARE MADE FROM 2219 ALUMINUM EXCEPT AS NOTED.

LONGERON ENVELOPE

FRAME ENVELOPE

SODA LIME GLASS - INNER WINDOW & WINDSHIELDS

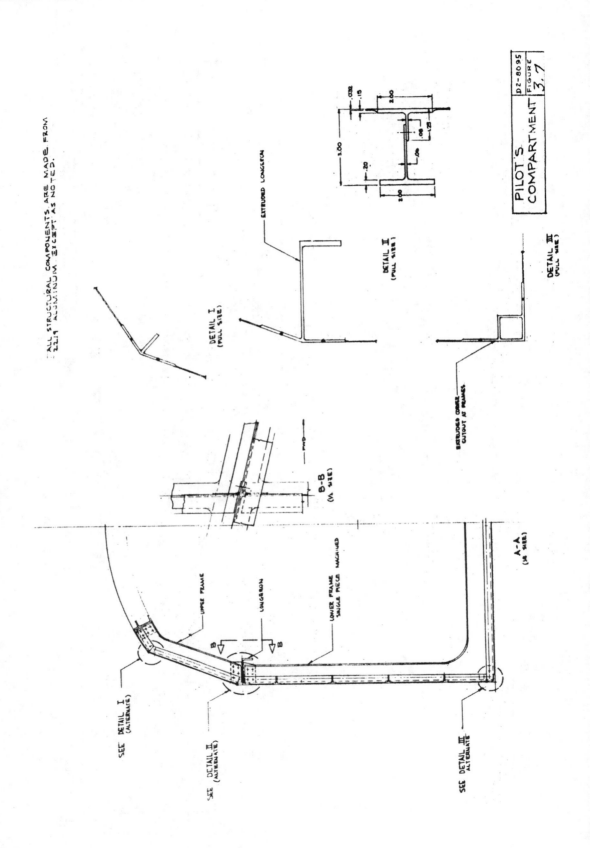

ALL STRUCTURAL COMPONENTS ARE MADE FROM
2219 ALUMINUM EXCEPT AS NOTED.

EXTRUDED LONGERON

DETAIL I
(FULL SIZE)

DETAIL II
(FULL SIZE)

DETAIL III
(FULL SIZE)

EXTRUDED CORES
CUTOUT AT FRAMES

B-B
(½ SIZE)

FWD

A-A
(¼ SIZE)

UPPER FRAME

LONGERON

LOWER FRAME
SINGLE PIECE MACHINED

SEE DETAIL I
(ALTERNATE)

SEE DETAIL II
(ALTERNATE)

SEE DETAIL III
ALTERNATE

D2-8095

FIGURE 3.7

PILOT'S COMPARTMENT

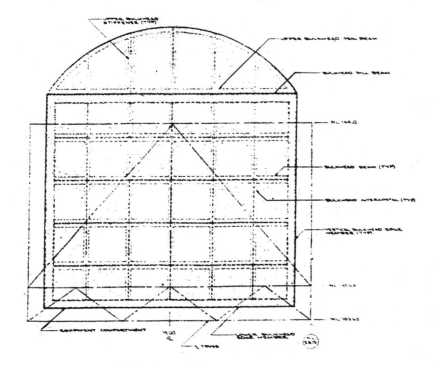

FRONT VIEW

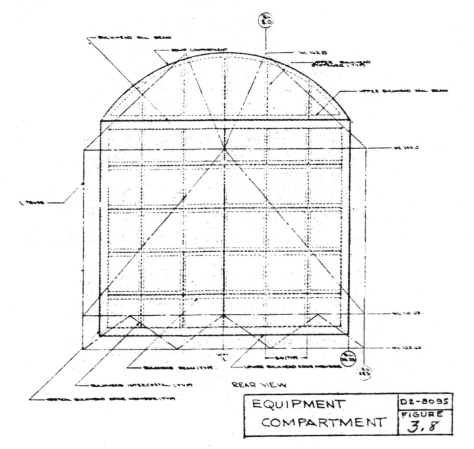

REAR VIEW

| EQUIPMENT | D2-8095 |
| COMPARTMENT | FIGURE 3.8 |

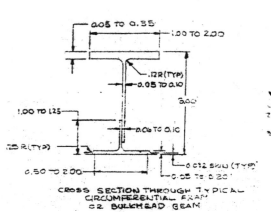

CROSS SECTION THROUGH TYPICAL
CIRCUMFERENTIAL FRAME
OR BULKHEAD BEAM

EQUIPMENT COMPARTMENT | D2-8095 | FIGURE 3.8

NOTES:
1. NO PENETRATIONS SHOWN ON THIS DRAWING.
2. COMPARTMENT TO BE FABRICATED OF 2.219 ALUMINUM
3. SEALANT & PNEUMATIC SEAL BETWEEN LOWER & UPPER ASSEMBLIES TO BE SILICONE

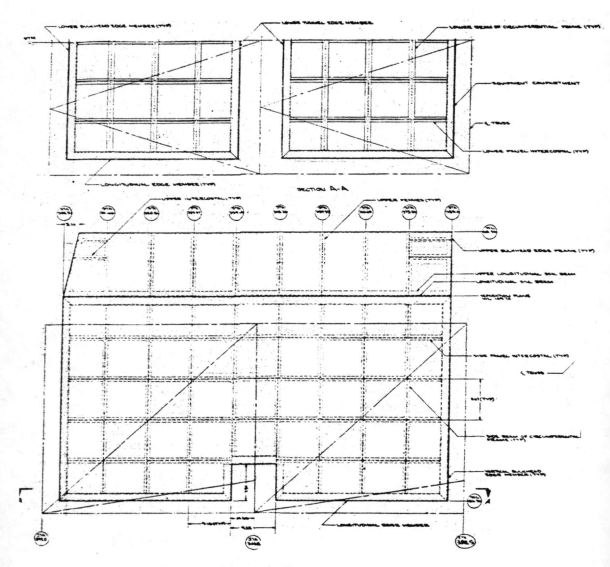

LEFT SIDE VIEW

FUSELAGE WINDSHIELD COVER TO BE ADDED LATER

SECTION 4 WING AND CONTROL SURFACES

4.0 WING AND CONTROL SURFACES - DESCRIPTION

4.1 MAJOR COMMENTS

4.1.1 The wing primary structure consists of an aft torque box and a forward wing structure. These two structures meet, and have in common, the torque box front spar running in a swept line from body station 388.50 at the body to body station 420.00 at the fin. The torque box consisting of, a rear spar, front spar, and inspar members provide support for the elevon, eleven actuator, fin, main landing gear, and wing air loads. The forward wing structure consisting of three spars, inspar members, and a leading edge member provide support for and transmits to the fuselage, the wing air loads. The leading edge member is comprised of a series of discontinuous, simply-supported beams located parallel to the wing leading edge. They provide support for the leading edge segments and beam the air load from the skin panels to the spars. Forward and aft wing skin support beams are located between spars and serve to beam skin panel air loads from the panels to the spar. The gap seal for the insulated panels and landing gear doors is located at the René' 41 corrugation plane and consist of multiple metallic foil leaves bridging the gap as shown on page 4.7. The seal configuration permits displacement of panels due to differential shear and thermal growth while maintaining minimum leakage rate consistent with thermal and aerodynamic performance considerations.

4.1.2 There are two fins with rudders and two elevons located at the aft portion of the glider symmetrically about the glider vertical centerline. The primary structure of the fin and control surfaces consists of ribs, spars and skin panels to carry torsional shear. The ribs and spars are fabricated of René' 41 chord members and corrugated webs. The skin panels are of two types, uninsulated and insulated. The insulated panels consist of a refractory alloy attached by structural clips to the René' 41 corrugated shear panel. Between the refractory alloy panel and corrugated panel are insulation blankets to protect the instructure from excessive heat. The uninsulated surface panels are fabricated of René' 41 skin and corrugation. Leading edges of refractory alloy are provided as required. The basic structure in the leading edge areas is protected from excessive heat by insulation blankets between the refractory alloy and the basic structure. The movable control surfaces are attached to the basic glider structure by René' 41 hinge brackets. The gap seal for the control surface insulated panels is similar to the wing insulated panel seals and is described in Section 4.1.1.

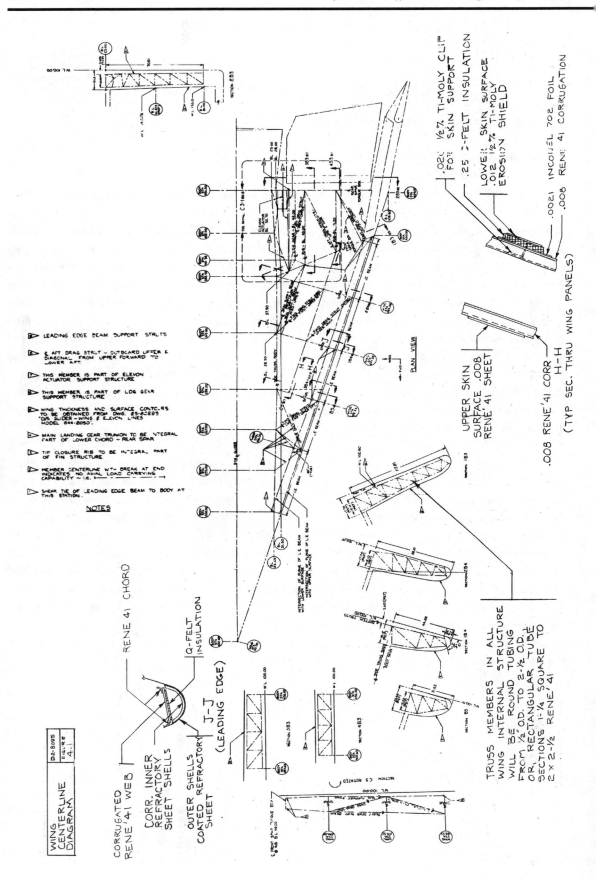

NOTES

▷ LEADING EDGE BEAM SUPPORT STRUTS

▷ & AFT DRAG STRUT – OUTBOARD UPPER & DIAGONAL FROM UPPER FORWARD TO LOWER AFT.

▷ THIS MEMBER IS PART OF ELEVON ACTUATOR SUPPORT STRUCTURE.

▷ THIS MEMBER IS PART OF LDG GEAR SUPPORT STRUCTURE.

▷ WING THICKNESS AND SURFACE CONTOURS TO BE OBTAINED FROM DWG. ES-8C293 "D/S GLIDER – WING & ELEVON LINES MODEL 844-2050.

▷ MAIN LANDING GEAR TRUNION TO BE INTEGRAL PART OF LOWER CHORD – REAR SPAR

▷ TIP CLOSURE RIB TO BE INTEGRAL PART OF FIN STRUCTURE

▷ MEMBER CENTERLINE WITH BREAK AT END INDICATES NO AXIAL LOAD CARRYING CAPABILITY – I.E.

▷ SHEAR TIE OF LEADING EDGE BEAM TO BODY AT THIS STATION.

WING CENTERLINE DIAGRAM | D2-8095 | FIGURE 4.1

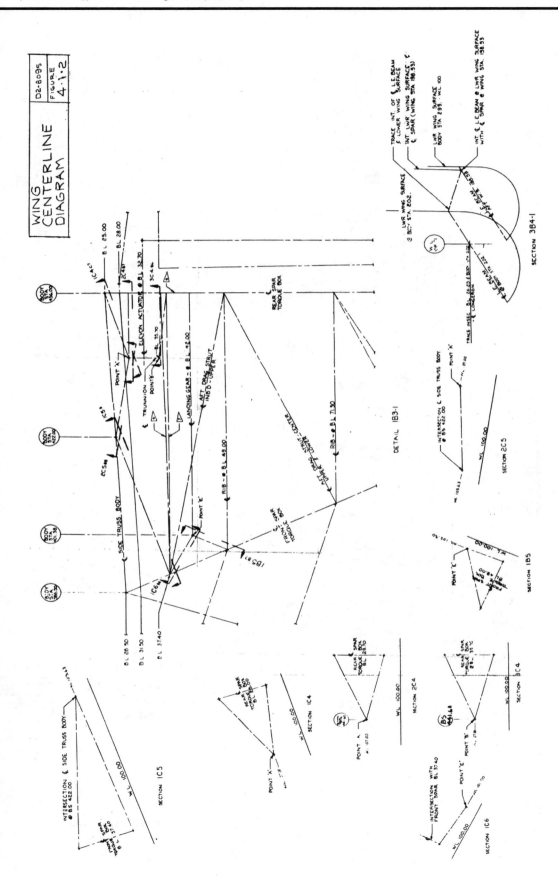

WING CENTERLINE DIAGRAM

D2-8095
FIGURE 4.1.2

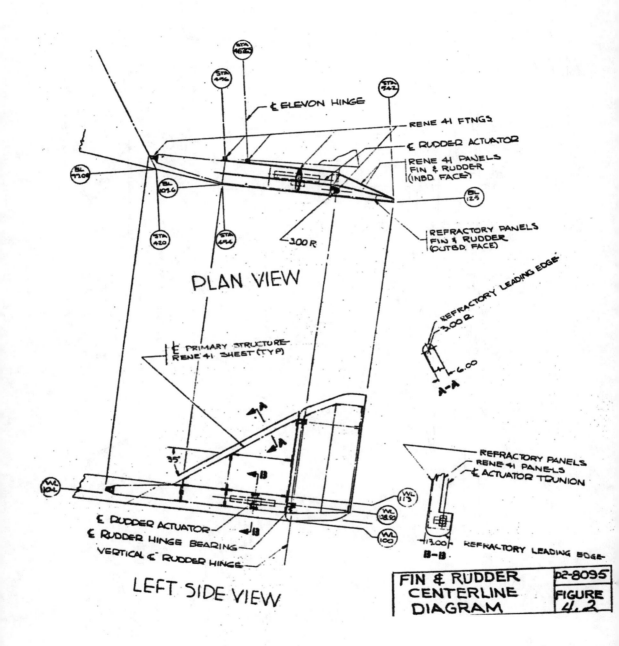

PLAN VIEW

LEFT SIDE VIEW

FIN & RUDDER CENTERLINE DIAGRAM	D2-8095
	FIGURE 4.2

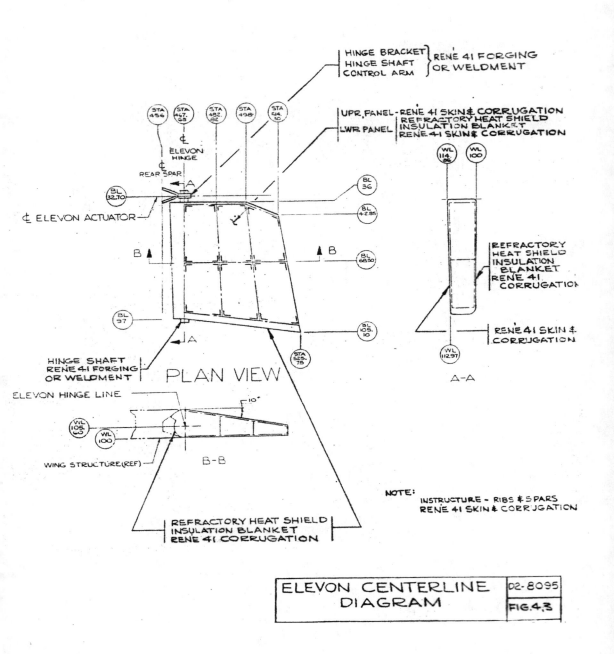

PLAN VIEW

B-B

A-A

NOTE:
INSTRUCTURE - RIBS & SPARS
RENE 41 SKIN & CORRUGATION

ELEVON CENTERLINE
DIAGRAM

D2-8095

FIG.4.3

WING & CONTROL SURFACES LEADING EDGES TO BE ADDED LATER

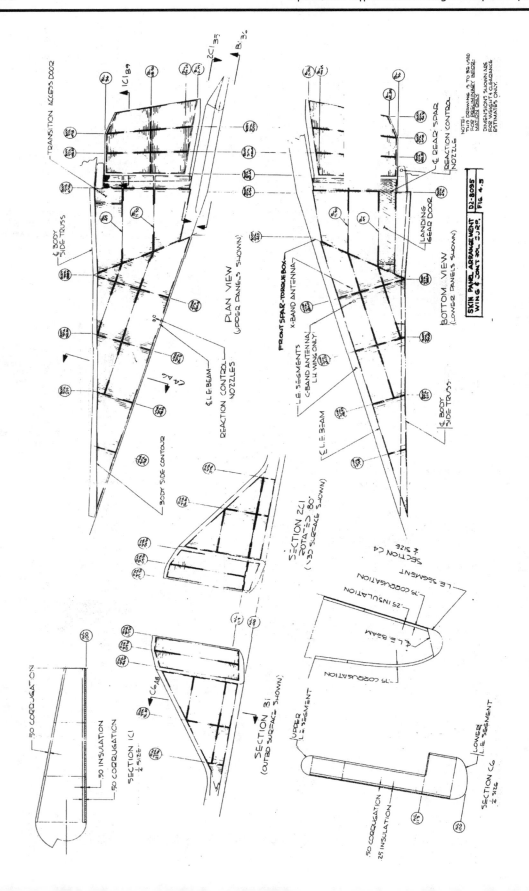

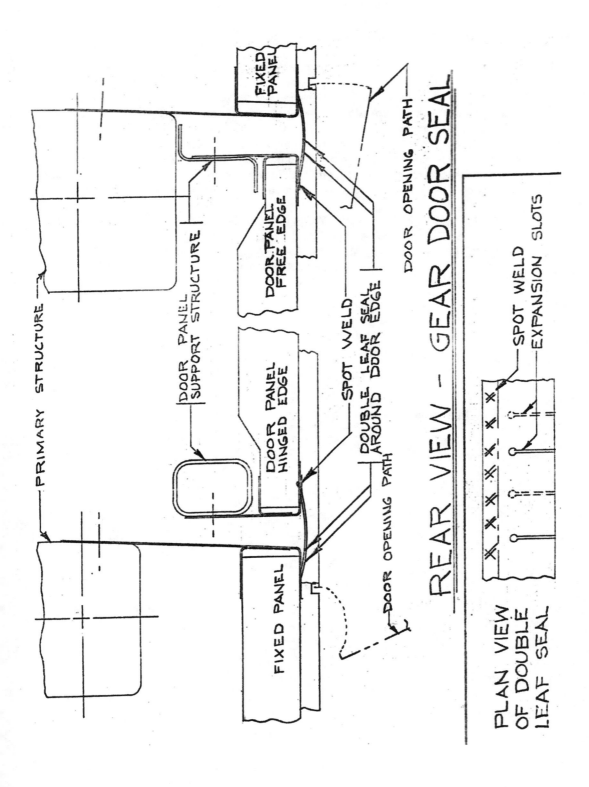

REAR VIEW – GEAR DOOR SEAL

PLAN VIEW OF DOUBLE LEAF SEAL

SECTION 5 LANDING GEAR AND MECHANISMS

5.0 LANDING GEAR AND MECHANISMS — Description

5.1 Landing Gear Description The landing gear is an all skid, uncooled, tricycle type utilizing yielding metal "energy strap" shock absorbers which operate as a yielding drag link between structure and gear strut. The main gear consists of a pair of high-drag wire brush skids. mounted on struts which are trunnion-mounted at the lower chord of the wing rear spar. The struts and skids are stowed within the lower wing surface and are extended aft to the landing position by pneumatic actuators. The landing gear doors are operated by separate pneumatic actuators. The nose gear consists of a cermetallic coated "dish pan" shaped skid mounted on a strut which is trunnion mounted near the forward end of the crew compartment. The skid and strut are extended aft to the landing position by a pneumatic actuator. The nose gear doors are operated by separate pneumatic actuators.

CONTROL SURFACE ACTUATION DESCRIPTION TO BE ADDED LATER
WINDSHIELD COVER ACTUATION DESCRIPTION TO BE ADDED LATER
CRYOGENIC & ELECTRICAL UMBILICAL HATCH ACTUATION DESCRIPTION TO BE ADDED LATER

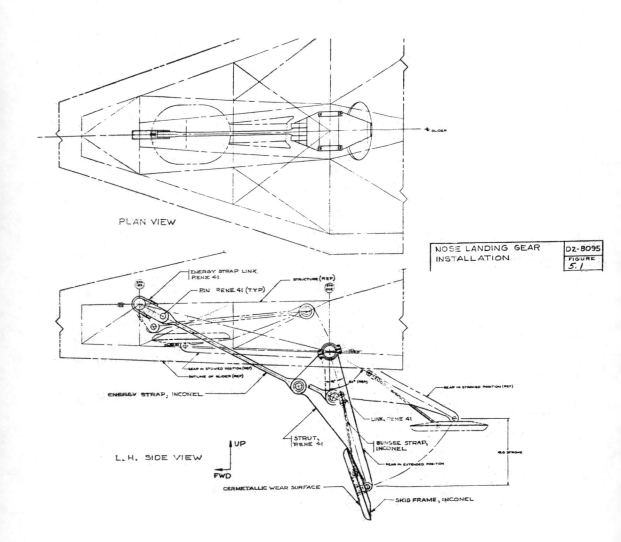

PLAN VIEW

NOSE LANDING GEAR INSTALLATION   D2-8095   FIGURE 5.1

ENERGY STRAP LINK, RENE 41
PIN RENE 41 (TYP)
STRUCTURE (REF)
GEAR IN STOWED POSITION (REF)
OUTLINE OF GLIDER (REF)
ENERGY STRAP, INCONEL
LINK, RENE 41
GEAR IN STOWED POSITION (REF)
STRUT, RENE 41
BUNGEE STRAP, INCONEL
GEAR IN EXTENDED POSITION
UP
FWD
L. H. SIDE VIEW
CERMETALLIC WEAR SURFACE
SKID FRAME, INCONEL

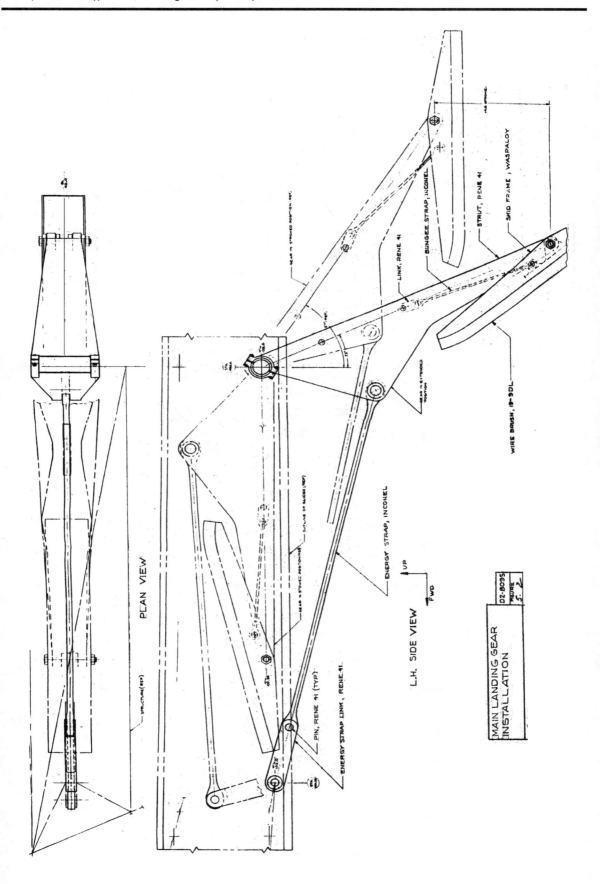

PLAN VIEW

L.H. SIDE VIEW

MAIN LANDING GEAR
INSTALLATION

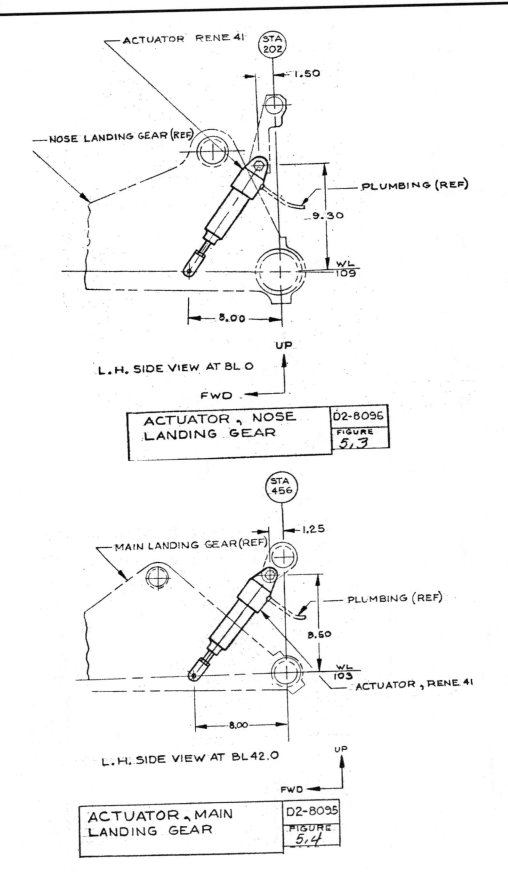

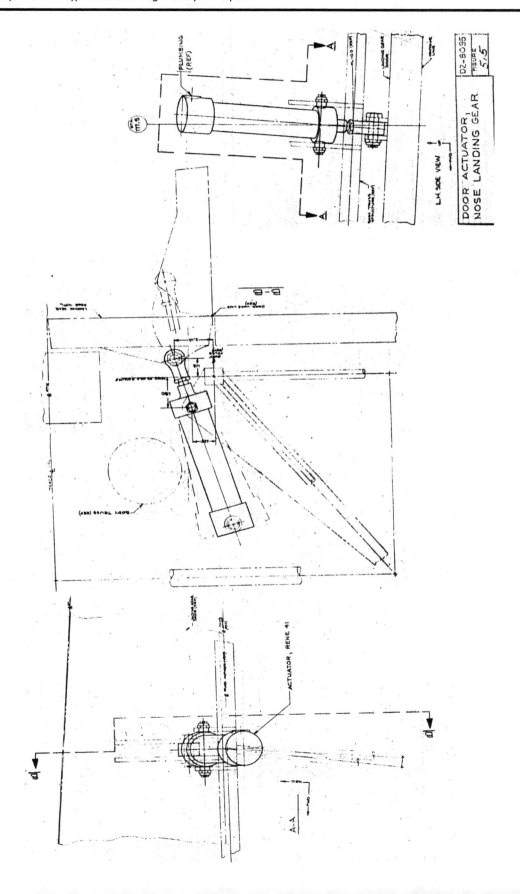

DOOR ACTUATOR,
NOSE LANDING GEAR

D2-8095

FIGURE
5.5

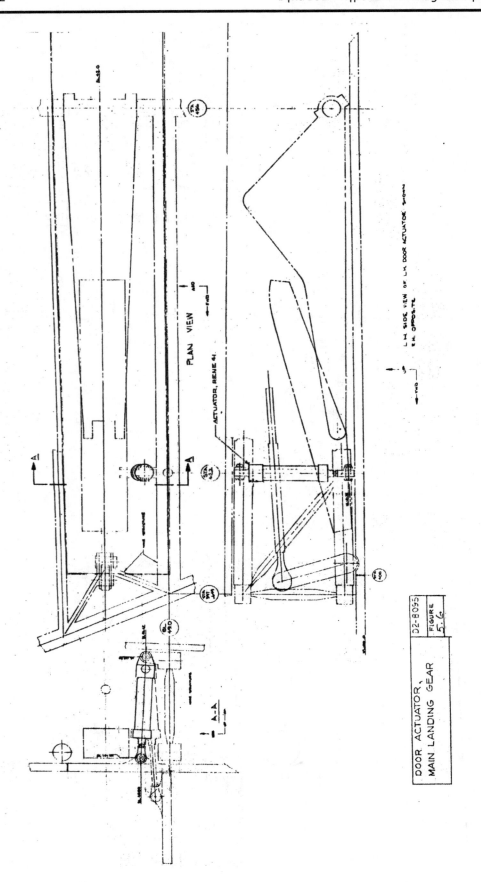

PLAN VIEW

ACTUATOR, RENE 41

L H SIDE VIEW OF L H DOOR ACTUATOR SHOWN
R H OPPOSITE

A-A

DOOR ACTUATOR, MAIN LANDING GEAR

D2-8095

FIGURE 5.6

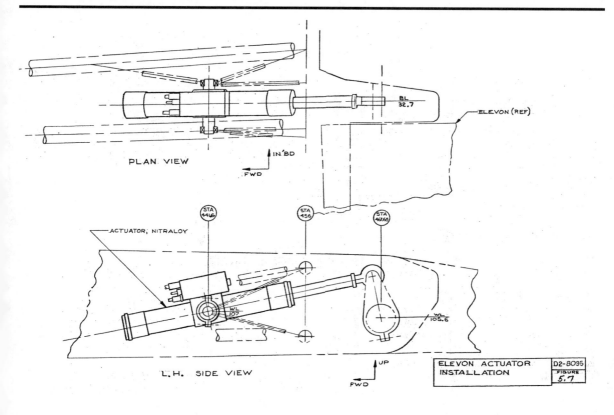

PLAN VIEW

IN'BD

FWD

ELEVON (REF)

BL
32.7

ACTUATOR, NITRALOY

STA
441.6

STA
456

STA
467.55

WL
107

WL
105.6

L.H. SIDE VIEW

UP

FWD

| ELEVON ACTUATOR INSTALLATION | D2-8095 |
| | FIGURE 5.7 |

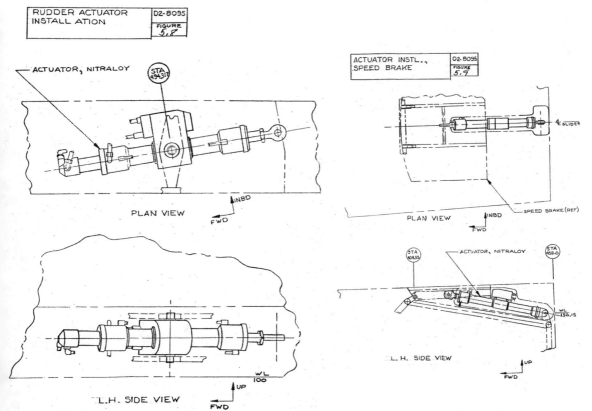

| RUDDER ACTUATOR INSTALLATION | D2-8095 |
| | FIGURE 5.8 |

ACTUATOR, NITRALOY

STA
494.31?

PLAN VIEW

INBD

FWD

L.H. SIDE VIEW

WL
100

UP

FWD

| ACTUATOR INSTL., SPEED BRAKE | D2-8095 |
| | FIGURE 5.9 |

GLIDER

PLAN VIEW

INBD

FWD

SPEED BRAKE (REF)

STA
468.55

ACTUATOR, NITRALOY

STA
458.0

WL
156.15

L.H. SIDE VIEW

UP

FWD

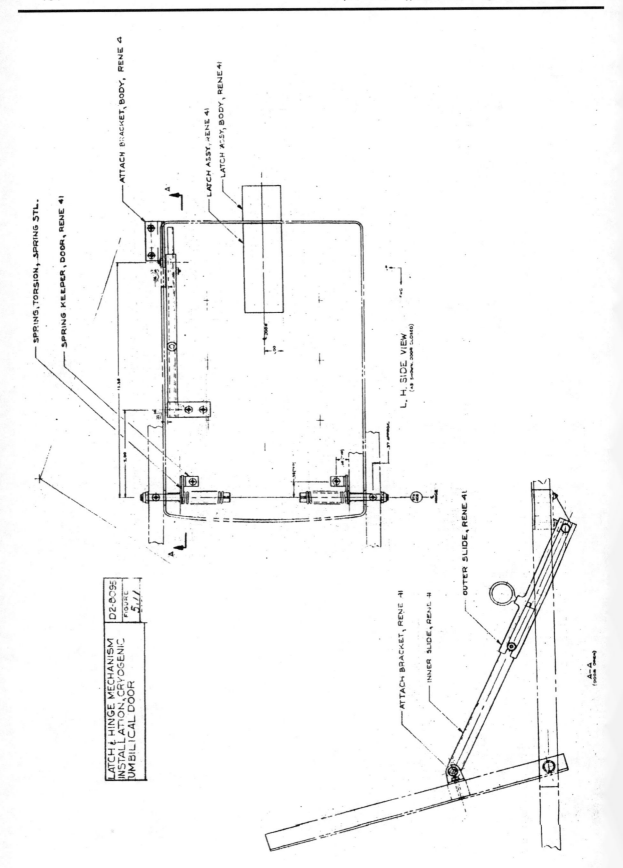

LATCH & HINGE MECHANISM
INSTALLATION, CRYOGENIC
UMBILICAL DOOR

D2-8095

FIGURE 5.11

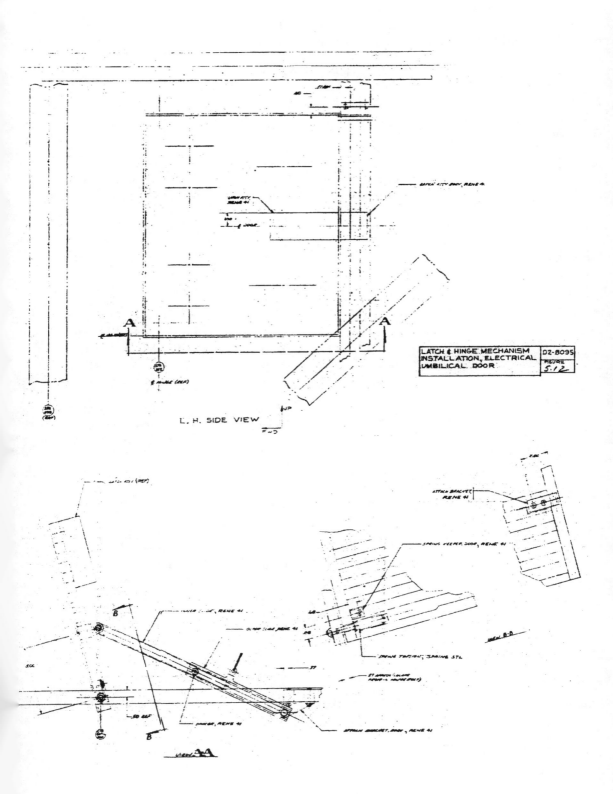

LATCH & HINGE MECHANISM
INSTALLATION, ELECTRICAL
UMBILICAL DOOR

D2-8095

FIGURE 5.12

L. H. SIDE VIEW

SECTION 6. TRANSITION SECTION, PROPULSION GLIDER INTERFACE

6.0 TRANSITION SECTION, PROPULSION GLIDER INTERFACE For detailed information on this section, refer to D2-8139, Transition Section Drawings Report.

SECTION 7. MATERIALS

7.0 MATERIALS

Glider Area	Material		Type & Use
Nose Cap	Zirconia	Grains	Pins or tile cement
	Graphite	Block	Shell - ATJ or RT 008
	Molybdenum	Forging	Support Ring, Retainers and Fittings
	René' 41	Sheet	Truss Linkage
Fuselage	Super Alloy René' 41	Sheet	Tubing - Primary Structure Formed Sections Primary Structure Upper Skin System Corrugation of Lower Skin
		Bar	Joints - Primary Structure
		Forging	Joints - Primary Structure
	Coated Refractory Alloy	Sheet	Shield - Lower Skin Panels Leading Edge Segments, Outer Fin Surface, Forward Body Sides
	Q-Felt	Blanket	Lower Surface Insulation Pilot & Equipment Comp. Insulation Secondary Power Bay, Insulation
	Hastelloy "X" or Inconel 702	Foil Screen	Sealing Insulation Retainer
Pilot's Compartment	2219 Aluminum	Plate	Skin Panels, Bulkheads Equipment Access Door
		Extrusions	Stiffeners, Frames, Longerons
		Sheet	Window Frames
	Super Alloy René' 41	Forgings or Bar	Compartment Supt. Fittings
	Silicone Rubber	Inflatable Seal	Equipment Access Door Pilot's Hatch
	Fused Silica	Glass.	Outer Windows
	Soda Lime	Glass	Inner Windows
Equipment Compartment	2219 Aluminum	Plate	Bulkhead, Sides, Bottom and Lid
		Sheet	Web Stiffeners
		Extrusions	Inboard Caps of Frames and Bulkhead Beams
	Super Alloy René' 41	Forgings	For Equipment Compartment Supports
	Silicone Rubber	Inflatable Seal	Between Lid and Lower Assembly and Sealant

Wing	Super Alloy René' 41	Sheet	Tubing - Primary Structure
			Formed Sections, Upper Skin Panels
			Corrugation - Lower Skin Panel
		Forgings	Joints - Primary Structure
			Actuator Trunnion Support
			Landing Gear Attachment
		Bar	Fasteners, Joints - Primary Structure
	Coated Refractory Alloy	Sheet	Leading Edge Panels
			Shield - Lower Skin Panel
		Bar	Fasteners - Rivets, Bolts
	Q-Felt	Blanket	Lower Surface Insulation
	Fiberfrax	Paper	Insulation Skin & Clips
	Hastelloy "X"	Foil	Sealing
	or Inconel 702	Screen	Insulation Retainer
Elevon & Tab	Super Alloy René' 41	Sheet	Upper Skin Panel
			Corrugation Lower Skin Panel
			Formed Sections, Spars and Ribs
		Forgings	Bearing Housings
		Bar	Fasteners – Rivets, Bolts and Nuts
		Tube	Hinge Shaft
	Coated Refractory Alloy	Sheet	Leading Edge Panels
			Shield Lower Skin Panels
			Formed Sections
		Bar	Fasteners - Rivets & Bolts
	Q-Felt	Blanket	Lower Surface Skin, Insulation
	Fiberfrax	Paper	Insulation Skin & Clips
	Hastelloy or Inconel 702	Foil	Sealing
		Screen	Insulation Retainer
Fin & Rudder	Super Alloy René' 41	Sheet	Inboard Skin Panels
			Corrugated Outboard, Skin Panel
			Formed Sections, Beams and Ribs
		Forgings	Bearing Housing
			Actuator Trunnion Support
		Bar & Wire	Fasteners - Rivets & Bolts
	Coated Refractory Alloy	Sheet	Shield Outboard Skin Panel
			Formed Sections
			Leading Edge Panels
		Bar & Wire	Fasteners - Rivets & Bolts
	Q-Felt	Blanket	Outboard Skin Insulation
	Fiberfrax	Paper	Insulation Skin & Clips
	Hastelloy "X"	Foil	Sealing
	or Inconel 702	Screen	Insulation Retainer

Date: Aug 1961   Title: DYNA SOAR- DATA- PROCESSING

SUGGESTED FRAMEWORK DYNA-SOAR DATA PROCESSING PLAN

21 AUGUST 61

DYNA SOAR- DATA- PROCESSING

1.0 INTRODUCTION

2.0 DATA CONVERSION

    2.1 GLIDER
        2.1.1 Blockhouse
        2.1.2 Airborne PCM
        2.1.2.1 Tape Duplication
        2.1.2.2 Quick Look
        2.1.2.3 Format Conversion
        2.1.2.4 Computer Conversion
        2.1.3 Airborne FM
        2.1.3.1 Tape Duplication
        2.1.3.2 Quick Look
        2.1.3.3 Format Conversion and Computer
        2.1.4 Telemetry
        2.1.4.1 PCM
        2.1.4.2 FM
    2.2 BOOSTER
        2.2.1 Blockhouse
        2.2.2 PCM
        2.2.2.1 Tape Duplication
        2.2.2.2 Quick Look
        2.2.2.3 Format Conversion and Computer Conversion
        2.2.3 FM
        2.2.3.1 Tape Duplication
        2.2.3.2 Quick Look
        2.2.3.3 Format Conversion and Computer Conversion
    2.3 RANGE
        2.3.1 Electronic Tracking
        2.3.2 Optical Tracking
    2.4 MISCELLANEOUS
        2.4.1 Meteorological
        2.4.2 Voice

1.0 INTRODUCTION

This report establishes the framework for the data processing plan to be used in support of the Dyna-Soar Step I flight test program. This plan will include detailed definitions of the operation, procedures and techniques required to convert raw test data into a form suitable for analysis by test engineers. The ultimate goal of the Dyna-Soar data processing plan is to provide a modus-operandi whereby all data user organizations receive test data for analysis in the shortest possible time after a test mission. The data processing operations required for the Dyna-Soar Step I flight test program can be functionally categorized by data source (i.e., Glider, Booster, Range, and Miscellaneous). In addition, these operations can be divided into general categories as follows:

  a. Data Handling
b. Data Conversion
    (1) Off-line
    (2) Computer
c. Data Computation

The data handling operation includes magnetic tape duplication, transcription of voice data, and assembly of meteorological conformation. Off-line data conversion includes the preparation of quick-look strip chart records and the conversion of raw data to computer entry format. Computer data conversion includes the application of calibration and correction information to raw data to provide useful engineering data for computation and/or analysis. In the data computation operation, those manipulations of engineering data necessary to determine system and/or subsystem performance characteristics are accomplished. This report considers the various data processing operations and establishes a framework for writing the detailed data processing plan. As a part of this framework, certain fundamental ground rules have been established. These are defined below.

    1. Data handling and off-line and computer data conversion will be accomplished for all data by the test range; AFFTC for the Air Launch program, AFMTC for the Ground Launch program.

    2. Data computation will be the responsibility of the individual data users. Data computation performed by digital computing machines at each user site will be done using standardized machine language and operating systems to facilitate interchange of programs and minimize duplicate programming efforts. The IBM 7090 has been selected as the standard data processing computer because of the availability of this machine at the test ranges. FORTRAN/FAP has been selected as the standard coding/operating system on the 7090.

    3. The test range will be solely responsible for all processing of data peculiar to the range (i.e., electronic and optical tracking data). Procedures and techniques for processing of all other data will be established by the data processing plan. The following paragraphs present a framework for the data processing plan. All phases of the data processing task are considered for each of the data sources. A functional block diagram of the data flow during data processing is shown in Figure 1.

2.0 DATA CONVERSION

Data conversion is defined to encompass all operations required to produce calibrated data in a form suitable for computation and/or analysis. This includes tape dubbing, quick-look editing, format conversion, and computer conversion (calibration, editing, correlation and updating). Because operations performed on a particular set of data depend upon its format, the data conversion operations are categorized in terms of the source of raw data. These data sources are: Glider (Blockhouse, Airborne PCM, airborne FM, telemetry PCM, telemetry FM); Booster (blockhouse, telemetry PCM, telemetry FM); Range (electronic tracking, optical tracking); and Miscellaneous (Meteorological, Voice).

2.1 GLIDER

2.1.1 Blockhouse
Strip chart records will be made of glider parameters prior to launch. These records will be primarily for assessment of system operation prior to booster lift-off. In case of system malfunction, the information contained on the strip chart records will be required by several user agencies for analysis of the system. Since duplication of these records is difficult, and since insufficient redundant recording equipment is available in the blockhouse, the data processing plan must establish the priorities and criteria whereby these records will be made available to interested user agencies.

2.1.2 Airborne PCM
Glider airborne PCM data will be recorded on magnetic tape and will be the prime source of glider

test data. The airborne data tape recorder will record both PCM and FM data on a single tape using a transverse, rotating head technique. Upon completion of a flight, the operations performed on the airborne recorded data tape become the responsibility of the test range (AFFTC, AFMTC). The three basic operations to be performed are: dubbing, quick look, and format conversion. The data processing plan will establish the detailed procedures to be followed to accomplish these operations. A basis for establishing these procedures is given below.

2.1.2.1 Tape Duplication
Prior to any data handling using the original airborne recorded data tape, at least one duplicate copy will be made for backup purposes. Additional duplicate copies will be made after high priority quick look and format conversion operations have been completed. These duplicate copies will be identical to the original airborne recorded tape and will be made available for quick-look editing to those user agencies who have transverse, rotating head playback facilities.

2.1.2.2 Quick Look
High priority quick-look records in the form of strip chart oscillographic, digital print out, etc., displays will be prepared after the backup duplicate tape (s) has been made. These records will be used for time editing and/or system malfunction analysis. The data processing plan will establish priorities, formats, and procedures for producing these records. Additional quick-look records will be prepared for user agency time editing and/or analysis format conversion has been completed. The data processing plan will also establish the procedures whereby user agency quick-look requests will be satisfied.

2.1.2.3 Format Conversion
After backup duplicate tape (s) and high priority quick-look records have been prepared, the raw data will be converted to computer input format. The output of the Format conversion operation will be a computer input data tape. This tape will not be distributed to user agencies but will remain under the control of the test range (AFFTC, AFMTC).

  All data channels will be processed on a single pass of the input tape (i.e., no channel editing will be accomplished). Since the entire airborne recorded data tape cannot in general be processed onto a single computer entry tape, time editing will be required. This time editing will consist only of determining convenient start and stop times for the computer entry tape (s) and will not result in deletion of data. The data processing plan will establish procedures computer identification formats to accomplish this start/stop time editing. The format of the converted data tape will be established by the data processing plan. This format will be based on standard IBM high density, binary tape formats and will be consistent with existing range format conversion equipments (TLM-62 Modified). This format definition will include prime channel and sub-commutator identification information as well as data and time words.

2.1.2.4 Computer Conversion (Calibration)
The Computer Conversion operation consists of combining the converted data tape with calibration and control information to produce a calibrated data tape. The calibrated tape will contain the same parameters as the converted data tape but in calibrated units. Editing of data on a channel and time basis per each data user's request will also be accomplished during the computer conversion process. The time and channel edited output data tape as well as listed information regarding the quality of the input data will be made available to individual data users. The data processing plan will establish the methods and techniques to be used in assessing the quality of the computer input data. The plan will also define procedures whereby each data user will be provided only that data in which he is interested. In addition the data processing plan will establish the calibration procedures, methods and techniques for converting raw data to engineering units. The data processing plan must also define methods whereby time correlation between the airborne recorded data and ground recorded data (i.e., tracking, &, telemetry, etc) can be accurately established and checked. This is necessary even though the airborne time code generator is synchronized to the ground generated time before lift off

since any errors in time correlation can manifest themselves as gross errors in the data computation and analysis phases.

2.1.3 Airborne FM

At present, requirements have been defined only for quick-look and tape duplication of FM data. A short discussion of the procedures for format conversion and computer conversion of these data is included in case such requirements are established. The FM sub-carrier frequencies recorded on the airborne tape are non-IRIG standard. Therefore, non-standard FM conversion equipment will be located in the Ground Station to provide FM playback capability. The output of the conversion equipment is a standard sub-carrier frequency for each FM parameter recorded. These standard frequencies are then available for processing through conventional FM Ground Station equipment.

2.1.3.1 Tape Duplication

Tapes containing FM data only will be prepared in the Range Ground Station. This operation will be performed after priority quick-look and format conversion operations have been carried out. Tape duplications of FM parameters will be made on IRIG standard 1 inch, 14 track magnetic tape. IRIG standard FM sub-carrier frequencies will be obtained from the non-standard sub-carrier conversion equipment. The data flow for this operation is illustrated in Figure 2. The IRIG standard FM tapes will be transmitted to requesting user groups. The process of duplicating the original airborne recorded non-standard FM data onto IRIG standard FM tapes will enable all user groups with standard magnetic tape playback equipment to perform quick-look and format conversion at their home sites. The data processing plan will establish procedures and priorities for the duplication and transmittal of FM data.

2.1.3.2 Quick Look

Low priority quick-Look records of FM parameters will be produced for requesting user agencies after duplication of the original tape. The standard FM sub-carrier signals will be fed through the conventional FM discriminators in the Ground Station, and the analog outputs thus obtained will then be recorded on strip chart recorders, oscillographs, etc. The data processing plan will define the procedures to be used in fulfilling user requests for FM quick-look data.

2.1.3.3 Format Conversion and Computer Processing

No requirement for conversion of FM data has as yet been established. However, the capability for multiplexing and conversion of a number of FM channels exists at the ranges (AFFTC and AFMTC) which can be used for conversion of these data if necessary. Processing of FM parameters presents a number of problems not associated with PCM data. Among these are multiplexing capacity, sampling rates, output formats, merging and calibration. The problem of merging these data is probably the most severe and requires considerable computer programming effort. The data processing plan will define the procedures whereby FM data can be converted for computer entry; however, no computer programs will be written until justifiable requirements for such conversions exist.

2.1.4 Telemetry

Both PCM and FM data will be transmitted from the glider via an RF communication and data link. The pre-detected output of this will be recorded on a wide-band recorder /reproducer on the ground. This recording will constitute a secondary source of data and will not be used for vehicle analysis unless the airborne tape is unusable, or the vehicle and airborne data tapes are lost. Duplicate copies of the telemetry pre-detection recorded tape will be made for those user groups with pre-detection playback facilities. Duplicate copies of the FM data will be made in the same manner as described for the airborne FM data for distribution to those user groups with standard IRIG quick-look playback facilities. Further processing of telemetered data will be necessary only in the event of an emergency and will be accomplished in an identical manner to processing of the airborne recorded data. The data processing plan will establish procedures for distributing telemetry quick-look records and duplicate tapes and will also define the procedures to be followed in event the telemetry data must be used as prime data. The plan will also establish techniques and methods for time correlation and selection of input data tapes from the multiple telemetry recorders used during the ground launch program.

2.2 BOOSTER

Booster data is transmitted via an RF link and recorded on the ground. This data is in the form of both PCM and FM information. The data processing operations to be performed on this data are identical to those for the glider airborne and/or telemetry data. For this reason, the discussion on glider data is merely referenced in the paragraph below, unless there exists some difference between the two sets of data.

2.2.1 Blockhouse Same as paragraph 2.1.1

2.2.2 PCM

Booster PCM data will be transmitted via an RF link for recording on the ground. This telemetered data is the prime source of booster test data. The conversion of this data to usable engineering information suitable for data computation is the responsibility of the test range (AFMTC). The data processing plan will define the procedures, methods, and techniques to be used in accomplishing these operations.

2.2.2.1. Tape Duplication

Prior to any data handling using the original telemetry data tape, at least one duplicate copy will be made for backup purposes. Additional duplicate copies will be made after high priority quick-look and format conversion operations have been completed. These duplicate copies will be identical to the original telemetry tape and will be made available to interested data user groups.

2.2.2.2 Quick Look
Same as paragraph 2.1.2.2

2.2.2.3 Format Conversion
Same as paragraph 2.1.2.3

2.2.2.4 Computer Conversion
Same as paragraph 2.1.2.4

2.2.3 FM

At present, requirements have been defined only for quick-look and tape duplication of the booster FM data. A short discussion of the procedures for format conversion and computer conversion of this data is included in case such requirements are established.

2.2.3.1 Tape Duplication
Same as paragraph 2.1.3.1

2.2.3.2 Quick Look
Same as paragraph 2.1.3.2

2.2.3.3 Format Conversion and Computer Conversion

2.3 RANGE

Range data consist of electronic (radar and telemetry) and optical tracking information.

2.3.1 Electronic Tracking

Trajectory information acquired by electronic tracking devices consists of range, azimuth, elevation, and velocity data from radar and telemetry tracking antennas. These data are recorded on magnetic tape along with range time. All operations required to convert these data into useful form - described below - will be performed by the Range. Tracking information received by the radar and telemetry tracking antennas will be converted to computer input format. These raw data will then be fed into the computer, together with calibration and control information, where computations will be performed to obtain a best estimate of trajectory. The output of this program is a magnetic tape containing the smoothed calibrated trajectory. A program, complete with necessary coefficients, will be available for interpolation of the estimated trajectory. The data processing plan will not be concerned with the detailed computations required to produce the best estimate of trajectory. The format of the calibrated range data tape w ill be specified. The format will include identification of each sample and identification of each file and record as well as time. Control information will be made available to permit interpolation of the computed range data. The magnetic tape will conform to IBM high-density format. The data processing plan will establish the formats and procedures to be used in user organizations with range electronic tracking data.

2.3.2 Optical Tracking
Optical tracking data will be recorded on photographic film. Operations required to reduce the film data recorded by the optical trackers will be performed by the Range. These data will be prepared for computer entry by punched cards on magnetic tape. Computer processing operations consist of computations required to produce a best estimate of trajectory. The output of the computer will be a magnetic tape containing the smoothed, calibrated tracking information. The format of this tape will be identical to that obtained for the electronic tracking devices as specified in 2.3.1 The data processing plan will not be concerned with the computations required to reduce optical tracking data but it will define the procedures and formats for distribution of this data to user groups.

2.4 MISCELLANEOUS

Miscellaneous types of information consist of meteorological and voice data.

2.4.1 Meteorological Data
Meteorological data consists of temperatures, winds, etc. made by various weather observations. These data will be required in some to correct the calibrated instrumentation data. These data will be entered into a computer from tapes or cards for processing. The output of the processing operation will be a set of correction information in a format specified by the data processing plan. Control information will also be available for use in applying the meteorological date to correction of the vehicle and range data.

2.4.2 Voice
Recordings of voice communications during test operations will be made available to user groups. The range will make transcriptions of these recordings.

August 31, 1961

MEMORANDUM: for Chief, Flight Research Division

Subject: Pilot's Report of Participation in ASD Centrifuge Program

1. An acceleration research program was accomplished on the centrifuge at the USAF Aero Medical Laboratory, Wright AFB, Ohio, from August 24 through 29, 1961, under the direction of the Dyna-Soar System Project Office. The purpose of the program was to determine the suitability of a one-position seat and one-position rudder pedals for use during Dyna-Soar boost. Description of the tests, results and conclusions is as follows:

PARTICIPANTS

The personnel participating as test subjects were: Maj. James W. Wood, AFFTC test pilot. Capt. Russell L. Rogers, AFFTC test pilot. Mr. William H. Dana, FRC research pilot. 1st Lt. Joseph P. Loftus, Aero Medical Laboratory Psychologist and rated jet pilot.

SEAT AND RUDDER CONFIGURATION

Pertinent seat dimensions are shown in Figure 1. Rudders were of the treadle type with the pivot approximately coincident with the ankle pivot. Maximum rudder deflections were ± 9 degrees.

TRACKING TASK

A rudder tracking task was provided for evaluation of tracking capability under stress of acceleration. The task required the centering of a randomly driven dot on a cathode ray tube. The dot was driven off-center by a square wave generator. Centering of the dot required virtually full rudder deflection until the square wave was removed, at which time the rudders were commanded returned to neutral. The error between rudder command and rudder position was time integrated for score. Mean time between square waves was an estimated three seconds. Tracking was commenced one minute before start of the run and continued for one minute after completion of the run for comparison of score under stress and without stress.

ACCELERATION PROFILES

Two acceleration profiles, representative of possible Dyna-Soar profiles, were used. Nominal acceleration versus time plots are shown in Figure 2; these profiles were modified slightly to meet centrifuge acceleration onset and reduction capabilities, and portions of the profiles shown at less than 1.0 g were simulated at 1.0 g. The gondola was rotated about its longitudinal axis to keep the acceleration vector as nearly back-to-chest (eyeballs-in) as possible.

TEST CONDITIONS

In addition to using two acceleration profiles (referred to as High and Low Profile), runs were performed both in summer flying suits and in AP-22s full pressure suits inflated to 3.5 psi differential pressure. Runs were made with no waiting period, and with a waiting period of one hour, simulating a delay on the launching pad. During waiting periods feet were not required to be on the rudder pedals; lower legs could be rested on the front of the seat with lower legs approximately horizontal. Table I gives a summary of runs accomplished.

RESTRAINT

Body restraint consisted of standard USAF seat belt and shoulder harness. The test subjects wore F-104 spurs attached to their flying boots. The knobs on the spurs were rested in holes in blocks of wood at the rear of the rudder pedals.

TEST RESULTS

All runs were made without significant pain or numbness in the feet. No changes in geometry of the seat or rudder were suggested except that this participant requested that rudder neutral position be rotated five degrees toes-toward-head, a request dictated by individual body geometry. The Low Profile was preferred to the High Profile by all participants in spite of the fact that the High Profile uses a mild acceleration for the final seven minutes. It was during this seven minutes of mild acceleration, and not the period of high acceleration, when the minor aches and numbness encountered by the participants was noticed. Breathing difficulty and chest discomfort during the higher accelerations were considerably less noticeable in a pressurized pressure suit than in a flying suit.

The rigorous exercise of little-used muscles required by the tracking tasks was responsible for some, and probably most, of the discomfort noted in the calves and arches. This discomfort was most noticeable in the pressurized pressure suit which restricted ankle mobility. Wearing of low-cut oxfords instead of flying boots during unpressurized runs caused less tiring of the legs and feet and appeared to improve tracking performance. All participants felt that translational rudders, rather than treadle pedals, would have presented no problems and would have allowed better tracking performance since this rudder design is conventional in aircraft and takes advantage of pilot experience.

CONCLUSIONS:

Dyna-Soar can very probably be adequately controlled during boost with a two-axis side controller and conventional rudders. A centrifuge program, using both the controller and the rudders and using a tracking task representing predicted D-S boost perturbations in all three axes, should be accomplished to verify this as soon as possible.

William H. Dana
Aeronautical Research Pilot

WHD:bc JRV Copies to: A. Brown, W. Stillwell, J. Walker, R. Brown - BAC (2)
N. Armstrong, Capt. H. A. Smedal - Ames (2), J. Weil

TABLE I

	Low Profile		High Profile	
	W/O FPS	With FPS	W/O FPS	With FPS
Without wait	9 runs all subjects	Dana	Dana Wood Rogers Loftus	Wood
With 1 hour wait	Rogers	Wood	Dana	Rogers
Subtotal	10	2	5	2
Total	12		7	

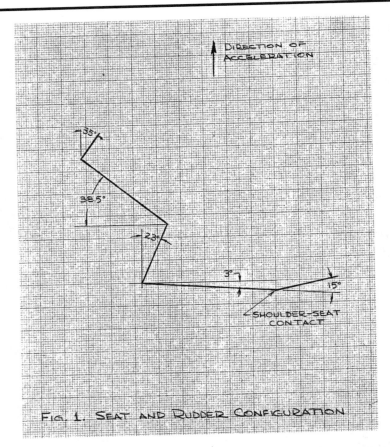

Fig. 1. Seat and Rudder Configuration

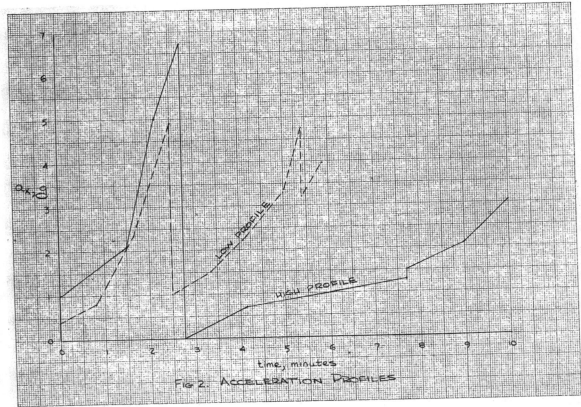

Fig. 2. Acceleration Profiles

Date: Sept 1961   **Title:** Memo: Flight Control & Development Proposals

MEMORANDUM for Chief, Flight Research Division

September 11, 1961

Subject: Dyna-Soar Flight Control Ground and Flight Development Proposals

1. A meeting was held at Dyna-Soar System Project Office on August 30 to discuss ground and flight development equipment for the D-S flight controller and control system. Personnel attending were:

C. J. Snyder, SPO	Maj. J. W. Wood, AFFTC	Lt. R. Johannes, FCL
L. L. Griffin, SPO	Capt. R. L. Rogers, AFFTC	A. J. Longiaru, FCL
G/L J. Henry, SPO	G. Graham, BAC	W. H. Dana, FRC
Capt. W. A. Ford, SPO	F. McRae, BAC	A. Murray, BAC

2. Mr. Snyder stated that, in light of the favorable results obtained in the recent centrifuge program using rudder pedals and a one-position seat, Boeing would be directed to stop development of a three-axis sidearm controller and to use seat geometry tested on the centrifuge (described in MEMORANDUM WHD to TAT, August 31, 1961).

3. Mr. Griffin stated that SPO hopes for approval of a follow-on contract to Minneapolis-Honeywell for continuing use of the Honeywell F-101A in flight control system development. The proposed contract, as envisioned by SPO, would include:

a. Refinement of the side controller to enable low L/D landings with the self-adaptive dampers engaged.

b. Installation of an electrically operated directional control system, similar to the longitudinal and lateral systems presently installed.

c. Installation of variable models in all three axes of the adaptive control system, the intention being to use the variable models for pseudo-variable stability.

SPO and AFFTC feel that this would allow use of the F-101A as a Dyna-Soar landing simulator for both dampers-on and dampers-off simulation. The proposed completion date of this program is February 1962.

4. Mr. Graham, of Boeing, stated that Boeing would commence static simulator testing of a prototype sidearm controller in September 1961.

5. Mr. Murray, of Boeing, stated that Boeing has done preliminary investigation of the possible modification of an F-102 aircraft to incorporate a Dyna-Soar type 2-axis side controller and a three-axis rate command flight control system (not self-adaptive). This modification is allegedly a simple one, involving only alteration of the F-102's damper system. This investigation is described in Boeing D-S Coordination Sheet 2-61 63-4-14. This program would (hopefully) give an adequate simulation of Dyna-Soar dampers-off landings, assuming that basic F-102 aircraft dynamics are similar to those of Dyna-Soar. The program would tentatively commence in January 1962 and first flight data would be available in April 1962.

6. A test-bed aircraft was proposed for flight development and qualification of the Dyna-Soar self-adaptive autopilot and flight control system. The most promising aircraft for this role appears to be an F-102 or F-106. This program would tentatively commence in April 1962.

7. Mr. Griffin stated that dynamic simulator testing of the sidearm controller is programmed for the Johnsville centrifuge starting in April 1962.

8. Figure I shows proposed time schedule of development program for Dyna-Soar flight control system.

9. The undersigned, representing FRC's Dyna-Soar pilot coordinating group, expressed the following positions to the meeting:

a. Concurred in refinement of present Honeywell side controller to allow low L/D landings in the F-101A for D-S dampers-on landing simulation.

b. Expressed disapproval of installation of electrically operated directional control system in F-101-A, and use of that aircraft as dampers-off landing simulator; because of inability of variable model system to simulate aircraft gust response and to avail itself of pilot's ability to damp oscillations.

c. Tentatively concurred with Boeing proposal for F-102 aircraft for D-S dampers-off landing simulator, contingent on modification being as inexpensive and as valid as described.

d. Concurred with proposal for flight control system development-and-qualification aircraft.

e. Concurred with edict to use two-axis controller, one-position rudder pedals, and one-position seat. All attendees of meeting agreed that centrifuge qualification of this flight control configuration must be completed expeditiously.

William H. Dana Aeronautical Research Pilot
WHD:bc JRV
Copies to: A. Brown, W. Stillwell, J. Walker, N. Armstrong, J. Weil

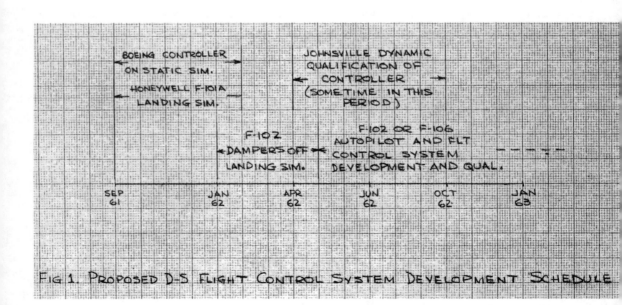

FIG 1. PROPOSED D-S FLIGHT CONTROL SYSTEM DEVELOPMENT SCHEDULE

Date: Sept 1961   Title: Space & National Security Panel

Space & National Security Panel

Convention Hall, Philadelphia PA
Panel: Dr. Dornberger, Dr. Welsh, Mr, Gardner, General Schreiver, Congressman Daddario

September 22 1961

CHAIRMAN GARDNER: What must we do to get us on the road to a military space capability? Please spell it out.

GENERAL SCHREIVER: I pointed out that I felt that from a program standpoint there were several areas that would require very great emphasis during the next few years. The first is in the area of large boosters and the ability to put large payloads into orbit and do this more economically. This certainly can be achieved as part of the lunar program. Secondly, I stated the need for man in space. Here we have not had a program that has had enough depth. The Mercury program has been a high-priority one, but there are many other things that can be added, other reentry techniques. The Dyna-Soar glider is one. There are many things that can and must be done, to a large extent as part of the lunar program.

We need to put a great deal of effort in certain techniques. Before we can have a defensive capability in space we must be able to rendezvous. We must be able to inspect. We should be able to dock, and we should be able to transfer either materiel or people from one space vehicle to another. These are specifics in terms of programs. Some of these will be a part of the lunar program and some perhaps will not be included and will have to be picked up separately.

CONGRESSMAN DADDARIO: In the background of all the remarks made here, there has been a partial answer to that question and no one could possibly spell out today in the time allotted a step-by-step program. In the first instance, we already have in the Air Force a forward look toward the ends involved. For example, two-thirds of the people in this country who have capacities in the life sciences, which General Schriever has christened bioastronautics, are in the Air Force; and these people are working presently with animals and with the capacities that they have available to them to create the environment in which man can live in space and become a useful part of the marriage of the man to machine as we reach that particular point.

There are also in the man-on-the-moon program the funds necessary to spread over a period of years so that when we do reach the moon we can, if this program is properly integrated, achieve not only the civilian goals involved but also the military goals which are in the intervening places between here and the moon.

We have the technical skills and the capacities both in and out of government, both military and civilian, properly bound together, to accomplish this end within the time allotted. It becomes necessary only to take those skills, capacities, and personnel we have and properly utilize them.

CHAIRMAN GARDNER: What is the panel's view of establishing another Von Neumann-type committee to ramrod a top priority for military control of space? Is this a job for the Space Council?

DR. WELSH: It could be a job for the Space Council. But I don't necessarily find a solution to problems coming from the appointment of another committee. We have several in Washington now.

CHAIRMAN GARDNER: How much should the military space budget be expanded?

DR. WELSH: It is my guess that the space budget that was obtained this year overall, military and civilian, was probably as much as could be obtained, and in the sense of introducing urgency into the program I believe that it's about what could be used. However, in the coming year it's going to be substantially larger. I don't think it's going to be a very significant portion of the total gross national product at any time. We've had a whole lot of comments about $20 billion, $30 billion, $40 billion a year for space. There hasn't been anything programmed or thought of or considered in that size or in that scope.

CHAIRMAN GARDNER: Why do we not have a single competent head over all space responsibilities and resources to do the job?

CONGRESSMAN DADDARIO: We have a civilian agency and we have a military agency, and we have got to take them and bind them together. I have suggested that we have, since the Air Force has been placed in charge of our military space program, someone who has the direct responsibility at an equal level with the director of NASA to utilize the full capacities of the Air Force together with the civilian agency to accomplish both military and civilian uses in space.

DR. WELSH: May I just add one point there? Time is the resource which is most scarce as far as our space program is concerned, and there are few things that are more time-consuming than to build up and develop an entire new agency or to break down already established lines of authority. We have made an effort to avoid this change in the structure of the government by setting up and making something out of this Space Council to try to bring the policies together, and I don't believe we can afford to waste time to set up just one agency to handle all the programs.

CHAIRMAN GARDNER: We seem to be optimistic about our chances of beating the Russians to the moon. Yet they are the ones already in space. How do we hope to overcome their lead?

DR. WELSH: The first thing we plan to do is to go forward much more rapidly on large boosters, liquid and solid. We are also planning to put a lot more attention on the rendezvous technique which may give us a step-up in acceleration. I don't know that anybody has said specifically that we're going to beat the Russians to the moon, but I am sure we won't be able to beat them if we don't start trying now.

CHAIRMAN GARDNER: I don't think anyone here has represented that we're going to beat the Russians to the moon. I for one do not believe that the manner in which we are presently organized will permit us to do so. We are not making enough use of the proven resources in the Air Force in connection with our lunar program, in my opinion; and unless we do so we can't buy back this precious commodity of time that we need.

GENERAL SCHREIVER: I have felt myself that we not only have the capability of catching up with the Soviets, but also the capability of getting ahead. After all, we have a nation that has a gross national product twice that of the Soviet Union. We have the resources. By that I mean the industry, the scientific, and the military. We have the know-how. I personally have seen large programs under-taken and I have seen them move forward very rapidly. From that point of view I have a great deal of confidence. The question is one of management. I have personally seen no indication on the part of the people that I have been dealing with in NASA at the top level that we aren't moving in the direction of creating the proper interface to work together closely in bringing all of the resources to bear, those that we have and those which are in NASA. Remember, this program has been approved for only a relatively short time. It's a massive program. The plan for Cape Canaveral where the launchings will take place has been in every respect a joint planning effort between the Air Force and NASA at the working level. My commander there, General Davis, working with NASA, worked up a detailed plan for Cape Canaveral just as an indication that we are working together in specific areas.

CONGRESSMAN DADDARIO: Enough monies have been appropriated by this Congress to do as much as we possibly can do at the moment to get this program on the road, to give it the kind of urgency it needs. There is no question in my mind that the Congress will in the succeeding years appropriate the additional funds that are necessary. I think you will find a space budget in the vicinity of $3.5 billion to $4 billion within the next couple of years.

CHAIRMAN GARDNER: Does the Air Force have even one approved man-in-space program today? If not, why not?

GENERAL SCHREIVER: Dyna-Soar is a man-in-space program, and that is approved.

CHAIRMAN GARDNER: Is our policy that space is for peaceful purposes fooling anyone but the US public?

DR. WELSH: It certainly isn't our policy to fool the US public. I'm afraid there has been, however, a good deal of confusion as to just what the policy was until fairly recently; and it has been spelled out more clearly, I believe, by the President in recent months than it had been at any time before that. I think it depends on what you mean by peaceful uses. The way I defined it in my initial statement was that the entire activity of the Department of Defense is for peaceful purposes in the sense of their effort to maintain the peace; and if we don't maintain the peace we won't have any chance to fly to the moon or any place else.

CHAIRMAN GARDNER: Is the USAF being denied authority money to go ahead in space in the way it wishes to do?

GENERAL SCHREIVER: All of the programs that we have approved, and there is a very significant program under way in the field of warning, navigation, weather, observation, and so forth, are, in my opinion, funded to the extent necessary to push them forward as fast as technology will permit. What I have pointed out was that there needs to be further emphasis as we go down the line now in additional areas.

CHAIRMAN GARDNER: Is it more important to have a space program which is aimed at getting national prestige or is it better to concentrate on those programs having a direct bearing on military power?

DR. DORNBERGER: The reasons for conquering space are interrelated-propaganda, political prestige, military, they're all interrelated. And maybe a prestige success can change our political situation in the world. The Russian success caused a lot of neutrals to go into the Russian camp because they are afraid that what the Russians can do with their scientific achievement in space they could do with weapons, too. Our first task should be to guarantee our survival and to prevent the Russians from getting supremacy in space. That, I think, is vital; and if the question ever comes up that we don't have enough money to do both, I would spend my money on the military supremacy in space.

CHAIRMAN GARDNER: Military tradition has often delayed the introduction of new weapons. Is this happening in the space field?

DR. WELSH: I don't think there is any evidence that the attitude of the military is one of delay. As a matter of fact, as we attempted to build up this new space program that was presented to the Congress, the first place we sought advice was from outstanding gentlemen like General Schriever; and we got it. We got good advice, and the advice was to go ahead and to go ahead fast. I have seen no evidence that the military was acting as a lag in this program. As a matter of fact, they're part of the catalyst in it.

CHAIRMAN GARDNER: Is spending so much money on space justified when we consider the other pressing needs which we have right here on earth? Cancer research is often mentioned in this connection. It would be nice to solve all of our earthly problems before we think about other problems. However, that isn't the way things are. The reality of life is that there is competition in the space field and there is opportunity there. The money we spend in space is not wasted. It will have large and regenerative civil fallouts for projects on earth. Our society is rich enough to take the space adventure and must spend money on the military space problem and still solve most of our problems on earth, or at least work on those problems.

CONGRESSMAN DADDARIO: I think we already have received a great deal of benefit from the space program. We have, for example, developed a valve small enough to be used in the human heart. There have been developed batteries with the capacity and ability to be utilized in operations so that they can help the heart to pulsate where it otherwise would not. There has been developed in the area of mental illness, which is one of our most costly problems and one of the ones we will be faced with in years to come, the ability to look into the brain, to map it out, to find out how it affects us and how it is affected by those things we do from day to day. Some scientists who have been working in this program have told me that if it were not for this help they could not do what they have done, and that in the last ten years they have learned more about the brain than they have in all of the previous history of study.

DR. WELSH: We didn't spend enough on cancer research before we ever had a space program, so that it isn't necessarily either/or; and it isn't that if we didn't spend the money on space we would spend it on cancer. We've got to do both and we can afford to do both.

GENERAL SCHREIVER: I'd like to make several points. Number one: already in space (and it's not even four years old) we have civilian programs which industry is falling all over itself trying to get in on. For example, communications. There are other immediate civilian applications, such as weather and navigation; and this is in the very early stages of this great experiment of ours. So who can say what we'll have ten years from now in terms of things that make our life better right here on earth because of our space program? The second point is that I don't quite understand why a nation that has a gross national product approaching $500 billion a year, at the same time has indeed a very large-scale unemployment problem. Many of our basic industries are not anywhere near full production. I think that we have to really harness our economy. We are in an all-out competition with the Soviet Union. One of the strongest things that we have is our economy, and anyone who tells me that we can't afford $5 or $6 billion a year for space I just would have to disagree with.

CHAIRMAN GARDNER: I would like to close with the observation that today the Soviets are ahead of us in big boosters and are testing larger ones. They are ahead of us in man in space and we don't know what else they're ahead of us in. I don't find a great deal of comfort in the fact that we have performed a few notable scientific experiments in space and allege that we're ahead in space science. I believe all of us must work and work diligently if any military space program of any major sort is to get born.

Date: Nov 1961   Title: Manned Operational Capability in Space

Needed: Manned Operational Capability in Space

November 1961

I believe the threat to our national security discussed by the other panel members is a major cause for concern. The development of effective functional capabilities in space will be a principal factor in the continuing struggle for leadership of the world. We have already seen that the Russians regard their space deeds with a militant pride. Their successes in space are presented as proof of the Communist credo that history is on their side and communism will rule the future world. We have the ability to prove them wrong, if we use our resources wisely.

It is not generally appreciated that in the last few years we have developed great resources which can be devoted to any one of several courses of action we may choose to develop a strong defense capability in space. Perhaps the best known of these resources are the missile test ranges which give us the ability to support space programs devoted to both military and civilian objectives.

In carrying out our ICBM programs we also have developed the capability to procure large modern systems which employ modern technology and, even more important, achieve reliable operational status with these systems after they have been delivered to the field. The experience gained in these programs covers the whole spectrum of research, development, procurement, transition of the systems from technical specialists to operational units in the field, and logistic support throughout all phases of the programs.

Perhaps the most invaluable asset of all is the corps of the Air Force officers and men in whom this experience is embodied. We have learned techniques of management which permit us to bring new systems into the inventory in a few short years as we did with our ballistic missiles.

But there is no short cut to the creation of a team of dedicated and experienced men with a tradition of accomplishment. The Air Force has a highly competent team trained in the skills required to develop the necessary military capabilities in space. These people are accustomed to working with the sense of urgency necessary in programs vital to the national security.

In carrying out our national space program, certain basic functional capabilities are necessary and fundamental to both military and civilian programs. The best approach to our military space program is a mixture of unmanned and manned space vehicles. More emphasis on manned spacecraft is required. We must be able to use space on a routine, day-to-day basis. In order to develop this ability we must begin by developing the ability to place large payloads in space, the ability to navigate and maneuver spacecraft, the ability to go into space and return to earth at times and places chosen to support a selected mission, the ability to rendezvous in space and accomplish refueling or cargo transfer; in short, to transport, use, and support man in space.

The unique abilities of man to observe and to use judgment are essential. Most important is the ability of man to exercise control based on his judgment. Man's abilities are necessary to support our national objectives and national security in the space age.

Along with development of the ability to place man in space in a functional role, a broad development program for components and techniques must be carried out. This will permit the combination of the experience and knowledge gained - in placing man in space in a functional-capacity - with a choice of developed components and techniques to give us national defense capabilities in space.

I believe it is clear today that we must be able to observe or inspect satellites and determine whether or not an unidentified satellite is hostile or friendly, just as we are able to identify unknown aircraft today. Should a satellite be determined to be hostile, then we must have a capability to neutralize it. If we cannot deal with such satellites, the ability of the nation to exist and. preserve its essential values will be severely compromised or nonexistent.

Finally, the key to rapid utilization of space by man for military or civilian purposes is flexibility.

We must not design our space vehicles and programs just to achieve those objectives which we can define now. We must design the vehicles with enough capacity to rapidly adapt to or incorporate the vast new knowledge which will flow from our space program. A manned space observatory should be designed so that the composition of the crew of observers can be modified or completely changed to accomplish new or additional missions. Such an observatory should have provisions for the addition or replacement of equipment simply and quickly in order to respond to new knowledge or new requirements.

The achievement of these capabilities in space will enable us to ensure that space is not used for aggressive purposes-provided we achieve them in time. This, I feel, is an urgent matter, and it may require that we rethink our traditional approach to the use of technology.

Historically, we have tended to overestimate what we could do on a short-term basis and to grossly underestimate what we could do on a long-term basis. As a result, the United States has been notably slow to recognize the military application of new inventions. This was an adequate philosophy for the days when oceans were real barriers and gave as time to mobilize after the outbreak of war. But today, technological surprise could be fatal.

Two of the most significant technical achievements of this century - the airplane and the liquid-fueled rocket - are American inventions. Yet in each case their first military application was made by other nations.

After the Wright brothers flew at Kitty Hawk, more than three years passed before the US Army decided to buy an airplane. Even then the military potential of the airplane was not realized. When World War I broke out, and when American pilots flew in battle, they flew foreign aircraft exclusively. No American-designed plane saw combat during World War I.

The work of Robert Goddard, inventor of the liquid-fueled rocket, was neglected in the US but was followed with interest overseas. By 1929, three years after Goddard's first successful test flight, Germany had embarked on a rocket program. Soviet rocket research started in 1933 and had recorded substantial accomplishments by 1935.

By 1939, Germany was spending one-third of her entire aerodynamic research budget at Peenemünde alone. She had 12,000 workers there and at least 1,000 qualified researchers in rocketry elsewhere in the country.

In the same year, Goddard had five technicians. In the whole of the United States there were probably not more than twenty-five people working with liquid rockets.

The United States finally was forced to recognize the military significance of the rocket after V-2s started to rain on London. Soon American scientists began to learn of Goddard's work through reading captured German documents. Nevertheless, little more than a decade ago the ICBM was called "a fantastic proposal." Early indications that the Soviets might be producing such a weapon was largely ignored.

I have cited these examples to indicate the danger that lies in inadequate military planning in the future. Now we may find ourselves in a similar position in regard to space. No one can predict what benefits its exploration will bring to mankind, but I am certain they will be immense. At the same time, space most certainly has potential military use.

We need to press forward vigorously with our national space effort. We should recognize that there is no inherent difference between basic military and non-military space technology. The same hardware and techniques used to send up a scientific capsule can also be used to orbit an early warning satellite. The same techniques that can send a man into space as a scientific observer may also send him there in a military role.

What really matters is not the technology but the intent. Our intentions as a nation are peaceful and are known to be so. We should not be afraid to develop whatever systems are needed for our legitimates self-defense and we should not be afraid to give them priority.

Only by being strong can we preserve the peace. This is the primary objective of the military forces today. When the Strategic Air Command was formed, the people of America understood that "Airpower Is Peace Power." We need a clear understanding today that "Spacepower Is Peace Power." Spacepower must become a vital part of our national strength and security.

GEN. BERNARD A. SCHREIVER, USAF

Date: Dec 1961   **Title:** Press Release

NEWS RELEASE

PLEASE NOTE DATE

DEPARTMENT OF DEFENSE
OFFICE OF PUBLIC AFFAIRS
Washington 25, D. C.

No. 1485-61
IMMEDIATE RELEASE
December 26, 1961
Oxford 75131

DYNA SOAR PROGRAM REVAMPED

SUBORBITAL FLIGHTS ELIMINATED

The Air Force has ordered a major revision of the DYNA SOAR program eliminating all suborbital flights employing the TITAN II as a booster. The development program will move directly from manned drops of the DYNA SOAR glider from a B-52 to orbital flight which will require an improved booster.

One approach to the design of this improved booster would be based on TITAN II technology and would incorporate large solid propellant rocket motors as the first stage.

In line with this action, the Air Force is terminating its contracts with the Martin Company for design work on the modified TITAN II as a DYNA SOAR booster. New contracts with the Martin Company will permit the contractor to proceed with initial work applicable to booster system considerations.

The primary purpose of the DYNA SOAR program is the exploration of manned, controlled lifting re-entry with a horizontal landing from orbital flight. The piloted DYNA SOAR glider will be launched into orbit from Cape Canaveral, Florida, and will use aerodynamic means for maneuvering during re-entry and descent to a landing at Edwards Air Force Base, California.

END

Date: Early 1962   Title: Cockpit Presentation

COCKPIT   ENVELOPE

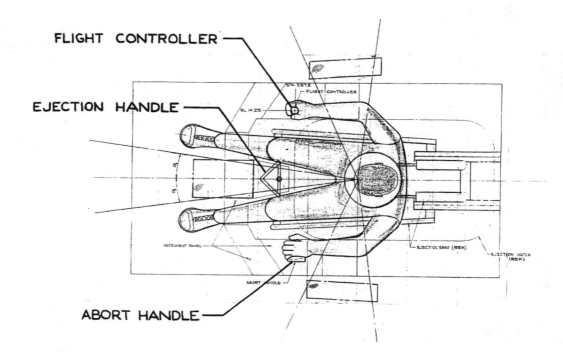

FLIGHT CONTROLLER

EJECTION HANDLE

ABORT HANDLE

COCKPIT   ENVELOPE

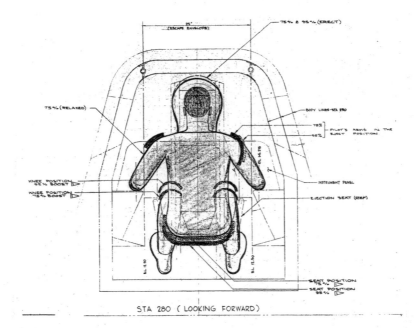

STA 280 (LOOKING FORWARD)

TIME— LINE ANALYSIS

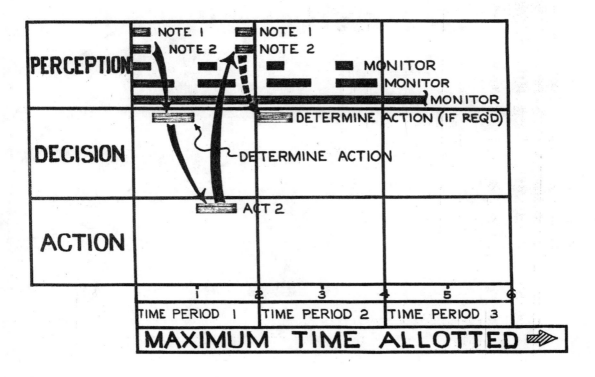

TERMINAL GLIDE PROFILE
$^L/_D$ MAX

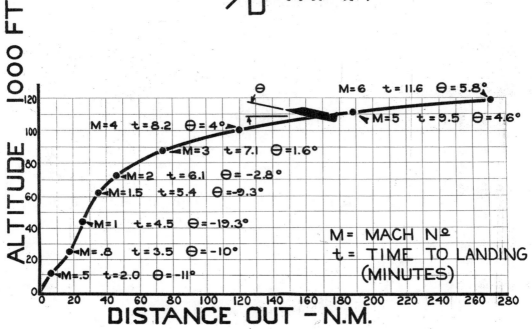

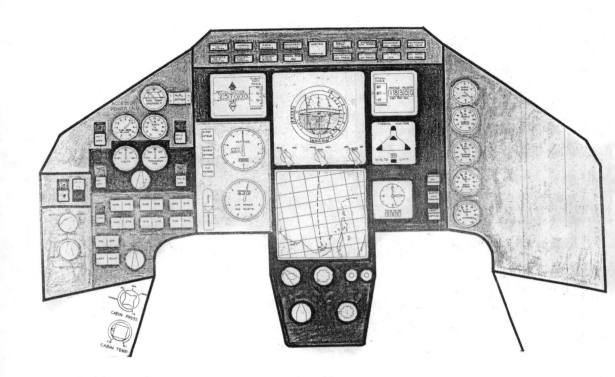

ABORT PROFILE

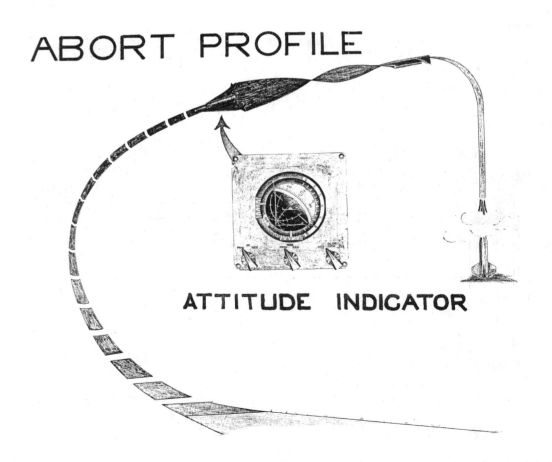

ATTITUDE INDICATOR

LANDING AREA DISPLAY

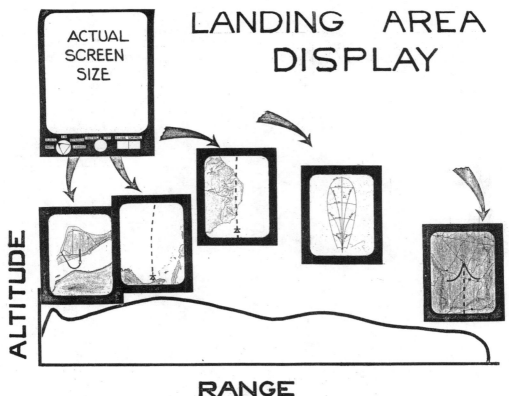

ALTITUDE/CLIMB RATE INDICATOR

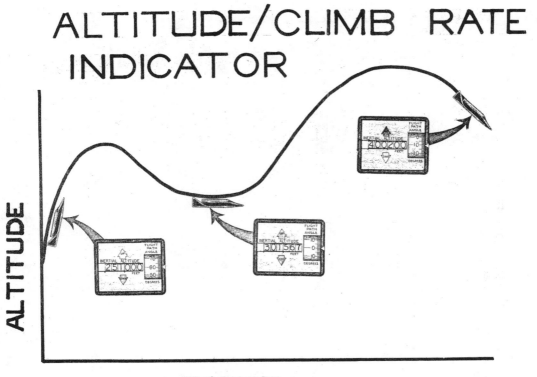

PILOT VISION FIELDS

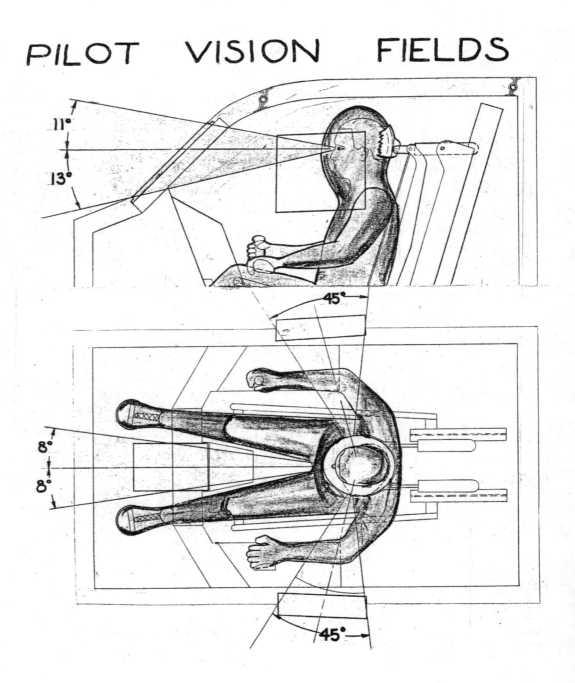

FWD VISION   64,000 FT. ALT.

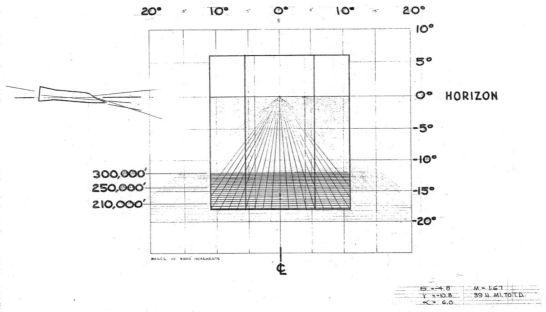

RANGE IN 5000' INCREMENTS

SIDE VISION   64,000 FT. ALT.

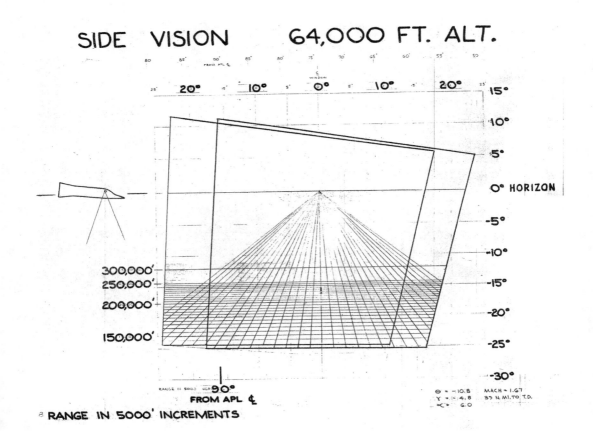

RANGE IN 5000' INCREMENTS

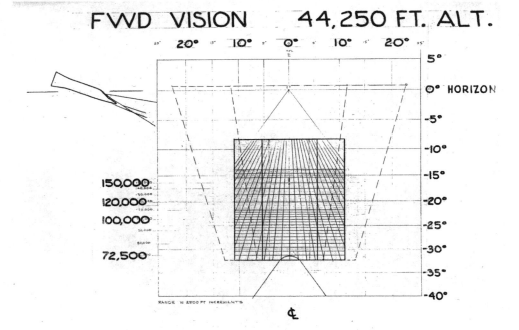

FWD VISION — 44,250 FT. ALT.

RANGE IN 2500' INCREMENTS

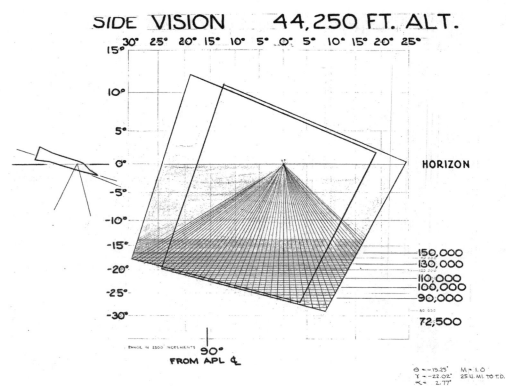

SIDE VISION — 44,250 FT. ALT.

RANGE IN 2500' INCREMENTS

FRONT VISION   20,000 FT. ALT.

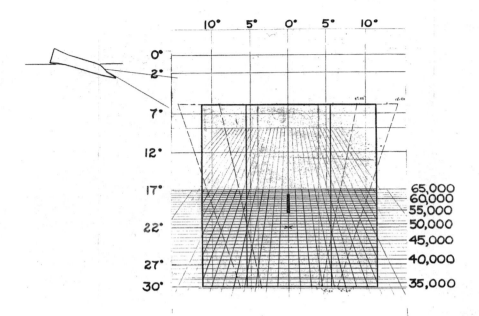

RANGE IN 1000' INCREMENTS

θ = 17.0°   M = .60
Y = 22.0°   9.3 N. MI. TO T.D.
α = 5.0°
DOTTED LINES REPRESENT 10° WIDE RECT. WINDOW

SIDE VISION   20,000 FT. ALT.

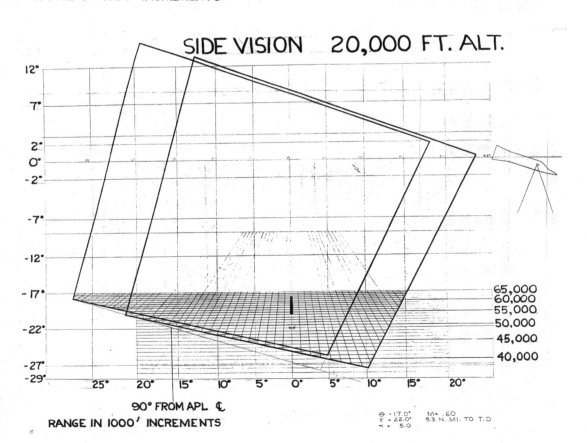

90° FROM APL ₵

RANGE IN 1000' INCREMENTS

θ = 17.0°   M = .60
Y = 22.0°   9.3 N. MI. TO T.D
α = 5.0°

FWD VISION   1,500 FT. ALT.

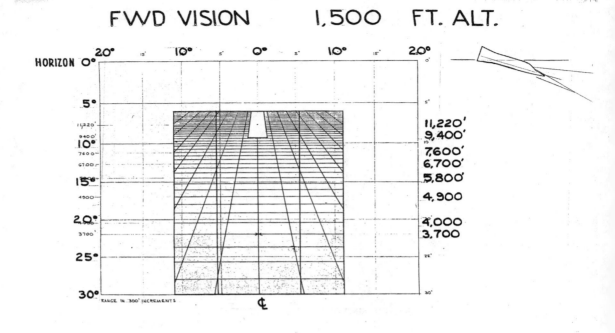

₈ **RANGE IN 300' INCREMENTS**

RESTRAINT SYSTEM

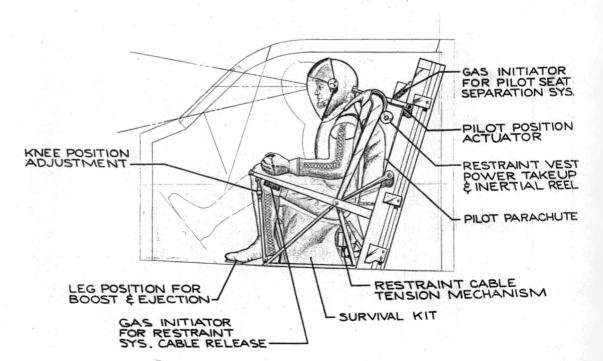

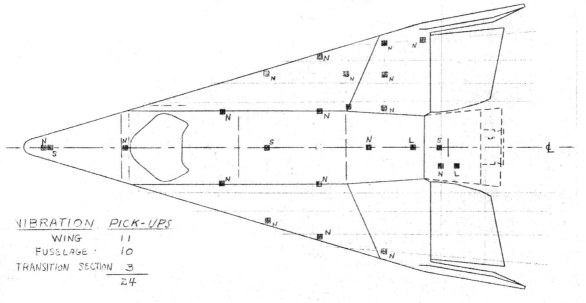

VIBRATION PICK-UPS
WING   11
FUSELAGE   10
TRANSITION SECTION 3
              24

UPPER SURFACE

N: NORMAL ACCELERATION
S: LATERAL ACCELERATION
L: LONGITUDAL ACCELERATION

FUSELAGE AREA

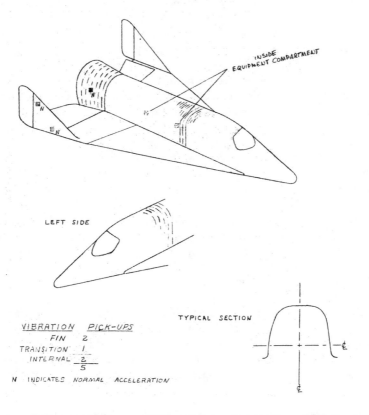

INSIDE EQUIPMENT COMPARTMENT

LEFT SIDE

TYPICAL SECTION

VIBRATION PICK-UPS
FIN   2
TRANSITION   1
INTERNAL   2
             5

N INDICATES NORMAL ACCELERATION

Date: Jan 1962   Title: Memo: Dyna-Soar Test Program

NASA-FRC

January 16, 1962

MEMORANDUM for Head, Manned Flight Control Branch

Subject: Dyna-soar Test Program

References:
(a) FRC Memo for Chief, Research Division, dtd 2-27-59, WHS, ADB, EJS: pm
(b) FRC Memo for DS-1 Files, dtd 2-12-60, WHS: dmo

1. This memorandum was originally prepared on December 17, 1960, for the Head of the Dyna-Soar Branch. Since it may contain a information of use to the Manned Flight Control Branch, it has been retyped for record purpose.

2. Several recent meetings and discussions have been held concerning the number of DS unmanned ground launched flights. As a result of some previous studies (references and (b)) and more recently the work on the corridor action study, the undersigned believes that an answer to this problem can be obtained by considering, in detail, the exact purposes for unmanned flights. Such consideration must include the complete vehicle system development and must take into account the effect of each phase of the program upon any other. It is not the purpose of this memorandum to present such a detailed evaluation, but to outline this viewpoint as an approach to the problem. Although it is not intended to review the purpose or function of the DS, a few comments will be made to provide the background for this memo.

3. It is believed that the DS is of importance to the NASA for the development of a winged entry vehicle designed for manned control only. The basic philosophy concerning the vehicle should be in direct contrast to the Mercury concept for control wherein the complete   mission can be flown without the pilot, in order to provide information on the complementary type of vehicle. The program is very ambitious in that it involves many new operational concepts and in addition, a new vehicle and vehicle-booster configuration mast be developed. It is felt that arguments can be strongly justified, on the basis of experience with ICBM test program, that it may not be possible to complete the development program in 16 flights. Recent experience with the Mercury launches serves to point out the need for more tests than were originally planned for that program. In addition, the number of equipment failures being encountered during the early tests of the X-15 is an indication of the problems to be expected with the DS.

4. It is felt that greater assurance of successfully completing the flight test program can be obtained by considering the DS as one complete test program in which the supporting tests are as important (and probably more important) to the successful completion of the program as the ground launched tests. In order to demonstrate the step 1 mission in 11 or 16 flights, it is critically important for each system to function properly and it is believed that a very extensive supporting test program must be planned. These test would include complete demonstration of all systems, subsystems, and components prior to using in the DS. The number and type of required supporting tests are estimated at an order of magnitude more complex than those for the X-15. With this approach, at the initiation of the ground launch tests, all systems would have been proven to a high degree of reliability and only operational concepts associated with ground launch, boost and corridor exploration should remain unproven.

5. Some light on the question of manned versus unmanned ground launch test program can be obtained by examining the respective risk. Not only risk to the pilot for any particular flight, but risk of failure of the complete program because of malfunctions and unknown factors (which are most certainly going to occur) encountered during the test program. A test program is desired which permits as many new phases of flight as possible to be approached in a gradual manner so that critical conditions can be approached with a

high degree of confidence. If it is agreed that all failures will occur because of unknown factors, then critical problems cannot be accurately evaluated, and hence corrected or avoided, until they have been explored. The missile test philosophy involves no such exploration and the encountering of problems invariably also involves the loss of a vehicle. It is believed that such a test program would involve a higher risk when compared to a manned build-up test program, for a test program involving only 11 or 16 flights. This is also based upon the belief that past history has amply demonstrated that the presence of a pilot greatly enhances the chances for successful completion of flights in the event of malfunctions.

6. Assuming a manned build-up test program, the question of the requirement for unmanned tests can be assessed through a mission analysis to determine areas where there is a high risk to the pilot and it is not considered safe to employ manned flights. Breaking the flight into a boost and glide phase and considering for the moment the glide; for this phase, risk to the pilot can be minimized through the use of a buildup of both velocity and lift type of test program. However, there does not appear to any technique with which the boost phase can be approached in a gradual manner. In addition, many phase of the boost operation will not have been previously tested or demonstrated for the DS booster vehicle configuration. Therefore, it is felt that a number of unmanned flights will be required to explore and demonstrate the boost operation. Again, a mission analysis, including escape system consideration, is required to accurately assess the types of unmanned tests required. It is noted that no tests of the escape system are presently contemplated by AF-BAC. This philosophy contrasts with the extensive demonstration program conducted by STG for the Mercury escape system.

7. A brief examination has indicated that conditions such as the following require unmanned tests: demonstration of fire-in-hole technique, including separation; separation of glider for escape at critical points such as, high q and second stage abort; separation of glider for normal boost trajectories; booster vehicle control and guidance demonstration for different types of boost trajectories; off-the-pad escape demonstration. It should be pointed out that not all of these tests involve the use of a complete Titan-DS configuration, and that none of the above tests require an unmanned vehicle for corridor exploration. However, the low altitude abort-boost recovery condition may impose a requirement for complete vehicle demonstration prior to manned flights. This problem is considered in more detail in reference (b).

8. In conclusion, a program has been proposed which presents one philosophy toward testing. This program can be altered, but the basic elements, i.e., extensive supporting tests, unmanned booster demonstration flights, manned ground launched build-up flights, and escape system demonstration tests, cannot be arbitrarily compromised without also compromising the basic objective of the program which is the development of a manned lifting reentry vehicle, using a new and complex booster-vehicle configuration, in a minimum number of flights.

Wendell H. Stillwell
Aerospace Technologist

WHS: dmo

Date: Feb 1962   **Title:** Visit to Boeing

NASA-FRC

February 23 1962

MEMORANDUM: for Head, Manned Flight Control Branch

Subject: Visit to Boeing Airplane Company to attend Dyna-soar Management review meeting

1. Messrs. A.D. Brown and W.H. Stillwell attended the Dyna-soar Management Review Meeting during the week of February 12, 1961. Since Mr Brown is preparing a separate memorandum on the entire meeting, the undersigned will confine his remarks to his immediate area of responsibility, i.e., control and guidance.

2. General comments:

    (a) Dyna-soar funding has been cut and BAC is attempting to fit the program to the costs. This creates serious problems and has necessitated many deletions of developmental test work.

    (b) The impact of change to the Titan III upon the program has not been evaluated, and it may be 6 months before such information is known. Significant changes may be involved because of acoustical problems, etc.

    (c) BAC plans to go ahead with the air-launch program as presently scheduled, but the ground launch program is expected to be delayed significantly beyond the present August '64 date.

3. Specific problems:

    (a) Gross weight is no 11,504 as compared with a design weight of 11,150 and the previous design weight (Oct. '61) of 10,600.

    (b) BAC is studying the use of elliptical orbit trajectories and vernier retro-rockets for orbital flight rather than hypersonic glide.

    (c) Consideration is being given to a separate reaction control system for the transition section, which will be used during the orbital portion of flight.

    (d) Serious problems have been encountered with the H2-O reaction control system. The problem arises from the low flow available from the liquid hydrogen system and the high demands of the reaction controls at low dynamic pressure which results in the pilot being able to easily demand more reaction control than the system can provide. This demand is so critical that BAC is studying a possible change to an H2O2, UDMH-nitrogen tetroxide, or hydrazine system. These alternatives mean either an increase in weight, cost, and development time or all three. BAC is not working on ways to decrease the demands on the system which appears to me to be a more basic solution if it can be accomplished.

    (e) A decision will be made by April 1 as to whether a drone recovery system will be used. An extensive effort will be required to develop the Sperry system (selected by the AF for use with Dyna-soar) and this must be weighed against the rather low probability of ever using the system to recover and unmanned glider at Edwards.

    (f) A decision will be made in April 1 as to the technique to use for a backup guidance system. Two methods are under study by BAC and SPO. One involves the development of an air-data system for angle of attack and temperature measurement, and in combination with an accelerometer should enable safe reentries to be made from any point during a flight. However, it does not appear that the system will provide any navigational capability and there are many problems associated with the development of the air-data system. The other method would consist of a dual inertial guidance system which has many advantages, but would require additional weight and space and would not provide safety for approximately 25% of the expected IGS failures which are of the type that you cannot tell which of a dual system is correct. This is also a very critical problem and the solution will undoubtedly require additional weight and cost.

Wendell H. Stillwell,
Aerospace Technologist

Date: Mar 1962   Title: Flight Simulated Off the Pad Escape & Landing

NATIONAL AERONAUTICS AND SPACE ADMINISTRATION
TECHNICAL MEMORANDUM X-637
FLIGHT-SIMULATED OFF-THE-PAD ESCAPE AND LANDING MANEUVERS
FOR A VERTICALLY LAUNCHED HYPERSONIC GLIDER

By Gene J. Matranga, William H. Dana, and Neil A. Armstrong

SUMMARY

A series of subsonic maneuvers simulating typical off-the-pad escape and landing procedures for a vertically launched hypersonic glider configuration was flown using a delta-wing airplane having a peak effective lift-drag ratio of 4.7.

None of the required maneuvers posed any particularly difficult or taxing situations for the pilots. Circular, overhead landing patterns flown at 240 knots indicated airspeed were relatively easy to perform and resulted in touchdown longitudinal dispersions of less than ±1,200 feet.

A reduction in the pilot's visibility from the cockpit did not noticeably impair his ability to navigate except when view of the area directly beneath the airplane was required. However, portions of the escape and landing maneuvers were adversely affected by the reduced visibility.

INTRODUCTION

From a safety-of-flight standpoint, take-off and landing are, generally, two of the most critical flight control areas. For a hypersonic glider, which is launched vertically from the ground atop a large booster rocket and is landed unpowered, these areas are particularly critical in the event of an emergency prior to launch. One proposal for providing for survival of the pilot and vehicle if a main booster malfunctions on the pad or shortly after lift-off is to propel the vehicle well away from the danger area by means of an auxiliary booster on the vehicle. If the vehicle is boosted high enough and fast enough, the pilot can right the airplane and land on a nearby runway. The problem of landing the glider, however, could be critical, since the lift-drag ratio of several proposed hypersonic-glide configurations is low (about 4), and the vehicle will land unpowered. To further complicate the landing problem, thermal-structural considerations generally dictate that window areas be minimized, thus limiting the pilot's view from the cockpit.

Low-speed X-15 landing maneuvers were successfully simulated in the study reported in reference 1. Based on this program, a series of analytical and flight-simulation studies is being conducted at the NASA Flight Research Center, Edwards, Calif., to investigate the subsonic, off-the-pad escape and landing maneuvers of a typical vertically launched hypersonic glider. This paper considers the results of a brief flight-test program in which an attempt was made to match predicted escape and landing maneuvers with flight data. Also considered is the effect on the maneuvers performed of limited pilot visibility from the cockpit.

SYMBOLS

an	normal acceleration, g units
C_L	airplane lift coefficient
g	acceleration due to gravity, ft/sec2
h	geometric altitude above touchdown point, ft
(L/D) '	effective lift-drag ratio
t	time, sec
V	true airspeed, ft/sec
Vdot	derivative of airspeed with time, dV/dt, ft/sec2
V_i	indicated airspeed, knots
V_v	vertical velocity, ft/sec
x	longitudinal distance from touchdown point, ft
y	lateral distance from touchdown point, ft
α	trim angle of attack, deg
γ	flight-path angle, deg

AIRPLANE

The test airplane is a single-place, delta-wing fighter-interceptor powered by a turbojet engine equipped with an afterburner. A three-view drawing and a photograph of the airplane are shown in figures I and 2, respectively. The physical characteristics of the airplane are presented in table I. The wing has an aspect ratio of 2.02 and varies in thickness from 5 percent at the root to 3 percent at the tip. The average wing loading during the tests was 36 lb/sq ft. Speed brakes located on the upper and lower surfaces of the wings were used in conjunction with the landing gear to provide the additional drag required for this investigation.

To further reduce the effective lift-drag ratio, the throttle was modified so that when idle power was selected the afterburner nozzle was forced to the full-open position. This reduced the idle thrust to slightly less than 200 pounds, compared with a normal idle thrust of about 500 pounds.

Longitudinal and lateral control of the airplane are provided by elevons located on the trailing edge of the wing. Directional control is provided by a conventional rudder. Two completely independent hydraulic systems operate the outboard elevons. The inboard elevons are electrically slaved to the outboard elevons and are actuated by electrohydraulic valves. Longitudinal-control forces are supplied artificially by a bungee and bobweight combination and are programmed as a function of Mach number. Lateral-control forces are supplied artificially by a bungee. The rudder is operated by a hydraulically powered system providing no external force feedback. Pedal forces are supplied artificially with a bungee. No artificial damping was provided during any of the maneuvers performed in this investigation.

INSTRUMENTATION

No internal recording instruments were used during the tests. All flight data presented were obtained from Air Force Flight Test Center Askania cinetheodolite cameras operating at I frame per second. Three-station solutions of these data determined the airplane position in space at any time. By differentiating the position data, forward velocity and vertical velocity were obtained. From a knowledge of airplane forward velocity, altitude, and wind conditions, indicated airspeeds were determined.

TESTS

Before attempting any simulated landing or off-the-pad escape maneuvers, a series of constant-speed, wings-level glides was performed to ascertain the lift-drag-ratio variation with airspeed. Tests were made at several engine power settings with only the gear extended and with the gear and the speed brakes extended. For most of the maneuvers discussed, the configuration consisted of gear and speed brakes in he extended position and engine at idle power with the afterburner nozzle open.

The escape maneuvers were entered into by executing a high-speed run about 1,000 feet above the ground n the clean configuration. At a predetermined point, the pilot performed a pull-up. At the vertical attitude position, engine power was reduced to idle and the speed brakes were extended. This position corresponds to glider auxiliary-booster rocket burnout and is where the simulation begins. As the pull-over continued, the gear was extended at the gear-extension-limit speed (260 KIAS). When the inverted, horizontal position was attained, the pilot rolled the vehicle to the erect, level attitude, accelerated to approach speed, and landed. Figure 3 aids in visualizing the relation between this maneuver and the escape maneuver of a vertically launched vehicle. This perspective sketch shows that the two trajectories merge when the test airplane reaches the vertical attitude and when the glider's booster rocket burns out; succeeding portions of the trajectories are coincident. It should be noted that two different speeds were utilized in the high-speed run, resulting in different speed and altitude combinations at the vertical attitude, thereby simulating a variety of auxiliary-booster-rocket capabilities.

To simulate the approach and landing maneuvers, a series of 360°-spiral, overhead landing patterns was performed at speeds from 180 KIAS to 290 KIAS and bank angles from 30° to 60°. Several straight in landing patterns were also performed at about 240 KIAS. All landings executed by the five participating pilots were made on the lake bed of Rogers Dry Lake at Edwards Air Force Base, Calif.

During most of the tests, the airplane canopy was fitted with an amber Plexiglas mask cut out to provide the pilot with a field of vision comparable to that of a currently proposed boost-glide vehicle. The pilot, with a blue visor in place, could see only through the cut-out portions of the mask. With the visor raised, the pilot could utilize the full field of view of the test aircraft.

RESULTS AND DISCUSSION

For convenience of presentation, the results of the various phases of this investigation are treated individually in the following discussion. First, the ranges of lift-drag ratios and angles of attack utilized are noted and compared with corresponding values anticipated for some hypersonic-glider vehicles. Succeeding subsections consider the off-the-pad escape maneuver; the landing, both following escape and during normal operations; the flare and touchdown; and the pilot's evaluation of the influence of cockpit visibility on his performance of the various maneuvers. Because of the lack of on-board instrumentation and the importance of qualitative evaluation in the critical flight areas being studied, pilot comments are relied upon heavily throughout this paper.

Glider Performance

The variations of angle of attack and effective lift-drag ratio with lift coefficient for the test airplane with the gear and the speed brakes extended, the engine at idle power setting, and the afterburner nozzle open are presented in figure 4. The angle-of-attack data were obtained from unpublished reports of the manufacturer's flight tests. The effective lift-drag ratios were determined from data obtained during constant-speed glides from an altitude of 20,000 feet to 5,000 feet by utilizing the forward speed and the rate of descent to determine the glide angle γ. Also, the rate of change of true airspeed was readily calculated, since the pilot flew a constant indicated airspeed during the approaches. These quantities were then combined in the following equation to obtain the effective lift-drag ratio

$$(L/D)' = \frac{g}{g \tan \gamma - Vdot}$$

The average lift coefficient was determined as a function of wing loading and the average dynamic pressure n the glides.

The faired data of figure 4 show that the peak lift-drag ratio of 4.7 occurred at a lift coefficient of 0.38 and an angle of attack of about 10°.

The trim angle-of-attack and effective lift-drag-ratio data from figure 4 are presented in figure 5 as a function of indicated airspeed at a wing loading of 36 lb/sq ft. For comparison, similar data for a proposed boost-glide vehicle at a wing loading of 28 lb/sq ft are included. The data for the two configurations are in good agreement, which indicates that the test airplane should closely simulate the performance of the proposed vehicle. In fact, the comparison is better than achieved in the successful simulation of the X-15 airplane discussed in reference 1.

Off-the-Pad Escape Maneuver

More than 40 simulated off-the-pad escape maneuvers were accomplished during this phase of the study. The maneuver is preceded by a low-altitude, high-speed run followed by a pull-up to a vertical attitude. At this point, power is reduced to idle and the speed brakes are extended. The pull-over is continued, and the landing gear is lowered at approximately 260 KIAS. When a horizontal, but inverted, attitude is reached, the pilot rolls to the erect, level attitude and accelerates to the proper approach speed.

Two sets of conditions exemplifying typical auxiliary-boost-rocket capabilities were considered. Figure 6 presents a typical trajectory of a high-energy escape and landing maneuver, and figure 7 presents a time history of the escape phase only. From an initial airspeed of about 500 KIAS during the low-level run, a 3.59 pull-up is performed. The vertical attitude is reached at an altitude of about 10,000 feet with a speed of less than 400 KIAS. With the pull-over continued at a reduced normal acceleration (approx. 2g), the peak altitude of 14,300 feet was realized with a speed of 172 KIAS about 32 seconds after the initiation of the maneuver. The approach pattern was flown at
240 KIAS with an average bank angle of 30°.

The trajectory and time history of a low-energy maneuver are presented in figures 8 and 9, respectively. With an initial speed of about 400 KIAS and a 4.5g pull-up, the airplane reaches the vertical attitude at an altitude of about 5,000 feet with a speed of about 320 KIAS. Normal acceleration is again reduced to about 2g, and the airplane goes "over the top" at an altitude of about 8,000 feet and an airspeed of 150 KIAS. Although the approach pattern shown in figure 8 is flown at 240 KIAS, it is considered to be a much tighter pattern than that of figure 6 because of the lower initial altitude on the downwind leg of the pattern. There is also much less margin for error, since little excess altitude is available anywhere around the pattern.

The following tabulation summarizes the average conditions for the more than 40 maneuvers performed;

Entry V_i, knots	h, ft	Pull-up a_n, g	Simulated burnout point V_i, knots	h, ft	Pull-over a_n, g	Over-the-top V_i, knots	h, ft
525	1,000	3.5	400	9,500	$=2.0$	190	15,000
400	1,000	4.5	325	5,000	$=2.0$	155	8,000

The pilots reported that the escape maneuvers, as performed in this investigation, were not particularly difficult or taxing and showed no significant difference in handling characteristics between the two types of maneuvers flown. This opinion is based upon the ability of the pilot to place the airplane on the downwind leg of the pattern at a more precise position and energy level in the escape maneuvers than is normally attained in power-off approaches.

Landing Patterns

In addition to the landings performed after each escape maneuver, a series of power-off landing approaches was performed over a speed range from 180 KIAS to 290 KIAS. Straight-in approaches as well as 360°-spiral, overhead patterns using from 30° to 60° of bank angle were made.

In these approaches it is most significant that the pilot was consistently able to position the aircraft at the approach end of the runway at the proper landing speed. Also, all five participating pilots agreed that a ircular, overhead pattern similar to that shown in figure 10 was easiest and most comfortable because it fforded a proper balance of excess energy without an excessively large rate of descent. This pattern (fig. 0), flown at 240 KIAS with a bank angle between 30° and 40°, is entered at a high-key altitude of about 5,000 feet. The average radius of turn was about 7,000 feet and resulted in a downwind-leg altitude of ,000 feet and a base-leg altitude of 3,500 feet. The average rate of sink in this pattern was about 120 :/sec, and the peak rate was about 140 ft/sec.

1 particular, these landing tests showed that, with an altitude of 7,000 feet and a lateral displacement of 2.5 autical miles on the downwind leg of the pattern (as with the lower-energy off-the-pad escape 1aneuvers), some indications of the limitations of the pattern could be determined. A velocity of 240 KIAS /as flown in the patterns so that sufficient energy would be available to successfully execute the flare and ouchdown. The combinations of geographic boundary conditions and minimum flying speed required that a ank angle of about 30° be utilized. Any further increase in speed would require increased bank angles with educed transit times and increased pilot response, thereby making the maneuver more taxing for the pilot.

s in the tests of reference 2, the pilots found that the slower approach speeds and sink rates associated vith a wing loading of 30 lb/sq ft to 40 lb/sq ft resulted in more comfortable patterns than experienced in imilar tests (ref. 3) using a vehicle with a wing loading near 80 lb/sq ft.

lare and Touchdown

"he advantages of low wing loading are particularly evident in the execution of the flare. The slow pproach speeds and low rates of descent available with a low-wing-loading vehicle allow the pilot to delay he flare to a lower range of altitudes where he can more accurately judge his position. The change of flight ath during the flare was definite and the rate of loss of airspeed during the flare was small enough to allow he pilot to take corrective action without decelerating to a dangerously low speed.

"he time history of a typical flare maneuver is presented in figure 11. The flare-initiation speed was 235 .IAS, and the altitude was about 500 feet. At the completion of the flare, the speed diminished to 205 .IAS. At this point, approximately 10 seconds remained for the final glide and deceleration to touchdown t a speed of 174 KIAS. During this final floating phase, no additional drag could be added to the test irplane to simulate the gear extension on the hypersonic vehicle. However, even if average speed bleed-off f 3 knots/second to 4 knots/second could have been doubled in these tests, the test airplane would be in a lass with the X-15 (see ref. 1), and the touchdown maneuver still would not be critical.

"he average touchdown speed during these tests was about 170 KIAS. The rate of sink at touchdown was 2ss than 3 ft/sec and averaged about 2 ft/sec. Touchdown longitudinal dispersions ranged between ±1,200 2et of the intended touchdown point, which is similar to that measured with the X-15 airplane (ref. 1).

ilot Vision-Field Simulation

During a portion of these maneuvers, the test-airplane canopy was restricted to give the pilot a field of iew comparable to that anticipated for a currently proposed boost-glide vehicle. Figure 12 is an illustration f the airplane canopy fitted with two different amber Plexiglas masks. With a blue visor in place, the pilot ould see only through cut-out portions of the mask; however, with the visor raised, the pilot was free to tilize the view field normally available from the test aircraft.

"he restriction of vision did not noticeably reduce the pilot's high-altitude-navigation capability except vhen it was necessary to locate some geographical landmark directly beneath the aircraft. The restricted ision did cause considerable difficulty because of the lack of a horizon reference during the portion of the scape maneuver from the vertical-attitude position to the point where the horizon reappeared. This egment of 3 to 4 seconds extended through approximately 45° of rotation in pitch. Once the horizon ppeared during the pull-over, the restricted vision posed no added hardship, however. It was noticed that he amber Plexiglas without the visor lowered was sufficiently restrictive to cause a noticeable eterioration in the quality of the maneuver.

Location of the high-key point was difficult with the restricted visibility, since the vehicle passed directly over this point. Other difficulties were encountered on the downwind leg of the approach pattern where lateral vision was insufficient, and during the 135° to 45° segment prior to rollout of the tight pattern where a view through the masked corner of the canopy was desired. Otherwise, vision was adequate after the aircraft had arrived within 45° of the runway heading on the final approach and remained adequate throughout the flare and landing.

CONCLUSIONS

A delta-wing airplane having a maximum effective lift-drag ratio of 4.7 was used to perform a series of subsonic, flight-simulated off the-pad escape and landing maneuvers of a vertically launched hypersonic glider. From this study the following conclusions can be made:

1. The off-the-pad escape maneuvers were not considered difficult or taxing by the pilot and were such that the pilot could position the airplane on the downwind leg more precisely than he could in normal power-off approaches.

2. After performing a series of straight-in and circular, overhead approaches, the pilots concluded that a circular pattern flown at 240 knots indicated airspeed was most desirable.

3. The flare maneuver was easy to judge and control. The touchdown longitudinal dispersions could be kept within ±1,200 feet without exceeding a touchdown rate of descent of 3 feet per second.

4. A reduction in the pilot's visibility from the cockpit did not noticeably decrease his high-altitude navigation capability except when it was necessary to observe the terrain directly beneath the aircraft. However, portions of the off-the-pad escape maneuver and landing approaches were adversely affected.

Flight Research Center,
National Aeronautics and Space Administration, Edwards, Calif., January 22, 1962.

REFERENCES
1. Matranga, Gene J.: Analysis of X-15 Landing Approach and Flare Characteristics Determined From the First 30 Flights. NASA TN D-1057, 1961.
2. Matranga, Gene J., and Menard, Joseph A.: Approach and Landing Investigation at Lift-Drag Ratios of 3 to 4 Utilizing a Delta-Wing Interceptor Airplane. NASA TM X-125, 1959.
3. Matranga, Gene J., and Armstrong, Neil A.: Approach and Landing Investigation at Lift-Drag Ratios of 2 to 4 Utilizing a Straight-Wing Fighter Airplane. NASA TM X-31, 1959.

TABLE I.- PHYSICAL CHARACTERISTICS OF THE TEST AIRPLANE

Wing:

Airfoil section, root	NACA 0005-1.1-30-6° (Modified)
Airfoil section, tip	NACA 0003-1.1-30-6° (Modified)
Area, sq ft	557
Span, ft	33.50
Mean aerodynamic chord, ft	18.25
Root chord, ft	25.08
Tip chord, ft	8.33
Aspect ratio	2.02
Taper ratio	.33
Sweep at leading edge, deg	52.50
Sweep at quarter chord, deg	46.50
Sweep at trailing edge, deg	16.50
Incidence, deg	0
Dihedral, deg	0
Geometric twist, deg	0

Outboard elevon:

Area (per side), sq ft	24.26

Span (normal to fuselage reference line), ft	11.73
Mean aerodynamic chord, ft	2.04
Maximum deflection, up, deg	40
Maximum deflection, down, deg	20
Inboard elevon:	
Area (per side), sq ft	9.04
Span (normal to fuselage reference line), ft	2.58
Mean aerodynamic chord, ft	3.75
Maximum deflection, up, deg	30
Maximum deflection, down, deg	5
Slat:	
Area (per side), sq ft	7.96
Span, ft	4.56
Mean aerodynamic chord, ft	1.10
Slat chord/wing chord	13
Vertical tail:	
Airfoil section, root	NACA 0005-1.1-25-6° (Modified)
Airfoil section, tip	NACA 0003.2-1.1-50-6° (Modified)
Area, sq ft	69.87
Span, ft	9.46
Mean aerodynamic chord, ft	7.85
Aspect ratio	1.28
Taper ratio	.46
Sweepback of quarter chord, deg	48.22
Rudder:	
Area, sq ft	9.29
Span (normal to fuselage reference line), ft	6.26
Mean aerodynamic chord, ft	1.23
Upper-wing speed brakes:	
Area (per side), sq ft	3.26
Span, ft	2.38
Maximum deflection, deg	45
Lower-wing speed brakes:	
Area (per side), sq ft	3.26
Span, ft	2.38
Maximum deflection, deg	60
Fuselage:	
Frontal area, sq ft	18.70
Length, ft	53.80
Fineness ratio	7.86
Wetted area, sq ft	466
Test center-of-gravity location, percent mean aerodynamic chord	23
Weight:	
Gross, lb	26,100
Empty, lb	17,100

Neil Armstrong Interview 2003.

Could you describe what was involved in your position as Pilot/Consultant?

The pilot consultant group (3 Air Force, 3 NASA) served the functions of representing operational and pilot opinion positions to the many design and development meetings at the Boeing plant and elsewhere. The pilot consultants served as pilots for the developmental simulators.

Could you explain how the F-102A and the F5D were used to simulate X-20 conditions and perhaps explain the abort procedure you devised for X-20?

The F-102A may have been used for Dyna-Soar landing approach studies, but I do not recall them.

In the case of a launch pad malfunction, the Dyna-Soar glider could be accelerated vertically to sufficient altitude and velocity to fly back to the landing strip that existed at Cape Canaveral. I believed we might be able to simulate that maneuver in free flight. The Douglas F5D had the ability to reproduce the lift/drag characteristics of the X-20, primarily because it had a high (306 Kt.) landing gear limit speed.

I devised a technique for obtaining a flight condition that would duplicate the vertical attitude and velocity at the termination of the escape rocket boost. At that point, the F5D was configured to match the X-20 flight performance and try to find an acceptable trajectory to make a safe landing at the skid strip.

Bill Dana continued the work after I transferred to Houston. The results of the program are documented in "Flight Simulated Off-the-Pad Escape and Landing Maneuvers for a Vertically Launched Hypersonic Glider", NASA TM X-637, 1962

Did that abort procedure accommodate different launch vehicles and does a variant survive today in the Shuttle's turn-around procedure?

The technique was developed specifically and uniquely for the X-20A. I do not believe it has much similarity to the Shuttle procedures.

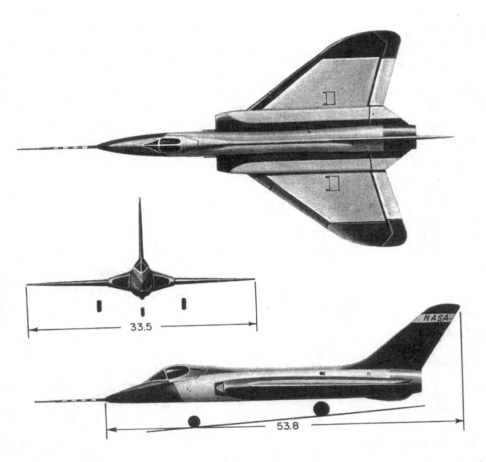

Figure 1.- Three-view drawing of the test airplane. All dimensions in feet.

Figure 2.- Photograph of the test airplane.

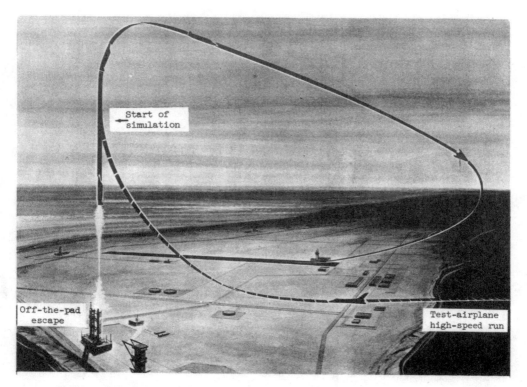

Figure 3.- Perspective drawing of test-airplane maneuver superimposed on hypersonic glider
off-the-pad escape and landing maneuver.

E-7598

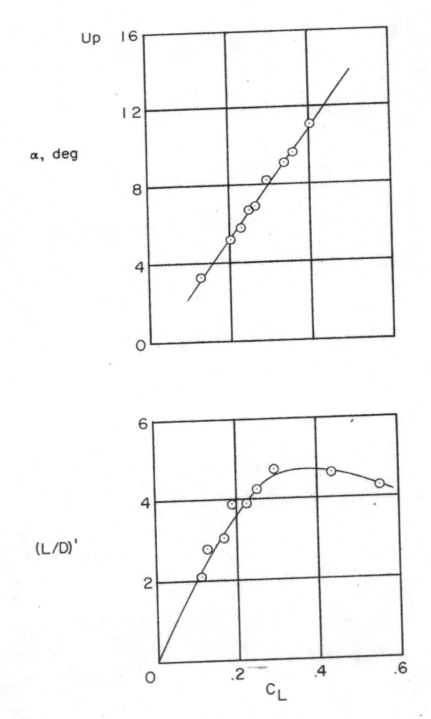

Figure 4.- Trim angle of attack and effective lift-drag ratio as a function of lift coefficient for the test airplane. Gear and speed brakes extended; engine at idle power with afterburner nozzle open.

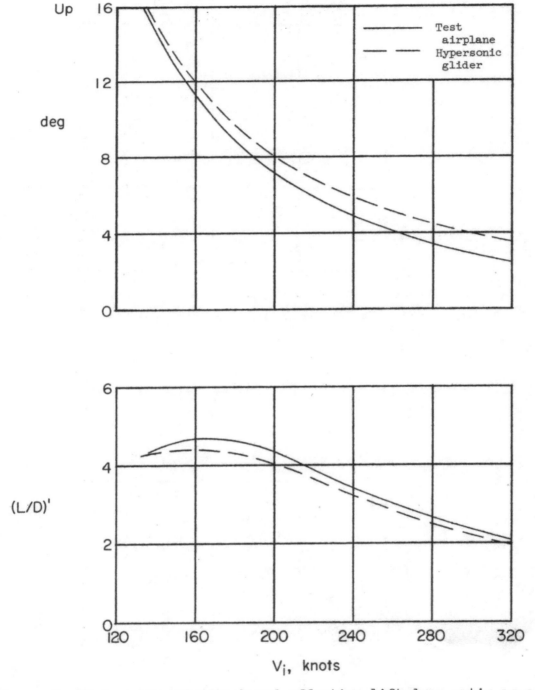

Figure 5.- Trim angle of attack and effective lift-drag ratio as a
function of indicated airspeed for the test airplane with a wing
loading of 36 lb/sq ft and the boost-glide vehicle with a wing
loading of 28 lb/sq ft.

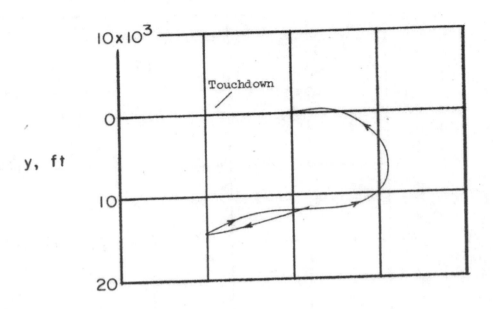

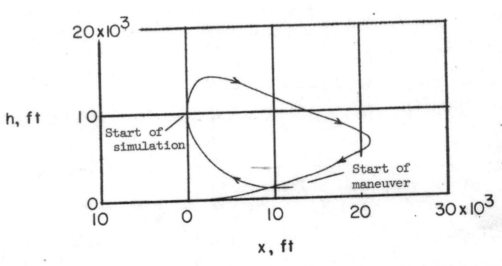

Figure 6.- Trajectory of typical high-energy off-the-pad escape and landing maneuver. Initial $V_i \approx 500$ KIAS; glide $V_i \approx 240$ KIAS.

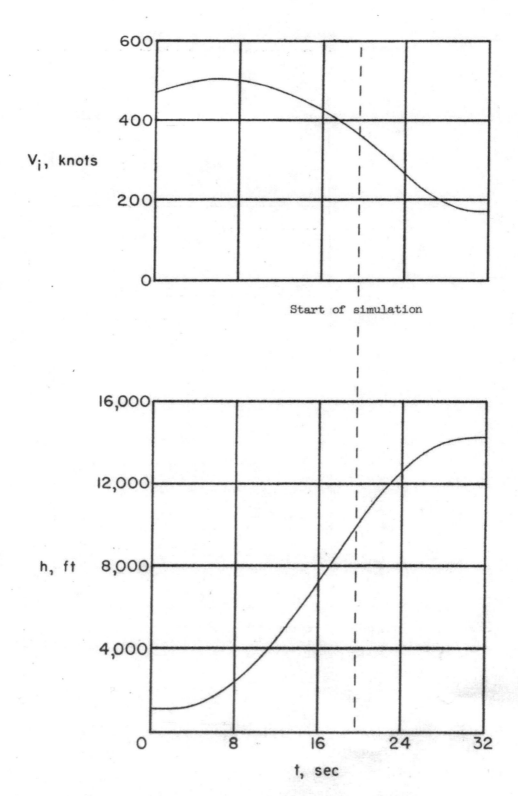

Figure 7.- Time history of typical high-energy off-the-pad escape maneuver.

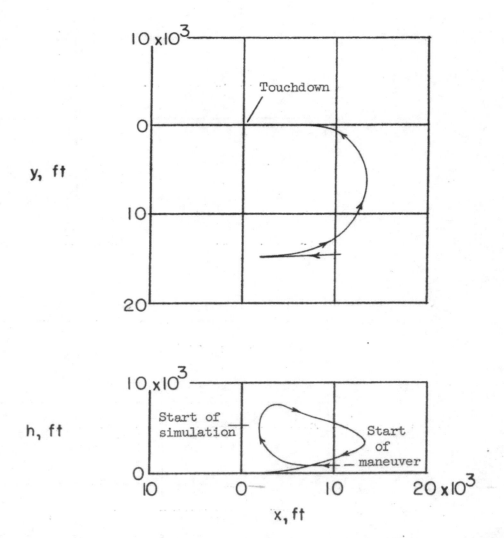

Figure 8.- Trajectory of typical low-energy off-the-pad escape and landing maneuver. Initial $V_i \approx 400$ KIAS; glide $V_i \approx 240$ KIAS.

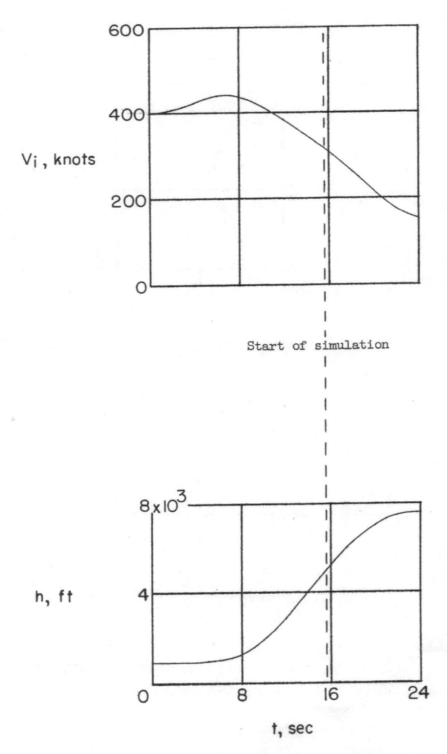

Figure 9.- Time history of typical low-energy off-the-pad escape maneuver.

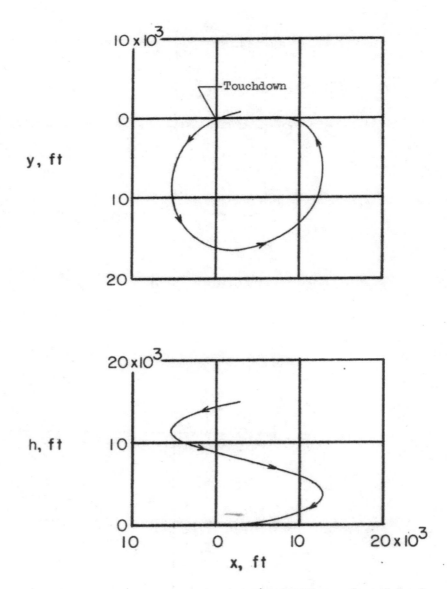

Figure 10.- Typical landing pattern. $V_i \approx 240$ KIAS; angle of bank $\approx 30°$; gear and speed brakes extended and engine at idle power with after-burner nozzle open.

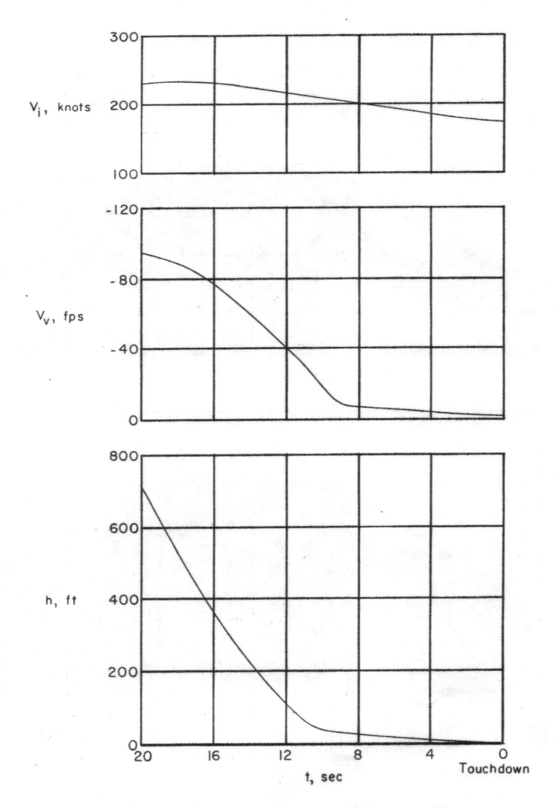

Figure 11.- Time history of typical flare maneuver.

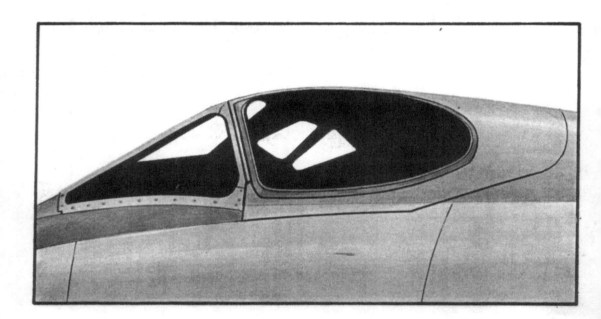

Figure 12.- Illustration of airplane canopy fitted with two different
amber Plexiglas masks.

Date: Mar 1962   Title: Press Release: Pilot Assignments

Press Release:

Thurs., Mar 15, 1962

Six Pilots Assigned To Dyna-Soar

DAYTON, Ohio – Four Air Force and two National Aeronautics and Space Agency test pilots have been assigned to the Dyna-Soar as pilot-engineer consultants, the Air Force announced today.

Two of the pilot-engineers are from Ohio. They are Air Force Capt. William J. Knight, 32, of Mansfield and Neil A. Armstrong, 31, of Wapakoneta, a NASA pilot.

Others selected are Maj, James W. Wood, 37, of Pueblo, Colo.; Cast. Henry C. Gordon, 36, of Gary, Ind., and Capt. Russell L. Rogers, 33, of Phoenix, Ariz., all Air Force officers. The other NASA pilot is Milton O. Thompson, 36, of Crookston, Minn.

Whether these men will actually test fly the Dyna-Soar spacecraft will be decided as the program develops, a spokesman for the Air Force aeronautical systems division at Wright-Patterson Air Force Base said.

The pilots are stationed at the Air Force flight test Center, Edwards AFB, Calif., and have been working as pilot-engineer consultants for the Boeing Co., Seattle, Wash. Boeing is system contractor for the manned orbital space glider.

The pilots also are working with the Minneapolis-Honeywell Regulations Co. of Minneapolis, contractor for Dyna-Soar flight control system development.

The Air Force announcement termed Dyna-Soar the most advanced piloted aerospace research system now under development. It is a winged space glider that will be boosted into orbital flight by a Titan launch vehicle from Cape Canaveral, Fla.

Date: Spring 1962   **Title:** Press Release and Chronology

Press Release and Chronology

<h2 style="text-align:center">THIS IS DYNA-SOAR</h2>

Men have flown to the edge of space in the X-15 rocket plane and have been launched like missiles into space in capsules.

In the case of the capsule, men travel through space with the speed of a ballistic missile, but must rely on parachutes to let them down gently, and on air-sea recovery units to get them home again.

Pilots of the X-15 rocket plane and other advanced research craft fly back to conventional landings on earth, but, to date, the speed and altitude they attain are not as impressive as the capsules.

An attempt to combine the best features of these two approaches – high speed flight into space and return with airplane like control – is the goal of the Air Force's Dyna-Soar program.

First objective of the program is to send an earth-launched, manned vehicle into space and, through a controlled re-entry into the atmosphere, bring it back to a conventional landing on earth.

The craft being developed for the job is a delta-winged glider. It is being manufactured by the Aero-Space Division of the Boeing Company in Seattle, Washington. It will look and behave more like an airplane than any space craft now built.

Rocketed into space by a powerful booster, the glider will be able to orbit the earth at speeds of more than 17,000 miles an hour. When the pilot is ready to return to earth, he will be able to fly his craft back into the atmosphere and band it at an airfield of his choice.

Lt. Gen. Roscoe C. Wilson, former Air Force deputy chief of staff for development, called Dyna-Soar the most important research and development project the Air Force has.

<h2 style="text-align:center">HOW WILL IT FLY?</h2>

The term Dyna-Soar is derived from "dynamic" and "soaring." It means that the vehicle will use both centrifugal force and aerodynamic lift.

Centrifugal force will sustain the glider when it attains orbital speed (about 18,000 miles an hour). At this speed, it will be flying just fast enough to offset the pull of the earth's gravity. The glider will remain is orbit like a satellite until the pilot decides to return. By firing small gas jets mounted on the glider, the pilot will be able to control the glider's attitude in space. Retro-rockets may be used to direct the craft out of its orbit and back iota the atmosphere.

The glider will enter the earth's atmosphere in a single long glide, contrary to the "skip-glide" technique first proposed by Dr. Eugen Sänger, originator of the Dyna-Soar concept (see History of Dyna-Soar, page 10).

The craft's wings will give it aerodynamic "lift" and maneuverability as it descends through the atmosphere. This combination of high speed, extreme altitude and maneuverability will permit the pilot to shorten or lengthen his range by thousands of miles and to maneuver far to the left or right of his flight path to reach his landing site. landing the Dyna-Soar glider should be no more complicated than landing the X-15 or a modern jet fighter.

DYNA-SOAR HAS UNUSUAL LANDING GEAR

The landing gear on Dyna-Soar, however, will be different than airplanes now in use. The Goodyear Tire and Rubber Company is developing main leading gear skids for the glider which will lack like wire brushes mounted on skis. Work on the unusual leading gear is being performed at Goodyear's Akron, Ohio, plant under a $45,000 subcontract to Boeing.

Dyna-Soar will not be equipped with brakes; the wire brushes and the friction they will create upon landing will bring the craft to a stop.

Bendix Corporation's Bendix Products Division of South Bend, Indiana, is developing a retractable nose gear for Dyna-Soar which will resemble a shallow kitchen dishpan. Work is being performed under a $75,000 subcontract to Boeing.

Searing temperatures which will be encountered by Dyna-Soar when it re-enters the earth's atmosphere rule out the use of rubber tires and lubricated bearings.

DYNA-SOAR WILL ENCOUNTER SEVERE HEAT

Parts of the Dyna-Soar surface will be heated in varying degrees from 2,000 to 4,000 degrees Fahrenheit when it glides trough the atmosphere on its way back from space. Its pilot, however, will remain comfortable in a cockpit kept at room temperature.

The air is front of the glider – the so-called stagnation point – will heat up to 20,000 degrees or more. This super-hot air, or "plasma," is expected to behave differently than air is its normal form.

A plasma – sometimes referred to as the fourth state of matter – is a good conductor of electricity and its flow is affected by a magnetic field. Air in its normal state has neither of these properties. Preliminary studies indicate that it will be very difficult to communicate through this plasma. It acts as a shield to radio waves.

Boeing scientists and others are experimenting with various techniques in an effort to solve this problem before the first manned Dyna-Soar is launched into space. A report by Dr. James E. Drummond of the Boeing Scientific Research Laboratories early in 1961 indicated that low frequency waves of electrically charged particles called "ions" can be used to bore holes in the plasma through which radio waves can travel. Other studies offer hope that certain very high radio frequencies may hold the secret to effective communication.

The plasma "sheath," flowing back over the craft as it re-enters the atmosphere, will afford a spectacular sight. It will look much like a shooting star blazing across the sky.

The Dyna-Soar glider will be constructed of high-nickel-alloy steel, molybdenum or columbium end ceramic materials highly resistant to heat. Unlike nose cones of ICBMs, which are coated with an ablative material which can boil off, the Dyna-Soar glider will radiate heat from its surfaces back into the atmosphere.

The ICBM nose cone plunges back into the atmosphere in a matter of seconds and must endure much higher temperatures, although for a relatively short time. The Dyna-Soar glider will come back in a more leisurely manner and will take a longer time (upwards of 30 minutes) to dissipate the heat.

Even this type of re-entry will scorch the surface of the glider until it looks like an old-fashioned wood stove, but it will be a simple task to prepare the glider for relaunching.

Much has been written of the high "G" loads which the passengers of Mercury-type capsules encounter when they fall back into the atmosphere in a deep, ballistic re-entry (up to 10 or 11 times the normal pull of gravity). Because of his shallow glide re-entry, however, the Dyna-Soar pilot will have to contend with no higher gravity loads than the pilot of a commercial jet airliner.

DYNA-SOAR WIND TUNNEL PROGRAM BIGGEST EVER

The most exhaustive wind tunnel program in the history of flight has been conducted on the Dyna-Soar space glider. By the time it is completed, it will triple the total time spent in similar data-gathering tests on the X-15.

Even the eight-jet Boeing B-52 global bomber, under whose wing the Dyna-Soar glider will nestle in early ,flight tests, required only half as much wind tunnel time.

Goal of the big effort is to gather information helpful in building a vehicle which will have to fly in every speed range from landing speed to orbital velocity.

The fact that the Dyna-Soar glider will be rocketed into space by a multi-stage booster and will cover a wide range of   speeds makes the extensive wind tunnel program necessary.

Every conceivable combination of the glider and its booster must be tested at various speeds – the glider atop the complete. booster, the glider and the booster after the first stage of the booster has fallen away, the glider in flight without the booster, and so on.

Wind tunnels used on Dyna-Soar include subsonic and transonic tunnels (from low speeds up to mach 1.4), supersonic tunnels (from mach 1.5 to mach 5.5), low hypersonic tunnels (from mach 6 to mach 10), and high hypersonic tunnels (from mach 12 to mach 25).

Even gentle breezes – simulating off-shore winds at Cape Canaveral – are directed at models of the Dyna-Soar glider-booster combination to determine how it will react while standing on a launch pad.

Of particular importance to the Dyna-Soar program are the so-called "hot shot" and shock tube tunnels which simulate speeds up to mach 20 and beyond. Although ICBM nose cones and other ballistic shapes have generated considerable Information in this speed range, Dyna-Soar is the first attempt to build a winged vehicle which will survive intact the blistering hypersonic speeds.

Started in earnest in early 1958, wind tunnel tests on Dyna-Soar models have been gathering data to answer questions of performance, stability and control, aerodynamic and structural heating loads and the like.

Virtually every mayor wind tunnel facility is the United States has contributed to the development of Dyna-Soar. Included in the wind tunnel program are facilities at Boeing, AVCO, Cornell Aeronautical Laboratory, General Electric, Martin-Marietta, the University of Washington. Ohio State University, and the University of Southern California.

Also included are the Air Force's Arnold Engineering Development Center cad the National Aeronautics and Space Administration's Ames Research Center, Langley Research Center and Jet Propulsion Laboratory. Altogether, about 30 wind tunnels and shock tubes have been involved.

EARLY BENEFIT SEEN BY AIR FORCE, N.A.S.A.

Both the Air Force and the National Aeronautics and Space Administration (which is participating in the technical development of the program) expect to accumulate valuable data from early tests of Dyna-Soar. The program will help determine what military uses of space are feasible and will aid NASA in space research.

As a test bed, the vehicle will furnish the opportunity to test military subsystems under actual space conditions and to determine the capability of man to operate them.

Early Dyna-Soar flights will be made at more than twenty times the speed of sound and will last for more than an hour. They will provide a means of conducting research and development tests in a true flight environment. Compared with the brief glimpse now available with the free-flight testing of scale models or parts mounted on rockets or nose cones, this is a long, leisurely look at the mysteries of space flight.

It is not possible at the present time to simulate simultaneously all of the environmental conditions of hypersonic and space flight merely by using ground facilities. Communication experiments with Dyna-Soar in actual hypersonic flight – just one of many tests planned – will contribute to the understanding of the problem of sending and receiving radio signals through the "plasma sheath."

PRINCIPAL ROLES IN DEVELOPMENT

Boeing, as system contractor for Dyna-Soar, is responsible for the manufacture of the glider. Under direction of the Air Force's Aeronautical Systems Division, Boeing also is responsible for tying in the vehicle subsystems, integrating vehicle and booster, and assembly and test.

Air Force management and engineering personnel of the Air Force Systems Command specifically assigned to the project at Wright-Patterson Air Force Base, Ohio, are managing the design work on Dyna-Soar.

The Space Systems Division is concerned with booster development.

The National Aeronautics and Space Administration is participating in the technical development of the program.

Both Air Force and NASA test pilots have been working with Boeing engineers since December, 1960, as consultants on certain design features of Dyna-Soar.

Associate contractors at work on the Dyna-Soar program include Martin-Marietta, suppliers of the Titan boosters which will rocket Dyna-Soar gliders into orbit; Radio Corporation of America, the communications system; and Minneapolis-Honeywell, guidance and secondary attitude reference.

BOEING'S MAJOR SUBCONTRACTORS

Boeing will spend more than $50 million on major subcontract items for the current Dyna-Soar program. Seven major subcontractors are involved. Their names and jobs are:

LING-TEMCO-VOUGHT, Dallas, Texas – nose cap for glider. Made of ultra-high temperature ceramic materials, the nose cap will protect the forward section of the glider from searing re-entry temperatures.

ELECTRO-MECHANICAL RESEARCH, INC., Sarasota, Florida – test instrumentation subsystem. Airborne equipment will be provided for collecting and transmitting Dyna-Soar test data to the ground. Also developed will be ground equipment for receiving, displaying, recording and processing the data.

GARRETT CORPORATION's AiResearch Manufacturing Division, Los Angeles, California – Dyna-Soar's hydrogen cooling system, a vital part of the vehicle's environmental control system. Hydrogen expelled from a hydrogen storage tank will enter a heat exchanger where it will absorb the heat extracted from the crew and equipment compartments.

MINNEAPOLIS-HONEYWELL's Aeronautical Division. Minneapolis, Minnesota – flight control electronics subsystem. This portion of the flight control system includes the electronic equipment necessary to achieve control of the glider through the use of automatic or manual commands.

SUNDSTRAND CORPORATION, Denver, Colorado – accessory power unit. Designed to power the vehicle's generator in flight, the gaseous hydrogen-oxygen unit will consist of a reaction chamber, prime mover, gear box, hydraulic pump, propellant shut-off valve, and metering valves and controls.

THIOKOL CHEMICAL CORPORATION, Elkton, Maryland – solid fuel acceleration rocket to be used either as escape rocket in case of emergency during launch or as a small booster rocket for additional acceleration after the last stage of Dyna-Soar's booster is expended.

WESTINGHOUSE ELECTRIC CORPORATION, Lima, Ohio – generator and control unit. Mounted on the accessory power unit, this will be the source of the glider's electrical power.

HISTORY OF DYNA-SOAR

In 1933, James R. Wedell flashed to a new world's speed record for land planes by averaging 305.33 miles an hour in the Phillips Trophy Race, and Lt. Cdr. Frank M. Hawks set a west-east non-stop record by flying from Los Angeles to Brooklyn's Floyd Bennett Field is 13 hours 26 minutes. News accounts of these deeds were carried in virtually every daily newspaper is the country.

That same year, at the University of Vienna, an obscure engineer and physicist wrote a book entitled, "The Technique of Rocket Flight.," in which he introduced the idea of a rocket airplane which would fly 50 times faster than Wedell's airplane and travel more than 10,000 miles beyond Hawks' non-stop mark. The book, by Dr. Eugen Sänger, caused scarcely a ripple of interest beyond a limited technical readership.

Three years later, the records of Wedell and Hawks had been erased and new aviation milestones added. By then, however, Sänger's work began to cause a stir. He was invited to Germany to continue his work under the auspices of the Hermann Goring Institute, the research organization of the Luftwaffe.

Sänger was assigned to a 10-year program for the purpose of developing his ideas on the long-range rocket aircraft. A specially built research center at Trauen was the site of his studies.

In formulating the design of his pioneer boost-glider, Sänger was joined by Dr. Irene Bredt, a brilliant mathematician who later became his wife. With a small team of technicians, various features of the design were painstakingly evolved.

The aircraft itself appeared in the designs as a low-wing vehicle with vertical stabilizers at the tips of the horizontal tail surfaces. The wing-section was that of a thin wedge with sharp leading and trailing edges. The water-cooled rocket engine of 100 tons static thrust was located in the tail.

Sänger's suggested launching procedure was unusual. He proposed to have the 100-ton craft take off from a railed track, almost two miles long, under powerful boost from a rocket-propelled sled. This tethered booster would bring the vehicle to a ground speed of more than one and a half times the speed of sound before release.

Leaving the track, the craft would climb under its own momentum at an angle of 30 degrees, reaching a height of some 5,500 feet before its rocket engine fired. Under propulsion, the aircraft would climb less steeply and, following the climax of thrust, would coast to a height of nearly 100 miles before failing back on a ballistic trajectory.

Instead of re-entering the atmosphere in a dive, however, it would return to earth along an undulated "skipping" trajectory, bouncing on top of the denser atmosphere like a flat stone skipped across the still waters of a pond. By this technique, Sänger proposed to achieve ranges of up to 14,600 miles.

(The "skip-glide" effect set Sänger's scheme apart from anything ever conceived and earned him recognition as the author of the Dyna-Soar concept, even though the "skip" feature was dropped by Dyna-Soar designers in this country.)

Target No. 1 in an early Sänger design exercise was New York City. The idea appealed to the German high command, but there were some major faults which detracted from the basic plan. The craft's tiny payload could not be overlooked, even if the problems of materials and propulsion could be solved. Over a range of 14,600 miles, its payload was only 672 pounds. In terms of chemical high explosives of the day, it was disproportionate to the overall weight.

TOO LITTLE, TOO LATE

Although Sänger and Bredt regarded their work as a purely preliminary study, it was continued right up to the summer of 1942 when, according to Sänger, "the long-term program came into conflict with the prosecution of the war." Hindered by the call-up of personnel, including those of participating industry, plus the acute shortage of such materials as nickel, copper and chrome, the project was faced with handicaps which could not be overcome. Germany, playing a losing hand in the war, chose to gamble its remaining resources on the V-1 and V-2 rockets. The skip-glide bomber project was shelved.

POSTWAR DEVELOPMENT

After World War II, several ideas for developing pilotless glide weapons were studied. In nearly every instance, the vehicle was designed to dive onto its target with the warhead. A great disadvantage was its approach speed to the target. It still was slow enough that its speed made it a sitting duck for high performance interceptors.

The work was by-passed when high-thrust rocket engines and light-weight rocket structures showed the way to ballistic missiles of outstanding range and performance, and which could embody a small-sized nuclear warhead.

The idea of the boost-glider was allowed to lapse until it became possible to think in terms of putting a man into space and bringing him back again. Even then, the first manned orbital craft officially conceived in the United States was not a winged vehicle but a ballistic capsule – closely akin to the present Mercury capsule.

Reasons why the ballistic approach was adopted in preference to lifting re-entry were basically these: (1) the available booster did not permit the orbital payload to exceed one ton; (2) the ballistic capsule was considered a relatively short-term development in view of progress with stabilized missile nose-cones and ablative heat shields; (3) the boost glider required a more complex structure for which there was relatively little practical experience even is the laboratory.

Calculations disclosed that the skipping procedure advocated by early Sänger studies resulted in considerably higher temperatures than a straight, gradually descending glide path. Severe heating would result from the "pull-ups" to generate the increased aerodynamic lift required to skip.

One study, called BOMI, called for boosting a glider to near orbital velocity and gliding to the target area thousands of miles away, arriving there at 15,000 feet or more per second and approximately 40 miles altitude. After dropping the bomb, the airplane would make a 180-degree turn and be boosted back up to the initial altitude and velocity conditions to make the return flight. This required carrying another rocket engine system for the return boost.

It soon was discovered that the aircraft would burn up in attempting to make the 180-degree turn because of the extreme increase in temperature resulting from this maneuver in the atmosphere. The additional rocket weight penalty also required a tremendous initial rocket booster. A far more efficient and practical method, it was decided, was to continue the flight path around the earth after dropping the bomb.

Studies in the United States have progressed through many phases involving different purposes and uses based on this same concept. In 1954, the government began to consider the concept more seriously. A series of studies by the Air Force, the civilian space agency (National Advisory Committee for Aeronautics) and industry followed. Included were Hywards, a winged hypersonic research and development system; 118P and Brass Bell, for various reconnaissance applications, and ROBO, a rocket bomber system requirements for which all companies were invited to study.

POST-SPUTNIK ERA

In November, 1957, one month after the Russians had launched an artificial satellite into orbit, the Air Force issued the first preliminary directives on Dyna-Soar. By March, 1958, a number of proposals had been submitted by industry members. In June of that year, the Air Force selected two major teams to prepare competitive studies of the Dyna-Soar. Boeing headed one team and the Martin Company and Bell Aircraft headed the other.

The unknowns which faced Boeing in 1958 were typical of those which confronted other members of the two teams. Dyna-Soar would fly at high Mach numbers and there were few men in the country who had much knowledge of hypersonics. Boeing was known for its experience in supersonics, engine inlets, and other related fields, but this was something else again.

Because Dyna-Soar would have to endure the torture of blast-off and reentry into the atmosphere, an entire new approach to materials and structures was demanded. The old ways weren't good enough.

One of the first steps toward solutions of these problems was to select preliminary design experts whose work in certain research areas fitted the needs of the Dyna-Soar program. Advance structures engineers who had conducted research into "hot frames" were added. Also brought into the program were engineers who had worked on the ROBO project (of which Dyna-Soar was a distant relative). While most of their experience had been with unmanned glide missiles, their ROBO background, it was felt, gave them an understanding of the flight regimes Dyna-Soar planners were talking about.

Aerodynamicists began wind-tunnel testing scores of models to gain data on hypersonic flight. Starting with simple, fundamental shapes, they tested them thoroughly. From these tests evolved the vehicle configuration Boeing proposed to the Air Force in 1959.

Typical of the contributions made by scientists and engineers at work on the program was that of Del Nagel, a young Boeing engineer who had been graduated not many years before from the University of Washington.

His discovery of the outflow phenomenon – a method of predicting flow and heat transfer characteristics – permitted Aerodynamicists to understand what heating problems they would face, and gave them a feel for predicting and correlating test model results to full scale.

As new materials were developed, they were tested at the company-owned 5,750-KVA radiant heat facility. Built for conducting heating tests on supersonic or space vehicles, while at the same time subjecting them to high loads, the device enabled engineers to duplicate high temperatures and loads they knew Dyna-Soar would encounter during re-entry.

So intense was the heat during some of the tests that insulation on the wires carrying electric current into the facility burst into flames.

Material which wouldn't take the punishment was thrown out and others were tried. In the end, engineers beat the so-called "heat barrier" and even the pessimists on the program began to smile.

Not all of the problems could be solved with exotic materials. Structures engineers finessed their way out of at least one tight corner by using triangular girder arrangements which permitted trusses to deform but virtually eliminated thermal stress – a technique used in building bridges, but seldom in aeronautics.

Banshee-like screams – produced by BOEING'S sonic testing facility – also were used to test Dyna-Soar developments. Skin panels were exposed to the energy produced by noise, similar to the severe exhaust and aerodynamic noise the space vehicle likely will encounter during flight at hypersonic speeds.

In November,1959, after intensive effort by both the Boeing team and the team headed by Martin and Bell, the Air Force made its decision. Boeing was designated system contractor and the Martin Company was named associate booster contractor.

Development work on the program, however, did not begin immediately. Because of the high costs involved plus serious doubts among many of the nation's top scientists and engineers that the program, as constituted, would be successful, the Air Force was ordered to perform a configuration verification study. This study, known as "Phase Alpha," began in December, 1959, and lasted until April of 1960.

All of the technical data which had been generated to support the program was collected and catalogued. All possible re-entry vehicle designs were reconsidered. When the study was completed, there was general agreement that the program could, indeed, be accomplished successfully.

WHERE ARE WE TODAY?

The materials and structural designs chosen for a radiation-cooled solution to the high temperature problem have been developed and demonstrated in the laboratory.

A review of a full-scale mockup of the Dyna-Soar glider and its related systems was carried out by a government inspection team in September, 1961. No major changes in the glider design were ordered.

The glider – the design of which is based on about three years of detailed studies – will be manufactured in Seattle at BOEING'S Missile Production Center.

Originally scheduled were sub-orbital flights down the Atlantic Missile Range with a modified Titan II ICBM booster. This was changed in December, 1961, with the announcement by the Air Force that a more powerful booster – one which would combine liquid and solid rockets – would be developed for Dyna-Soar. This new booster, based on Titan II technology and employing solid propellant rockets, will propel Dyna-Soar to orbital velocities. As a consequence, the sub-orbital flights have been dropped from the test program.

Here is the sequence of tests:

Air-drops of powered gliders from a B-52 mother ship at Edwards Air Force base to check the craft's stability and control at slow speeds, and to give the pilots opportunities to perfect landing techniques. Later, gliders equipped with rocket engines will be flown faster than sound to see how they handle in the supersonic regime.

At the conclusion of the Edwards tests, unmanned and manned flights around the world will be launched from Cape Canaveral. These tests will check every phase of Dyna-Soar's operation, including stability and control, performance, and the effects of aerodynamic heating on the craft during re-entry.

The Air Force has not said publicly how much time will be saved in Dyna-Soar's development by the decision to give it an orbital booster. The initial announcement only said that the new booster will assure "early attainment" of manned orbital flight. It did not disclose when the first flight would take place. No official schedule for Dyna-Soar development ever has been announced.

Main purpose of the sub-orbital flights was to gain data on hypersonic flight, a regime never before sampled by manned winged vehicles. This is the speed range beyond mach 6. The new program will explore this area and accomplish orbital flight as well.

The orbital flight is expected to pose fewer problems for the pilot than the short one. For one thing, the extra hour involved in the around-the-world flight will give the pilot more time to adjust to his chores in space, set up his re-entry conditions, and make the necessary preparations before beginning his descent through the atmosphere.

DYNA-SOAR FUNDING

In his budget message to Congress in January, 1962, President Kennedy asked for $115,000,000 to be spent on the Dyna-Soar program during Fiscal Year 1963. This represented an increase of $15,000,000 over the amount earmarked by the Kennedy Administration for the program during FY 1962.

The spending level for Dyna-Soar during FY 1962 was $100,000,000, which was $30,000,000 more than that proposed by the Eisenhower Administration. Even this, however, was less than Congress was willing to spend to accelerate the program.

On the recommendation of the appropriations committee of the U.S. House of Representatives in 1961, Congress appropriated an additional $85,800,000 (bringing the total to $185,800,000) and urged that the Dyna-Soar program be speeded up in FY 1962. Secretary of Defense McNamara decided against this and the additional funds were not spent.

Testifying before a House subcommittee of the appropriations committee in the spring of 1962, Secretary McNamara said:
"I have personally reviewed the project (Dyna-Soar) and concluded that, while we cannot say categorically it will yield an important military weapon, we do believe that its potential is sufficiently great to warrant the expenditures we have proposed ($100,000,000).

". . . The military applications of manned orbital flight of the type toward which Dyna-Soar is directed are likely to be great. We can conceive of a number of such applications, although we have not developed them specifically, because we are trying to achieve the objective of the Dyna-Soar program (research)."

Following is a history of funding for Dyna-Soar since the program was given an official go-ahead early in 1960:

FY 1963 — $115,000,000
FY 1962 — $100,000,000 (an additional $85,800,000 was not spent)
FY 1961 — $58,000,000

QUOTES

"The choice of flight paths available to the Dyna-Soar pilot will be almost infinite. By combining the high speed and extreme altitude of his craft with his ability to maneuver, he will be able to pick any air field between Point Barrow, Alaska, and San Diego, California, with equal ease."
—   George H. Stoner
        Dyna-Soar program manager for Boeing September 22, 1960

"The Air Force considers Dyna-Soar the most important research and development project it has . . . The Dyna-Soar will open a new era . . . It is the first step towards practical man-in-space flights."
—   Lt. Gen. Roscoe C. Wilson
        Deputy Chief of Staff-Development, USAF   September 22 , 1960

"The Dyna-Soar, in effect, is the first vehicle which will combine the advantages of manned aircraft and missiles into a single system . . . It is interesting to note that although the Dyna-Soar will attain peak speeds of over 15,000 miles an hour during flight, its proposed landing speed is to be less than that of some of our present-day combat aircraft."
—   Gen. Thomas D. White, USAF Chief of Staff,
        in AIR FORCE Magazine, September, 1960

"So far, Dyna-Soar has been programmed solely as an experimental craft for research purposes. However, as the first piloted military space system planned by the United States, Dyna-Soar has important operational potentialities which are now being explored by the Air Research and Development Command. A factor contributing to increased confidence in the Dyna-Soar concept has been the encouraging progress made in flight tests of the rocket-powered X-15. This experimental aerospace craft has been designed to study environmental conditions at the border of the atmosphere where Dyna-Soar will operate."

— Lt. Gen. Bernard A. Schriever, Commander,
ARDC, in AIR FORCE Magazine, September, 1960

"Dyna-Soar . . . offers an enormous potential for future maneuverable capability in space and the atmosphere . . ."

— Lt. Gen. Roscoe C. Wilson, Deputy Chief of
Staff-Development, USAF, September 22, 1960

"We are quite impressed with the philosophy of this manned, maneuverable space vehicle which is recoverable under pilot control. My impressions are favorable."

— Major Robert White, Air Force test pilot of X-15 rocket plane, January 18, 1962

"The committee foresees the need for an operational, manned military space vehicle over which the pilot has the greatest possible control and believes that the Dyna-Soar concept provides the quickest and best means of attaining this objective."

— Appropriations Committee, U.S. House of Representatives, May, 1961

DYNA-SOAR MILESTONES

November, 1957	— Development directive issued; development plan approved.
March, 1958	— Proposal received from seven contractors.
June, 1958	— Selection of Boeing and Martin to compete for selection of source for development of Dyna-Soar.
April, 1959	— Evaluation of Boeing and Martin proposals begun.
November 9, 1959	— Selection of Boeing and Martin as contractors. Boeing to be responsible for the manufacture of the vehicle portion of the system, and for integration of vehicle subsystems, integration of vehicle and booster, and assembly and test. Martin to manufacture booster portion of Dyna-Soar.
December 11, 1959	— Boeing receives letter contract for Step 1 of Dyna-Soar.
December 11, 1959	— Phase Alpha study ordered by Air Force, Phase Alpha is an intensive comparative study and re-definitization of all Dyna-Soar technology to date.
March 28, 1960	— Boeing submits recommendation of system configuration to Air Force.
March 28-30, 1960	— Aerospace Vehicles Panel of the Air Force Scientific Advisory Board meets to review findings of Phase Alpha study.
April 11-14, 1960	— Dyna-Soar Symposium, Langley Field, Va., held to acquaint industry with problems and requirements.
April 25, 1960	— Air Force okays Phase Alpha study report, gives go-ahead to permit actual design of Dyna-Soar glider immediately.
September 20, 1960	— Boeing holds first of a series of bidder's conferences as step toward selecting major subcontractors for work on Dyna-Soar.
September 22, 1960	— Air Force reveals first details of Dyna-Soar glider configuration at Air Force Association convention in San Francisco.
December 13, 1960	— Minneapolis-Honeywell Regulator Co. of St. Petersburg, Fla., named associate contractor for Dyna-soar primary guidance subsystem.
December 16, 1960	— Air Force announces selection of Radio Corporation of America as associate contractor for Dyna-Soar communications package.
January 6, 1961	— Boeing announces award of its first major Dyna-Soar subcontracts to Chance Vought Corp. (for nose cap) and Minneapolis-Honeywell Regulator Co. of Minneapolis, Minn. (for flight control electronics).
January 13, 1961	— Air Force announces decision to substitute Titan II for Titan I as booster for Dyna-Soar glider and subsystems.
September 22, 1961	— Government review team completes inspection of full-scale mockup of Dyna-Soar glider end related systems.
December 28, 1961	— Air Force announces decision to develop new booster for Dyna-Soar. Combining both liquid and solid fuel rockets, it will be capable of hurling glider into orbit.

Date: Jun 1962   Title: DOD Press Release: X-20 Designation

NEWS RELEASE
PLEASE NOTE DATE

DEPARTMENT OF DEFENSE
OFFICE OF PUBLIC AFFAIRS
Washington 25, D. C.

IMMEDIATE RELEASE   June 26, 1962   No. 1057-62   Oxford 75131

DYNA SOAR DESIGNATED X-20
BY THE AIR FORCE

The Air Force has designated its DYNA SOAR manned space glider the X-20.

The new designator describes the experimental nature of the program to test the manned spacecraft and follows the long standing procedure of giving each craft a letter - number designator.

The X-20 is a one-man, piloted, delta winged, space glider to be launched into orbit by a TITAN III rocket booster. The TITAN III will be a modified liquid-fueled TITAN II which will use large solid propellant rockets as the first stage.

The X-20 is a combination aerodynamic and space vehicle. Its design will permit the pilot to choose his point of reentry into the atmosphere from orbit. The wings of the X-20 will provide the means for aerodynamic maneuvers by which the pilot can extend or shorten his flight path and can turn to either side over distances of several thousand miles. This maneuverability will permit great flexibility for landing at a suitable air base on return from earth-orbit with the inherent reliability and accuracy of controllable, piloted aircraft.

The National Aeronautics and Space Administration is assisting the Air Force in the development and testing of the X-20. The NASA role stems from its interest in the exploration of winged hypersonic flight phenomena up to orbital conditions.

Several years of study and design competition preceded the beginning of work on the X-20 by the Boeing Airplane Company in early 1960. Radio Corporation of America and the Minneapolis-Honeywell Corporation are also Air Force contractors on the X-20 program.

END

Date: Jun 1962   **Title:** DOD Press Release: X-20 Acceleration Test Fire

NEWS RELEASE
PLEASE NOTE DATE

DEPARTMENT OF DEFENSE
OFFICE OF PUBLIC AFFAIRS
Washington.25, D. C.

IMMEDIATE RELEASE   June 26, 1962   No. 1066-62   Oxford 75131

X-20 ACCELERATION ROCKET
SUCCESSFULLY FIRED

The acceleration rocket being developed for the Air Force X-20 manned space glider was successfully static test fired recently.

The test, by the Thiokol Chemical Corporation in Elkton, Maryland, demonstrated that the solid propellant grain design can deliver the performance required for two missions. The acceleration rocket will provide quick escape for the X-20 in the event of booster malfunction during orbital launch as well as accelerate the glider to supersonic speeds in early airdrops from a B-52 mother ship.

The rocket engine configuration fired in this test was equipped with low-cost, fixed immovable nozzles and an interim rather than flight-weight case. In order to minimize costs, tests of the flight weight case and of nozzles which will control direction of the rocket's thrust are being conducted separately.

The X-20 is a piloted, delta-wing glider being developed by the Air Force with technical advice from the National Aeronautics and Space Administration. The X-20 combines the maneuverability of the piloted X-15 aerospace craft with the orbital capability of the manned Mercury capsule.

END

Date: Aug 1962   Title:Memo Development of Dyna-Soar Configuration

MEMORANDUM to Chief, Research Division

Subject: Development of the Dyna-Soar Configuration

August 23, 1962

1. The attachment is a comparison of some of the major changes in the Dyna-Soar configuration, and the corresponding aerodynamic characteristics of these configurations. The Boeing documents used in this study are listed on the following page and the models are shown in chronological order.

2.   Model 814-1050 came out in March, 1959, and was the final version of the 814 series.

Model 844-2005 was proposed in July, 1960 and was the glider used in the Phase Alpha Study. The wing thickness was reduced to eliminate pitching moment problems.

The 844-2035 configuration came out in November 1960 and had larger elevon control surfaces for more effective hypersonic control.

The 844-2050 configuration is dated August, 1961.The slope of the windshield was decreased to eliminate a hot spot.

Model 844-2050-E came out in December, 1961 and is expected to be the final configuration. A ramp has been added to the top of the aft section of the fuselage for stability reasons.

3. This attachment illustrates the development from an idealistic to a realistic vehicle design while working at the fringe of the state-of-the-art.

Ronald W. Du Val Student Trainee (Aeronautical Engineering)

RWD: gee DRB
JW TAT Attachment: As stated

REFERENCES

D5-439911	DS-1 Aerodynamics Report - Glider Performance and Aerothermodynamics Model 814-1050	3/21/59
D2-6909	Preliminary Systems Design Report Model 844-1005	7/ 13/60
D2-6909-1	Interim System Description of the Dyna-Soar Step I Model 844-2035	11/29/60
D2-8080-1	Glider Performance Characteristics Report Model 844-2050	8/28/61
D2-7326-11	Program Progress Report Dyna-Soar Step I	April-June 1961
D2-80065	Aerodynamic Stability and Control Data Model 844-2050-E	12/20/61

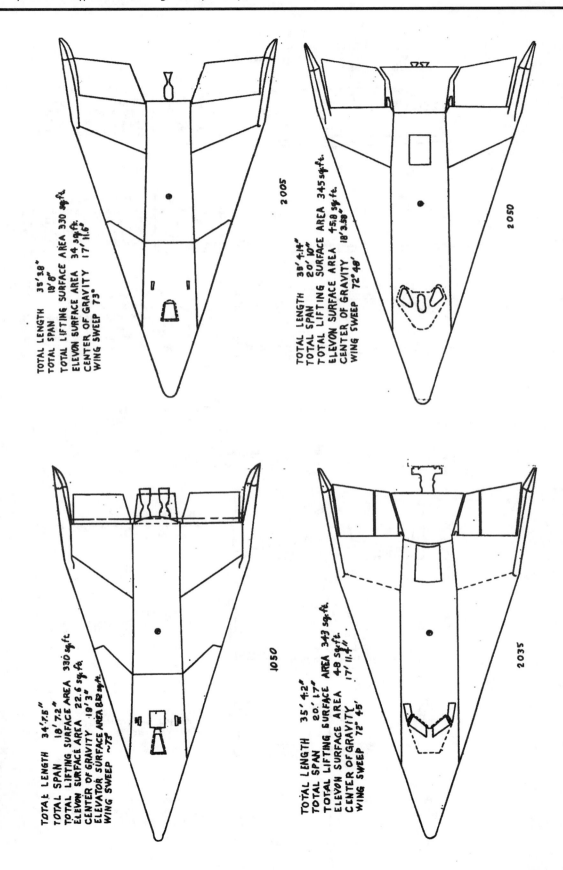

TOTAL LENGTH   35'3.8"
TOTAL SPAN   18'6"
TOTAL LIFTING SURFACE AREA 330 sq.ft.
ELEVON SURFACE AREA   34 sq.ft.
CENTER OF GRAVITY   17'11.6"
WING SWEEP 73°

2005

TOTAL LENGTH   35'4.14"
TOTAL SPAN   20'10"
TOTAL LIFTING SURFACE AREA 345 sq.ft.
ELEVON SURFACE AREA   45.8 sq.ft.
CENTER OF GRAVITY   18'3.58"
WING SWEEP 72°48'

2050

TOTAL LENGTH   34'7.5"
TOTAL SPAN   18'7.2"
TOTAL LIFTING SURFACE AREA 330 sq.ft.
ELEVON SURFACE AREA   22.6 sq.ft.
CENTER OF GRAVITY   18'3"
ELEVATOR SURFACE AREA 8.82 sq.ft.
WING SWEEP ~73°

1050

TOTAL LENGTH   35'4.2"
TOTAL SPAN   20.17"
TOTAL LIFTING SURFACE AREA 343 sq.ft.
ELEVON SURFACE AREA   4.8 sq.ft.
CENTER OF GRAVITY   17'11.4"
WING SWEEP 72°45'

2035

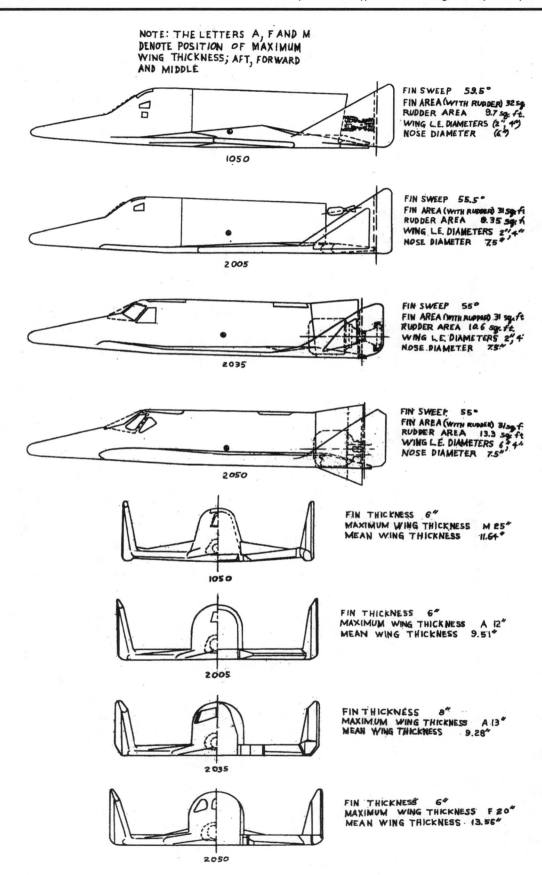

NOTE: THE LETTERS A, F AND M
DENOTE POSITION OF MAXIMUM
WING THICKNESS; AFT, FORWARD
AND MIDDLE

1050

FIN SWEEP   59.5°
FIN AREA (WITH RUDDER) 32 sq.
RUDDER AREA   9.7 sq. ft.
WING L.E. DIAMETERS (2", 4")
NOSE DIAMETER   (6")

2005

FIN SWEEP   55.5°
FIN AREA (WITH RUDDER) 31 sq. ft
RUDDER AREA   8.35 sq. ft
WING L.E. DIAMETERS 2", 4"
NOSE DIAMETER   7.5", 4"

2035

FIN SWEEP   55°
FIN AREA (WITH RUDDER) 31 sq. ft.
RUDDER AREA   10.6 sq. ft.
WING L.E. DIAMETERS 2", 4"
NOSE DIAMETER   7.5"

2050

FIN SWEEP   55°
FIN AREA (WITH RUDDER) 31 sq. f.
RUDDER AREA   13.3 sq. ft
WING L.E. DIAMETERS 6", 4"
NOSE DIAMETER 7.5"

1050

FIN THICKNESS 6"
MAXIMUM WING THICKNESS M 25"
MEAN WING THICKNESS   11.64"

2005

FIN THICKNESS 6"
MAXIMUM WING THICKNESS A 12"
MEAN WING THICKNESS   9.51"

2035

FIN THICKNESS 8"
MAXIMUM WING THICKNESS A 13"
MEAN WING THICKNESS   9.28"

2050

FIN THICKNESS 6"
MAXIMUM WING THICKNESS F 20"
MEAN WING THICKNESS   13.56"

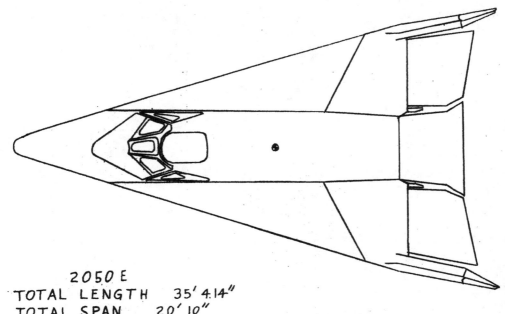

2050 E
TOTAL LENGTH   35' 4.14"
TOTAL SPAN   20' 10"
TOTAL LIFTING SURFACE AREA 345 sq.ft.
ELEVON SURFACE AREA 45.8 sq. ft.
CENTER OF GRAVITY 19' 7"
WING SWEEP   72° 48'

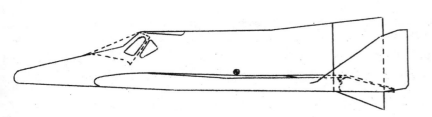

2050 E
FIN SWEEP 55°
FIN AREA (with rudder) 31.4 sq.ft.
RUDDER AREA 10.67 sq. ft.
WING L.E. DIAMETERS 6", 4"
NOSE DIAMETER 7.5"

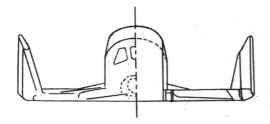

2050 E
FIN THICKNESS 8"
MAX. WING THICKNESS F 20"
MEAN WING THICKNESS 13.56"

Date: Sept 1962   Title: DOD Press Release: Unveiling X-20

DEPARTMENT OF DEFENSE
OFFICE OF PUBLIC AFFAIRS
WASHINGTON 25, D. C.

NO. 1510-62
OXford 7-5131

HOLD FOR RELEASE
UNTIL DELIVERY OF ADDRESS
EXPECTED ABOUT 9-00 A.M. (PST)

REMARKS BY
THE HONORABLE JOSEPH V. CHARYK
UNDERSECRETARY OF THE AIR FORCE
AT THE UNVEILING OF THE X-20
AIR FORCE ASSOCIATION CONVENTION
LAS VEGAS, NEVADA
THURSDAY – 20 SEPTEMBER 1962

During the past year we have been thrilled and filled with pride at the wonderful exploits of the X-15. It has explored new regimes of altitude and speed in the aerospace environment, and Major Bob White has taken this aircraft to the fringes of space.

The X-15 has provided invaluable information on structures, materials, aerodynamics, human factors and the interaction of all the various subsystems in this new regime. It has confirmed the engineering choices that were made in its design and added new confidence for future applications. Despite all this, it is sometimes hard to realize that it has sampled only a very minute portion of the environment from the ground to space.

But today we stand on the threshold of a new experience with a new potential. The X-20, the vehicle you are about to see, is designed to explode the explored narrow band of speed and altitude into a complete corridor to space – a corridor within which man will be able to exit and re-enter from space under his own control, using the atmosphere to arrive and land at a place of his own choosing.

The X-20 does not represent a vehicle for a specific military job, but neither did the Wright Brothers' airplane nor Robert H. Goddard's early rockets, and yet, without the pioneering effort, without the vision, without the search for new knowledge and without the determination to explore the unknown, the fruits of these efforts could never have been realized. And so it is, I believe, with the X-20. It will be a milestone in the conquest and utilization of space and the men who fly it will be the artisans who will convert space flight from an experiment to a new medium for exploitation – hopefully for peace but, if need be, for the preservation of our national security.

END

Date: Oct 1962   Title: DOD Press Release: Officer Selection

DEPARTMENT OF DEFENSE
OFFICE OF PUBLIC AFFAIRS
Washington 25, D. C.

NEWS RELEASE
PLEASE NOTE DATE
IMMEDIATE RELEASE

October 22, 1962

NO. 1717-62 Oxford 75131

OFFICERS SELECTED FOR THIRD CLASS
AT US AIR FORCE AEROSPACE RESEARCH PILOT SCHOOL

Names of ten Air Force officers selected for the third class at the U. S. Air Force Aerospace Research Pilot School of the Advanced Course at Edwards Air Force Base; California, were announced today by the Department of the Air Force.

Those named and their birth places are:

Captain Alfred L. Atwell, North Garden, Virginia; Captain Charles A. Bassett, Berea, Ohio; Major Tommie D. Benefield, Jefferson, Texas; Captain Michael Collins, Alexandria, Virginia; Captain Joe M. Engle., Chapman, Kansas; Major Neil R. Garland, Hicksville, New York; Captain Edward G. Givens:, Quanah, Texas; Captain Francis G. Neubeck, Washington, D.C.; Captain James A. Roman, Paris; France; Captain Alfred H. Uhalt, New Orleans, Louisiana.

The eight-month post-graduate course at the Air Force School is designed to develop space pilots, engineers and program managers for aerospace projects such as the X-20 (DYNA SOAR) and follow-on aerospace programs. It is operated by the Air Force Flight Test Center of the Air Force Systems Command. Colonel Charles E. Yeager, first man to fly faster than the speed of sound, is Commandant of the school.

Those selected met rigid educational, physical, and psychological criteria. Each selectee has a degree in engineering, one of the physical sciences, or mathematics. Prior to selection each underwent intensive medical and psychological examinations to determine character, motivation, and emotional factors at the U.S. Air Force Aeromedical Division, Brooks AFB, Texas.

Curriculum of the school combines classroom instruction, space simulator flights, and Proficiency flying in high performance specially-modified jet aircraft.

Classroom instruction includes mathematics, engineering and mechanics, an introduction to atomic physics, celestial and space navigation, computers, aerospace environment, Einstein's Theory of Relativity, orbital mechanics, trajectories and other subjects pertinent to aerospace flight.

Laboratory work includes computer operation and extensive training in static, dynamic, orbital rendezvous, and ballistic control simulators at Edwards AFB and the Naval Air Development Center at Johnsville, Pennsylvania. Other laboratory work will be done at the U. S. Air Force School of Aerospace Medicine at Brooks AFB, Texas; Ling-Temco-Vought Company at Dallas, Texas; and Griffith Observatory at Los Angeles, California.

END

Date: Nov 1962   Title: AIRFRAME STRUCTURAL DESIGN

X-20 (DYNA-SOAR) AIRFRAME STRUCTURAL DESIGN CRITERIA

(This document supersedes AIR FORCE DYNA-SOAR MILITARY TEST SYSTEM, STEP I, STRUCTURAL DESIGN CRITERIA, 61WWDR-570, (ASNRF-4001) dated May 1961).

6 November 1962

STRUCTURES AND WEIGHTS BRANCH
FLIGHT MECHANICS DIVISION
DYNA-SOAR ENGINEERING OFFICE
DYNA-SOAR SYSTEM PROGRAM OFFICE

Aeronautical Systems Division Air Force Systems Command United States Air Force Wright-Patterson Air Force Base, Ohio

FOREWORD

The original issue of this document, entitled AIR FORCE DYNA-SOAR MILITARY TEST SYSTEM STEP I, STRUCTURAL DESIGN CRITERIA, dated May 1961, and further identified as ASNRF 4001, STRUCTURES DIRECTIVE #I, Control Number 61WWDR-570, has been extensively revised to reflect the December 1961 Dyna-Soar System program re-direction.

A list of individuals who cooperated in developing the criteria presented herein would be too long to reasonably include in the document, however, special acknowledgement is made to Lt. William F. Walkenhorst of the Structures and Weights Branch, Flight Mechanics Division, Dyna-Soar Engineering Office, who prepared the original issue of the document, and the personnel of The Boeing Company, The Martin Company, Aerospace Corporation, the Air Force's Space Systems Division and the Meteorological Development Laboratory of the Air Force Cambridge Research Laboratories who provided such invaluable assistance in contributing to the original issue.

Further acknowledgement is made to The Boeing Company for their assistance in preparing the document in its present form.

R. B. Baird

TABLE OF CONTENTS

.0 SCOPE

his document contains the structural design criteria for the Dyna-Soar glider and transition section rframes. This document further contains the structural design requirements applicable to the integrated 20A/624A air vehicle and is applicable to the degree specified in Tabs A and B to the Dyna-Soar System pecification.

1.1 PRECEDENCE
In the event of conflict between the requirements specified herein and the requirements specified in Tabs A and B to the Dyna-Soar Specification, ASNR 62-4, the Tabs to the Dyna-Soar System Specification shall take precedence. In the event of conflict between the criteria specified herein and the requirements of the military specifications cited in Paragraph 2.2, this document shall take precedence.

.0 APPLICABLE DOCUMENTS

2.1 STRUCTURAL DESIGN CRITERIA DOCUMENTS

2.1.1 GLIDER AND TRANSITION SECTION DOCUMENTS
The detailed structural design criteria for the glider and transition section developed in accordance with Dyna-Soar System program requirements and as approved by the Dyna-Soar System Program Office, shall be applicable.

2.1.2 STANDARD SPACE LAUNCH VEHICLE (SSLV) DOCUMENTS
The detailed structural design criteria for the SSLV developed in accordance with the SSLV program requirements and as approved by the Air Force, and which ensures that the SSLV airframe is structurally adequate to carry the Dyna-Soar payload and to meet the performance requirements specified in Tab B to the Dyna-Soar System Specification shall be applicable.

2.1.3 ENVIRONMENTAL/GEOPHYSICAL SPECIFICATION THE ENVIRONMENTAL/ GEOPHYSICAL SPECIFICATION, X-20 (DYNA-SOAR), ASNR 62-22, dated October, 1962 developed in accordance with Dyna-Soar System program requirements, sets forth the geophysical environmental data to be used in the X-20 program including structural design.

2.2 MILITARY SPECIFICATIONS
Military specifications and other publications, with the modifications and deviations of APPENDIX A incorporated, are applicable to the Dyna-Soar program as specified in APPENDIX A.

.0 STRUCTURAL DESIGN REQUIREMENTS

.0.1 GENERAL
he design, criteria specified in this document establish the general structural design requirements for the

Dyna-Soar glider and transition section airframes and for the integrated air vehicle airframe. Detailed structural design criteria developed for the glider and transition section airframes, or components thereof shall be in conformance with the criteria specified herein and shall be sufficiently comprehensive to be the basis for the development of integrated Structures of sufficient capacity to meet the performance requirements specified in Tabs A and B to the Dyna-Soar System Specification, ASNR 62-4. All primary structure used in the Dyna-Soar System shall withstand without excessive deformation or failure, the largest forces resulting from all critical combinations of limit loads resulting from normal flight and ground loading conditions. The structural design shall provide for the effects of cosmic radiation, vacuum temperature extremes, and ionization and dissociation of gases on insulations, seals, subsystems, etc. The abort vehicle shall be capable of accomplishing a successful abort from the dispersed boost trajectory assuming correct functioning of the Malfunction Detection System so as to meet the requirements for pilot safety as specified in ASNR 62-4A and within the limits specified herein. The integrated air vehicle will not be designed for loads resulting from manual control by the pilot during boost. Unless otherwise specified all load factors are limit values and all load and load factor formulae are based upon yield limit values.

3.1 GEOPHYSICAL ENVIRONMENT

3.1.1 GENERAL
The geophysical environments applicable to the X-20 system are specified in ASNR 62-22, Environmental/Geophysical Specification, X-20. All systems and subsystems, or components thereof, shall be designed for the most severe environmental conditions and combinations of conditions to the extent specified herein for the applicable geographical locations.

3.1.2 SPHERE OF OPERATIONS
Geographical locations applicable to operation of the X-20 System are as follows:

1. All of the continental United States for purposes of research, development, shipping handling, and storage operations.

2. Launch sites are:
a. Cape Canaveral, Florida for ground launches b. Edwards Air Force Base (EAFB), California for air launches

3. Landing sites are:
a. Edwards Air Force Base, California (normal)
b. Cape Canaveral, Florida (abort)

3.2 GROUND PHASE, GLIDER AND TRANSITION SECTION
Structural design of the glider and transition section shall include consideration of all environments to which the structure and its component parts are exposed during manufacturing, handling transportation, erection and storage. All structural assemblies shall be capable of withstanding the ground loads specified herein.

3.2.1   HANDLING AND ERECTION

3.2.1.1 Towing
The glider landing gear will not be designed for towing loads.

3.2.1.2 Jacking
Jacking loads for the glider and transition section separately, or in combinations as applicable, shall be as specified in the table below. The vertical load shall act singly and in combination with the longitudinal, lateral, and combined longitudinal and lateral loads. The horizontal loads at the jack points shall be reacted in such a manner as to cause no change in the direction or magnitude of the vertical loads at the jack points.

COMPONENT	LANDING GEAR TRUNNION 3-POINT ATTITUDE (STROKED) LIMIT (MIN DESIGN WT)	OTHER JACK POINTS LEVEL ATTITUDE LIMIT (MAX DESIGN WT)
Vertical	1.35F*	2.0F
Longitudinal	0.4F	0.5F
Lateral	0.4F	0.5F

* F is the static vertical reaction at the jack point.

3.2.1.3   Hoisting

The hoisting loads shall be based on the glider and transition section maximum design weights. The longitudinal axis of the glider and transition section shall be considered in both the vertical and horizontal. positions with relation to the ground plane. Vertical hoisting loads shall be equivalent to a limit load factor of +2.0g. Horizontal loads with relation or to the ground plan shall be zero.

3.2.2 TRANSPORTATION

The glider and transition section shall be capable of withstanding the combinations of limit load factors specified in the following diagram during transportation at the minimum design weight.

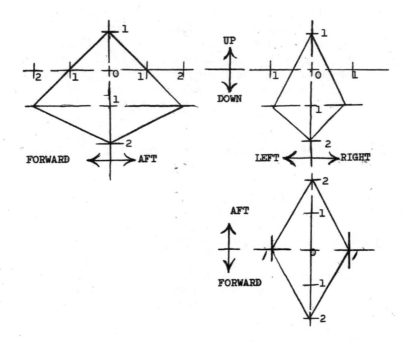

3.2.3 CLOSED OR SEALED COMPONENTS

The effects of pressure loads caused by temperature and/or atmospheric pressure changes shall be considered in the design of sealed or closed components. When required, venting shall be provided to prevent damage to the normally closed components. Components requiring sealing to prevent contamination shall be equipped with provisions to maintain the seal but to prevent damage from pressure differentials.

3.3 PRE-LAUNCH AND LAUNCH PHASES

The glider/transition section and air vehicle shall be capable of sustaining all pre-launch and launch loads. The full design load conditions shall include consideration of ground winds, gusts, engine thrust transients and differentials, stand misalignments, control system operation, and such other conditions as may be experienced during the pre-launch and launch operations.

3.3.1   GROUND WINDS
The ground wind environment for the launch site shall be as depicted in Figure 3.3.1-1. The gust increment shall be critically phased. The air vehicle shall be designed to withstand oscillations induced by the above specified ground winds.

3.3.2   LAUNCH LOADS
The effects of winds, gusts, control system operation, and thrust shall be included in combination with the following factors in determining the launch dynamic loads;

    a. Engine thrust transients and differentials

    b. Vehicle launch stand coupling, misalignments, and dynamics

    c. Engine gimballing and/or thrust vectoring

    d. Acoustics and vibrations

3.3.3   ABORT REQUIREMENTS
The glider/transition shall withstand the loads associated with the pad abort requirements specified in Paragraph 3.1.1.1.6 ASNR 62—4A. Quoted here in for reference only: "During abort from the pad, the glider shall have a structural capability of recovery within design limit conditions. Touchdown and slide-out shall be accomplished within design ultimate conditions, assuming a landing strip can be attained within the performance capabilities of the glider."

3.4 BOOST PHASE
The air vehicle shall be capable of withstanding all loading conditions specified herein during the boosting of the glider to orbital flight conditions. The Dyna-Soar flight phases are as defined in ASNR 62-4A.

3.4.1   BOOST DESIGN TRAJECTORIES
The design trajectory is as specified in ASNR 62-4A & 4B.

3.4.2   BOOST DESIGN LOAD CONDITIONS

3.4.2.1   Design Weights and Center of Gravity Positions
Boost loads shall reflect air vehicle mass distributions and magnitudes for all critical weights from air vehicle maximum design weight to air vehicle minimum design weight.

3.4.2.2   Transient Loads
Dynamic and elastic effects shall be considered in the determination of external loads. The effects of reduced modulus of elasticity due to increased temperatures, reduction in stiffness due to local buckling under load, fuel slosh, and other significant effects shall be included in the dynamic analyses as applicable.

3.4.2.3   Off-Nominal Conditions
Variations and tolerances in aerodynamic characteristics, positions of fixed surfaces, vehicle flexibilities, guidance and control system characteristics, thrust level, etc., shall be considered in load derivations. The parameters will be considered independently and combined statistically with nominal loads except where demonstrated correlation exists, in which case the actual measured or estimated correlation coefficients may be used.

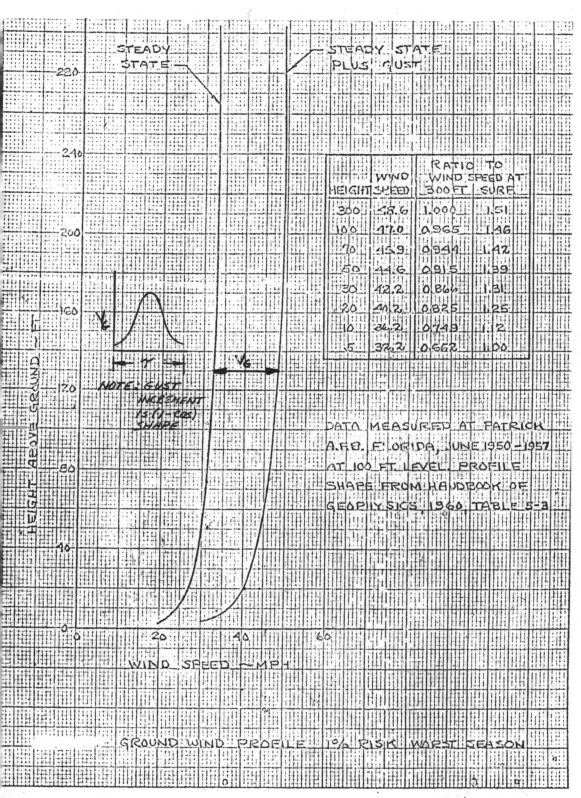

FIG. 3.3.1-1

3.4.3   FLIGHT LOADS

The air vehicle shall be capable of withstanding the following loads acting either separately or in any combination and in both the pitch and yaw planes. Loads shall be computed considering body dynamic and aeroelastic conditions. The design loads shall be the maximum which occur at any altitude with gust loads superimposed.

3.4.3.1   Axial Loads

Axial loads produced by engine thrust (including transients).

3.4.3.2   Maneuvering Loads

Maneuvering loads produced by Tip over to gravity turn and other attitude changes required to follow the design trajectory.

3.4.3.3   Wind Shear Loads

Wind shear loads resulting from flights through the design winds specified in ASNR 62-22 (shown on Figures 3.4.3.2-1 through -5 herein—for reference only), Loads derivation will include consideration of the winds occurring from any direction (with the directional characteristics accounted for), air vehicle launch azimuth, and with the shear spikes assumed to occur at any altitude within the boost phase.

3.4.3.4   Gust Loads

Gust induced loads as determined by dynamic analyses and considering a critically phased horizontal (1- cosine) gust with a true velocity of 11 feet per second acting anywhere along the boost design trajectory through 50,000 feet altitude.

3.4.4   ABORT REQUIREMENTS

The glider/transition shall withstand the loads associated with the abort requirements specified in ASNR 62-4-A. These are as follows: (See ASNR 62-4A—quoted herein for reference only) "The glider shall be capable of accomplishing a successful abort within the ultimate load capability of the glider transition section from the dispersed boost trajectory, assuming; correct functioning of the Malfunction Detection System, so as to meet the requirements for pilot safety—" stated in Paragraph 3.2.9 of ASNR 62-4A.

3.5 ORBITAL FLIGHT PHASE

The orbital vehicle as defined in ASNR 62-4A shall be capable of withstanding the geophysical environment and loading conditions encountered during the orbital phase of the design mission as defined in ASNR 62-4A.

3.5.1   GEOPHYSICAL ENVIRONMENT

The geophysical environment for orbital flight shall be in accordance with applicable portions of ASNR 62-22.

3.5.2   REACTION CONTROL

Maximum available reaction control forces in roll, pitch, and yaw and deorbiting reaction forces, in any combinations shall be considered.

3.5.3   TRANSITION SEPARATION

Transient loads resulting from glider and transition separation shall be considered.

3.6 RE-ENTRY, HYPERSONIC GLIDE, AND APPROACH AND LANDING PHASES

The Dyna-Soar flight phases are as defined in ASNR 62-4A. The glider shall be capable of withstanding the environments and loading conditions associated with (1) establishing equilibrium glide conditions on the nominal trajectories, (2) flying the angle of attack schedule, and (3) accomplishing an approach and landing, all as specified in ASNR 62-4A.

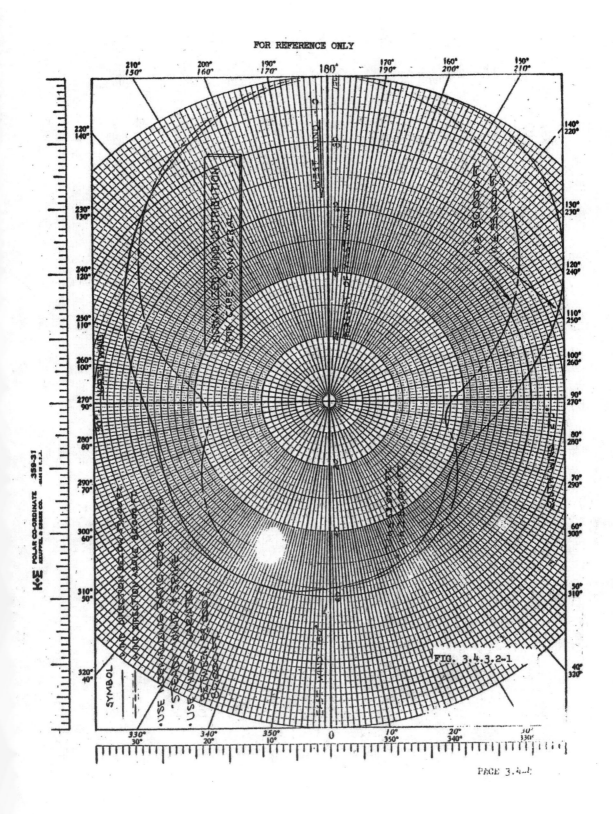

FOR REFERENCE ONLY

FIG. 3.4.3.2-1

FOR REFERENCE ONLY

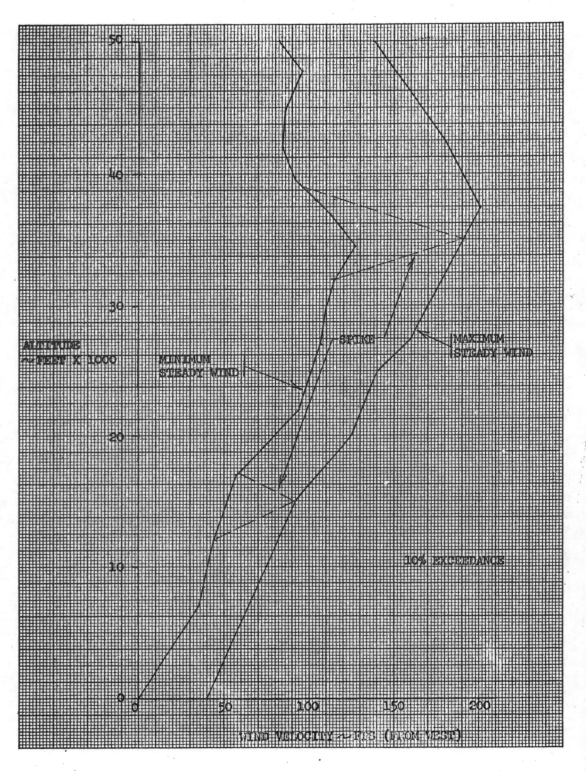

WIND PROFILE
0 TO 50000 FEET

FIG. 3.4.3.2-2
PAGE 3.4-4

FOR REFERENCE ONLY

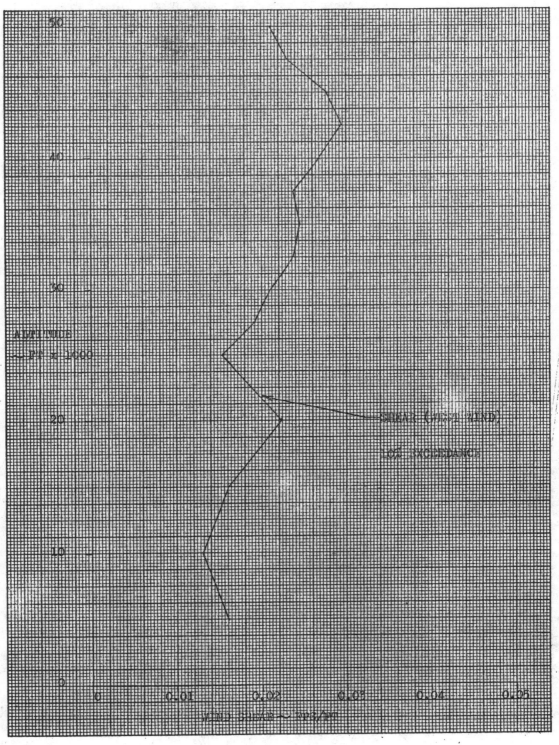

WIND SHEAR
0 TO 50,000 FT

FIG. 3.4.3.2-3
PAGE 3.4-5

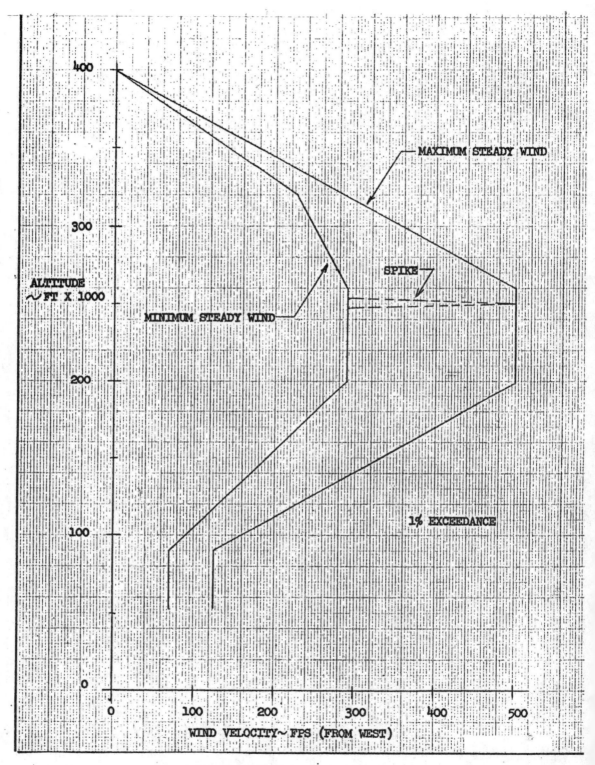

WIND PROFILE, 53,000 TO 400,000 FT

FIG. 3.4.3.3-4
PAGE 3.4-6

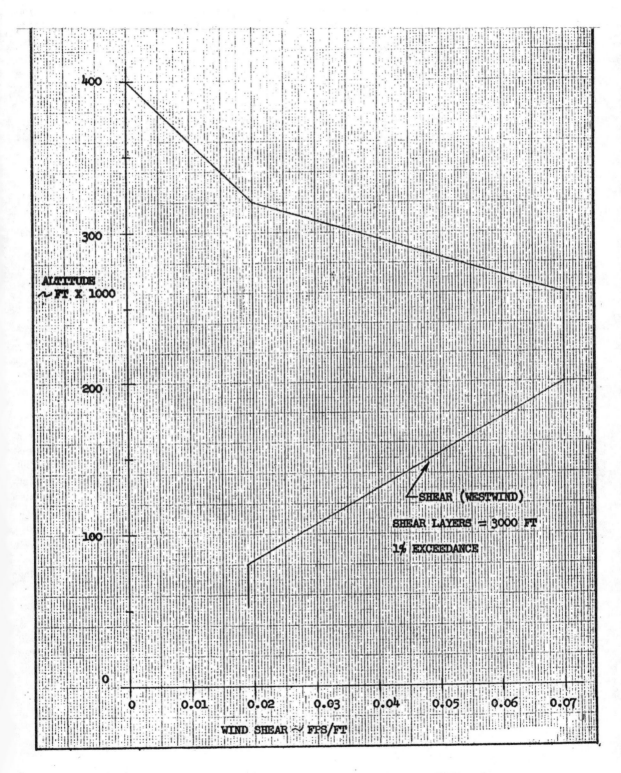

WIND SHEARS, 53,000 TO 400,000 FT

FIG. 3.4.3.2-5
PAGE 3.4-7

3.6.1   STRUCTURAL DESIGN REQUIREMENTS

The glider will have design limits compatible with the altitude margin requirements as specified in ASNR 62-4A. The glider design shall provide for transient maneuver conditions considering at least the following: a. Maneuver capability for adjustment of an equilibrium glide C_L to accomplish range control or increase flight safety. b. Oscillations about the equilibrium glide lines due to tolerances in the flight control and guidance systems.

3.6.1.1   Nominal Re-Entry

During nominal re-entry the glider shall have a structural capability of achieving the nominal range-velocity trajectory from a dispersed boost trajectory prior to initiation of the hypersonic glide phase and with an angle of attack modulation as specified in ASNR 62-4A.

3.6.1.2   Hypersonic Glide

During hypersonic glide and to high key, the glider shall have a structural capability of achieving (1) the minimum range based on C_L max with nominal bank angle of zero degrees, (2) the maximum range based on L/D_{max} with nominal bank angle of zero degrees, and (3) lateral and longitudinal range modulations — all as specified in ASNR 62-4A.

3.6.2   DESIGN LOAD CONDITIONS

3.6.2.1   General Requirements

General requirements which shall apply to all design load conditions are as follows: a. Design weights employed shall be those appropriate to the flight regime to be analyzed. b. The center of gravity positions at each design weight shall be the critical maximum-forward or maximum-aft position. c. The augmented flight control subsystem will be considered as operating at all times

3.6.2.2   Symmetrical Flight Conditions

In the velocity range above M = 5.0, the minimum symmetrical maneuver limit load factor shall be as determined by requirements for glider performance to meet Mission and Pilot Safety requirements as specified in ASNR 62-4A. The minimum maneuver capability for flight conditions below M = 5.0 shall be as set forth in Figure 3.6.2.2-1.The minimum maneuver capability, for all velocities, shall be that to satisfy the requirements of paragraph 3.6.1.2. The design of the glider shall be based on, but not limited to, consideration of the following symmetrical flight conditions: a. Balanced Maneuver b. Symmetrical Maneuver with Pitch c. Banked Pull-Out d. Approach Pull-Out

3.6.2.3   Unsymmetrical Flight Conditions

The glider shall be designed for at least the following unsymmetrical flight conditions: a. Rolling Pull-Out b. Roll in Approach c. Sideslips and Yawing Maneuvers

3.6.2.4   Thermal Conditions

The glider shall be capable of withstanding the thermal environments associated with the design trajectories of paragraph 3.6.1 in conjunction with the load factor requirements of paragraph 3.6.2.2.

3.6.2.5   Gust Conditions

During the approach phase the glider shall be capable of withstanding the gusts specified below in conjunction with the following flight conditions.

3.6.2.5.1 Equilibrium Glide Flight

The vertical and horizontal gust load factors to be used for equilibrium glide shall be determined as specified in paragraph 3.6.2.5.3.

3.6.2.5.2 Maneuvering Flight

The vertical and horizontal gust load factors to be used in combination with symmetrical maneuvering flight shall be determined as specified in paragraph 3.6.2.5.3. The maneuver limit load factors to be used are +2.0 and +1.0.

3.6.2.5.3 Gust Load Factors

Gust velocities shall be as specified in ASNR 62-22 (shown herein as Figure 3.6.2.5.3-1 for reference). The gust load factors are calculated by means of the equation

$$\Delta n_g = \frac{\pm p(Vtrue)\ (U_{true})\ mK_w}{2\ W/S}$$

and the gust alleviation factor, $\underline{K_w}$, shall be taken from the following table.

V_{true}	Vertical Gust K_w	Horizontal Gust K_w
Subsonic $M \leq 1.0$	$\dfrac{0.88\ \mu}{5.3 + \mu}$	1.0
Supersonic $1.0 < M < 5$	$\dfrac{\mu\ 1.03}{6.95 + \mu\ 1.03}$	1.0
Hypersonic $M \geq 5.0$	1.0	1.0
Where $\mu = \dfrac{2W/S}{g\bar{c}m\rho}$		

Gust Alleviation Factors

3.6.2.5.4 Power-Spectral Analysis

A power-spectral density gust analysis based on 1% probability of exceedance atmospheric turbulence using the data specified in ASNR 62-22 (shown herein as Figures 3.6.2.5.4 - 1 through 3.6.2.5.4 - 3 for reference) shall be used to insure that there are no critical frequency responses for the glider structure.

3.6.2.6   Landing Loads

3.6.2.6.1 Normal Landing

The glider shall be designed to withstand the maximum loads induced by landings on concrete and asphalt runways and the Edwards Air Force Base dry lake bed with a sink speed of 8 feet per second for the follow ing configurations: a. The glider shall be at minimum design weight, air launch configuration, and the gear and airframe shall be at ambient temperature. b. The glider shall be at a minimum design weight, ground launch configuration and the critical residual temperature resulting from the reentry phase.

Pilot imposed elevon deflections during or after touchdown shall not exceed those required to accomplish a safe tangential landing in accordance with Paragraph 3.1.1.1.5 of ASNR 62-4A. The vertical impact velocity of the nose gear shall be determined for the worst combinations of e.g. location, attitude, and velocity as specified in the glider detail specification D2-80600.

3.6.2.6.1.1 Ground Winds
The design ground wind shall be a steady state wind of 20 knots magnitude considered from any direction.

3.6.2.6.2 Emergency Landings
The ability of the glider to withstand landings under emergency conditions, specifically, as required by ASNR 62-4A and by paragraph 3.3.3 herein shall be based upon ultimate strength as appropriate.

FOR REFERENCE ONLY
NON-STORM TURBULENCE PROBABILITIES

Altitude: Feet
0 - 2,000	.320
2,000 - 10,000	.080
10,000 - 20,000	.045
20,000 - 30,000	.060
30,000 - 40,000	.065
40,000 - 50,000	.023
50,000 - 60,000	.020

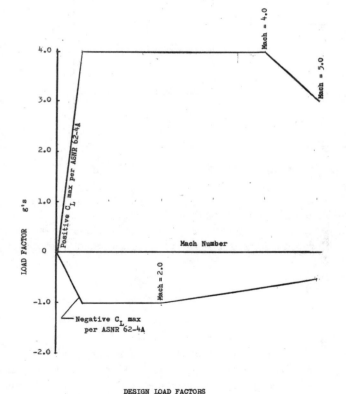

DESIGN LOAD FACTORS

FIG. 3.6.2.2-1
PAGE 3.6-7

3.7   ESCAPE REQUIREMENTS
The glider and escape system for the crew shall be designed to withstand all loads encountered in ejecting the crew under the conditions specified in ASNR 62-4A.

3.7.1   CRASH LOADS
The structural design shall consider crash loads on the crew compartment and equipment contained therein, the failure of which could result in injury to the crew or prevent egress from the vehicle. Crash loads shall be those loads resulting from the ULTIMATE load factor combinations listed below. The crew compartment shall be assumed depressurized and all other loads shall be assumed zero.

a. N_x = 40g, a forward load.
       N_y = 0
       N_z = 0

b. N_x = 37.6g, a forward load.
       N_y = 13.7g, a lateral load.
       N_z = 0

c. N_x = 0

$N_y = 0$

$N_z = 20g$, a down load.

d. $N_x = 0$

$N_y = 0$

$N_z = 10g$, an up load .

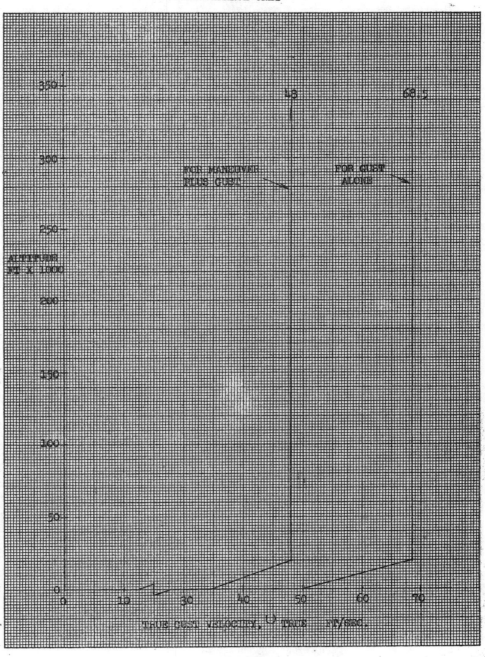

TRUE GUST VELOCITY VS. ALTITUDE   FIG. 3.6.2.5.3-1
PAGE 3.6-8

FOR REFERENCE ONLY

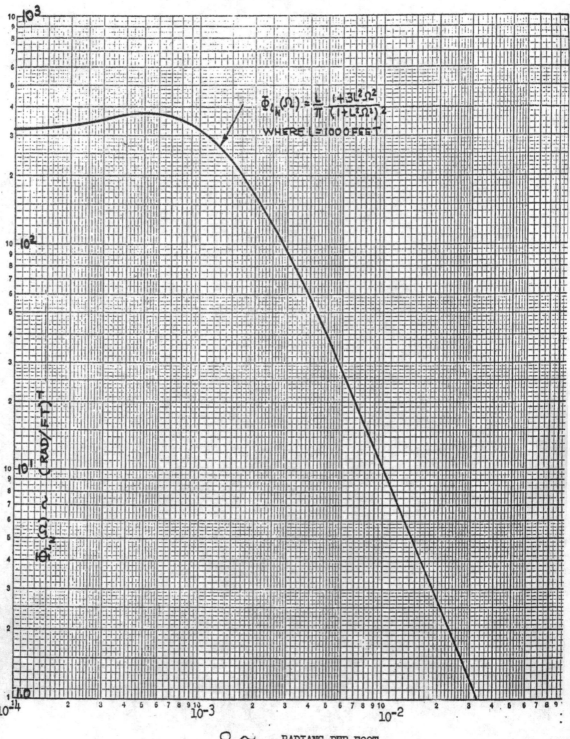

$$\Phi_{4_h}(\Omega) = \frac{L}{\pi} \frac{1+3L^2\Omega^2}{(1+L^2\Omega^2)^2}$$
WHERE L = 1000 FEET

$\Omega \sim$ RADIANS PER FOOT

NORMALIZED POWER SPECTRUM OF
ATMOSPHERIC TURBULENCE

FIG. 3.6.2.5.4-1
PAGE 3.6-9

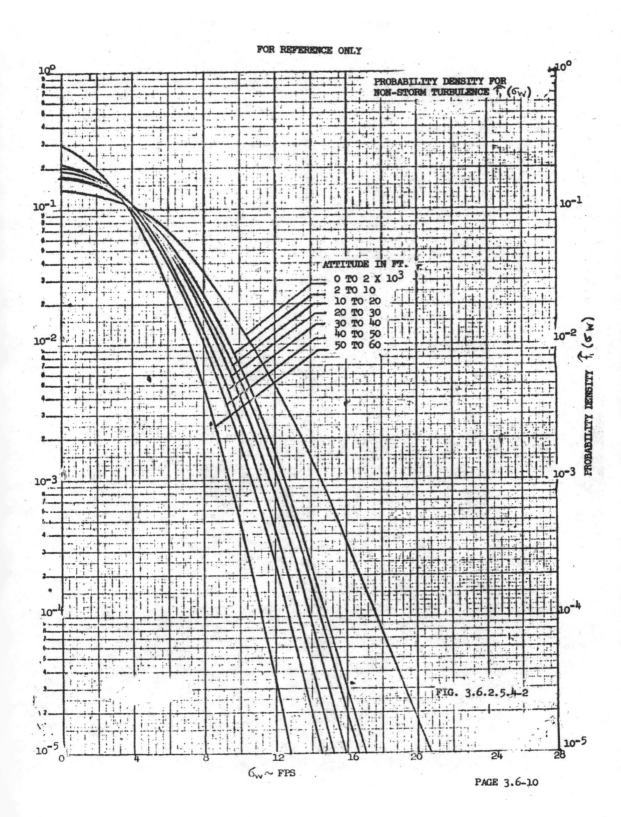

FOR REFERENCE ONLY

PROBABILITY DENSITY FOR
NON-STORM TURBULENCE $\hat{f}_i (\sigma_w)$

ATTITUDE IN FT.

0 TO 2 X 10^3
2 TO 10
10 TO 20
20 TO 30
30 TO 40
40 TO 50
50 TO 60

PROBABILITY DENSITY $\hat{f}_i (\sigma_w)$

FIG. 3.6.2.5.4-2

$\sigma_w \sim$ FPS

3.8 AIR LAUNCH/FERRY PHASES

The glider structure shall be such that it is capable of withstanding the environments imposed during captive flights except that the carrier aircraft may be subject to such flight restrictions as are practical and necessary to prevent undue glider structural penalties. The design shall include consideration of the aerodynamic and inertia loads resulting from response of the carrier aircraft to gusts and maneuvers and the leads incurred during take-off. Vibration and acoustic environments imposed on the glider when attached to the carrier aircraft shall be considered. The glider shall be designed to withstand any special ground handling requirements incidental to captive flights.

3.8.1   AIR LAUNCH FLIGHT CONDITIONS
The air launch glider flight trajectories shall be as specified in ASNR 62-4A.

3.8.2   AIR LAUNCH DESIGN LOAD CONDITIONS
The air launch design load conditions, while the glider (or the glider and transition section) is attached to the carrier aircraft, shall be those of the carrier aircraft as modified and as flight restricted for the Dyna-Soar captive flights. The glider design load conditions after release from the carrier aircraft shall be within the limits of paragraph 3.6. The power or thrust of the acceleration rocket motor (ARM) applicable to the appropriate phases of the air launch missions shall be as specified for the ARM.

3.9 SAFETY FACTORS AND/OR DESIGN FACTORS
Safety factors shall be applied to loads, pressures, service life, operating energy, etc., for the purpose of assuring structural reliability.

3.9.1   GLIDER AND TRANSITION SECTION

3.9.1.1   Primary Airframe and Unpressurized Subsystems

3.9.1.1.1 Loads
The primary airframe structure and unpressurized subsystem components shall be designed to withstand a minimum of 1.5 times the limit loads resulting from the normal ground, pre-launch, launch, boost, exit, orbital, re-entry, hypersonic glide, and approach and landing loading conditions.

3.9.1.1.2 Thermal
Safety factors (except as specified herein shall not be applied to thermal strains as determined from design temperatures and temperature gradients. Safety factors shall not be applied to heat transfer coefficients used in the derivation of design temperatures and/or temperature gradients.

3.9.1.2 Pressure Vessels
All pressure vessels shall withstand, without excessive deformation or failure, the largest forces resulting from all critical combinations of limit loads resulting from flight and ground loading conditions as applicable. The design of pressure vessels shall be such that the application of proof pressures will not cause the yield strength of the vessel materials to be exceeded. Design shall be based on ultimate load conditions except that design conditions shall be based on nominal pressures without factors applied where pressure increases the structural allowables or produces relieving stresses.

3.9.1.2.1 Cabin and Pressurized Equipment Compartments

a. The glider cabin and pressurized equipment compartments shall be designed to withstand a limit pressure of 1.33 times the maximum expected positive internal pressure applied under 1g loading conditions and an ultimate pressure of 1.5 times limit pressure applied under 1g loading conditions.
b. The glider cabin and pressurized equipment compartment design factors for zero, maximum

expected positive internal, or maximum expected negative differential pressure in combination with all other flight loads shall be 1.0 for limit loads and 1.5 times limit for ultimate loads.

c. The cabin and pressurized equipment compartments proof pressure factors shall be 1.2 times the maximum expected positive internal pressure applied under 1g loading conditions.

3.9.1.2.2 Subsystems Tanks

All subsystems tanks shall be designed to have a life equal to a minimum of 1.5 times the service life specified in Tab A to ASNR 62-4.

a. All subsystems tanks shall be designed to withstand limit pressure as determined by:
(1.) the effects of transient peaks 3 sigma variations in pressure, and vehicle accelerations or
(2) maximum relief valve setting times a factor of 1.5 except that limit pressure shall be at least 1.5 times the nominal pressure.

b. All subsystems tanks shall be designed to withstand an ultimate pressure of 1.33 times the limit pressure specified in (a) above.

c. The subsystems tanks proof pressure shall be the limit pressure specified in (a) above applied under 1g loading conditions.

d. The subsystems tanks burst pressure shall be at least 1.33 times the limit pressure specified in (a) above applied under 1g loading conditions.

3.9.1.2.3 Subsystems Pressure Vessels (Other Than Tanks)

The proof and burst factors for components mounted on or in the glider and transition section shall be as specified in the following table. The factors shall, be applied to the nominal pressures under 1g loading conditions.

The design of all components shall include consideration of any loads that may result from transients, accelerations, vibrations, and shall further include consideration of operating temperatures.

Systems	Proof	Burst
Hydraulic System		
Lines and Fittings	2.0	4.0
Accumulators	2.0	4.0
Actuators & Other Components Subject to System Pressure	1.5	2.5
Components Subject to Back Pressure Only (except lines and fittings)	1.5	3.0
Passive Cooling Components	2.0	4.0
Reaction. Control System		
Lines and Fittings	2.0	4.0
Other Components	1.5	2.5

*These factors shall be applied to maximum expected pressure.

Systems	Proof	Burst
Pneumatic System		
Lines and Fittings	2.0	4.0
Actuators and Other Components	1.5	2.5
Pressure Vessels	1.67	2.5

APU and Environmental Control System

Lines and Fittings	2.0	4.0
other; Components	1.5	2.5
O2, N2, and H2 Tanks ①	.1.5	2. 0

A.R.M. Nozzle Actuation System

Lines and Fittings	2.0	4.0
Accumulators (Precharged)	1.67	2.5
Actuators and Other Components	1.5	2.5

① - Tank factors are applied to maximum relief valve setting.

3.9.1.2.4 Solid Rocket Motors (Case Only)

The solid rocket motor shall be designed to withstand limit load conditions of 1.0 times the maximum expected differential pressures applied in combination with all other flight loads and ultimate load conditions of 1.5 times limit except, that where the critical load condition are not aerodynamically induced, this factor may be 1.4 times limit. The solid rocket motor proof pressure shall be 1.1 times the maximum expected differential pressure that occurs at the maximum stabilized temperature and shall include 3 sigma variations. The proof pressure shall be applied under 1g loading conditions. The solid rocket motor shall be designed to withstand a burst pressure of 1.4 times the maximum expected differential pressure as defined above applied under 1g load conditions.

3.9.1.2.5 Liquid Rocket Engines

Liquid rocket engines shall be designed to withstand limit load conditions of 1.0 times the maximum expected differential pressure applied in combination with all other flight loads and ultimate load conditions of 1.5 times limit. The liquid rocket engine proof pressure shall be 1.2 times the maximum expected-chamber differential pressure applied under 1g loading, conditions. The liquid rocket engine shall be designed to withstand a burst pressure of 1.33 times proof pressure applied-under 1g, loading conditions.

3.9.1.3 Equipment Support Structure

The glider and transition section equipment support structures shall be designed to withstand a minimum of 1.5 times the limit loads resulting from all flight and ground loading conditions except as specified in paragraph 3.7.1.

3.9.1.4 Landing Energy Absorption System

The glider landing energy absorption system shall be designed to withstand 1.3 times the energy resulting from the normal landing conditions. The design factors for all other parts of the landing gear and its support structure shall be as specified in paragraph 3.9.1.1.

3.9.2 SPACE LAUNCH SYSTEM

3.9.2.1 Primary Airframe

All primary airframe structure shall be designed to withstand without permanent deformation or failure, the largest forces resulting from all crtical combinations of loads resulting from normal flight and ground loading conditions.

3.9.2.1.1 Loads

The booster primary airframe structure shall be designed to withstand a minimum of 1.4 times the forces resulting from the pre-launch, launch, and boost loading conditions. The design of the structure shall include consideration of operating temperatures.

3.10 GENERAL LOAD REQUIREMENTS

3.10.1 AERODYNAMIC LOAD PARAMETERS

The aerodynamic load parameters shall be as determined from applicable wind tunnel data. Where applicable wind tunnel data does not exist, these parameters shall be as calculated by rational and conservative methods.

3.10.2 EFFECT OF DEFLECTION ON LOADS

The effects of aeroelastic and thermoelastic deflections on the distribution and intensity of loads, derived by rational methods, shall be considered.

3.10.3 ACOUSTIC AND VIBRATION LOADS

3.10.3.1 Acoustic

The glider and transition section structures shall be capable of withstanding the applicable acoustic environments encountered during all phases of Dyna-Soar operations except that the ground launch glider shall, not be penalized by protective systems required by captive-flight tests. Noise from the carrier airplane during captive-flight tests, rocket noise during ground launch as specified in ASNR 62-4A, boundary layer noise, and localized aerodynamic noise during boost and re-entry will be considered in the analytical analyses.

3.10.3.2 Vibration

The glider and transition section structures shall be capable of withstanding the applicable vibration environments encountered during all phases of Dyna-Soar operations including boundary layer noise, fuel slosh, buffet, and other mechanical and aerodynamic phenomena. Transmitted vibration from the SSLV are as specified in ASNR 62-4A.

The equipment mounting structures shall incorporate provisions for supporting the equipment with minimized deflections and minimized amplification of the structural vibrations.

3.10.4 Transient Response

The magnitudes and distributions of loads shall include the effects of the dynamic response of the structure resulting from the transient or sudden application of loads, such as abrupt maneuvers, gusts; landing, etc.

3.10.5 Additional Subsystem Loads

All subsystems and components that are subjected to structural or other loads not of the particular subsystem or component origin, shall be capable of withstanding, without excessive deformation, such loads when applied simultaneously with the applicable limit pressure. In addition, these loads shall not cause deformations which detrimentally effect the operation of the subsystems or components.

3.11 SERVICE LIFE

The equipment and structure shall be capable of withstanding the levels of shock, vibration, and noise which are applicable to the particular vehicle location. Ground handling and transportation conditions shall be taken into account. The structure shall be such that fatigue failures and structure creep failures will not occur from exposure to the spectra of repeated loads and temperatures consistent with the specified service life.

3.11.1 GLIDER

Service life requirements, but excluding replaceable components as identified below, are specified in ASNR 62-4A.

3.11.1.1 Replaceable Components

The following glider structural components shall withstand at least one design mission prior to replacement: refractory metal erosion shields and windshield cover, surface panel insulation, leading edges, nose cap, landing gear skids, energy straps and bearings, nose gear knuckle joint, elevon and rudder bearings, and gas actuated actuator linings.

3.11.2 TRANSITION SECTION
The transition section and its components shall be capable of withstanding the design mission plus the equivalent total accumulated static firing time anticipated for the transition section or the individual components. The effects of rocket plume and pressures resulting from rocket motor and/or engine firings shall be considered.

3.12 STIFFNESS REQUIREMENTS

3.12.1 FUNCTIONAL
Structural rigidity shall be sufficient for the designed functioning of the system and all of its component elements.

3.12.2 FLUTTER AND DIVERGENCE
The glider, transition section, and integrated air vehicle structures shall not flutter, buzz, or exhibit other related dynamic instability or divergence phenomenon at any point along the dispersed design trajectories as specified in ASNR 62-4A and -4B. A 32 per cent increase in dynamic pressure anywhere along these trajectories, both at constant Mach number and separately at constant altitude, shall not result in flutter or divergence. Additionally, the damping coefficient, g, shall be at least 3 per cent for all important flutter and vibration modes.

3.13 THERMAL REQUIREMENTS
Thermal effects shall be considered in the design of the glider, transition, and booster airframes. Thermal effects on the structure, including heating rates, temperatures, thermal stresses and deformations, mechanical   and physical property changes, shall be based on critical 3   design heating environments without additional factors of safety, except for certain conditions where δ specific factor of safety may be required.

3.13.1 HEAT TRANSFER COEFFICIENTS
Heat transfer coefficients for aerodynamic heating shall be based on wind tunnel tests applicable to the configuration or on recognized and verified theoretical techniques.

3.13.2 EXTERNAL SURFACE HEATING
Design heating rates shall be the most severe resulting from analyses of laminar, transitional and turbulent continuum flow.

3.13.3 GLIDER REQUIREMENTS
The glider design shall be based on the following thermal conditions:
a. Maximum temperature resulting from thermal equilibrium or transient conditions.
b. Temperatures resulting from transient conditions due to maneuver imposed on steady-state thermal equilibrium conditions existing at the time the maneuver is initiated.

3.13.3.1 External Surface Temperatures
The external surfaces will be assumed to operate at radiative equilibrium temperatures except where it can be shown by analysis that significantly lower temperatures exist due to transient conditions. Radiative equilibrium temperatures shall be based upon surface emissivities as substantiated by test.

3.13.3.2 Internal Structure Temperatures
The time-temperature histories of the internal structure during boost, orbit, re-entry glide and maneuver flight shall be based on analyses of transient conditions. Thermal properties of the structural and heat protection systems shall be as established by test. Total panel conductance will

be as determined by tests of actual configurations.

3.13.3.3 Glider Thermal Protection

The complete thermal protection system in each area of the glider, including external erosion shield and insulation, structure, internal store insulation, and cooling shall be considered as a composite system for analysis. The effects of re-radiation from internal stores and heat sinks shall be included in the analysis.

3.13.4 THERMAL PROTECTION SYSTEMS

3.13.4.1 Ablation

Where an ablative material is used as the thermal protection system, the ablative material thickness shall be based on recognized ablative heat transfer theory as substantiated by suitable material property tests. The effect of thermal conductance and heat capacity shall be included in the thermal analyses. A factor of safety of 1.5 shall be applied to the nominal thickness required for ablation. If additional thickness of material is included as insulation only, a factor of 1.0 shall be applied to the nominal thickness required for this purpose.

3.14 MATERIAL-PROPERTIES AND ALLOWABLES

Material strengths and other physical properties shall be selected from authorized sources of reference such as MIL-HDBK-5 and from contractor test values when appropriate. Strength allowables and other mechanical properties used shall be appropriate to the loading conditions, design environments, and stress states for each structural member. Materials properties and allowables are to be consistent with overall vehicle reliability requirements.

Allowable material strengths used in design shall reflect the effects of load, temperature, and time associated with the design environment. Allowable yield and ultimate properties for common (MIL-HDBK-5) materials are as follows:

    a. For single load path structures, the minimum guaranteed values (A values in MIL-HDBK-5) shall be used.

    b. For multiple load path structures, 90 per cent probability values (B values in MIL-HDBK-5) may be used.

Determination of materials properties, including emissivity, stress-strain relationships, creep, fatigue, hardness, tendency to cold weld, shall include consideration of the effects of reduced pressures and the resulting out-gassing.

4.0 TEST REQUIREMENTS

Test requirements for the glider/transition section shall be as specified in ASNR 62-18.

5.0 DOCUMENTATION REQUIREMENTS

Reporting requirements shall be as specified in the Dyna-Soar System Statements of Work and/or other contractural documents.

6.0 DEFINITIONS AND NOMENCLATURE

6.1 DEFINITIONS

Air Vehicle - The air vehicle consists of the glider and transition section, SSLV, and all subsystems that become airborne upon launch.

Glider - The glider is the forward winged portion of the air vehicle and all subsystems thereof.

Standard Space Launch Vehicle (SSLV) - The SSLV is the portion of the air vehicle that provides the primary propulsive boost force to the glider and transition section and all subsystems thereof.

Transition Section - The transition section is the structural tie between the forward portion of the

SSLV and the aft end of the glider. The transition section may be utilized as a container for equipment and/or subsystems. All or part of the transition section may be carried with the glider during orbital flight but will normally be jettisoned before re-entry.

Limit Load - Limit load is the maximum load, or combination of loads, which a structure may be expected to experience during the performance of specified missions in specified environments.

Ultimate Load - Ultimate load is the maximum load for which the structure is designed. It is obtained by multiplying the limit load by the ultimate factor of safety.

Factor of Safety - The factor of safety is a multiplication factor used in design to assure adequacy of the structure.

Abort Vehicle - See Figure 3.0.2, ASNR62-4A.

Orbital Vehicle - See Figure 3.0.2. ASNR62-4A

Load Factor - Load factor is the ratio of the external forces acting on a mass to the weight of the mass. A subscript designates the direction of the load and of the load factor.

Critical Conditions - A critical condition is a loading condition for which the structure is to be designed.

Failure - Failure is the condition when a structure can no longer perform its intended function. Failure of a structure may result in the loss of the vehicle, or any part thereof, and/or which may present a hazard to operating personnel.

Material Yielding - Material yielding exists when either the yield stress or the yield strain is realized. For materials which do not exhibit a definite yield point, the yield stress shall be as defined below:

The yield strain is a permanent deformation of .002 inches per inch. The yield stress is determined by the intersection of the stress-strain curve with a straight line parallel to the original slope of the straight portion of the stressstrain curve which passes through the point of zero stress and .002 in/ in strain.

Hazard to Personnel - Any condition which may precipitate a structural failure resulting in endangerment to the safety of personnel is considered hazardous.

Excessive Deformations - Deformations, either elastic or inelastic, resulting from application of loads and/or temperatures, are excessive when any portion of the vehicle structure exceeds its design limits and which may preclude the vehicle from performing its intended mission.

Basic Glider Configuration - The basic glider configuration has all devices such as cockpit enclosures, landing gear, speed limiting devices, and extendable aerodynamic surfaces in their closed or retracted positions.

High-Drag Glider Configuration - The glider high-drag configuration is the same as the basic glider configuration, except that the speed limiting devices are in the fully-open position or any critical intermediate position as limited by the available actuation force and/or the external surface temperature limitations.

High-Lift Glider Configuration - The glider high-lift configuration is the same as the basic glider configuration, except that the extendable aerodynamic surfaces and speed limiting devices, if applicable, are in the maximum extendable position or any critical intermediate position.

Landing Glider Configuration - The glider landing configuration is the same as the basic configuration except that the landing gear and any other devices that are extended or open for landing are in their landing positions.

Design Weight - The weight as specified in ASNR62-4A on which performance, loads and temperature calculations are based:

a. Maximum Design Weight - That weight which is representative of the vehicle with maximum internal and external loading for which provisions are made.

b. Minimum Design Weight - That weight which is the maximum design weight (in (a) above) less the estimated expendables but including the reserves.

c. Re-entry Design Weight - That weight which is the maximum design weight (in (a) above) less estimated expendables used to the initiation of re-entry.

Pressure Vessels - Pressure vessels are defined as those containers which must be designed to withstand internal operating pressures; i.e., pressurized compartments, propellant tanks, storage bottles, plumbing, actuators, etc., but excluding components such as structural tubing, fins, etc., even though the components are acted upon by internal or external pressures.

Limit Pressure - Limit pressure is the maximum expected operating pressure (MEP) as determined by (a) the effects of transient peaks, 3 sigma variations in pressure, and vehicle accelerations; or (b) maximum relief valve setting times a specified factor of safety; or (c) the nominal pressure times a specified factor of safety.

Nominal Pressure - Nominal pressure is the maximum pressure to which a pressurized system, subsystem, or component is subjected under steady state conditions.

Proof Pressure - Proof pressure is the pressure applied to all pressure vessels, systems, subsystems, and components, as a singular load condition and at specified temperatures, to demonstrate adequacy prior to installation. Proof pressure is the product of the limit or nominal pressure and the appropriate proof pressure factor.

Burst Pressure - Burst pressure is the product of the limit or nominal pressure and the appropriate burst factor of safety.

Relative Velocity - Relative velocity is the velocity of the vehicle relative to a non-rotating earth.

Relative Wind Velocity - Relative wind velocity is the velocity of the vehicle relative to the free airstream.

Ground Phase –

    a. SSLV - The ground phase for the SSLV begins when a specific SSLV, identified to be used in conjunction with the Dyna-Soar payload is assembled on the launch pad and continues through glider mating with the SSLV and air vehicle servicing in readiness for flight.

    b. Glider - The ground phase for the glider includes all phases of ground operations, but excluding pre-launch and landing, from the time the glider leaves the assembly line until it is returned for refurbishment.

Pre-Launch Phase.- The pre-launch phase is the period following removal of the gantry or other support and terminating at lift-off.

Launch Phase - The launch phase is the period beginning at liftoff and extending to the time when the launch transients have damped out.

Boost Phase - The boost phase is the period beginning when the launch transients have damped out and ending when the final propulsive unit has expended.

Flight Phase - The flight phases include all flight operations wherein the glider has separated from the SSLV or carrier aircraft and ending when the glider has come to rest after landing.

Staging - The programmed physical separation of the component parts from the remaining portion of the vehicle.

Creep Failure - A structural creep failure is any structural deformation due to creep which is detrimental to the utility or integrity of any structural component.

6.2 NOMENCLATURE

a	= sonic velocity - (feet per second)
c	= average chord - (feet)
C_L	= coefficient of lift (non-dimensional)
$C_L\alpha$	= lift curve slope - (per radian)
g	= 32.2 fps^2
g	= flutter and vibration damping coefficient
h	= altitude - (feet)
K_w	= gust alleviation factor (non-dimensional)
L/D	= lift over drag ratio (non-dimensional)
m	= slope of curve for C_N vs alpha (per radian)
M	= Mach number (non-dimensional)
MAC	= mean aerodynamic chord (feet)
n_g	= gust load factor (non-dimensional)
$n_{z,y,x}$	= load factor (non-dimensional subscript denotes axis along which load factor acts)
q	= dynamic pressure - (pounds per square foot)
S	= wing area - (square feet)
t	= time - (seconds)
U_{true}	= true gust velocity - (feet per second)
V	= velocity - (feet per second)
V_g	= gust velocity - (feet per second)
V_s	= steady state wind velocity - (feet per second)
T	= gust time base - (second)
V_{true}	= true velocity - (feet per second)
w	= weight (pounds)
α	= angle of attack (degrees)
β	= yaw angle (degrees)
δ	= control displacement (degrees)
Δ	= incremental change
μ	= $\dfrac{2W/S}{g\,c\,m\,p}$
ρ	= density (slugs per cubic foot)
δ	= density ratio - (non-dimensional)
σ	= standard deviation (non-dimensional)
ϕ	= bank angle (°)
γ	= time base of gust (secs)

Subscripts x- roll axis, y - pitch axis, z - yaw axis

Date: Jan 1963   Title: Information Fact Sheet

AIR FORCE
INFORMATION
FACT SHEET

X-20 DYNA-SOAR

OFFICE OF INFORMATION
OFFICE OF THE SECRETARY OF THE AIR FORCE
INTERNAL INFORMATION DIVISION
WASHINGTON 25. D C

X-20 · DYNA-SOAR

"It looks like a cross between a porpoise and a manta ray. It is jet black, and its stubby nose leads a thick, round-shouldered, flat-bellied body about 35 feet long which is abruptly sawed off at the read end. The entire body sits atop a great triangle of a wing whose outboard edges point straight up in the shape of an airplane's vertical stabilizer. It is a manned space glider -- and one of the most important things to have happened in aviation since the Wright Brothers' first flight..." - John G. Hubbell in the September 1962 issue of Reader's Digest.

INDEX

The information contained in this fact sheet is for use in the Air Force Internal Information Program.

JANUARY 1963

I - BACKGROUND OF DYNA-SOAR

Definition

The X-20 Dyna-Soar is a manned aerospace glider combining characteristics of missile, airplane and satellite. It will have multi-orbit capability. Dyna-Soar will be a true aero-spacecraft — a vehicle that can operate both within and beyond the sensible atmosphere. An aircraft flies within the atmosphere and a true spacecraft flies principally in the space environment.

The name Dyna-Soar is an abbreviation of dynamic soaring, a term that describes the aerodynamic principles of its operation, thrust, lift and drag. The booster's thrust will hurl Dyna-Soar into orbit, building up a store of energy, the kinetic energy of motion and the potential energy of its altitude. The pilot uses retro rockets to terminate orbit and then maneuvers the aerodynamic shape through the atmosphere by the use of the control surfaces of the craft. The delta-winged aerospace glider is to be launched into orbit by the Titan III rocket booster.

The X-20 design will permit the pilot to choose his point of re-entry into the atmosphere from orbit. The wings of the X-20 will provide the means for aerodynamic maneuvers. The pilot can extend or shorten his flight path and can turn to either side. This maneuverability will permit landing at any suitable air base on return from Earth-orbit.

Significance

While the bulk of Dyna-Soar's military and peaceful significance may lie beyond a test period, this research phase is one of the most vital steps ever taken in either military or scientific exploration of space. The manned orbital launches towards which the program is aimed will demonstrate that controlled re-entry into atmosphere, and normal landing selection by a manned aerospace vehicle is reliable and safe to the point it could be called routine. Gen. Bernard A. Schriever, chief of Air Force Systems Command, has said: "The nation that first achieves the ability to maneuver, to communicate and to carry out military missions in and from space will enjoy a strategic advantage that could be decisive if not effectively countered."

History

Commencing in 1954 the Air Research and Development Command sponsored several studies on boost glide systems. These were finally combined into a single plan, accepted by the Air Force in 1957. The Air Force interest in a boost glide system was shared by the National Advisory Committee on Aeronautics, predecessor to the present NASA.

In November 1957, the Air Force issued the first preliminary directives on Dyna-Soar, and in June 1958, selected two major teams to prepare competitive studies. Boeing headed one team; Martin Company and Bell Aircraft the other.

Two years later, in November 1959, the Air Force selected Boeing as system contractor and the Martin Company as associate booster contractor. Progress has been uniform since an intensive study ordered by the Air Force ("phase Alpha") was concluded in April 1960.

A government inspection team, in September 1961, reviewed a full scale mockup of the Dyna-Soar glider and its related systems, ordering no significant changes.

In December 1961, the Air Force announced the development of the Titan III booster, which will propel Dyna-Soar to orbital velocities rather than the originally scheduled sub-orbital flights.

Titan III Facts

*Greater payload capability and versatility with increased launch rate.

*In addition to launch vehicles, Titan III program calls for development of associated aerospace ground equipment designed to reduce substantially the time-on-launch-pad and the number of pads required.

*Titan III (as part of the National Launch Vehicle Program) will meet requirements in the 5,000 to 25,000 pound payload range for relatively low altitude orbits, accommodating payload capabilities ranging from placing 10 tons in a 100-nautical-mile orbit to orbiting 13,900 pounds at 1,000 nautical miles.

*Development and management for Titan III is the responsibility of the Space Systems Division (SSD), Air Force Systems Command.

*Under contract to SSD, Aerospace Corporation will provide systems engineering and technical direction.

*Titan III will employ a modified version of the Titan II guidance system. Martin-Marietta Corporation, Denver Division, is system integrator. Martin also produces the Titan II core.

II - RECENT DEVELOPMENTS

Joint USAF/NASA Understanding
The following is a verbatim reproduction of a joint memorandum signed by Secretary of the Air Force Eugene M. Zuckert and National Aeronautics and Space Administration (NASA) Administrator James E. Webb.

MEMORANDUM OF UNDERSTANDING

Subject: Principles for NASA participation in Dyna-Soar program, (Supersedes Memorandum of Understanding dated November 14, 1958.)

1. The Dyna-Soar glider is a manned re-entry vehicle with unique capabilities of high maneuverability at hypersonic speeds, and conventional tangential landing. One of the objectives of the Dyna-Soar program is to explore the problems and potentialities of this type of highly maneuverable manned re-entry vehicle,

2. NASA endorses this objective as necessary to the national aerospace program.

3. The following principles of conduct will be applied in the remaining phases of the project:
    (a) Dyna-Soar is an Air Force program.

(b) Technical support (consulting and ground-facilities testing) will continue to be provided by NASA Centers as initially agreed upon with the Air Force.

(c) Instrumentation and flight-test support will be provided by NASA Flight Research Center to an extent to be mutually agreed upon with the Air Force.

Signed:
Eugene M. Zuckert James E. Webb
Dated: August 7, 1962

Designation of X-20

The Dyna-Soar piloted aerospace glider has been designated the X-20. This describes the experimental nature of the program to test the piloted aero-spacecraft and follows the long standing procedure of giving each craft a letternumber designator.

Although "X-20" is the official primary designator for the piloted aerospace glider, and the program for its development and test, it is not intended to replace the project nickname, "Dyna-Soar."

Continued reference to the nickname is authorized for general program identification. In this case, it is proper to use the two terms separately or in combination; for example, the "X-20," the "X-20 Dyna-Soar," or the "Dyna-Soar."

From Sub-Orbital to Orbital

In December 1961, the Department of Defense announced the Titan III program to provide a standardized space launch system for performing a variety of manned and unmanned booster missions.

The Titan III, replacing Titan II as the launch system for the X-20, eliminates all sub-orbital flights and gives impetus to the Dyna-Soar program by changing it from sub-orbital to orbital.

The Air Force estimates that the Titan III will accelerate the orbital Dyna-Soar program by two years, permitting the scheduling of unmanned Dyna-Soar flights in 1964 and manned flights a year later.

X-20 Acceleration Rocket Fired

The acceleration rocket being developed for the Air Force X-20 manned space glider has been successfully static test fired.

The test, by the Thiokol Chemical Corporation in Elkton, Maryland, demonstrated that the solid propellant grain design can deliver the performance required for two missions. The acceleration rocket will provide quick escape for the X-20 in the event of booster malfunction during orbital launch, as well as accelerate the glider to supersonic speeds in early airdrops from the B-52 mother ship.

The rocket engine configuration fired in this test was equipped with low cost, fixed nozzles and an interim rather than a flight-weight case. In order to minimize costs, tests of the flight-weight case and of nozzles which will control direction of the rocket's thrust are being conducted separately:

III - DYNA-SOAR PILOTS

Five out of six pilots selected to fly the X-20 Dyna-Soar piloted orbital glider are Air Force officers with extensive experience in supersonic aircraft and experimental flight test techniques.

These pilots, who will be the first to fly the 18,000 mile per hour aerospace craft, are all graduates of the Air Force's Experimental Test Pilot School, or Aerospace Research Pilot School at Edwards AFB, Calif.

They are pilot-engineers who handle supersonic aircraft or slide rules with equal ease. All pilots are checked out in "Century Series" aircraft, and will complete courses in aerodynamics, flight test techniques, astronomy, astrophysics, meteorology, star navigation, communications, computer theory, rocket engines and fuels, orbital mechanics, heat transfer, and space medicine before their initial flight.

The five Air Force space pilots are;

Capt. Albert H. Crews, 33, an experimental test pilot at the Air Force Flight Test Center. He is one of eight test pilots attending the Aerospace Research Pilot course at Edwards AFB, Calif. Captain Crews is a native of El Dorado, Ark.

Maj. Henry C. Gordon, 36, also an experimental test pilot at the Air Force Flight Test Center, and air combat veteran of Korea. Major Gordon was born in Valparaiso, Ind.

Capt. William J. Knight, 32, another member of the test pilot team at Edwards AFB, was born in Noblesville, Ind. He is qualified to pilot all the Air force's "Century Series" aircraft.

Maj. Russell L. Rogers, 34, flew 142 missions during the Korean conflict. He was born in Lawrence, Kans. and has been a test pilot at the Air Force Flight Test Center since April 1959.

Maj. James W. Wood, 38, has been a test pilot at Edwards AFB since 1957. Until his selection as one of the Dyna-Soar pilots, he was assistant chief of fighter operations in the Air Force Flight Test Center's Directorate of Flight Test.

Sixth member of the Dyna-Soar team is Milton O. Thompson, a test pilot with the National Aeronautics and Space Administration who is a WW II Naval veteran.

Prime training facilities for Air Force piloted orbital flight of the X-20 program are at Edwards AFB.

IV - DYNA-SOAR TERMS

BOOSTER - That portion of the Dyna-Soar vehicle which provides orbital boost force to the spacecraft transition and all systems thereof.

ENERGY MANAGEMENT - The control of the Dyna-Soar vehicle flight path is such that the potential and kinetic energy of the vehicle is expended in a manner permitting a pilot controlled subsonic landing at a predetermined point on the earth's surface.

FLIGHT CORRIDOR - The altitude depth or envelope within which the glider is capable of an equilibrium or maneuvering flight. The altitude depth at a particular velocity is determined by the maximum lift and weight of the glider for the upper limit and the capability of the structure to withstand thermal heating and/or aerodynamic loads at the lower limit.

GLIDER - The winged portion of the Dyna-Soar vehicle, which reenters the earth's atmosphere after hypersonic or orbital flight.

GO-AROUND CAPABILITY - The ability of a powered air vehicle to make more than one landing approach.

INTERSTAGE - The structural tie that transmits the loads from the one stage to the next stage of the booster.

MAN-RATED - Any vehicle, glider, test apparatus or equipment designed and approved by recognized authority to accommodate an operator, pilot or other personnel; also the approval of a man-made environment for safe human occupancy.

RE-ENTRY DEVICE - Any man-made device intended for re-entry into earth's atmosphere at extreme speeds.

V - QUOTATIONS

'If we are to have spacecraft with the flexibility required in military systems we must develop the capability to land these craft at points and times of our choosing, and to land them in a condition in which they can be readily turned around and re-used. We cannot be satisfied with ballistic type re-entry in which we parachute our craft into the ocean and recover them by salvage-type operations. The special technology necessary to satisfy the military requirement will not be produced as a fallout of present or planned NASA programs."

Secretary of the Air Force Eugene M. Zuckert

"The Dyna-Soar may be regarded as a logical follow-on to the X-15 in the exploration of aerospace. This military test system has the capability for manned, maneuverable, hypersonic re-entry from orbital altitudes and velocities with a normal landing at conventional air fields."

Under Secretary of the Air Force Joseph Charyk

"Eventually I see the development of a manned vehicle that can take off from existing runways, go into orbit, maneuver into a parking orbit, de-orbit, maneuver while re-entering the Earth's atmosphere and land at
an air base in the conventional manner. A manned vehicle of this type could have many important military applications in space. Since space systems are extremely expensive, one of the first tasks of a manned space vehicle would be to repair equipment operating in our unmanned satellites. The first nation to develop a manned space vehicle with complete mission flexibility could possibly dominate the space above the atmosphere...."

Air Force Chief of Staff Gen. Curtis E. LeMay Oct. 26, 1961

"The choice of flight paths available to the Dyna-Soar pilot will be almost infinite. By combining the high speed and extreme altitude of his craft with his ability to maneuver, he will be able to pick any air field between Point Barrow, Alaska and San Diego, California with equal ease."

George H. Stoner Dyna-Soar program manager for Boeing Sept. 22, 1960

"The committee foresees the need for an operational, manned military space vehicle over which the pilot has the greatest possible control and believes that the Dyna-Soar concept provides the quickest and best means of obtaining this objective."

Appropriations Committee, House of Representatives May, 1961

"It may be asked, why send a man aboard a Dyna-Soar into space, when significant scientific information on space flight can be obtained from unmanned probes? The answer is that man always has been a major contributing factor to the development of any type of equipment. He has the ability to think, to change plans, to take corrective action in the event of emergency, and to analyze."

Robert L. Twiss, aviation editor of The Seattle (Washington) Times Ordnance, March-April, 1962

VI -BIBLIOGRAPHY

Haugland, Vern "Manned Space Flight" Ordnance, January-February, 1963
Hubbell, John G. "Here Comes the Dyna-Soar" Reader's Digest, September, 1962 (Condensed from Air Facts)
Twiss, Robert L. "Dyna-Soar—The Space Glider" Ordnance, March-April, 1962
Butz, J. S. Jr. "Dyna-Soar Plus Titan III" AIR FORCE/SPACE DIGEST, February, 1962
Blair, Edison T. "This is Dyna-Soar" The Airman, October, 1960
Yoler, Usaf A. "Dyna-Soar - A Review of the Technology" Aerospace Engineering, v.20, no.8

Date: Jan 1963   Title: AFSC Press Release

AFSC NEWS SERVICE

OFFICE OF INFORMATION HEADQUARTERS
AIR FORCE SYSTEMS COMMAND
UNITED STATES AIR FORCE
ANDREWS AIR FORCE BASE, WASHINGTON 25, D. C.

TELEPHONE (Area code 301) 981-4137
4138

RELEASE NO. 31-R-5

For Release: 21 January 1963

X-20A DYNA-SOAR GUIDANCE SYSTEM
IN FIRST USAF FLIGHT TEST

ANDREWS AFB, Md. – The first test flight of the X-20A Dyna-Soar's inertial guidance system recently was conducted over the Eglin Gulf Test Range, Air Proving Ground Center, Eglin AFB, Fla., the Air Force Systems Command revealed today.

A "test bed" NF-101B aircraft, piloted by Jim Bailey, a Minneapolis-Honeywell engineering pilot, flew the first in a series of test flights over the EGTR. The inertial guidance system was carried in the nose of the jet "Voodoo" aircraft for the flight.

Minneapolis-Honeywell, contractor for the inertial guidance system, will conduct developmental testing at APGC for about six months.

The program will provide information on operation of the inertial guidance system and its ability to provide necessary navigational data. About 24 flights are planned over the EGTR.

Don R. Hanford, an electrical engineer from Minneapolis-Honeywell, accompanied Bailey as an engineering observer.

Actually there are two facets to the program: Flight testing at the APGC; and rocket sled testing at Holloman AFB, N.M,

The X-20A Dyna-Soar is a hypersonic, lifting re-entry, orbital vehicle currently under development by the Air Force.

The inertial guidance system is designed to permit the pilot to guide the Dyna-Soar space glider – known as the X-20A – in orbit and to bring it back to earth at selected landing sites. The pilot will use the manual controls of the X-20A, re-entering the atmosphere at a speed of around 15,000 miles per hour. He will make a conventional landing at a base of his choice.

Minneapolis-Honeywell began to set up shop at Eglin AFB in mid-October with Russell Thompson serving as site supervisor. Thompson's staff will eventually grow to about 25 people.

APGC is furnishing considerable physical facilities, fuel and equipment to Minneapolis-Honeywell for conducting the tests. The Deputy for Aircraft and Missile Test headed by Col. Henry A. Orban monitors the program with Maj. William A. Henderson serving as Air Force project coordinator.

Contract management for the Dyna-Soar is handled by AFSC's Aeronautical Systems Division at Wright-Patterson AFB, Ohio. The Titan III booster for the system is under management of the Space Systems Division, Inglewood, Calif.

X-20 PERFORMANCE SPECIFICATION
ASNR 62-5
GLIDER/TRANSITION-LAUNCH VEHICLE MALFUNCTION DETECTION SUBSYSTEM

30 January 1963

This Specification Tab is a part of the Dyna-Soar Statement of Work, dated 26 January 1962
ASNR 62-5 Revised 1 November 1962

1.0 DEFINITION AND SCOPE OF SYSTEM

1.1 Objective

1.1.1 General

In recognition of the inherent reliability limitations of large boosters and the associated hazards facing the pilot, the X-20A program requires that an abort capability be provided for the unmanned and manned glider. The glider/transition-launch vehicle malfunction detection subsystem, described herein, is intended to detect all malfunctions of sufficient importance to require pilot corrective action or immediate abort, In performing the foregoing functions the Malfunction Detection Subsystem (MDS) shall supply appropriate signals to the glider. Through the integrated action of the MDS, crew and glider subsystems a safety goal approaching 100 per cent is attainable. It is essential to recognize that air vehicle instrumentation, pilot recognition and manual corrective action are complementary to the MDS. Since many launch vehicle failures are of such a nature that the glider/transition limits will be exceeded within the reaction time of crew response, it is necessary that an automatic abort capability be provided to initiate timely abort action for those launch vehicle failures which will be catastrophic.

1.2 Basic Operating Principles

Recognizing that a malfunction may exhibit many failure modes, and that the rate and consequences will vary with many parameters, the MDS may sense the result of the malfunction rather than the initial failure. Corrective action, other than automatic engine shutdown, if required to insure escape, shall not be incorporated. Differentiation between "minor" and "catastrophic" type failures shall be accomplished by utilization of suitable sensors in the MDS such that rapid failure progression will result in immediate automatic abort action,

1.3 Description of System

The MDS shall consist of a series of sensors located throughout all stages of the launch vehicle and in the transition section, a crew display, a separation sequence programmer located in the glider, and the required connecting circuits and enabling circuits. The glider/transition-launch vehicle MDS block diagram is presented in Figure 1.

1.4 Scope and Purpose of Specification

1.4.1 Scope

The definitions, objectives and performance requirements established herein encompass the performance, testing and glider/transition-launch vehicle interfaces associated with the design of the MDS.

1.4.2 Purpose

This specification will provide documentation of the launch vehicle and glider/transition requirements that must be considered in the integration and development of the MDS.

2.0 APPLICABLE DOCUMENTS

The following Government specifications and publications of the exact issue specified form a part of this specification to the extent noted herein. In case of conflict between requirements specified herein and the specifications and publications referred to below, this specification shall govern. ASNR 62-4A Dyna-Soar System Specification, Glider/Transition Section, dated 6 July 1962 ASNR 62-4B Tab B, X-20A System Specification, System 620A Requirements Document for System 624A, dated 31 January 1963

3.0 REQUIREMENTS - PERFORMANCE AND DESIGN

3.1 Performance

3.1.1 Performance - Functional

3.1.2 MDS - Glider/Transition-Launch Vehicle

The glider/transition-launch vehicle MDS shall detect potentially catastrophic malfunctions in flight and generate signals to initiate appropriate action which will result in a maximum probability of successful glider abort. The MDS shall monitor parameters in which variations indicate dangerous malfunctions in glider/transition-launch vehicle subsystems. The choice of parameters to be monitored and the response characteristics of the sensors performing such monitoring shall be consistent with reliability and safety requirements.

3.1.2.1 MDS Ground Rules

3.1.2.1.1

During Prelaunch, the malfunction detection subsystem shall be rendered passive up to 5 seconds prior to lift-off, and shall be monitored by ground control. During Prelaunch, after abort system arming, launch vehicle explosion, structural failure or inadvertent destruct initiation shall result in immediate automatic glider abort. The MDS shall become active 5 seconds prior to lift-off and continue to function throughout boost flight.

3.1.2.1.2

Sense malfunctions with sufficient lead time to initiate abort. The MDS reaction time shall be sufficiently short to;
a. Be compatible with the structural and mechanical capability of the glider and transition section immediately prior to the initiation of abort and also, after abort vehicle separation from the launch vehicle.
b. Not exceed separated glider control authority.
c. Filtering time shall be 30 milliseconds. Sensor dead-band shall be adequate to minimize the probability of unwarranted aborts. d. Sensors dead-band shall be short enough to provide an abort sequence which is compatible with glider or glider/transition-launch vehicle structural and control limits.

3.1.2.1.3

Initiate automatic launch vehicle engine shutdown and glider abort for impending catastrophes. (No launch vehicle shutdown for first 32 seconds of flight). For the first stage, solid engines "engine shutdown" shall be construed to mean thrust termination and/or thrust reversal.

3.1.2.1.4 MDS need not identify specific malfunctions nor correct malfunctions in flight. The MDS shall only detect malfunctions or their effects and initiate thrust termination and glider abort.

3.1.2.1.5 MDS should be functionally independent of other launch vehicle subsystems. A permissible exception to this rule shall be utilization of launch vehicle power for energizing the MDS.

3.1.2.1.6 MDS design goal is to approach 100 per cent reliability.

3.1.2.1.7 MDS equipment must be more reliable than equipment being monitored. "Reliable", is defined as having the ability to function repeatedly within prescribed limits and tolerances. The limits and tolerances shall be assigned in consonance with the MDS mission.

3.1.2.1.8 Design redundancy shall be provided where dictated by sensor reliability or area criticality. Consideration shall be given to the placement of sensors and wiring to prevent simultaneous failure of redundant elements as a result of their environment.

3.1.2.1.9 The MDS shall be designed to minimize the possibility of unnecessary abort consistent with the other ground rules.

3.1.2.1.10 The glider/transition-launch vehicle MDS shall not cause any subsystem malfunction or a false abort. A single component failure of the subsystem shall render only that part of the subsystem inoperative.

3.1.3 MDS - Launch Vehicle
The characteristics of those sensors which are intended to monitor launch vehicle subsystem malfunctions or their effects shall be determined by launch vehicle subsystem performance characteristics. The settings of these sensors shall be the responsibility of the 624A SPO. The choice of launch vehicle subsystem parameters and the response characteristics of the sensors performing such monitoring shall be consistent with the reliability and safety requirements contained in document ASNR 62-4B,

3.1.3.1 Sensor settings shall be adjustable prior to flight. Provisions shall be made for locking and sealing of the settings to insure that setting cannot be changed without authorization.

3.1.3.2 The launch vehicle portion of the MDS shall be powered by the launch vehicle power supply. This voltage shall energize relays which maintain the level of glider MDS voltage. Loss of launch vehicle voltage (through the opening of an MDS sensor) shall cause de-energizing of the relays, loss of glider MDS voltage and, thus the initiation of an abort.

3.1.3.3 When a malfunction is detected, the launch vehicle MDS signal to the glider shall drop and remain below 10 volts in a maximum of 50 milliseconds to signal an abort.

3.1.3.3.1 The launch vehicle engine shutdown relay shall initiate shutdown at any time MDS voltage drops below 10 volts.

3.1.3.4 The launch vehicle subsystem shall contain provisions for disarming the glider/transition-launch vehicle MDS while on the launch pad and arming the subsystem prior to lift-off.

3.1.3.4.1 After arming of the abort system, provisions shall be made for automatic abort while on the pad in the event an inadvertent launch vehicle explosion occurs.

3.1.4 MDS - Glider

The glider portion of the glider/transition-launch vehicle MDS shall consist of wiring, connectors, and switching devices to provide sequence control of abort or normal glider/booster separation operations. A glider initiated abort causes the voltage to the Separation Sequence Programmer and launch vehicle shutdown relays to drop below 10 volts, thus initiating launch vehicle shutdown and abort. (No launch vehicle shutdown for first 32 seconds of flight.)

3.1.4.1 Glider-launch vehicle interface requirements shall be as presented in document ASNR 62-4A.

3.1.4.2 Separation Sequence Programmer - Abort occurs when the abort initiation discreets drop to zero from the nominal 28 volt signal. However, the operating points of the Separation Sequence Programmer are:

a. Below approximately 23,000 fps when the launch vehicle MDS signal drops below 10.5 + volts for a period exceeding 0.3 milliseconds.
b. Above approximately 23,000 fps when the launch vehicle MDS signal drops below 8.5 + 0.5 volts for a period exceeding 1.0 + 0.1 seconds. Abort signals shall not be generated in the presence of an external or glider applied blocking voltage.

3.2 Reliability The following requirements shall be demonstrated prior to the first manned ground launch:

3.2.1. As a requirement, the pre-injection reliability of the launch vehicle MDS when detecting malfunctions shall be a minimum of .9991. The probability of the subsystem giving a false warning shall be .0009 or less.

3.2.2 As a requirement, the reliability of the glider portion of the glider/transition-launch vehicle MDS when detecting malfunctions shall be a minimum of .9999. The probability of the glider portion of the MDS giving a false warning shall be .0001 or less. 3.3 Human Factors Cockpit presentation shall consist of an abort warning light visible to the pilot and energized by the glider separation sequence programmer at the instant of abort initiation. The pilot is also supplied with a manual abort switch. Recognizing that some launch vehicle malfunctions will result in glider/transition-launch vehicle divergence rates too low to trip the MS sensors, it is considered that, where adequate pilot reaction time is available, the pilot will be utilized as an active part of MDS.

4.0 REQUIREMENTS - TESTING, CHECKOUT AND EQUIPMENT DESIGN

Testing, performance, calibration and environmental testing provisions are as provided in ASNR 62-4A. In recognition of the glider/transition launch vehicle MDS objectives stated in paragraph 1.1, it is mandatory that all testing and all checkout operations or procedures be designed and conducted in a manner that will preserve MDS functional qualities and leave it unimpaired to accomplish its intended mission.

4.1 Capabilities

4.1.1 The glider/transition-launch vehicle malfunction detection subsystem shall be so designed as to provide plug- in test connections for checkout and fault isolation. Establishment of test connections shall not interrupt the normal circuits nor require their disconnection. Access shall be in accordance with accented design practices.

4.1.2 The glider/transition-launch vehicle malfunction detection subsystem shall be, monitored by the checkout equipment during pre-launch operations-to insure proper functioning at lift-off. An indication of improper MDS or other subsystems operation shall prevent launch of the glider/transition-launch vehicle.

4.1.3 Further provisions shall be defined in D2-80045. "Performance Requirements Ground Checkout Equipment Subsystems."

4.2 Equipment The glider/transition-launch vehicle malfunction detection subsystem checkout equipment shall be integrated in the countdown through the Master Operational Controller (MOC). The MOC shall integrate these inputs with those from the Pad Safety Officer, Flight Safety Officer, and superintendent of Range Operations to determine countdown continuance, hold-fire, or to initiate glider abort.

4.3. Procedures

4.3.1. Qualification testing of the MDS shall be conducted to insure that equipment will perform its intended function under X-20A environmental conditions, demonstrating the reliability specified in paragraph 3.2.

4.3.2 During early launch vehicle R and D test flights the MDS shall be monitored. Data shall be telemetered to verify design adequacy and functional design.

4.4 Monitoring

4.4.1 The glider/transition-launch vehicle MDS shall be monitored for diagnostic-purposes and reliability assessment.

4.4.2 Pre-Launch Phase The MDS shall be monitored during pre-launch as previously noted in Paragraph 4.1.2.

4.4.3 Operational Telemetry channels to be made available for monitoring of performance of the MDS under the operational mode may be classified as follows: a. Those will provide a quantitative check where feasible of parameters which are being monitored by the MDS. b.. Those which will provide a qualitative check (operation, no operation) of all MDS sensors.

5.0 NOTES

5.1 Definitions

Active: Able to respond and to a signal and activate other systems without approval from another monitoring system or source.

Catastrophe: A condition which:
a. Destroys the glider and/or launch vehicle, or
b. Incapacitates the pilot so that he cannot act in his own behalf, or
c. Places the glider in a position from which successful abort cannot be achieved.

Passive: Able to respond to a signal and activate other systems only after approval from another monitoring system or source.

Reaction Time: The interval from the detection of a subsystem malfunction or glider-transition-launch vehicle flight path divergence, at a predetermined critical value, to glider separation from the launch vehicle.

MDS Mission The MDS mission is defined as monitoring parametric variations in the booster and transition which are considered indicative of impending catastrophe and initiating the glider separation sequence. The mission begins as soon as launch preparations allow the system to be armed and ends with transmission of the separation sequence programmer signal to separate the glider from the launch vehicle.

Launch Vehicle Limits: Limits concerning the launch vehicle, (environment, trajectory, attitudes, temp., guidance, control, structure, etc,) which, if exceeded, would constitute a catastrophe.

Glider-Transition Recovery Limits All limits concerning the glider and transition section which define the structural and control capability of the glider- transition section during an abort maneuver.

Glider/Transition Launch Vehicle Limits Limits concerning the glider/transition-launch vehicle which are an integration of both the launch vehicle and glider-transition limits as applicable to the total vehicle.

Normal Variation: Non-critical fluctuations of parameters being monitored.

Dead-Band: The time interval preceding failure detection. Its origin is the tolerance allowed in a sensor alarm setting and it includes sensor reaction time.

Transient Filtering: (Approximately 30 milliseconds) time from generation of failure signal (by sensor) to completion of filtering and transmission of signal to signaling relays.

Signaling Relay Reaction: Time from receipt of signal from filters to transmission of signal to launch vehicle shutdown relays and separation sequence programmer.

Separation Sequence Programmer Reaction: Time from receipt of abort signal malfunction signaling relays to transmission of signal to fire acceleration rocket motor and start separation sequence.

Abort Engine Thrust Build-up: Time from receipt of signal from separation sequence programmer to the point where glider separation is possible. (Separation is possible when glider acceleration is equal to launch vehicle deceleration.)

Separation Device Reaction: . Time from receipt of signal (from delay timer in separation sequence programmer) to firing of separation device.

Glider Recovery Period: This period begins with glider separation and ends at the glider recovery limit. Sensor alarm settings shall be such as to permit release of the glider in time to complete the recovery maneuver prior to arrival at the glider limit.

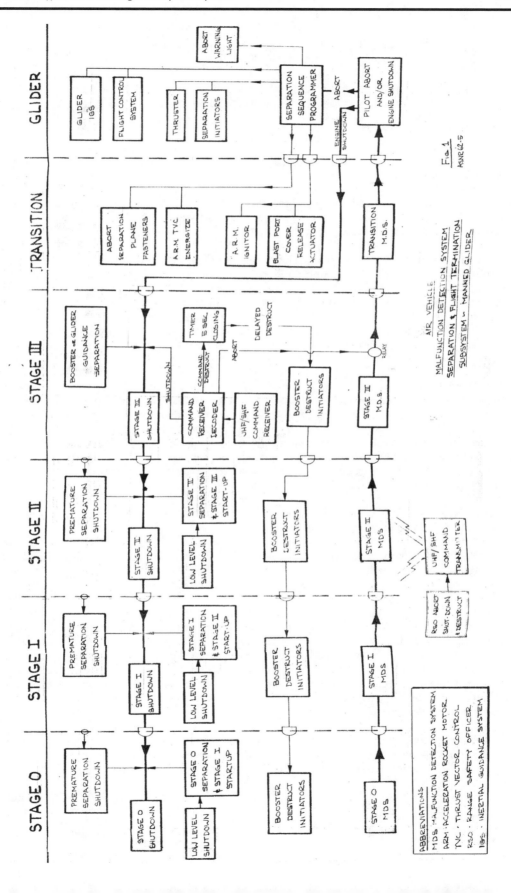

Fig. 1
ASNR 62-5

AIR VEHICLE
MALFUNCTION DETECTION SYSTEM
SEPARATION & FLIGHT TERMINATION
SUBSYSTEM ~ MANNED GLIDER

ABBREVIATIONS
MDS - MALFUNCTION DETECTION SYSTEM
ARM - ACCELERATION ROCKET MOTOR
TVC - THRUST VECTOR CONTROL
RSO - RANGE SAFETY OFFICER
IGS - INERTIAL GUIDANCE SYSTEM

Date: Mar 1963   **Title:** Memo: From James Webb to Col. Young

NATIONAL AERONAUTICS AND SPACE ADMINISTRATION
WASHINGTON 25, D. C.

OFFICE OF THE ADMINISTRATOR

March 1, 1963

MEMORANDUM for Colonel Young

So as not to have to rely on memory but preserve some note in a manner that does not get any wide distribution, the following is set down as the subjects of discussion with Mr. McNamara and Mr. Gilpatric by Dryden, Seamans, and myself today.

(1) We talked about the fact that our series of agreements, culminating with the Gemini 1, might be extended to include an additional one on Dyna-Soar, one on the B-70, one on the TFX. With respect to the B-70, it was pointed out by Dr. Seamans that we were already working with the Air Force and had agreed to put about a million dollars into instrumentation. It seemed best to continue on this program rather than endeavoring to have a formal exchange of letters, and Dr. Seamans was to talk with Mr. Rebel as to any further steps that might usefully be taken. However, it was generally agreed that there might come a time when we would want to have an exchange of letters and work the B-70 program much as the X-15.

(2) With respect to Dyna-Soar, Mr. McNamara expressed his strong desire to know precisely what benefits this would have for us and whether or not those benefits could not be obtained from some other program or from additions to one of our programs rather than having to spend the 600 million dollars involved in finishing the Dyna-Soar project. He expressed a desire to be briefed on the Gemini projects and it was agreed that he would be willing to go to Houston for this briefing. Dr. Seamans is to get in touch with Dr. Brown to arrange this; perhaps as Mr. McNamara travels to Seattle to examine Dyna-Soar he can stop for three hours at Houston. Dr. Seamans is to arrange this briefing and is to be present.

(3) With respect to the TFX, no conclusion was reached, but inasmuch as we have a letter on file, I believe this will come forward to completion.

(4) I mentioned that in connection with communications satellites, there were some indications that both in Congress and in industry an understanding of why NASA should continue work in this field was being questioned. We pointed out that the decisions necessary to set up a system, either military or civilian, could undoubtedly be made better eighteen months from now if we could wait that long, but there seems some urgency in connection with the military procurement. It was generally agreed, particularly in view of the low estimates of revenue from such a system, that our first endeavor should be to determine exactly what could be accomplished with a system, and then determine whether or not a military requirement over and above the commercial system existed. I pointed out the need for going forward with our multiple launch experiments and the other facets of our program that would be of great benefit in determining what really could be done with a system. I also pointed out that the continuance of our program was important to prevent assumption of control in this area by AT&T by default of any other competence.

(5) In a general discussion as to what military missions might be envisioned and the areas of technology which should be developed to make them possible, certain conclusions were arrived at which Dr. Dryden and Dr. Seamans both understand.

James E. Webb
Administrator

Copy to:AA - Dr. Seamans, AD - Dr. Dryden

Date: Mar 1963   Title: X-20 Program Report

X-20A PROGRAM REPORT

D2-80852   March, 1963

TABLE OF CONTENTS

X-20 PROGRAM REPORT

The X-20 is the most advanced aerodynamic test vehicle under development anywhere in the free world. It is intended to explore the region of flight beyond that planned or contemplated for the X-15, and is a logical and timely extension of its capabilities in this field. This is a flight region which it is necessary to explore, and one which will he explored by someone at some time either in this country or elsewhere.

Many of the X-20 conditions of flight cannot be simulated on the ground because:
(1) Wind tunnels or their equivalents capable of duplicating the environment above about Mach 5 simply do not exist. The X-20 must fly to Mach numbers of 27.
(2) Control of attitude and aspect of the vehicle must he achieved from orbital speeds down to landing. The interactions that occur between the structure, control systems, pilot, and subsystems because of loads, heat, elasticity, and other factors can be estimated or computed from derived theory, but they cannot be verified short of flight test.

The X-20 has been designed conservatively to meet the expected requirements. Even that design poses many problems of configuration, materials and techniques of construction, structural concepts to meet complex load and temperature situations, and control of the vehicle. These problems are identified and are being solved. The design is well along, materials of construction have been chosen, processes and techniques of construction have evolved from the engineering laboratory to the factory.

The industrial team now combined to produce the X-20 broadens the base of technical and manufacturing skills throughout then country, improving the national capability to produce space equipment. The manufacturing stage is now underway with a solid technical foundation.

X-20A OBJECTIVES

Note from the stated objectives, the number of things to be learned from the X-20 Program.

First, information will be obtained giving guidance to decisions on future re-entry systems, manned or unmanned, and covering a broad spectrum of maneuver capability and landing capability.

Second, at least one additional way to return from orbit will be tested. However, in this case it will permit return to a base or equivalent in a conventional airplane landing fashion.

Third, enough will be learned about navigation control and communication to permit application of this re-entry method to either a manned or unmanned routine utility system.

Fourth, a great deal will be learned about how to use a pilot aboard re-entry vehicles for decisions and for reliability.

Later on the X-20 might be used to rendezvous with some satellite and bring it or some piece of it back, or to simulate other systems.

The X-20 is unique in an early ability to do these things and because of the margin in size, weight that can be returned from orbit, and on-board electric and hydraulic power, it will provide substantial capability for direct follow-on application for future testing.

BOEING X-20 PROGRAM MANAGEMENT

The organization chart shows the management of the X-20 program within this branch of the Aero-Space Division Of The Boeing Company. Under George Snyder, Vice-President and Manager of the X-20 Branch, are A. M. Johnston, Program Manager and focal paint of contact for the Air Force; J. H. Goldie, Manager of Engineering Design and Systems Test; E. G. Czarnecki, Manager of the Support Technologies to the Engineering Department; W. E. Ramsden, Manager of Manufacturing, Quality Control and Materiel operations; E. W. Reischl, Facilities Manager; administrative supporting services in the area of Contract Administration and Financial control are managed by B. K. Carr, Jr., and L. B. Ludford, respectively.

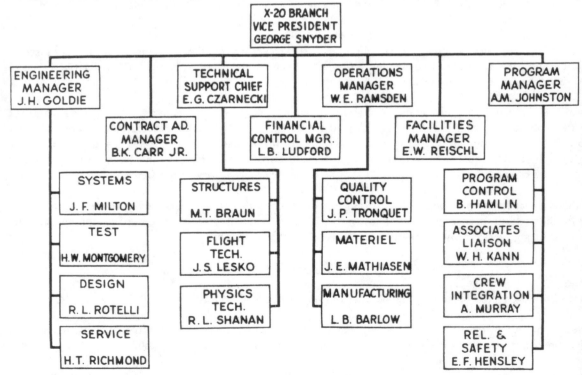

BOEING PROGRAM CONTENT AND TIMING

The air launch program consists of 20 launches from a B-52 at Edwards AFB four of which will achieve low supersonic speeds by using the acceleration rocket for power. Ten ground launch flights are planned from Cape Canaveral, 8 of which are single orbit and 2 multi-orbit missions with landings at Edwards AFB. The first two ground launches will be unmanned and automatically controlled to be landed at Edwards. The subsequent 8 piloted flights will explore the re-entry corridor and obtain research information under a wider variety of flight conditions.

Eight flight gliders including four refurbishments after ground launch flights and 15 transition sections are required to conduct this program.

A full-scale glider mock-up inspection was completed in September 1961, resulting in Air Force approval of the glider design configuration. In August 1962, the Air force redirected the program to utilize the Titan III Launch vehicle. The first air launch flight is scheduled for January 15, 1965. In November the first unmanned ground launch will be made at Cape Canaveral and the first piloted flight is scheduled for May of 1966. Flights are planned at three-month intervals.

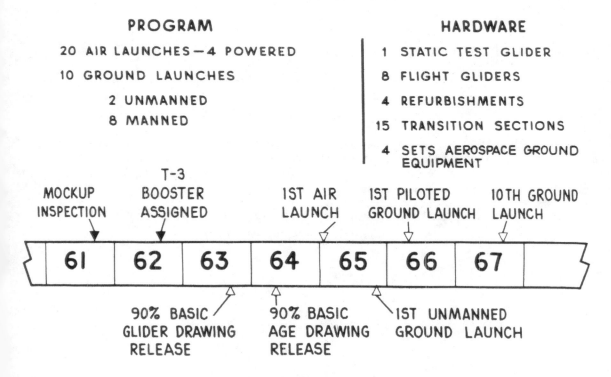

BOEING PROGRAM CONTENT AND TIMING

PROGRAM

20 AIR LAUNCHES — 4 POWERED

10 GROUND LAUNCHES

    2 UNMANNED

    8 MANNED

HARDWARE

1 STATIC TEST GLIDER

8 FLIGHT GLIDERS

4 REFURBISHMENTS

15 TRANSITION SECTIONS

4 SETS AEROSPACE GROUND EQUIPMENT

MOCKUP INSPECTION — T-3 BOOSTER ASSIGNED — 1ST AIR LAUNCH — 1ST PILOTED GROUND LAUNCH — 10TH GROUND LAUNCH

61 | 62 | 63 | 64 | 65 | 66 | 67

90% BASIC GLIDER DRAWING RELEASE — 90% BASIC AGE DRAWING RELEASE — 1ST UNMANNED GROUND LAUNCH

DEVELOPMENT PLAN

The development plan shown here is monitored by Program Management and by the Air Force, using PERT techniques. There are 23 program elements monitored for schedule control and 135 elements for cost control. Weekly reports are made to the program managers and weekly operations meetings attended by functional heads or their delegated representatives to solve identified problems. As of this date, Boeing has met all major milestones on schedule.

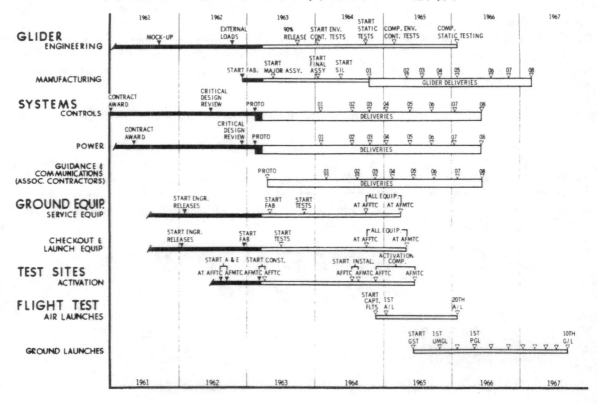

BOEING X-20 FUNDING REQUIREMENT

Shown here are the Boeing fiscal funding requirements necessary to support the development plan. A cost incentive contract has been negotiated on this program. The shaded portion of the bars indicate funds allocated to date, which include $70.4 millions for Fiscal year 1963. As of March 1, 1963, $64 millions of the Fiscal Year 1963 allocated funds were committed.

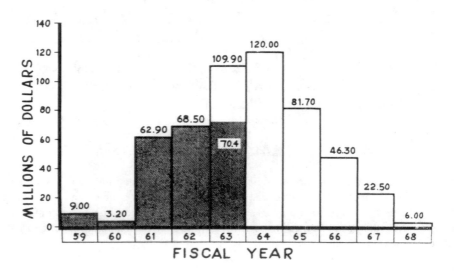

GLIDER DEVELOPMENT AND HARDWARE STATUS

This chart shows the completion status of the major hardware items of the glider, the subsystems in the glider, and the ground support equipment. Engineering development and design are well along. Sources for materials and equipment have been established, and manufacturing processes and tools have been developed to the point where fabricated parts are beginning to flow. Planning for qualification tests and flight tests is nearing completion and qualification testing has begun on a number of subsystems.

The Aerospace Ground Equipment design is based on use of existing equipment and state-of-the-art. As a result, AGE design is proceeding rapidly with development testing being used essentially to confirm the design criteria.

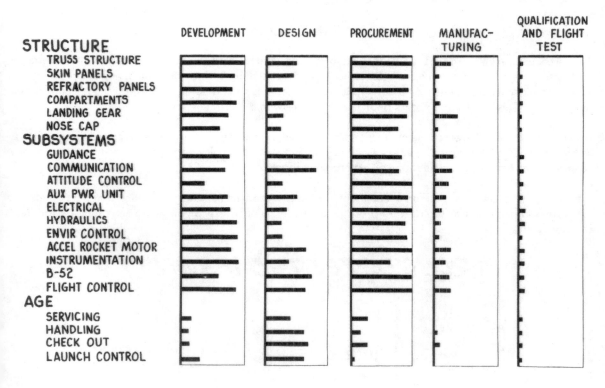

TECHNICAL PROGRESS

This section of the Boeing presentation covers the Technical Progress on the X-20 Program since the inception of the contract. A wide variety of material is contained, and in some cases this material is discussed in some depth of detail.

The reader should consider, in following this material, that the X-20 Technical Progress not only is substantial for the development itself, but, in many cases, it will be applicable in the future to a wide variety of other types of space and maneuverable re-entry vehicles.

AERODYNAMICS

	DEVELOPMENT	CONFIGURATION	VERIFICATION
STATUS	90%	90%	10%

- ## WIND–TUNNEL HOURS
- ## HEATING
- ## CONFIGURATION TAILORING

AERODYNAMICS

This area embracing aerodynamic heating, stability and control, and configuration tailoring is essentially complete except for flight verification. Since no ground facility exists which can simultaneously produce the significant flight environment, complete verification of the design and configuration must be accomplished by flight test. This verification program for the X-20 is now in the planning stage.

A major portion of the aerodynamic tests were devoted to tailoring an aerodynamic configuration. This is stable and controllable over the entire flight regime and that obeys all restraints imposed by performance requirements, heating, and limitations of structural materials.

The following charts present some details of the X-20 aerodynamic effort including a wind tunnel test program summary and a description of advances made through use of thermal paints applied for heat transfer testing.

GROUND TEST CAPABILITY

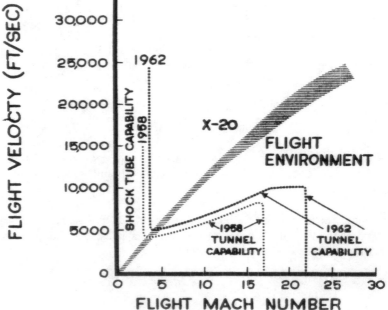

GROUND TEST CAPABILITY

The X-20 hypersonic flight environment shown here as a shaded curve is completely new. Ground test facilities cannot simultaneously duplicate flight Mach number and flight velocity in a hypersonic flight environment. Shock tubes can produce a very high velocity at low mach numbers, while shock tunnels can produce high Mach numbers at low velocities. The X-20 Program and its testing requirements have extended wind tunnel capabilities, as shown by the difference between the 1958 dotted curve and the 1962 capability curve. Despite this increase in capability in the past 4 years, the ground facilities still fall far short of duplicating the X-20 flight conditions. Theories and analyses have been used to correlate wind tunnel capability to the flight regime. This means that high Mach number data which is obtained at velocities less than 10,000 ft/sec. has to be extrapolated to flight velocities up to 24, 500 ft/sec. Similarly, high velocity data obtained in shock tubes at Mach numbers less than 4 have to be extrapolated to flight Mach numbers up to 29. The only way to carbine these parameters is through a flight test program with the actual vehicle.

WIND TUNNEL TESTING

The X-20A design is based on approximately 17,000 hours of wind tunnel testing over the range of Mach numbers from low subsonic and up to 22. This is a greater number of hours than were used in designing the B-52 or the X-15. The requirement for the large number of test hours stemmed from the number of flight configurations that had to be examined, the increased speed regime, the large angle of attack range and the conflicting requirements on configuration detail that exists between the various speed and altitude regimes. In addition, the X-20A had to be carefully tailored to withstand severe heating at high re-entry velocities. About one-third of the wind tunnel program was devoted to heat transfer testing.

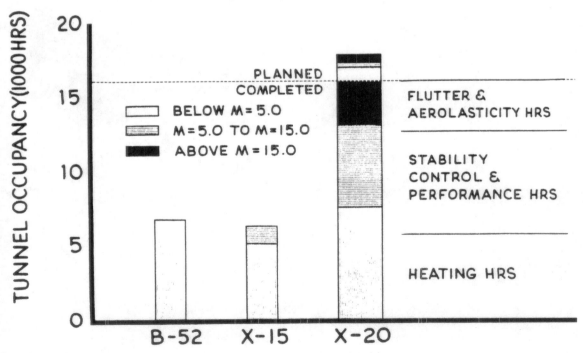

TEMPERATURE DISTRIBUTION – PAINT MODELS

The X-20 Program has pioneered some new testing techniques as illustrated by this chart. Obtaining thermal data from a conventional model requires the building of a steel model with thermocouple installations at points which are anticipated to be critical in the heat transfer region. A thermal paint technique has been developed whereby an inexpensive model is covered with a paint which is extremely sensitive to temperature. The painted model is tested and the temperatures existing across the model in the painted area are deduced from the color of the paint after the test. A continuous distribution of temperature in a critical area is thereby obtained using a cheap model which can be molded or carved from wood. This makes thermal paint a very useful and economical method for discovering they heating effects of joints, manufacturing irregularities and shock wave impingement.

This chart shows the use of the thermal paint technique for obtaining temperature distribution over the entire X-20 model. The coloring on the model ranges from deep purple in the very hot areas to lavender and lighter shades in the cooler areas. The results of two tests are shown. On the left is an angle of attack of 40°, consistent with a medium L/D glide attitude. On the right is an angle of attack of 15°, corresponding to flight near maximum L/D as would be required for large lateral range maneuvers or for long range glide during re-entry.

TEMPERATURE DISTRIBUTION
PAINT MODELS

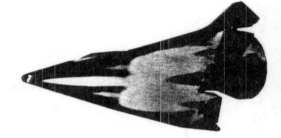

MACH NO.	10	MACH NO.	10
ANGLE OF ATTACK	40°	ANGLE ATTACK	15°
YAW	0°	YAW	0°

(Editor's Note: A better copy of this illustration was unavailable. Unfortunately the purpose of the illustration is negated by the poor quality.)

SHOCK IMPINGEMENT – FIN LEADING EDGE HEATING

The X-20 heat transfer testing has uncovered a few surprises. Fin leading edge heating was one of these. The fin projecting above the X-20 glider wing was originally predicted by leading edge theory to have a heat rate normalized to the stagnation pointy as shown by the lower curve on this chart. Wind tunnel test data showed fin leading edge heating rate considerably higher than theory and changing with angle of attack as shown by the test data points on this plot. These increased heating rates resulted from impingement of the shock wave from the main wing surface on the fin leading edge. The location and character of this impingement changed as a function of angle of attack. As a result of these data the glider design was changed to the upper line shown on this chart and labeled current design. This design line is well outside the experimental heating rates recently measured for fin leading edge.

The data shown on this chart have been gathered and compared from several different facilities. The data show the effect to be independent of Mach Number and Reynolds lumber over a wide range when normalized as done here by a stagnation point heat rate value.

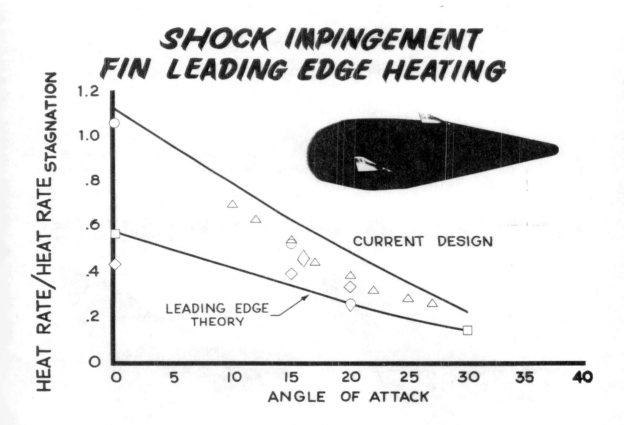

STRUCTURES AND MATERIALS

In the structures area, X-20 progress has also been substantial. The "% complete" bars shown include technical processes as well as the percent completion of the end product. Development testing of pieces and components is essentially complete. Detail design decisions and layouts leading to them are nearly complete even though a smaller percentage of the fabrication drawings have been released. A significant amount of planning, fabrication technique development, tooling, and detail part shaping have been accomplished even though a smaller portion of the manufactured parts have been completed. "Verification" (accomplished only by flight testing) indicates that a significant start has been made by detail planning of flight operations and facilities even though no flights have been made.

The X-20 program has significantly advanced the state-of-the-art throughout the entire spectrum of materials required for construction of high temperature manned vehicles. The material capability improvements have been aggressively backed up by detail design studies and fabrication skills development.

Refractory metal coatings have been advanced from the laboratory-scale coatings of inconsistent quality available at the start of the program to higher performance consistent coatings applied at high rates in manufacturing shops. Development testing of the X-20 nose cap is well under way.

STRUCTURES AND MATERIALS

STATUS

DEVELOPMENT	DESIGN	FABRICATION	VERIFICATION
90%	30%	10%	10%

- MATERIALS PROGRESS
- COATINGS

X-20A RE-ENTRY GLIDE TEMPERATURES

These pictures show the temperature profiles on the X-20A glider. The environment corresponds to the design lateral range mission of 1500 N. M. This picture portrays the thermal distribution at the critical velocity of 20,700 ft/sec. As developed by test, substantial margins of safety exist between the temperatures shown and the ultimate capability.

X-20A RE-ENTRY GLIDE TEMPERATURES

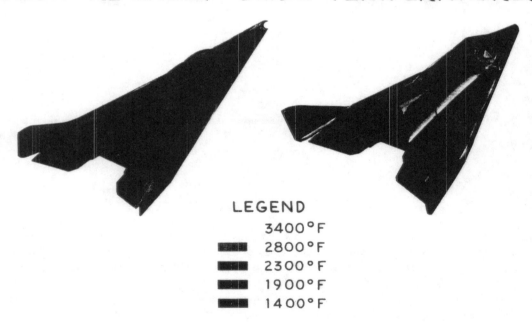

LEGEND
3400°F
2800°F
2300°F
1900°F
1400°F

(Editor's Note: A better copy of this illustration was unavailable. Unfortunately the purpose of the illustration is negated by the poor quality.)

MATERIALS PROGRESS

The key to the X-20 program has been the development of aircraft quality materials capable of withstanding the thermal environments. The bulk of the effort has been towards qualification of "off the shelf" materials for the more severe environment. Examples of this would be the increase in the temperature capability of transparent materials from 1000°F to 1800°F; for thermal insulation from 2000°F to 3000°F.

Conventional structural materials i.e., aluminum, titanium, stainless steel, have been replaced by a nickel "super alloy" René 41, originally developed for jet engines. This alloy retains significant strength up to 2000°F and forms a self protective high emittance oxide. It has allowed extending the maximum temperature of the primary structure to 1800°F. Bearings are required to operate in the same environment as the primary structure. Titanium carbide, a ceramic, has been qualified as an aircraft bearing to meet the maximum temperatures.

The upper temperature capability of the exterior surfaces has been extended to 3000°F on lees edges and skin panels and up to 4500°F on the nose cap. The former has been accomplished by development of refractory alloys columbium to 2750°F and molybdenum to 3000°F. The use of refractory alloys required development of protective coatings and close cooperation between Boeing and raw material suppers to upgrade the quality of the raw material to meet aircraft standards. Careful considerations of materials capability and limitations have allowed use of zirconia as a structural ceramic for the nose cap. A significant point has been reached by a nose cap which successfully passed a series of ground tests simulating a complete flight.

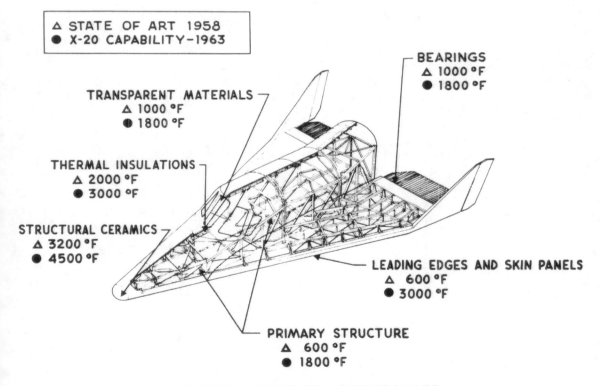

MATERIALS PROGRESS

△ STATE OF ART 1958
● X-20 CAPABILITY-1963

BEARINGS
△ 1000 °F
● 1800 °F

TRANSPARENT MATERIALS
△ 1000 °F
● 1800 °F

THERMAL INSULATIONS
△ 2000 °F
● 3000 °F

STRUCTURAL CERAMICS
△ 3200 °F
● 4500 °F

LEADING EDGES AND SKIN PANELS
△ 600 °F
● 3000 °F

PRIMARY STRUCTURE
△ 600 °F
● 1800 °F

REFRACTORY COATINGS – OXIDATION LIFE

Since 1959 the scatter of oxidation life test data or coated molybdenum alloys has been reduced by a factor of 5 and the median of the data increased by over 700°F. This was accomplished by an intensive development program which emphasized reproducibility and quality control. The coatings were evaluated under test conditions which simulated typical X-20 boost and re-entry trajectories, including simulation of temperature, time, pressure and mass flow.

The design curve for coated molybdenum generated from the resulting data is shown. This curve represents conservative use of the data. In one recent test a specimen has withstood twelve re-entry cycles to a maximum temperature of 2800°F. This and other similar data indicates that increase in the design oxidation life can be made in the near future.

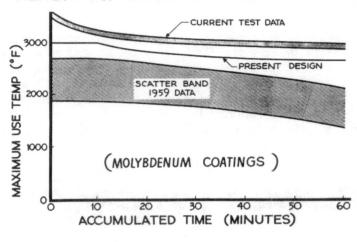

MOLYBDENUM COATINGS – EMITTANCE

A high emittance is required for coatings to hold exterior temperatures to a minimum. As a result, emittance has been a prime consideration in the development of coatings for refractory alloys. Two curves are shown. The present design curve and the recently improved curve. This data shown as recent coating improvements represents the results of current tests on coated molybdenum. These results are reproducible and the values are currently being qualified for design use.

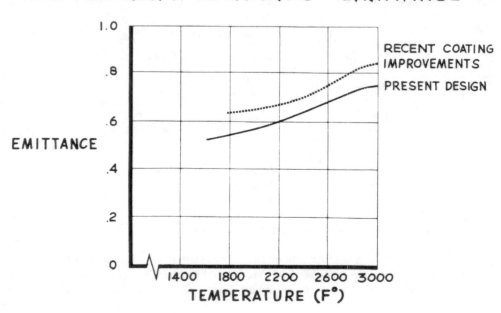

GLIDER RE-ENTRY WEIGHT

With present design material emittance there is a lateral range capability of approximately 1700 N. Mile at the design weight of 11,150 lb. With the improved emittance shown on the preceding page the lateral range can be extended to allow use of nearly the full aerodynamic turn capability of the glider at the glider design weight. Conversely, at the design lateral range the allowable re-entry weight can be increased by several thousand pounds.

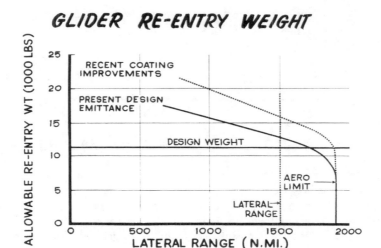

ASSOCIATE CONTRACTORS
MAN-MACHINE INTEGRATION
FLIGHT TEST PROGRAM X-20 POTENTIAL

ASSOCIATE CONTRACTORS

The Air Force has requested Boeing to discuss two of the subsystems which are not a direct responsibility of Boeing. These are the associate contracts for Communication and Tracking equipment and for Primary Guidance Equipment.

The design of the X-20 system inherently frees the pilot and glider from dependence on ground tracking and communications. However, flight testing experience, both with airplanes and spacecraft, has shown that improvement in mission success and data acquisition warrant both two-way communication and vehicle tracking at least during the critical portions of the flight.

ASSOCIATE CONTRACTORS
RCA - COMMUNICATION AND TRACKING
MINNEAPOLIS-HONEYWELL - PRIMARY GUIDANCE

COMMUNICATION PROFILE

Two of the more critical regions needing communication and tracking are near booster burnout and during the maximum re-entry heating period from about 22,000 feet per second to about 12,000 fps. In both these regions, the vehicle passage through the air generates a shock wave which heats the air until it ionizes. The ionized air surrounding the glider significantly interferes with normal radio communications. Complex analyses and limited ground experiments have indicated use of super-high frequencies (SHF) (above 10,000 Mc) will penetrate this "ion sheath." Because the precise extent of SHF penetration remains to be shown by flight test, ground installation of this equipment is being planned only in needed regions. Normal radio frequencies (UHF) is planned elsewhere and is often available.

COMMUNICATIONS PROFILE

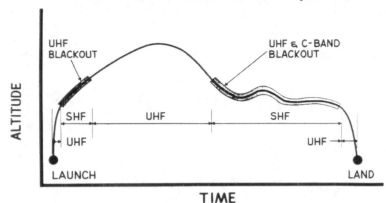

RE-ENTRY COMMUNICATIONS & TRACKING

Completely dual SHF transmitters and receivers are provided in the glider. In addition, both top- and bottom-mounted antennas are used to reduce the ionization effects. These will provide voice, command, telemetry, and range safety communications between the Glider and the ground for slant ranges up to 500 miles. Digital techniques are employed in determining slant range to facilitate computation. Somewhat different frequencies are used for transmission from air-to-ground than those from ground-to-air to eliminate self-interference. A search and rescue transceiver will be packed in the pilot's survival kit. This unit has provisions for both two-way voice and direction-finding tones and is capable of operating over distances up to 200 miles using international standard frequencies.

RE-ENTRY COMMUNICATIONS AND TRACKING

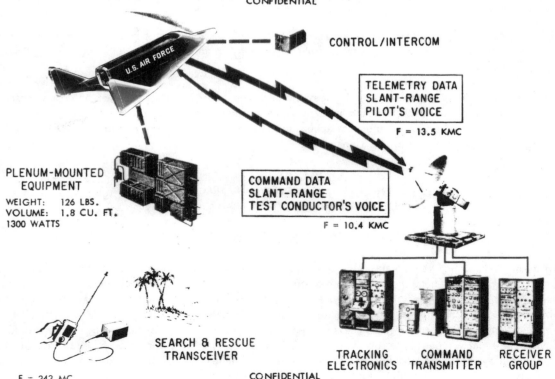

CONFIDENTIAL

STATUS

The RCA drawings are 100% released. Equipment development is 95% complete. Four (4) sets or airborne development hardware are being provided for accomplishing (a) subassembly Testing at Camden, (b) Flight Test Integration at Newcastle, Delaware, (c) Design Evaluation Tests at Camden, and (d) Tests at Seattle to insure compatibility of the RCA equipment with the rest of the glider subsystems.

The Newcastle test program will install airborne equipment in a KC 97G aircraft for flight integration with the surface tracking and communication hardware.

The first air launch of the X-20 is scheduled for January 1965 and will provides final system verification prior to orbital flights.

All activity is on schedule.

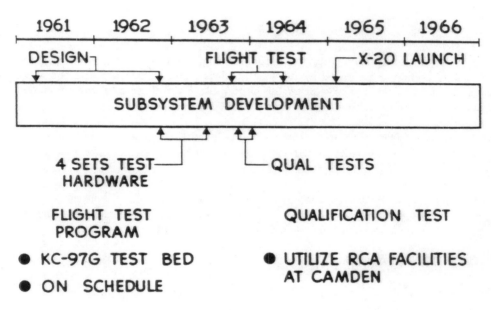

STATUS OF RCA EQUIPMENT

FUNCTIONAL BLOCK DIAGRAM
(INERTIAL GUIDANCE SYSTEM)

An inertial guidance system was chosen for the X-20 because it provides an autonomous capability without dependence on ground tracking facilities. It was also chosen because of the availability and flight-proven status of most components. Minneapolis-Honeywell (St. Petersburg) is the contractor.

Data sensed by the inertial guidance system is processed by the signal data converter and displayed on the energy management display in the form of up to 10 landing sites that can be safely reached by the glider. The coordinates of the site selected by the pilot are then processed in the inertial guidance computer and displayed for use by the pilot.

Data are also sent to the telemetry system for on-board recording and for transmission to the ground. These data include instantaneous speed, altitude, attitude and angle-of-attack, for post-flight data reduction.

FUNCTIONAL BLOCK DIAGRAM

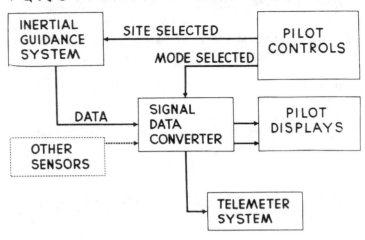

UNIQUE FEATURES

A novel approach to the determination of altitude during re-entry is being used. Decelerations and horizontal velocity components sensed by the Inertial Measurements Unit are inserted as variables in a mathematical model of an air density/velocity/altitude formula. From these data, the computer can reduce the computed guidance system altitude error from ±400,000 ft to ±10,000 ft.

Minneapolis-Honeywell has incorporated detectors in the inertial guidance system to identify malfunctions. These will warn the pilot of guidance system failures and alert him to switch reference to a less accurate emergency system.

A recent development by Minneapolis-Honeywell reduces the drift rate of the integrating gyros used on the X-20. This increase in accuracy is accomplished by the simple technique of interrupting power to the spin motor periodically in a precisely controlled cycle.

ALTITUDE COMPUTATION

SUBSYSTEM MALFUNCTION DETECTION

GYRO SPIN-MOTOR INTERRUPTION TECHNIQUE

INERTIAL GUIDANCE COMPONENTS

The X-20's navigation and guidance equipment consists of the three components shown here. The tolerance inherent in a large landing footprint is such that X-20 guidance accuracy requirements are less stringent than those of many present-day weapon systems.

The Inertial Measurement Unit is a miniaturized 4-axis platform originally developed for the IM99B program. It is currently used in the Centaur vehicle and is very similar to that planned for Gemini. The Guidance Computer (Verdan) is used in the GAM-77; over 1500 units have been built.

Only minor modifications were required to adapt the remaining units to meet X-20 requirements. In every case, the performance accuracy of this subsystem exceeds requirements.

PRIMARY GUIDANCE SYSTEM COMPONENTS

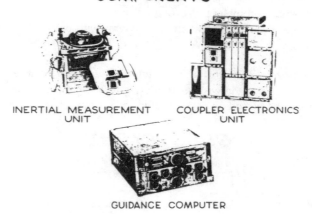

INERTIAL MEASUREMENT   COUPLER ELECTRONICS
          UNIT   UNIT

GUIDANCE COMPUTER

STATUS

All aspects of the guidance system are on schedule. As of this date 5 sets of prototype hardware have been fabricated. As shown here, one set has been installed in an F101B for the purpose of verifying the system concept and the ability of the system to provide necessary navigation data. To date, 8 flights have been made; the F101B flight test program is scheduled for a total of 24 flights lasting through May 15, 1963.

A second set will be sled tested on the high speed track at Holloman AFB to verify subsystem performance under high "G" forces. This will be a 6 month test program commencing in August 1963.

A third set will be delivered to Boeing for engineering evaluation.

A fourth set is scheduled for Environmental Tests and a fifth set is to be used for Design Evaluation at the factory.

Starting February 15, 1963, environmental tests are being performed by Minneapolis-Honeywell at their St. Petersburg facility.

All test programs are scheduled for completion by February 15, 1964.

STATUS

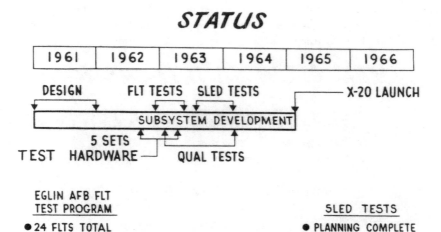

SUBSYSTEMS

For all subsystems, the development activities are almost complete, the design is well along, fabrication and fabrication know-how is over 50% complete. Prototype flight hardware verification tests have been started on all of the subsystems. The task of integrating the subsystems with each other and with the glider and the pilot is underway at Boeing.

There will undoubtedly be further problems with subsystems – every program has these. The basic choices of hardware and contractors were made wisely, attempting to avoid new developments unless essential, to permit the maximum effort to be placed on aerodynamic and structural advances.

SUBSYSTEMS

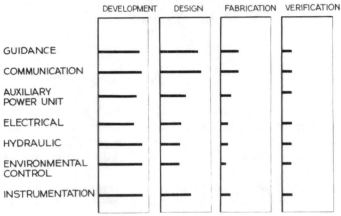

MAN-MACHINE INTEGRATION

The presence of a pilot in the X-20 offers substantial potential to improve reliability, test flexibility, and vehicle recovery. To take full advantage of this, the pilot has been provided with monitors and controls for the subsystems previously mentioned, with displays, with vision for landing, and with a stable, controllable, "forgiving" airplane.

You will note that actual flight testing has been accomplished in this area. For example, the side stick controller configuration has been developed in close coordination with the team of Air Force and NASA consultant pilots, including a development test program by Minneapolis-Honeywell in an F-101 with adaptive flight control system. Acceptable vision for approach and landing have been demonstrated in simulated X-20 landing flight tests using modified F-104 and F-5D aircraft.

MAN-MACHINE INTEGRATION

	DEVELOPMENT	DESIGN	FABRICATION	FLIGHT TEST
STATUS	80%	85%	25%	5%

- ● HANDLING QUALITIES
- ● DISPLAYS AND VISION
- ● ENERGY MANAGEMENT

PILOTING THE X-20

In each phase of flight, as is shown on the chart, the pilot is required to perform functions that are both necessary and useful to the success of the test flight.

The pilot is an essential part of the pre launch operations and the final countdown. In case a serious malfunction requires mission abort after engine ignition, an abort rocket is ignited and the pilot controls and navigates the vehicle back to the Cape or to one of several emergency landing sites. At least one site is within rate at any boost termination point. As these fields are often fairly small, the glider has been designed to land on an 8000' x 150' runway of asphalt, concrete or compact earth. In a multi-orbit flight, the injection and de-orbit maneuvers will be controlled entirely by the pilot. In orbit, the pilot maintains attitude of the X-20 by utilizing the same set of controls to operate the reaction control nozzles that he uses to operate the aerodynamic control surfaces in the atmosphere.

The pilot can deviate widely from the standard re-entry trajectory, by observing his instruments and making proper judgements. Thus, he will accelerate the process of gathering hypersonic research data and still be able to navigate to Edwards AFB.

Finally, the pilot, by visual observation of landmarks through the X-20 windows, will land the X-20 in the conventional manner using the techniques established on other airplanes.

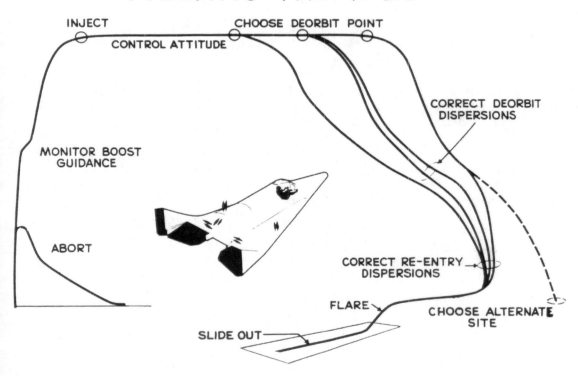

X-20A DEVELOPMENT SIMULATOR ACTIVITY

The majority of the development flight simulator work has been accomplished; the remainder will be completed in 1963,. A total of approximately 8000 development simulator test hours will have been conducted on the X-20 by the end of 1963. Over half of those hours include one of the assigned USAF or NASA pilots in control of the simulated vehicle.

The extensive simulation program is required because of the unusual features of the flight environment and tailoring of configuration and flight controls to attain a stable, controllable airplane with good handling qualities.

Piloted simulator studies have included the abort configuration (glider plus acceleration rocket) and the air vehicle (booster plus glider). These studies were also conducted at the Johnsville centrifuge and demonstrated that abort and boost accelerations do not degrade the pilot's control capability. The studies have shown that the pilot can control the entire air vehicle successfully throughout the boost phase, if desired.

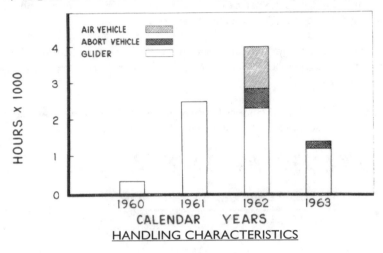

X-20A DEVELOPMENT SIMULATOR ACTIVITY

HANDLING CHARACTERISTICS

The handling qualities of any airplane are not defined by any known analytical technique. Therefore, pilot opinions must be used. These opinions are somewhat standardized by having pilots checked out in standard airplanes. Further, the data on this chart combines the opinion of seven pilots over a large number of simulator runs.

Recently, the simulator has been modified to incorporate the final cockpit configuration, actual hydraulic servos, and a prototype set of flight control electronics, using analog computers to simulate only the aerodynamics. The pilots report even better ratings than shown here.

HANDLING CHARACTERISTICS
(WITH AUGMENTATION)

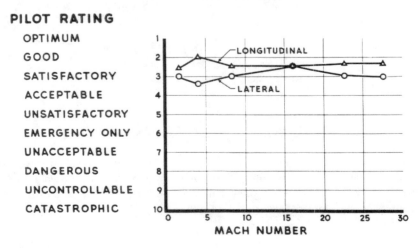

X-20 COCKPIT

The instrument panel has been designed to make the fullest use of the pilot's ability to control the glider, monitor subsystem performance and select subsystem operation.

Switches and knobs used in switching from one subsystem to an alternate or in performing discrete functions required in normal operation are located within reach of the left hand since the pilot grips the sidearm control stick with his right hand.

Five windows are provided. The two outboard windows are uncovered at all times for vision in boost, orbit, re-entry, and larding. The three forward windows are protected until the worst re-entry heating is past by a heat shield which is then jettisoned. The size, shape, and location of each window has been chosen to allow the pilot to see the runway at any time during his 360° landing pattern starting at about 31,000 feet and ending with slideout completion.

The primary flight instruments are located centrally and provide the flight data for manual control in normal and emergency situations. The flight simulation programs, "flown" 'by the consultant pilots have contributed strongly to the type and layout of the instruments selected. The flight instruments include the array of velocity, altitude and attitude data and energy management required to reach the landing site. The latter display, (lower center of the panel) is described on the following pages.

PILOT'S DISPLAY

The energy management display is a primary instrument used by the pilot in navigating and controlling the glider to reach a selected landing site. The display (on this chart) showing the location of the landing site with respect to range capability and how to fly the glider to reach the landing site.

The display consists of a film overlay and a cathode ray tube both driven from the primary guidance system. The landing site is displayed on the scope as a small dot and the relationship of this dot to the grid on the precomputed film overlay defines the angle of attack and bank angle required to reach the landing site. The overlay is changed at velocity increments of 1000 fps.

The pilot can determine quickly that the selected site is within the range control capability and how to fly the glider to reach the site. Any of 10 precomputed landing sites may be selected by the pilot.

As indicated earlier, the pilot has a very large range of control available to him by varying pitch and bank angles. He must still observe structural heating limits in the critical velocity regime. To provide simultaneous monitoring of limits as well as navigation, the current display design is shown on the following chart.

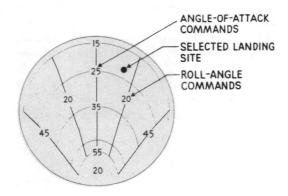

PILOT'S DISPLAY
V = 20, 000 FPS

PILOT'S DISPLAY

Structural limits are shown on the film overlay in angle of attack-altitude coordinates. The scales have been left off to keep markings on the display to a minimum. The second dot on the scope shows the current glider angle of attack and altitude. The dot moves horizontally to the right as angle of attack is increased, and moves up or down slowly as altitude changes. Since angle of attack can be changed quite rapidly without immediate change in altitude this display materially assists the pilot in monitoring the extent of attitude change he is permitted to make at any given condition. By observing the movement of this dot, and its position with respect to the structural limit, the pilot is cognizant, of his safety margins.

Using this display the pilot can:
  a)   Maneuver from one glider condition to another
  b)   Perform a recovery following boost abort
  c)   Maneuver the glider within structural limits, to reach the landing site
  d)   Perform special flight test maneuvers to intentionally approach the limits on later flights

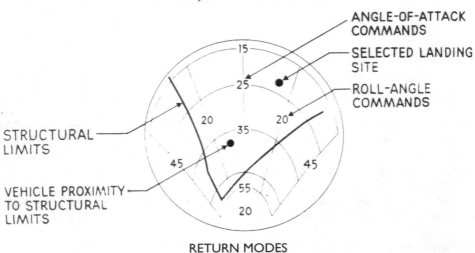

PILOT'S DISPLAY V = 20,000 FPS

RETURN MODES

Using the energy management techniques previously described, the pilot can accommodate a very wide tolerance on de-orbit times or orbit positions and still reach his desired landing site.

A comparison of the re-entry windows for the X-20A and a semi-ballistic vehicle (L/D = .25) which are desired to land at the same site shows the higher lift vehicle to have the greater range control. Both vehicles obtain range control capability by developing and modulating the coefficients of lift and drag. The greater tolerance of the X-20A results from the large variation in L/D that is obtained by modulating angle of attack between 18 and 52 degrees. Control is achieved by reaction jets and aerodynamic control surfaces. The smaller semi-ballistic vehicle tolerance is in correspondence with its relatively small control over L/D.

The increased tolerance to errors of the winged X-20A continues to be important, though smaller, as the speed and altitude decrease. This allows easy final corrections and increased tolerance to winds.

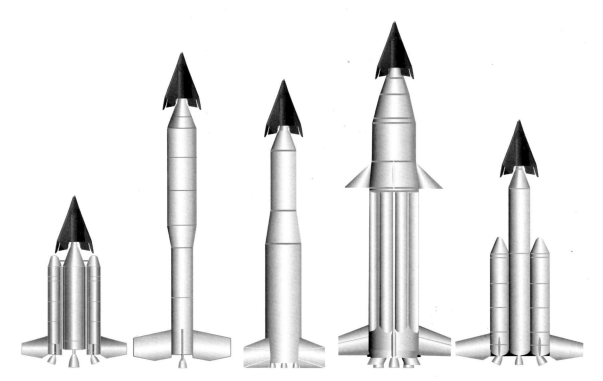

Above from left to right is the Space Systems Division's Phoenix A388 which was proposed in July 1961. A brand new launch system it was to use solid strap-ons and a single J2 LOX/LH2 main stage amounting to 536,000 lbs of thrust. It would have flown in late 1963. Next is the Martin Plan C which was a Titan II first stage with a LOX/LH2 second stage and an overall thrust of 409,230 lbs. It was eliminated because of payload limitations and because solids were considered more reliable for the 1st stage. Center is the Convair Astro IV with 613,630 pounds of thrust this was an all LOX/LH2 booster and was disqualified for the same reason as the Titan Plan C. Next is the Saturn C-1 with a set of upper stabilisers which was disqualified due to cost and complexity, finally at right is the Titan II with solid boosters (later known as Titan IIIC) which was the final choice for Dyna-Soar before the program was cancelled. (Renderings by the Editor.)

A group shot of the six assigned astronauts in front of a model of the Titan IIIC. (above)
A huge model of Dyna-Soar atop the Titan IIIC with the proposed launch structure, note there are no stabilisers (right)

Martin's enormous Boosted Arcturus a conglomeration of seven Titan first stages attached to a pair of North American engines capable of sending 60,000 lbs to the moon.(left)

Von Braun's A9-A10 conceived at Peenemünde during World War II.

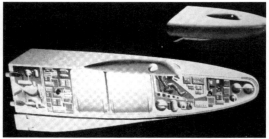

Avro Canada's STV from June 1958(left) it was to have been a rocket/ramjet hybrid. This design came nine months behind Ames Mach 10 demonstrator. Note the similarity to the BoMi model opposite.

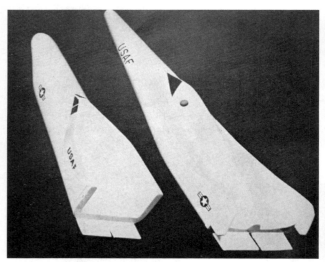

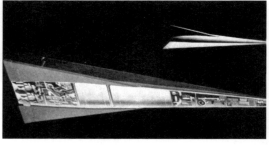

A host of ideas (clockwise from top left) two USAF hybrids crossing over between Dyna-Soar and the lifting body concept WADD II at left, an NACA boat shaped "skip" semi-ballistic reentry vehicle based on Alfred Eggar's ideas, an NACA delta-wing concept, an artist concept of Bell's BoMi orbital stage, a Dyna-Soar/M2 abort sequence.

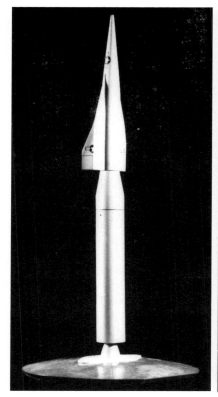

Clockwise from above, Bell's BoMi/ROBO in its final configuration. The three core version would have a take-off weight of 740,800 lbs. with the glider weighing in at 24,000 pounds. Thrust was by an ammonia and fluorine propellant. A Krafft Ehricke two-stage design at right was later turned into a Revell model kit for kids. An artist's illustration showing something akin to the early Bell BoMi designs alongside a Mercury capsule and a modified Mercury.

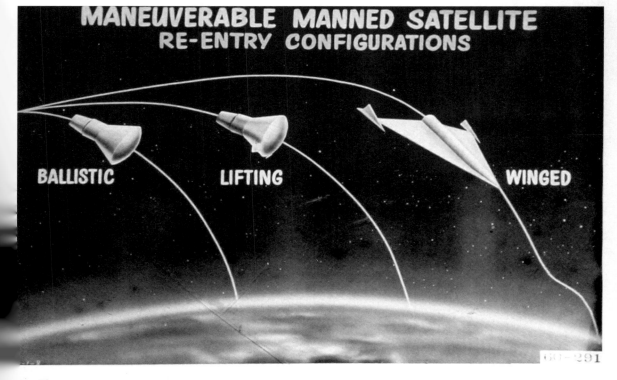

MANEUVERABLE MANNED SATELLITE
RE-ENTRY CONFIGURATIONS

BALLISTIC   LIFTING   WINGED

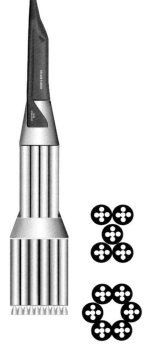

A Martin engineer holds a model of something similar to the enormous Arcturus proposal. This launch vehicle has only two large engines on the first stage. (far left)

A conceptual three stage booster using solid rockets. Six on the first stage, four on the second stage and a third nestled inside. (left)

Below are the six men assigned to fly Dyna-Soar. Clockwise are James W. Wood, Milton O. Thompson, Albert H. Crews, Henry J. Gordon, Russell L.Rogers and William "Pete" Knight. Crews was brought in to replace Neil Armstrong who was one of the original six chosen in March 1962

X-20
Dyna-Soar

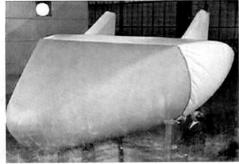

Dozens of contractors were involved in the early research including Goodyear. At left is an artist rendering of an inflatable reentry vehicle and a mockup above. It clearly shared some similarities with the Ames M1-L.

Below is a Goodyear rendering showing something similar to Langley's HYWARDS design. Overleaf are more Goodyear models showing their proposals.

LIGHT WING LOADING RE-ENTRY VEHICLE GOODYEAR

DYNA-SOAR TYPE RE-ENTRY VEHICLE

GLIDE RE-ENTRY VEHICLE

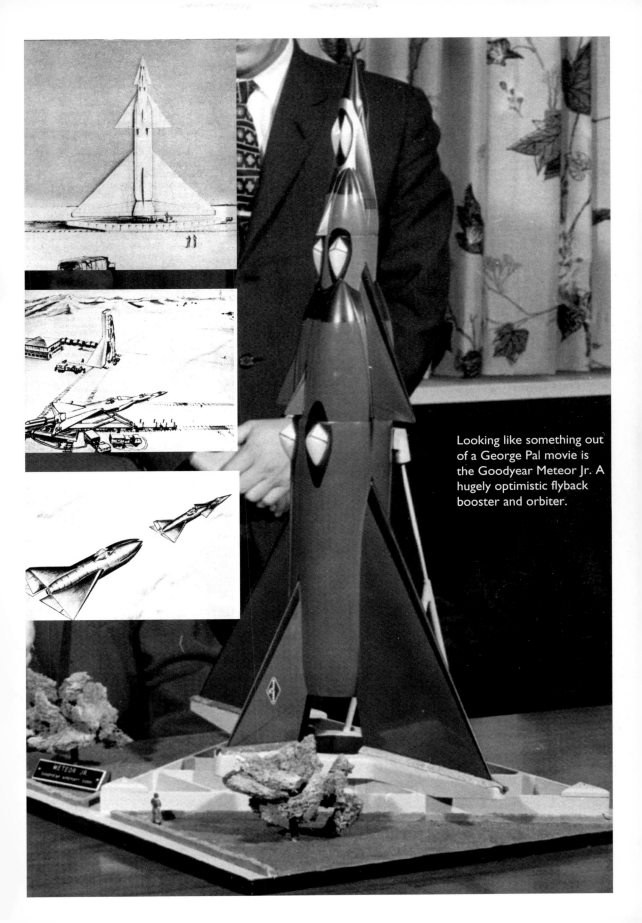

Looking like something out of a George Pal movie is the Goodyear Meteor Jr. A hugely optimistic flyback booster and orbiter.

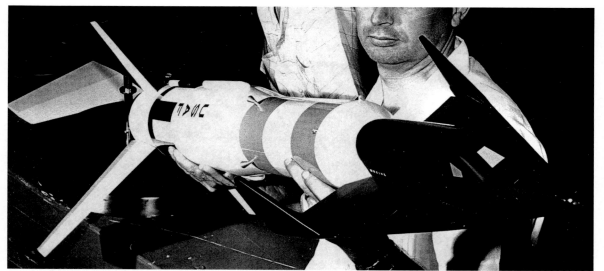

Above and at left two Martin engineers assemble a large model of the Titan I /Dyna-Soar.

Centre is a landing skid from the Boeing mock-up and some models being used to test the skids.

At right is a diagram showing the projected flight path of the sub-orbital Dyna-Soar, Brazil was ruled out due to politics. By creating another obstacle this may have contributed to the demise of the program.

The six assigned pilot/consultants discuss the Dyna-Soar
The escape hatch below. An earlier design had the entire crew cabin ejecting
and deploying a parachute. (See DVD)

The mock-up placed on wheels showing the window arrangment and at right with the window cover in place and the six pilots. Below the mock-up is seen next to a simulator and an unidentified booster.

Side view of the engineering mock-up 814-2050 at the factory in Seattle.
Below is a Boeing Configuration development diagram showing the speed at which
the design changed for the glider.

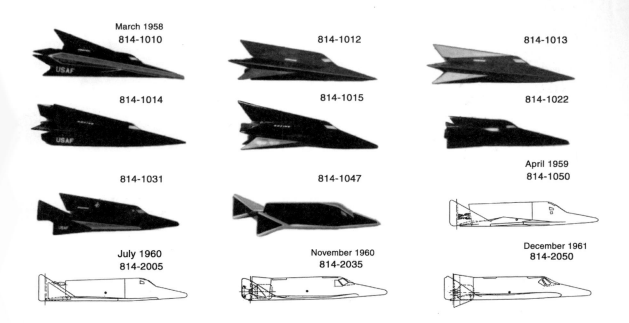

March 1958
814-1010

814-1012

814-1013

814-1014

814-1015

814-1022

814-1031

814-1047

April 1959
814-1050

July 1960
814-2005

November 1960
814-2035

December 1961
814-2050

A nose-on view at the factory showing some of the internal wing frame. Below a version of the cockpit.

The rear of the mock-up, showing a cross section inside the glider.
Below a wing is put through a thermal test at Boeing.

5A10326

11-30-62   CONFIDENTIAL   2A128541

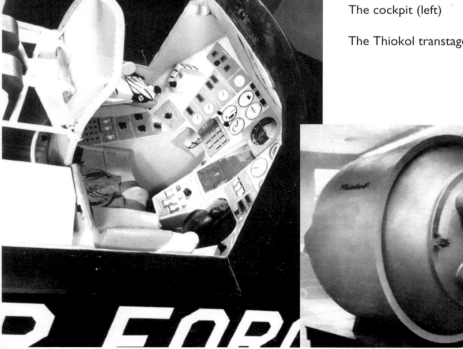

The air-frame for Dyna-Soar (above)

The cockpit (left)

The Thiokol transtage motor (below)

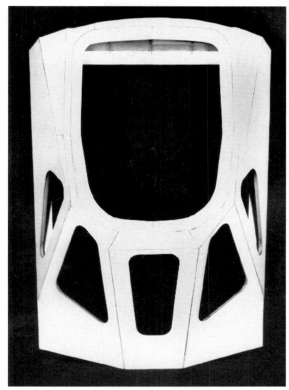

The proposed window arrangement. (above)

An engineer works on a wind-tunnel model just before cancellation (above right)

Artist's concept of ASSET (below)

The Martin World's Fair X-23A version in Oklahoma (bottom)

The ten man advanced Lockheed proposal (below).

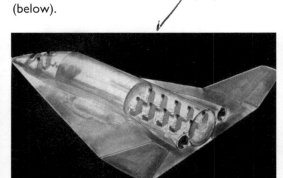

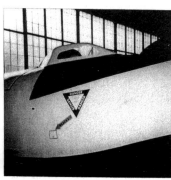

The Martin X-23A mock-up is moved from the Cradle of Aviation in
New York to the Oklahoma Omniplex (above)

The advanced X-20X with a five person crew capacity.(below)

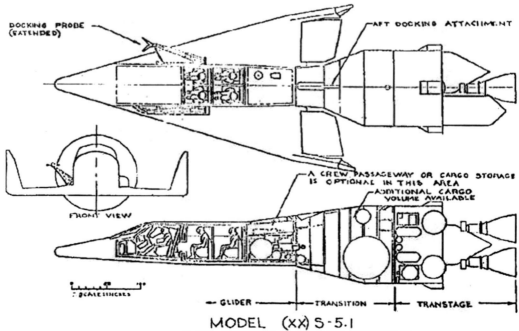

DOCKING PROBE
(EXTENDED)

AFT DOCKING ATTACHMENT

FRONT VIEW

A CREW PASSAGEWAY OR CARGO STORAGE
IS OPTIONAL IN THIS AREA

ADDITIONAL CARGO
VOLUME AVAILABLE

SCALE INCHES

GLIDER   TRANSITION   TRANSTAGE

MODEL (XX) S-5.1
X-20 SHUTTLE SPACE VEHICLE

RETURN MODES

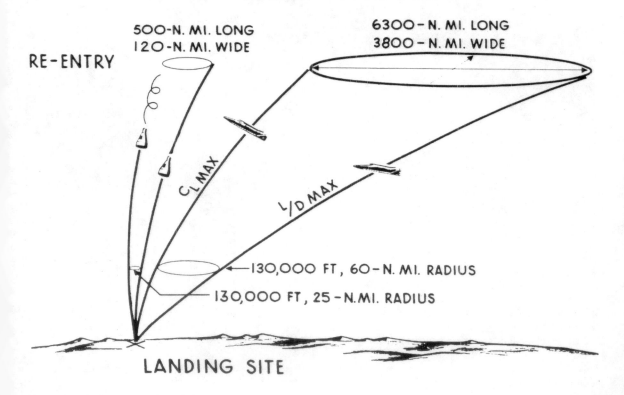

500-N. MI. LONG
120-N. MI. WIDE

6300-N. MI. LONG
3800-N. MI. WIDE

RE-ENTRY

CL MAX

L/D MAX

130,000 FT, 60-N. MI. RADIUS

130,000 FT, 25-N.MI. RADIUS

LANDING SITE

PILOT'S MANEUVER CAPABILITY

The range control capability of the X-20A during the once-around mission from Cape Canaveral to Edwards Air Force Base is shown on the accompanying chart. For a re-entry north of Australia, the landing area available to the pilot is centered at Edwards Air Force Base and covers the entire United States. By flying at high attitude the pilot could land in Hawaii or flying at low attitude he could land in Bermuda.

Maximum cross range distances of ±1700 N Mile can be reached by banked flight (at 45° bank) at low angle of attack.

The landing area available for a semi-ballistic vehicle is much less than the X-20A because of the lower L/D. The landing area can be extended along the ground track by varying the time that the de-orbit rockets are fired, but the maximum cross range is about ±60 N. miles.

As can be seen by the chart, the X-20A landing sites accessible, decrease as the velocity and altitude decrease during the re-entry. When the velocity is 23,000 fps only 33 minutes remain to landing and nearly the entire USA is accessible to the pilot. Even when the velocity is only 10,000 fps and only 15 minutes remain to touchdown, he still has about 500 miles by 400 miles available for alternative landing site selection.

MANEUVER CAPABILITY

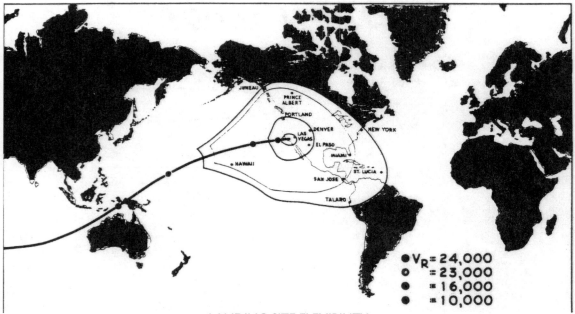

VR = 24,000
= 23,000
= 16,000
= 10,000

LANDING SITE FLEXIBILITY

Since the X-20A pilot can choose virtually any air field in the Continental United States, his ability to land will not be restricted by weather conditions. If he discovers his planned landing site is weathered in, there are other clear fields that he can select.

The map indicates the cloud cover situation at 64 out of approximately 200 major Air force bases on a recent typical winter day, (March 3rd). Much of the country was cloudy to some degree but air strips along the California coast and in the south central states were clear. Many of the remaining sites were only partly cloudy with relatively high clouds and could have been used.

True all-weather capability requires on-board landing aids plus assistance from the ground. The remote control recovery system to be used in the two unmanned flights may well be the forerunner of such a system.

Thus, from pre-launch to landing, every attempt has been made to exploit the unique capabilities of the pilot and to provide safety consistent with X-aircraft philosophy. The integrated man and machine must be tested in the real mission to verify whether this has actually been done.

LANDING SITE FLEXIBILITY

EDWARDS AFB △

● AIR FORCE BASES
    CLEAR
//// PARTLY CLOUDY
    CLOUDY

FLIGHT TEST PROGRAM

The theme of this discussion must now be switched from what has been done to what is planned to be done. The detailed plans for each test flight are not completed yet; it is normal to revise these continually. To properly design the X-20A, a significant amount of planning has been done. The basic purposes of each flight have been defined, flight paths determined, range requirements established and partially implemented, the launch complex and flight control facilities defined, and specific instrumentation is being purchased to allow 750 items to be measured during each flight.

Time doesn't permit a very thorough discussion of this planning but a brief progress report and some of the expected benefits from the flight test program are included.

PHILOSOPHY OF X-AIRPLANE

The approach to designing and testing an X-airplane is significantly different from an operational aircraft. Very little is known about the actual aerodynamic environment between M=6 and orbital speeds. Whenever there is a conflict in theories, for example, the X-20A has been designed to the one predicting the worst results. In addition, a safety factor has been added beyond that prediction. The only things limiting the size of the safety factor are weight, cost, and the state of the materials art.

Initial flights of X-aircraft are also programmed at lower speeds, gradually and hesitantly approaching the design limits. The cost of non-recoverable boosters and of multiple landing sites prohibits this philosophy for the X-20A. Instead, the first flight is programmed at full velocity but at a pitch attitude of minimum heating. This is expected to reduce the peak heat input on this flight to 63% of that designed for. This is still a larger initial step forward than is usual, so the first flights are unmanned.

Later flights are planned to explore the full environment with pitch and bank maneuvers gradually increasing to determine the true vehicle limits.

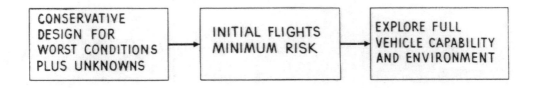

PHILOSOPHY OF X-AIRPLANE

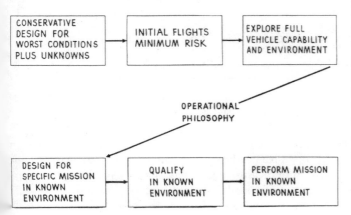

If a later operational vehicle has had the advantage of a preceding X-aircraft thoroughly exploring the environment in which it must operate, its design can be much less conservative. (Estimates have been made showing up to 20% weight reduction attainable.) As the operational vehicle never needs to be flown outside this known environment, its development cost and timing should be significantly reduced.

X-20 TEN FLIGHT PROGRAM

The currently planned X-20 Flight Test Program includes a twenty-flight Air Launch Program at Edwards Air Force Base, plus ten ground-launches with the Titan III booster. The ten-flight ground-launch program is designed to provide the maximum amount of research data on piloted orbital flight and winged re-entry without undue risk to the pilot in exploring an untried flight regime.

The first two ground-launch flights are currently planned to be unmanned. They will provide confidence in the adequacy of the booster and glider combination and will demonstrate that the glider is capable of a minimal safe re-entry from orbital flight. They are programmed to land at Edwards, using a ground control landing system, after a "once-around" flight from Cape Canaveral.

Later flights will be piloted and will provide data on manned control of the glider as previously discussed. As can be seen from the chart, the re-entry tests are the essence of the program; orbital tests are a bonus which will be exploited to the extent they don't interfere with the primary tests.

X-20 TEN-FLIGHT PROGRAM

PRIME OBJECTIVES	ORBITAL TESTS	RE-ENTRY TESTS
AIR VEHICLE COMPATIBILITY	SYSTEMS EVALUATION	NOMINAL RE-ENTRY PITCH = 36°
MINIMUM FLIGHT MARGIN VERIFICATION	SYSTEMS EVALUATION	ATTITUDE PROGRAM PITCH = 15°-50°
PILOTED LIFTING RE-ENTRY & LANDING	PILOT CONTROL & SYSTEMS EVALUATION	NOMINAL RE-ENTRY PITCH = 36°
MANEUVER IN ORBIT	ATTITUDE MANEUVERS ± 90°—VISUAL REFERENCE CHECK	CONTROL MANEUVERS PITCH = 10°-50°, 10,000 - FT MARGIN
MANEUVER DURING RE-ENTRY SYSTEMS EVAL	PILOT CONTROL OF TRANSTAGE ENGINE— SPACE MANEUVER DEMONSTRATION	CONTROL MANEUVERS PITCH = 10°-50°, ROLL = ± 20° 6000 -FT MARGIN
	SUBSYSTEM EVAL— ALTERNATE CONTROL MODES	DYNAMIC STABILITY & AEROTHERMAL MANEUVER PITCH & ROLL PULSES 2°-5° PER SECOND
MANEUVER IN ORBIT VEHICLE REUSE	VISUAL REFERENCE CONTROL & NAVIGATION	DYNAMIC STABILITY & AEROTHERMAL MANEUVER MIN REFURBISHED GLIDER
MANEUVER IN ORBIT	AERODYNAMIC RANGE CONTROL DURING EARLY RE-ENTRY	DYNAMIC STABILITY & AEROTHERMAL MANEUVER PITCH, YAW, ROLL
MANEUVER DURING RE-ENTRY PRECISION RECOVERY	MULTIOBIT FLIGHT APOGEE ADJUSTMENT DEMONSTRATION	SUSTAINED FLIGHT NEAR THERMAL LIMITS PITCH = 40° ROLL = ± 30°
	MULTIORBIT FLIGHT ORBITAL MANEUVER DEMONSTRATION -PLANE CHANGE 3°	SUSTAINED FLIGHT NEAR THERMAL LIMITS LATERAL RANGE - 1000 NAUTICAL MILES

TEST INSTRUMENTATION SUBSYSTEM SENSOR LOCATIONS

Three different configurations of test instrumentation are currently planned; one for air launch, one for early ground launch flights, and one for the later ground launch flights. The one shown is the early ground launch arrangement and hence, includes more subsystem measurements than later. Each flight will carry at least 750 channels; about 1000 different measurements will be made during the entire program.

Temperature and pressure sensor instrumentation has been located at various points on the vehicle to provide research flight data. This instrumentation will measure the magnitude, the distribution of surface temperature and pressure resulting from the local flow environment which is a function of vehicle attitude, relative velocity and density altitude.

1) Nose region
2) Wing and fin leading edge – stagnation line to shoulder value
3) Lower wing surface
4) Heat shield and vicinity of pilots compartment
5) Elevon and rudder control surface – variation of pressure and temperature with deflection will provide data on mutual interference effects
6) Base area region

Pressures, along with temperatures (from which heating rates can be derived), represent the only data which can be measured, within practical feasibility, that will give insight to the local environment within which the vehicle operates. Knowledge of these parameters will provide comparative data for evaluating the validity of current analytical prediction methods and experimental data obtained from ground facilities throughout the entire re-entry profile.

SENSOR LOCATIONS

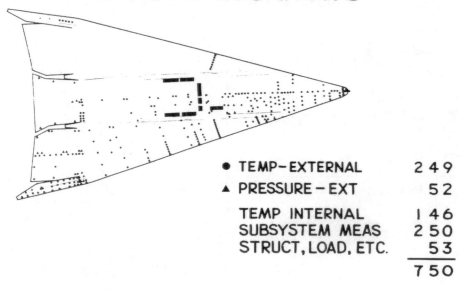

• TEMP-EXTERNAL	249
▲ PRESSURE – EXT	52
TEMP INTERNAL	146
SUBSYSTEM MEAS	250
STRUCT, LOAD, ETC.	53
	750

BENEFITS FROM AN X-20 FLIGHT TEST PROGRAM

The full-scale flight test program will provide the link between the ground derived data and the actual flight environment and will result in the realization of the benefits shown. Succeeding charts will amplify each of these except for the first which has already been discussed.

BENEFITS FROM X-20 FLIGHT TEST PROGRAM

• SCALING FROM WIND TUNNEL TO FULL-SCALE FLIGHT

• PROVIDE SIMULTANEOUS ENVIRONMENTS

• DISCOVERY OF UNANTICIPATED TRENDS

• REQUIREMENTS FOR OPERATIONAL RE-ENTRY SYSTEMS

• DEVELOPED TEST BED

SIMULTANEOUS ENVIRONMENTS

Flight test provides the demonstration and proof of an integrated workable system. Theory and ground testing provide the basis for design of all systems. However, ground testing is limited in that it provides only partial simulation of a single parameter or, in some cases, limited combinations of parameters. Even this testing is confined to a specific area of a vehicle or to scale models. Only in flight do all of the parameters occur simultaneously in the proper sequence, integrated with respect to size, shape, distribution and interactions. It is through this integration that theories can be modified or verified, concepts confirmed, and confidence for interpolation and extrapolations established.

The example shown applies only to the structure of the airplane. Not only is this tested for the first time with the proper combination of environments but also the entire vehicle is validated including pilot and subsystems as well.

SIMULTANEOUS ENVIRONMENTS

PREFLIGHT	FLIGHT TEST
INDIVIDUAL PARAMETERS: • TEMPERATURE • PRESSURE • LOADS • FLUTTER • HEATING RATES	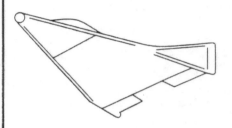
SIMULATED SEPARATELY OR IN LIMITED COMBINATION	DEMONSTRATION AND PROOF OF INTEGRATED WORKABLE SYSTEM

UNANTICIPATED TRENDS – LOW TURBULENT HEATING, X-15

Every X-airplane has yielded some surprising results from actual flight. These are of benefit to the aeronautical designer on future systems in avoiding costly mistakes. The X-20A flight test will certainly produce such surprises, but obviously they cannot yet be predicted.

As an example, however, turbulent heat transfer data obtained from X-15 flights were from 10% to 30% lower than were predicted by various theoretical methods. Each of these theoretical methods had been supported by wind tunnel data over a limited range of test conditions; however, none of these wind tunnel tests completely simulated the conditions of the X-15 flights. The X-15 test results thus indicated that existing theoretical methods for predicting turbulent heating rates are valid only over a limited range of flow conditions.

An interesting situation has developed from this trend. One theory advanced which explains the low heating on the X-15, when extrapolated to higher Mach numbers, predicts higher heating for the X-20A. This theory is now being used to evaluate the X-20A performance and safety margins.

UNANTICIPATED TRENDS

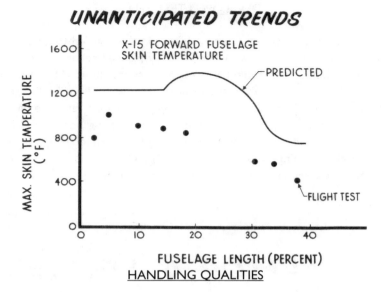

FUSELAGE LENGTH (PERCENT)

HANDLING QUALITIES

An example of how the X-20A flight test program will contribute to defining the "General Requirements for Operational Manned Re-Entry Systems" is shown here. Handling quality specifications for aircraft have been based on detailed studies of many missions for a wide variety of aircraft. These studies include: theoretical analysis, fixed and moving base simulator evaluations, and actual flight testing of both experimental and operational aircraft.

By analyzing and compiling these results, a series of standard government specifications have been developed over a period of years. These specifications provide a good design tool which result in an aircraft having satisfactory handling qualities without unnecessarily compromising the aircraft design.

Simulator tests of the X-20 with the Air Force X-20 consultant pilots indicated an anomaly in current handling qualities requirements. The chart shows current handling qualities requirements plus X-20 handling qualities ratings by the consultant pilots. As shown by the X-20 ratings, a configuration having a low pitch frequency and low damping was rated acceptable for normal operation even though current requirements would consider it unacceptable. A partial explanation for the difference between current requirements and pilot ratings lies in the pilot control task. At hypersonic speeds, a pilot has no need for a snappy vehicle response to control commands; hence does not consider it necessary. Accordingly, vehicle dynamics unacceptable for contemporary aircraft may be satisfactory for re-entry vehicles during hypersonic flight. During flight tests, the stability augmentation of the X-20 will be disengaged to permit evaluation of unaugmented handling qualities as an aid to developing requirements for future vehicles.

Similar examples could be discussed on vision requirements, crew station specifications, performance prediction techniques, and pilot aids for all-weather capability.

HANDLING QUALITIES

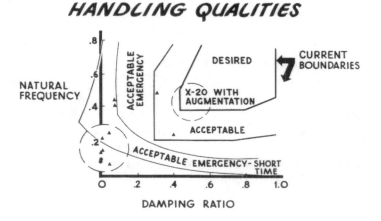

DAMPING RATIO

DEVELOPED TEST BED

Valuable materials and re-entry cooling development testing in the real flight regime cap be accomplished using the X-20A glider without interfering with the currently planned aerodynamic performance, control, and communication testing. Some of the attractive areas for test are shown. Materials and cooling systems for test could be either for research purposes or those desired for specific future vehicles. A major advantage is that the significant variables such as local pressure, heating rate, shearing force and velocity may be controlled over the entire range by selection of location on the glider or of the re-entry trajectory. The radiation-cooled vehicle introduces no contaminants into the surrounding gas to disturb chemistry or heating rates as do ablative cooling systems. The value of these tests is increased since the glider experiences flow regimes including laminar and turbulent continuum, slip, transition, free molecular and ionized flow.

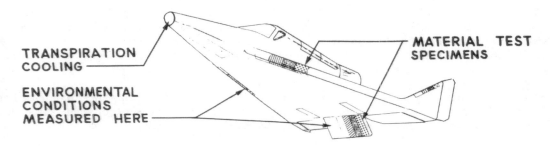

DEVELOPED TEST BED

TRANSPIRATION COOLING

MATERIAL TEST SPECIMENS

ENVIRONMENTAL CONDITIONS MEASURED HERE

- ● MATERIALS TESTS COMBINED WITH GLIDER TEST PROGRAM

- ● DUPLICATES ENTIRE RE-ENTRY ENVIRONMENT
- ● RADIATION-COOLED STRUCTURE GIVES UNCONTAMINATED FLOW
- ● IONIZATION & SLIP FLOW REGIMES CAN BE INVESTIGATED

EQUIPMENT COMPARTMENT

This is a view of the Equipment Compartment with the lid removed to show the payload which for the ten flight program is the Test Instrumentation Subsystem (TIS). The TIS includes equipment for sensing, recording, processing and transmitting flight data. This equipment occupies 18.9 cubic feet of the 75.6 cubic feet available for payload. The compartment is provided with a controlled temperature and pressure environment. (Pressure is 10 ± .2 psia, temperature is about 115°F).

This available volume plus the re-entry weight increase permitted by the new emittance values allows experimental hardware to be flight tested in addition to the full re-entry testing.

EQUIPMENT COMPARTMENT

PAYLOAD FLEXIBILITY

These insets are examples of various alternative payload installations that have been studied. This work was done on a separate study contract which extended from November, 1960 through June, 1962, expending $1.7 million. The basic purpose of the study was to examine several potential military space missions, to evaluate the extent to which the X-20 could perform these, and to outline an experimental flight test program (beyond the current 10-flight program) which could demonstrate the feasibility of the hardware actually performing such missions.

The missions examined included satellite inspection, logistic support, three forms of reconnaissance, and subsystem tests. Available military subsystem equipment were used in each of these examples (for example, the catadioptric camera lenses are developed). The Titan III will provide adequate weight into orbit to handle the payloads discussed plus additional glider endurance expendables in the shaded area of the transition still leaving enough fuel in the transtage to demonstrate rendezvous.

The results of these studies have caused considerable Air Force interest; however, no effort is currently planned to extend or to implement them. Additional explanatory material is available to you if desired. The X-20A / TIII capabilities are unique in providing 75 cubic feet of conditioned volume for returnable payload, plus over 12 kva of electrical power, plus a pilot to control the demonstrations, plus adequate orbital propulsion and expendables.

PAYLOAD FLEXIBILITY

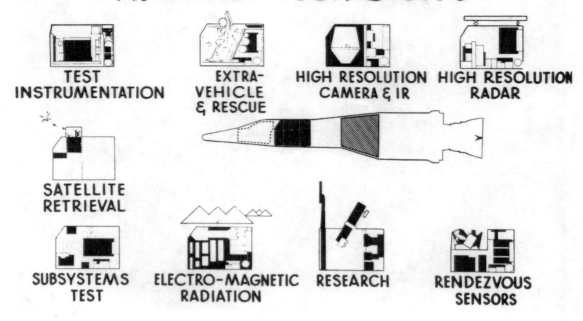

TEST INSTRUMENTATION   EXTRA-VEHICLE & RESCUE   HIGH RESOLUTION CAMERA & IR   HIGH RESOLUTION RADAR

SATELLITE RETRIEVAL

SUBSYSTEMS TEST   ELECTRO-MAGNETIC RADIATION   RESEARCH   RENDEZVOUS SENSORS

X-20 POTENTIAL (MULTI-ORBIT)

The X-20 possesses growth potential obtainable through vehicle refinement and advanced operating techniques. This chart shows some of these potentials in the form of increased space maneuver and re-entry payload capabilities. The chart is based on a 3-orbit mission.

1. Improve T-III Boost Capability – The booster "thrown weight" (defined as the weight above the transtage) is expected to increase by approximately 4,000 lbs. by 1967. Presently quoted weights are intentionally conservative.

2. Glider Inert Weight Reduction – Upon completion of the flight test program there will be a potential glider structural and subsystem weight reduction through design improvement and state of the art advances. This weight saving is estimated to be 750 pounds.

3. Considerable performance potential is available through the refinement of the abort rocket system and the transition multi-orbit provisions. These areas were initially designed with primary regard for cost and only small attention on optimizing weight.

4. Improved Emittance – Current tests indicate a molybdenum design emittance of .85 at high temperature as compared to the present design value of .75. This will allow an increase in returnable payload weight of 2500 lbs. and in available immediately.

5. The previous discussion is based on obtaining the space maneuver by use of the transtage. If this weight is utilized for "synergetic maneuver" the capability shown by the dotted line should be obtained. The synergetic maneuver is discussed on the following chart. As can be seen, if the full growth shown here were attained, a space plane change of 24 degrees could be accomplished.

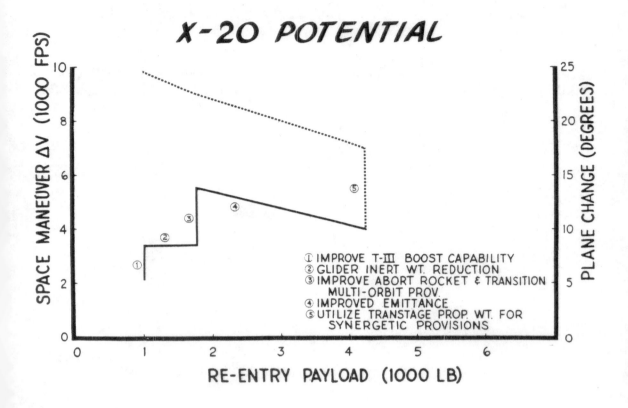

X-20 POTENTIAL

① IMPROVE T-III BOOST CAPABILITY
② GLIDER INERT WT. REDUCTION
③ IMPROVE ABORT ROCKET & TRANSITION
   MULTI-ORBIT PROV.
④ IMPROVED EMITTANCE
⑤ UTILIZE TRANSTAGE PROP. WT. FOR
   SYNERGETIC PROVISIONS

SYNERGETIC PLANE CHANGE CONCEPT

The definition of the word "synergetic" is: working together, cooperating. The synergetic plane change maneuver is the mating of propulsive and aerodynamic forces with a resultant angular plane change that is greater than the sum of the changes that could be realized from the individual forces. The maneuver shown on the accompanying chart can be separated into seven sequences.

1) De-orbit thrust to generate an elliptic trajectory that will intersect the atmosphere.
2) Unpowered descent to perigee altitude within the atmosphere. As the atmosphere is entered the vehicle is banked and pitched to optimum attitude to start the turning maneuver. During the descent velocity losses due to aerodynamic drag result in perigee occurring at sub-orbital speeds.
3) An acceleration period to attain orbital velocity. During this period the thrust is canted to the side to produce some additional turning.
4) 90° banked powered turn at orbital velocity and temperature limit altitude. Vehicle is pitched to 18° for high lift-to-drag.
5) An acceleration period to overcome aerodynamic drag and permit climb to orbital altitude. The thrust may be canted to obtain increased turn capability.
6) Unpowered ascent from perigee to orbit attitude. Optimum pitch and roll attitudes are maintained to continue turning while in the sensible atmosphere.
7) Injection impulse to recircularize at desired orbit altitudes.

Although this concept has not been thoroughly studied, the capabilities of the X-20A (high lift-to-drag ratio and radiation-cooled structure) uniquely suit it to this type of maneuver.

SYNERGETIC PLANE CHANGE

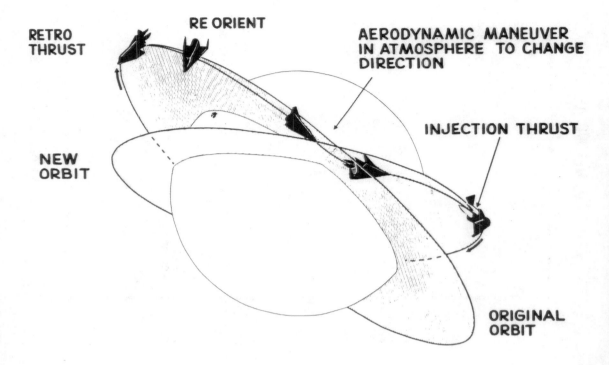

RETRO THRUST

RE ORIENT

AERODYNAMIC MANEUVER IN ATMOSPHERE TO CHANGE DIRECTION

INJECTION THRUST

NEW ORBIT

ORIGINAL ORBIT

SPACE AND RE-ENTRY MANEUVER FLEXIBILITY

The chart makes this comparison between a semi-ballistic vehicle and an X-20 with and without synergetic capability. The semi-ballistic vehicle was assumed to have a basic weight of 7000 lbs. plus 1000 lbs. of payload.

The X-20 configuration has a lateral range potential of 1900 NM if the turn is initiated at reentry. For a thrown weight of 28,000 lbs. sufficient fuel is carried to make an in-orbit propulsive turn of 15.8° or a synergetic turn of 20.3°. If these turns were used to accomplish a space mission the X-20 would have a 1900 NM lateral range capability from the new orbital plane. If a lesser amount of plane change were required for the orbital mission the remaining plane change can be transformed into additional lateral range at 60 NM per degree of plane change. As an example, the X-20 could make a synergetic plane change of 10° using only part of its available fuel. After completing its mission in the new plane it could re-enter, perform another turn and then maneuver laterally to a landing site displaced 2500 NM from the second orbital plane.

The semi-ballistic: configuration shown has limited lateral range from aerodynamic means because the lift to drag ratio (L/D) is limited to 0.25. The lighter basic weight of the ballistic vehicle allows more fuel to be boosted into orbit with a resultant greater propulsive plane change capability.

FLIGHT REGIMES

The aerodynamic and structural research resulting from the X-20 Program will have an early and direct benefit to the recoverable booster, should the government choose to develop one.

SPACE AND RE-ENTRY MANEUVER FLEXIBILITY
TOTAL THROWN WEIGHT = 28,000 LBS

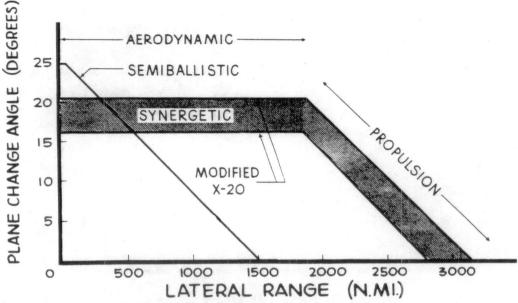

Reusable boosters will generally fly in the speed-altitude regimes shown on the accompanying chart. They will use materials and encounter structural limits which are quite similar to the X-20. These re-entry trajectories and flight limits were established from a wide range of Boeing design studies on reusable launch systems.

Some launch systems under consideration use air breathing propulsion and may encounter an even more severe environment in the Mach 6 to 12 region than that experienced by the X-20. The data from the X-20 Program should permit reasonable extrapolation to this region.

MISSION OPERATING ENVELOPE
TITAN III 28,000 LBS PAYLOAD

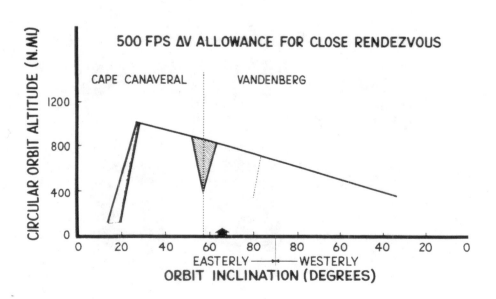

FLIGHT REGIMES

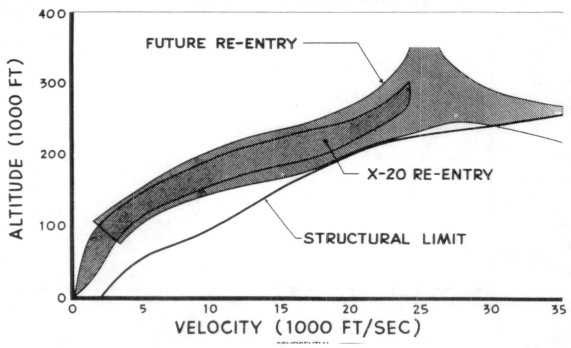

REUSABLE LAUNCH SYSTEMS TYPICAL CONFIGURATION

On the following page are shown photographs of models of configurations which have resulted from some of the Boeing reusable launch system studies. The first stage units generally have wings and aerodynamic controls. Studies of upper stages have covered the entire spectrum from very blunt bodies to slender winged gliders which look very much like the X-20.

Much of the X-20 technology, particularly in the areas of aerodynamics, structures and materials, flight controls, guidance and navigation, and operational techniques, is directly applicable to the reusable booster program.

REUSABLE LAUNCH SYSTEMS TYPICAL CONFIGURATIONS

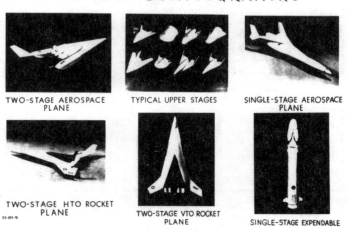

TWO-STAGE AEROSPACE PLANE

TYPICAL UPPER STAGES

SINGLE-STAGE AEROSPACE PLANE

TWO-STAGE HTO ROCKET PLANE

TWO-STAGE VTO ROCKET PLANE

SINGLE-STAGE EXPENDABLE

SPACE SYSTEM EVOLUTION

There appear to be two main streams of evolution being followed in the flight to higher speeds and altitudes. This chart showy the growth of manned ballistic systems. The current Mercury vehicle has leaned very heavily on the technology developed by the extensive ballistic missile program. The Mercury technology is almost exactly that of Gemini. This same family then grows to much higher speeds and becomes Apollo.

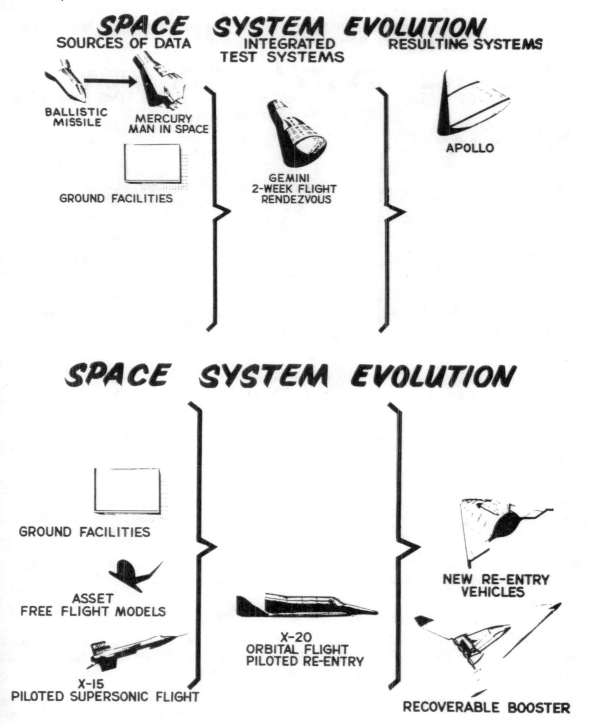

SPACE SYSTEM EVOLUTION

This chart depicts the growth of winged vehicles as a further step beyond the capabilities of current aircraft. The X-15 vehicle was designed to test the capability of winged technology for still higher altitudes and velocities and the Asset free flight models for testing higher temperature materials and aerothermodynamic theory. Both of these systems have or will contribute to the development of the X-20.

The X-20A does not appear to be the end of airplane technology any more than the Gemini appears the end of ballistic technology. There will be piloted, controllable vehicles eventually either at higher speeds or, such as the recoverable booster, at the same or lower speeds.

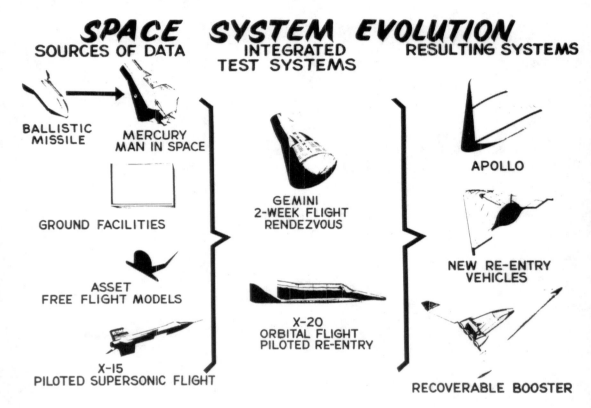

SPACE SYSTEM EVOLUTION

SOURCES OF DATA   INTEGRATED TEST SYSTEMS   RESULTING SYSTEMS

BALLISTIC MISSILE   MERCURY MAN IN SPACE   APOLLO

GROUND FACILITIES   GEMINI 2-WEEK FLIGHT RENDEZVOUS

ASSET FREE FLIGHT MODELS   X-20 ORBITAL FLIGHT PILOTED RE-ENTRY   NEW RE-ENTRY VEHICLES

X-15 PILOTED SUPERSONIC FLIGHT   RECOVERABLE BOOSTER

SPACE SYSTEM EVOLUTION

There has been in the past, and will continue to be in the future, much cross fertilization of technical knowledge stemming from developments in each of these two families. Both system types must be developed to assure the availability of technical data to cope with future unanticipated events.

The optimum configuration for all missions in all environments is unlikely to exist. If it does, it may resemble neither, or may be a combination of the best features of both.

SPACE PROGRAM GOALS

Because of the uncertainty of the eventual usefulness of space to the national security, and because of the great gaps in existing technology, we understand the government has established certain "building block" capability goals which are relatively independent of any particular mission. Some of these are shown here.

The X-20, in its current configuration, provides special capability for the starred items on this list; these are discussed in more detail on the next page. It should be noted here, however; that the X-20 program will and/or can contribute substantially in the other four areas as a bonus value if desired:

a) Experience in launch readiness will be gained with the X-20 / Titan III Air vehicle.

b) Addition of rendezvous electronic subsystems (which can be incorporated readily) would allow X-20 demonstration of rendezvous with an uncooperative target. This may be a desirable duplication with the Gemini system.

c) Operations independent of ground control, pilot use of space maneuver, and quick-return experience will improve mission performance in a hostile environment.

d) Long duration space tests could be accomplished by the X-20 space vehicle through improvement of glider subsystems and use of the volume available in the glider transition.

SPACE PROGRAM GOALS

- LAUNCH READINESS
- RENDEZVOUS WITH UNCOOPERATIVE TARGET
- ★ MANEUVER IN ORBIT
- ★ MANEUVER DURING RE-ENTRY
- MISSION PERFORMANCE IN A HOSTILE ENVIRONMENT
- ★ PRECISION RECOVERY
- ★ VEHICLE REUSE
- LONG-DURATION SPACE TESTS
- ★ TESTING OF COMPONENTS AND SUBSYSTEMS

KEY X-20 APPLICATIONS

These five Space Program Goals can be developed in considerable depth during the current X-20 Program of ten flights.

The current X-20 Program provides for space testing with the glider, transition and transtage combined. Thorough experimentation with piloted re-entry, maneuver and landing will develop the system and the reusable glider.

The program can proceed, if desired, into many areas of exploration, including:

a. Extended orbital maneuver capability,
b. Extended operational recovery systems,
c. Advanced testing of materials, components and subsystems.

As indicated earlier, some of these features can be included in the current program without affecting the primary objectives.

In 1960 an Industrial Survey was conducted by the Air Force and Boeing to establish industry capabilities to manufacture components from various superalloys and refractory alloys. The results of this survey follow.

1960 INDUSTRIAL SURVEY

Principal Values

Summarizing the principal values expected to come from the X-20 Program, the aerodynamic exploration will provide guide lines for design of aircraft operating over a wide range of lifting conditions in these speed regimes.

KEY X-20 APPLICATIONS

GOAL	KEY TECHNOLOGY	X-20 PROVIDES FROM	
		CURRENT PROGRAM	EXTENSION
MANEUVER IN ORBIT	SPACE PROPULSION VEHICLE CONTROL SYNERGETIC PLANE CHANGE	3200 FPS WITH TRANSTAGE INTEG SPACE VEHICLE	UP TO 8000 FPS RENDEZVOUS SYSTEM SYNERGETIC TEST CAPABILITY
MANEUVER DURING RE-ENTRY	HI-TEMP STRUCTURE AERO HEATING ADAPTIVE CONTROL FLUTTER	RADIATION-COOLED STRUCT FLIGHT-TEST VERIF PROVEN HARDWARE & TECH FLUTTER-FREE AIRCRAFT	ABLATIVE STRUCT TESTS
PRECISION RECOVERY	MODULATED ATTITUDE CONTROL CONTROL SURFACE EFFECT HIGH L/D CONFIG AERO ENERGY MGT & DISPLAY GOOD HANDLING QUALITIES LIFTING SUPERORBITAL	FULL PILOT CONTROL CONFIRM. OF GROUND DATA VERIFIED PERFORMANCE PROVEN METHODS & HDWRE PILOT EXPER & RATING	ALL-WEATHER CAPABILITY IMPROVED GLIDER CONFIG LANDING AT BASE OF PILOT'S CHOICE SUPERORBITAL TESTING
VEHICLE REUSE	NONABLATIVE STRUCTURE HORIZONAL LANDING RELIABILTY MAINTAINABILITY	REUSABLE STRUCTURE CONVENTION. LDG & RECOVERY DESIGNS, STRUCTURE, COMPON DESIGNS, TECHNIQUES, EXPER	ADVANCED MAT'LS CAPAPAB IMPROVED SUBSYSTEMS ADVANCED DESIGNS
TESTING OF COMPONENTS & SUBSYSTEM	PAYLOAD ADAPTABILITY ONBOARD POWER RETURN PAYLOAD CAPAB ONBOARD COOLING RE-ENTRY TEST CAPAB	ADAPTABLE CONFIG 2-12 KW SYSTEMS 75-100 CU FT UP TO 4000 LBS DEVEL LARGE CAPACITY	USE OF 800 CU FT TRANSITIONS READY INTERCHANGEABILITY OF LARGE RETURNABLE GLIDER PAYLOAD TESTING OF ADVANCED COMPONENTS & MAT'LS TRANSPIRATION COOLING

New concepts of structural arrangement will be developed capable of withstanding the re-entry environment.

New materials and manufacturing processes will be developed to allow the ready manufacture, use and reuse of this type of hypersonic aircraft.

Guide lines or techniques will be developed for the guidance and control of the vehicle and for the use of man in making decisions and improving reliability.

The flight test program will provide a test bed with sufficient volume, payload, weight, and, available power to permit testing of new materials, subsystems and components in an environment which cannot be duplicated in ground facilities. Last but not least will be the wealth of X-20 data that will be used to validate or correct the existing theories used in this design and that also will, contribute to extrapolation to the step beyond the X-20.

Bonus Values

Other benefits will accrue from this program.

First we will have another concept of recovery of vehicles and equipment from space that may be safer, more reliable, and versatile than ballistic recovery systems.

We may develop a means of obtaining greater maneuver in orbit than is obtainable through propulsion alone. The vehicle can be used as a space test bed as well as a re-entry test bed.

The experience in developing the X-20 with its new materials, processes and techniques will broaden the national industrial base as well as providing the Department of Defense with an increased technical and operational capability.

Date: Mar 1963   Title: Symposium Proceedings (Extract)

PROCEEDINGS OF 1962 X-20A (DYNA-SOAR) SYMPOSIUM

GENERAL TESTING AND GROUND SUPPORT, AND SUBSYSTEMS

Technical Documentary Report No. ASD-TDR-63-148, Volume I

March 1963

Aeronautical Systems Division
Wright-Patterson Air Force Base Ohio

FOREWORD

The 1962 X-20A Symposium, held In November at Aeronautical Systems Division, Wright-Patterson Air Force Base, Ohio, had the following objectives:

(a) summarize the current X-20A program and its status;
(b) provide a report on technical progress achieved since the April 1960 conference held at Langley Air Force Base, Virginia; and
(c) stimulate thinking, coordination, and exchange of technical information. The symposium theme was "Progress In X-20A Technology" and to support this theme, speakers and session chairmen in particular related the current state of knowledge to that available in April 1960.

The symposium included a general session to summarize and relate the X-20A program, Titan III program, X-20A test planning, overall technical progress and the current X-20A configuration. Sixty technical papers were presented in ten technical sessions of approximately 3-1/2 hours each. While numerous additional papers would have been desirable for more complete coverage of the program, it is believed the more significant developments have been included or referred to. Time limitations and the desire to clarify the essential messages of each paper for the wide representation of technological disciplines in the audience led to the plan to include more complete data and greater detail in the written papers published in this, and the additional volumes of the proceedings. Proceedings of this symposium are contained in five separate volumes. In addition, a summary voume, Volume VI, containing the abstracts of all papers, has been published and made available to all organizations requesting the proceedings.

Volume I - General, Testing and Ground Support, and Subsystems Volume II - Flight Mechanics and Guidance Volume III - Structures and Materials Volume IV - Instrumentation and Communications Volume V - Bioastronautics Volume VI – Abstracts

Acknowledgement is due the contractors involved, The Boeing Company, MinneapolisHoneywell, and Radio Corporation of America, and the several Governmental agencies, the Air Force Flight Test Center, NASA, and Wright- Patterson AFB for the support received in planning and conducting the symposium.

Special acknowledgement is due the authors of the papers presented herein, the symposium co-chairman, J. H. Goldie of the Boeing Company the session chairmen and cochairmen, the executive review committee which included Mr. P. Korycinski of NASA Langley Research Center, and G. Johnson of the X-20 SPO and J. H. Goldie, T. S. Liu who effectively handled many of the technical arrangements, T. Lonchar of the SPO illustration group and the art departments and administrative support groups at the SPO and contractors.

ABSTRACT

The Proceedings of the 1961 X-20A (Dyna-Soar) Symposium are presented in this report. The X-20A program developing the pilot controlled lifting reentry glider and conducting full-scale hypersonic flight research is one of the most extensive advanced developments under way in the Air Force. To insure the timely dissemination of the data obtained by the X-20A program, the 1962 X-20A symposium was held to present new and significant knowledge gained during this development period to the aerospace industry, scientific and Government agencies. Materials were presented in ten technical areas consisting of flight mechanics, structures, materials, guidance, communications, instrument power and environment subsystems, bioastronautics, testing and ground support. Papers in two of these areas are assembled In each of five volumes of this report. A separate volume, Volume VI, contains the abstracts of all papers in Volumes I to V.

PUBLICATION REVIEW

The publication of this report does not constitute approval by the Air Force of all the findings or conclusions contained herein. It is published only for exchange and stimulation.
WALTER L. MOORE Colonel, USAF Director, X-20 System Program Office

INTRODUCTION

The X-20A (Dyna-Soar) program, to develop and demonstrate piloted research vehicle capable of orbital flight, controlled maneuverable reentry from orbit, and extensive research of the reentry and hypersonic flight regimes, represents one of the most extensive advanced developments under way in the Air Force. The X-20A program has naturally become of interest to many organizations and personnel engaged in the advancement of technology and the development of advanced systems. Although a prime objective of the X-20A program is the development of basic capabilities important to potential military systems requiring operation in space and the flexibility of controlled and maneuverable reentry, it is obvious that the extensive research, development, and testing required to achieve such advanced objectives provides a host of technological advancements as by-products.

The Air Force considers it important to insure the dissemination of the knowledge gained by the X-20A program to the aerospace industry, scientific and Governmental agencies to permit and encourage use of this knowledge in facilitating development and reducing cost of other projects and systems being developed for the Government. To assist in accomplishing the objective of disseminating data on the technical progress and achievements of the X-20A program and related research, the Air Force and NASA jointly sponsored the "Joint Air Force-NASA Conference on Reentry and Hypersonic Vehicles at the Langley Research Center of NASA in April 1960. This conference was found highly effective, and it was therefore decided to conduct subsequent Government-industry conferences or symposia as additional X-20A progress became available. The 1962 X-20 Symposium was held early in November at the Aeronautical Systems Division, Wright-Patterson Air Force Base, Ohio. In the following volumes are the proceedings of all the sessions of the symposium. Having established the theme of "Progress," the symposium schedule did not permit "state-of-the-art" reviews. Technology baseline of 1960 was used for all measures of progress in various disciplinary areas of the 1962 symposium. Some new ideas of particular interest to the X-20, but not necessarily being incorporated in the current program, were also included in these proceedings for the purpose of stimulating thinking and exchanging information.

THE X-20A (DYNA-SOAR) CONFIGURATION

INTRODUCTION The X-20 System has been developed to demonstrate piloted maneuvering reentry from orbit with controlled glide landing at a preselected landing site and to gather research data on lifting reentry from orbital flight. The present configuration is the result of two and one-half years of design and development effort directed at meeting these requirements. The X-20A Program requires the design and analysis of the six configurations shown on Figure 1. Each of these configurations must be analyzed for structural loads, performance, stability and control and vehicle interfaces. This paper briefly discusses the launch orbital and abort vehicles. However, primary emphasis is on the X-20A glider.

FIGURE I
X-20 FLIGHT CONFIGURATIONS

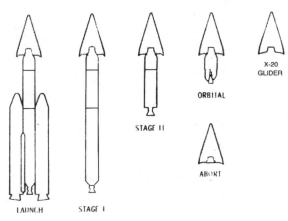

X-20
GLIDER

ORBITAL

STAGE II

ABORT

LAUNCH   STAGE I

FIGURE 2
GENERAL ARRANGEMENT
Air Vehicle Model 853-1001 Titan IIIC/X-20

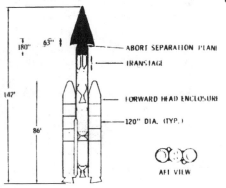

7/180"   63"

142'

86'

ABORT SEPARATION PLANE

TRANSTAGE

FORWARD HEAD ENCLOSURE

120" DIA. (TYP.)

AFT VIEW

AIR VEHICLE AND ORBITAL FLIGHT VEHICLE CONFIGURATIONS

The air vehicle consists of the X-20A Glider mated to a Titan III booster by a transition section as shown in Figure 2. The Titan III booster is made up of two 5-segment 120-inch solid rockets and three liquid rocket stages. The upper liquid stage is referred to as the transtage. The transtage is a multipurpose stage that remains with the transition section and X-20 glider at boost burnout. This configuration is the orbital flight vehicle as shown on Figure 3. The transtage contains the primary boost guidance system and a reaction control system for attitude control of the orbital flight vehicle. The transtage engines may be restarted for velocity vector trim during the coast to apogee, and on multiorbit flights these engines will provide orbit injection and de-orbit thrust. The engines could also be used for limited orbit shifting.

A solid propellant acceleration rocket motor for escape off the pad or escape during portions of boosted flight is installed In the transition section forward of the abort separation plane. On multiorbit flights additional expendables for the pilot and vehicle subsystems will be installed in the transition section aft of the abort separation plane. A typical once-around flight profile is shown in Figure 4. The orbital flight vehicle is boosted eastward from Cape Canaveral to a velocity of 24,470 feet per second and an altitude of 320,000 feet at boost burnout. Apogee altitude is 480,000 feet. The X-20 Is separated from the transition section and transtage at the initiation of reentry and proceeds to a landing at Edwards Air Force Base. The critical reentry heating regime occurs over the Pacific Ocean between 17,000 and 24,000 feet per second. A typical three-orbit flight is shown in Figure 5. For this flight the orbital vehicle is injected into a circular orbit with an altitude of 600,000 feet. At boost burnout, velocity is 24,536 feet per second and altitude is 322,000 feet. Retrothrust rockets are fired over Angola. As in the case of once-around flights, the X-20A separates from the transition section and transtage, re-enters, and lands at Edwards Air Force Base.

X-20A GLIDER CONFIGURATION

The external configuration of the X-20A Glider as shown in Figure 6 was developed on the basis of five major design requirements.

1. The glider must withstand the thermal environment associated with lifting and maneuvering reentry.
2. The glider is required to fly and maneuver in a flight regime of dynamic pressures of from approximately 0 to 500 psi and velocities of from 300 to 25,000 feet per second.
3. Adequate space and vision had to be provided for the pilot.
4. A requirement exists for adequate volume for the test data system necessary to fulfill the X-20's role as a research vehicle.

FIGURE 3
ORBITAL VEHICLE CONFIGURATION

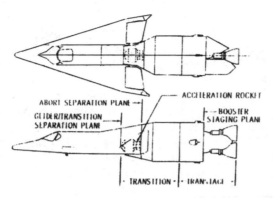

ABORT SEPARATION PLANE
GLIDER/TRANSITION SEPARATION PLANE
ACCELERATION ROCKET
BOOSTER STAGING PLANE
TRANSITION — TRANSTAGE

FIGURE 4
ONCE-AROUND TRAJECTORY
& GLOBAL RANGE

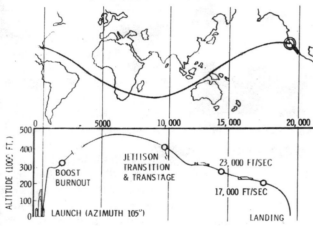

FIGURE 5
MULTIORBIT
Launch Azimuth 67°

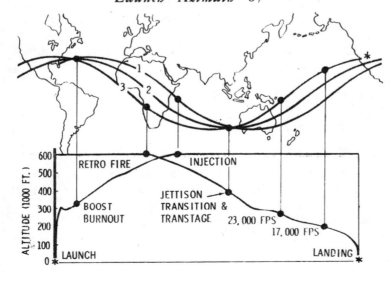

5. The glider must be designed for loads associated with boost and landing as well as reentry flight loads.

Early studies initiated because of the first requirement resulted in a radiation-cooled structural design using refractory and superalloy materials. Because of the temperature limits of these materials, "trade-offs" involving structural temperature and glider performance and stability were required in the development of the configuration. Some significant influences of these "trade-offs" on the external configuration are as follows:

a. The radii of the nose cap and wing and vertical tail leading edges were established by the trade between component heating load and L/D max. A larger radius reduces both heating load and L/D max.

b. The flat-bottom surface of the wing provides better heat distribution, higher CL max and higher L/D max than some of the earlier V-bottom configurations.

c. The wing sweep angle of the 72°48 ft was determined as a result of the trade between leading edge heating load, L/D and the need to maintain a proper relationship between the aerodynamic center and vehicle center of gravity.

An increase in sweep would reduce leading edge heat load, increase L/D, and would have an adverse effect on pitch static stability. Structural temperatures also limit rudder and elevon deflections. This contributed to elevon and rudder size requirements. Significant changes have been made in the external configuration as a result of the second requirement. These changes are based on wind tunnel tests conducted since 1960 and include the following:

a. The elevon size was changed from 10.3 percent to 13.3 percent of the wing area to improve pitch trim and control at hypersonic speeds.

b. A ramp was added to the aft body for directional stability at transonic speeds. This change also reduced elevon hinge moments.

c. The wing chord section was changed to approximate a low

FIGURE 6
GENERAL ARRANGEMENT
D-S Model 844-2050

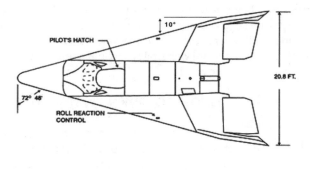

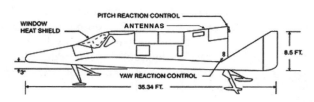

speed air foil in order to improve low-speed pitch stability.

d. The vertical tail wedge angle was changed from 8° to 10°. This wedge angle resulted from a trade study of the wedge angle versus the area required for directional stability at hypersonic speeds. In this case 1° of wedge angle is equivalent to 4 square feet of fin.

The third requirement established that the pilot be placed well forward for adequate vision and determined the depth and cross-section of the body at the pilot's station. The forward placement of the pilot and local body contours in this area has a destabilizing effect. This required tailoring of other portions of the glider to maintain required stability characteristics. The fourth requirement resulted in a payload volume of 75 cubic feet. This was a factor in determining the size of the body. The last requirement contributed directly to the weight of the glider by setting design conditions for structure. The boost dynamic pressure of 1020 psf established wing and body panel flutter design requirements. Bending moments due to wind shears during boost, and loads due to landing determined design criteria for portions of the wing and body trusses.

Each of the other requirements also contributed either directly to the weight of the vehicle or indirectly by establishing size and weight requirements for supporting subsystems. Since wing loading, maximum C_L capability and allowable structural temperatures determine the vertical height of the reentry corridor, the above five design requirements had a major influence in establishing the design weight of 11,150 pounds and a wing area of 345 square feet: The present X-20A configuration has a C_L max of 0.7 and an L/D ratio range of 0.8 to 1.9 at hypersonic speeds. The landing L/D ratio is 4.4. Flight between L/D max and C_L max at hypersonic speeds is accomplished at angles of attack between 18° and 52°. At 4000 feet per second, on a nominal reentry, a transition is made from the high angles of attack used at hypersonic speeds to angles of attack ranging from 0° to 15° for approach and landing. Conventional delta wing aerodynamic control surfaces are provided. Elevons are used together for pitch control and differentially for all control. The dual rudders operate together for yaw control and each is displaced 13° outboard for use as speed brakes. Control surfaces are hydraulically actuated in response to pilot or guidance system commands.

The pilot has four modes of control - an automatic mode, a manual direct mode, and two manual augmented modes. All controls are "fly-by-wire" wherein all pilot or guidance commands are in the form of electrical inputs at the control surface hydraulic actuators. During reentry, aerodynamic controls are augmented by two redundant hydrogen-peroxide reaction control systems. At dynamic pressures above 10 psf, aerodynamic control is adequate so the reaction control systems are shut down and the remaining fuel is dumped Both front and side vision are provided for the pilot. During orbital flight and reentry, the forward windshield is protected with a heat shield. This heat shield is jettisoned at Mach = 4 to 6 to provide forward vision for approach and landing. A tricycle landing gear using skids and an all metal shock absorbing system allows the pilot to land at the lake bed at Edwards Air Force Base or a conventional asphalt or concrete runway. Landing velocity at touchdown varies from 80 to 220 knots.

GLIDER STRUCTURAL ARRANGEMENT

The basic hot structure of the X-20A is shown in Figure 7. Primary loads are carried in the body and wing trusses. The trusses are determinate structure constructed from René 41 material and are designed to compensate for thermal expansion. The upper wing, body, and inboard fin surfaces are covered with René 41 panels; the leading edges of the wing and fin are TZM molybdenum, the bottom surface of the wing and outboard surface of the fins are covered with D36 columbium heat shields; and the nose cap is zirconium. Insulated RT 0029 graphite. During the maximum lateral range reentry maximum structural temperatures are 3650°F on the nose cap, 2825° on the wing leading edges, and 2450°F on the wing lower surface. The body truss provides space for four compartments or bays; the forward pilot's compartment, a central equipment compartment, an aft equipment bay and a secondary power bay.

FIGURE 7
X-20 TRUSS STRUCTURES

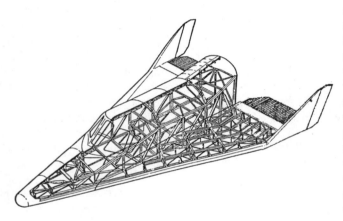

GLIDER EQUIPMENT INSTALLATIONS

All equipment except the reaction control thrust nozzles, the control surface hydraulic actuators, antennas and wave guides, and some of the test instrumentation is installed in one of the four compartments or bays shown on Figure 8. Equipment installed outside these areas is either insulated, as in the case of hydraulic and reaction control components, or is capable of withstanding the environment such as in the case of wave guides and antennas. Each of these compartments or bays is passively cooled by water walls. The water walls drop the external maximum temperature of approximately 1800°F to below 200°F on the compartment surfaces allowing the use of aluminum compartment structure.

The pilot's compartment is pressurized to 7.35 psia and has an atmosphere of 43.5 percent oxygen and 56.5 percent nitrogen by weight. This is equivalent to a cabin pressure altitude of 18,000 feet and oxygen partial pressure altitude of approximately sea level. The equipment compartment Is pressurized to 10.0 psia and has a 100 percent nitrogen atmosphere. The aft equipment bay and the secondary power bay are unpressurized.

FIGURE 8

COMPARTMENTS

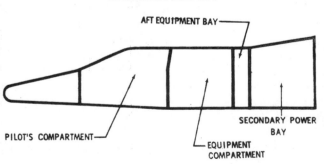

To minimize fire hazards in case of subsystem malfunctions, both the aft equipment bay and the secondary power bay are equipped with a nitrogen purging system. An environmental control system provides cooling within the pilot and equipment compartments, as well as cooling of the hydraulic system and power generation equipment This system uses cryogenic hydrogen as a heat sink with redundant glycol water loops for transferring heat from the compartment and equipment heat exchangers to the primary hydrogen cooler. Access to compartments is provided by removable panels and doors in the hot structure as shown in Figure 9.

With the exception of the pilot's hatch and umbilical door, access to equipment and service panels within compartments is attained by removal of the hot structure panel and a corresponding inner panel in the

FIGURE 9
GLIDER ACCESS

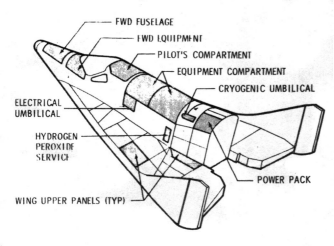

- FWD FUSELAGE
- FWD EQUIPMENT
- PILOT'S COMPARTMENT
- EQUIPMENT COMPARTMENT
- CRYOGENIC UMBILICAL
- ELECTRICAL UMBILICAL
- HYDROGEN PEROXIDE SERVICE
- POWER PACK
- WING UPPER PANELS (TYP)

FIGURE 10
INBOARD PROFILE
D-S Model 844-2050

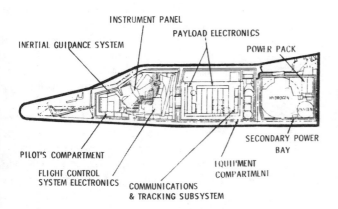

- INSTRUMENT PANEL
- INERTIAL GUIDANCE SYSTEM
- PAYLOAD ELECTRONICS
- POWER PACK
- HYDROGEN
- OXYGEN
- SECONDARY POWER BAY
- PILOT'S COMPARTMENT
- EQUIPMENT COMPARTMENT
- FLIGHT CONTROL SYSTEM ELECTRONICS
- COMMUNICATIONS & TRACKING SUBSYSTEM

compartment cold structure. 'The hot and cold structure on the pilot's hatch is an integral assembly that is removed or installed as a unit. The umbilical doors are separate hinged panels of hot structure and water wall that close and latch automatically at umbilical disconnect. Access to equipment outside the compartments is attained directly, by removing wing or body panels.

PILOT'S COMPARTMENT

Major equipments installed in the pilot's compartment as shown in Figure 10, are the guidance system; the flight control electronics, gas bottles for the landing gear extension, heat shield jettison, and pilot extension pneumatic systems; pilot's instrumentation and controls; pilot's seat; the compartment cooler; the compartment pressure regulating equipment; the pilot's breathing and ventilating equipment; and the test data recorder. The pilot's hatch on this compartment is equipped with ordnance-actuated fasteners and thrusters for hatch jettison to allow pilot escape with the ejection seat. This escape system is designed to operate at velocities up to Mach .9 in flight and down to 70 knots in the landing slide-out. A mockup photo of the pilot's cockpit is shown in Figure 11.

The side arm controller on the right hand side of the cockpit is for pitch and roll control of the glider. Conventional rudder pedals are used for yaw control. Controls for voice communications, hatch jettison, and flight control trim are on the left hand console. The handle on this console is used for initiating manual separation of the glider from the orbital flight vehicle. This handle also is used for manually firing the acceleration rocket and initiating abort separation. Figure 12 is a close-up of the instrument panel.

The left hand side of the panel contains glider subsystem controls, pilot's suit controls, the landing gear extend handle, the cabin pressure dump lever, and controls for the UHF-SHF communications and tracking subsystem; the center portion contains the flight instruments; and the right hand portion

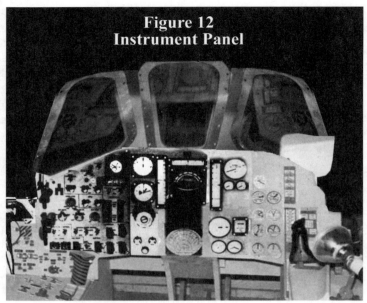

Figure 12
Instrument Panel

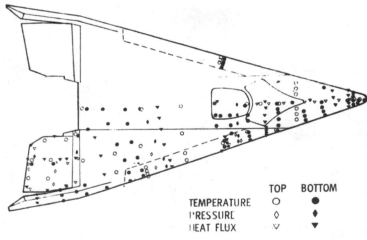

FIGURE 13
TYPICAL THERMOCOUPLE & PRESSURE SENSOR LOCATIONS

	TOP	BOTTOM
TEMPERATURE	O	●
PRESSURE	◊	◆
HEAT FLUX	∇	▼

FIGURE 14
AFT EQUIPMENT BAY
Rear View

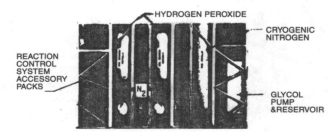

APU & GENERATOR CONTROLS

HYDROGEN PEROXIDE

CRYOGENIC NITROGEN

REACTION CONTROL SYSTEM ACCESSORY PACKS

N₂

GLYCOL PUMP & RESERVOIR

contains the subsystem instruments. The long panel on the right is the annunciator panel. This annunciator panel provides automatic warning for various glider subsystem malfunctions. The instrument on the extreme right-hand side is for monitoring reaction control system operation and glider trim at very low dynamic pressures. A separate indicator light is provided in this panel for each thrust nozzle.

Each indicator light is energized by thrust chamber pressure when the nozzle it represents is producing thrust. By monitoring the panel the pilot can pinpoint a malfunctioning nozzle and take corrective action. Also, the pilot can detect and make corrections for errors to aerodynamic trim by monitoring this panel for unsymmetrical use of reaction control thrust in each control axis. Piloted reentries on the flight simulator have shown this technique enables the pilot to reduce reentry reaction control fuel consumption by 36 pounds. The instrument in the lower center of the panel is the energy management display. This display provides the pilot with boost program monitoring information; the glider's position with respect to the selected and alternate landing sites; information on the angle of attack and bank angle program to reach the landing site; and information on the angle of attack and sink rate required to remain within the reentry corridor. Use of the energy management display allows the pilot to fully exploit the glider's maneuver capability in obtaining hypersonic research data.

EQUIPMENT COMPARTMENT AND AFT EQUIPMENT BAY

Major items installed in the equipment compartment, as shown in Figure 10, include the test data system electronics, communication and tracking subsystem electronics, compartment environmental control equipment, the electrical power

FIGURE 15
SECONDARY POWER BAY
Left Side View

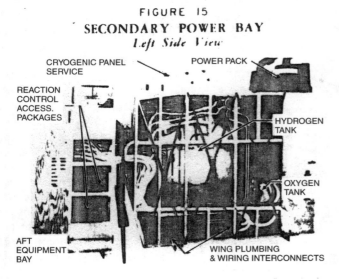

CRYOGENIC PANEL SERVICE

POWER PACK

REACTION CONTROL ACCESS. PACKAGES

HYDROGEN TANK

OXYGEN TANK

AFT EQUIPMENT BAY

WING PLUMBING & WIRING INTERCONNECTS

FIGURE 16
SECONDARY POWER BAY
Right Side View

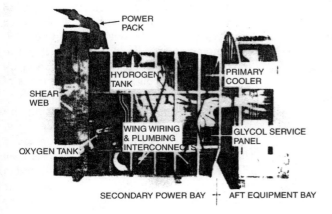

POWER PACK

HYDROGEN TANK

PRIMARY COOLER

SHEAR WEB

WING WIRING & PLUMBING INTERCONNECTS

GLYCOL SERVICE PANEL

OXYGEN TANK

SECONDARY POWER BAY — AFT EQUIPMENT BAY

distribution system, and the electrical umbilical connection. A volume of 75 cubic feet has been provided in the equipment compartment for installation of the test data system components. Three test data system configurations, each capable of 750 measurements, will be used during the flight program. Each configuration will weigh 1000 pounds including recorders, wiring, plumbing and sensors In the course of the total program, thermal pressure, subsystem performance, and bioastronautic data will be obtained from a total of 1034 sensors by use of the three test data configurations. All data will be recorded on board and selected data will be telemetered. Figure 13 shows typical sensor locations.

AFT EQUIPMENT BAY

The aft equipment bay contains the cryogenic nitrogen supply; the generator and auxiliary power unit (APU) controls; the glycol water reservoir and pump; and propellant tanks and service provisions for two redundant reaction control systems. Water walls on this compartment are supported by an aluminum frame structure. Figure 14 is a photograph of the mockup of this installation.

SECONDARY POWER BAY

The secondary power bay contains the cryogenic hydrogen tank, two redundant cryogenic oxygen tanks, the primary cooler, the power generation equipment and the hydraulic system accessories. Power generation is provided by two power packs consisting of a hydrogen-oxygen APU driving a 12 KVA 400 cycle AC generator and an 8.5 GPM hydraulic pump. This equipment supplies power to the redundant electric and hydraulic systems. All of the functional units are accessible for maintenance directly from either the top or sides. All of the components that lie inboard or on the bottom of the secondary power bay are basically the nonfunctioning minimum maintenance elements, such as tubing, wire runs and structural supports. Servicing of the cryogenic systems is accomplished through the cryogenic umbilical at the upper left hand side of the bay. The glycol-water system service panel is on the right hand side of the bay and hydraulic system service provisions are on the upper portion of the powerpack. Water walls on this bay are also attached to an aluminum frame structure. Figures 15 and 16 are mockup photos of the secondary power bay installation.

In the secondary power bay, it is obvious that there are a large number of components in a relatively small volume. To accomplish this installation similar or related components are assembled into single units. Each unit is designed to minimize the number of connections that must be broken for replacement of the unit. This type of installation reduces weight, reduces the number of innerconnections by such techniques as manifolding and increases the packaging density, thus allowing more efficient use of the available volume. A typical unit is the power pack, shown in Figure 17. This unit contains components from 7 subsystems and 27 functionally related parts. Another feature of the design is the multiple use of structure. For example

FIGURE 17
POWER PACK

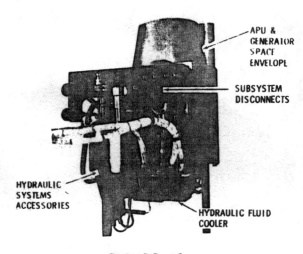

APU &
GENERATOR
SPACE
ENVELOPE

SUBSYSTEM
DISCONNECTS

HYDRAULIC
SYSTEMS
ACCESSORIES

HYDRAULIC FLUID
COOLER

FIGURE 18
RE-ENTRY FLIGHT CORRIDOR

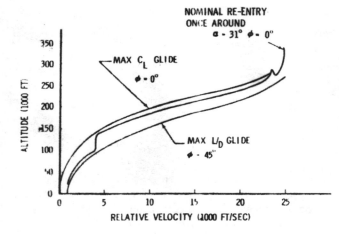

NOMINAL RE-ENTRY
ONCE AROUND
α - 31° ϕ - 0°

MAX C_L GLIDE
ϕ - 0°

MAX L_D GLIDE
ϕ - 45"

ALTITUDE (1000 FT)

RELATIVE VELOCITY (1000 FT/SEC)

FIGURE 19
GLIDER RANGE CAPABILITY
Footprint

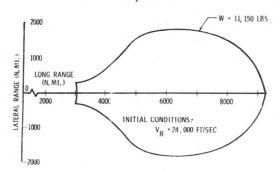

W - 11,150 LBS

LONG RANGE
(N. MI.)

LATERAL RANGE (N. MI.)

INITIAL CONDITIONS:
V_R - 24,000 FT/SEC

the structure that provides the packaging supports for the components of the power pack, when bolted in place, provide a necessary shear panel across the top of the equipment area. In a similar fashion the primary cooler support rack, when bolted in place provides a shear panel in this part of the compartment structure.

Although details of the installations in the other compartments are not presented here. this same design philosophy is being applied throughout the glider. The X-20A Glider is capable of exploring a wide range of hypersonic flight and maneuver conditions in the corridor shown in Figure 18. Flight at L/D max is used to attain maximum longitudinal range and maximum lateral range is attained by flying at L/D max with a nose angle of 45°. From an initial velocity of 24,000 feet per second, the X-20A has a maximum lateral range capability of 1800 nautical miles and is capable of a maximum of 6250 nautical mi. longitudinal range variation, as shown in Figure 19. Maneuvering to a landing site within the envelope illustrated is accomplished by varying L/D and roll angle within the limits of the reentry corridor. Lesser maneuvers may also be accomplished by originating the maneuver at a lower velocity. Since April, 1960, the X-20A Program has progressed from the study phase to the hardware phase. The X-20A Glider configuration presented in this paper is in the detail design stage and is being released for fabrication. The present status of the design is as follows:

1. The engineering design is approximately 23 percent complete.
2. The development program supporting the design is over 75 percent complete.
3. Procurement specifications for 73 subcontracted subsystems or components have been prepared and released.
4. Major subcontracted subsystems have been under development since early in 1961 and designs have progressed to the prototype hardware or qualification test stage.
5. Release of the airfraMe for manufacturing has been underway since early this year.
6. All areas of the design are progressing and all major releases to contractors will be completed by mid-1963.

Date: Oct 1963   Title: History of the X-20A DYNA-SOAR

AFSC Historical Publication Series 63–50–I   K243.042-22
                                                                    October 1963

History of the
X-20A DYNA-SOAR

Volume I
(Narrative)

63 ASE–94

by
Clarence J. Geiger

Historical Division Information Office
Aeronautical Systems Division
Air Force Systems Command

Approved by:

R. G. RUEGG
Major General, USAF
Commander, ASD

PHILIP H. POLLOCK
Chief, Historical Division
Office of Information

ERNEST L. SMITH
Lieutenant Colonel, USAF
Director of Information

TABLE OF CONTENTS

LIST OF ILLUSTRATIONS

CHRONOLOGY

1944	August	Drs. Eugen Sänger and Irene Bredt of the German air ministry completed their calculations for a manned, rocket bomber.
1945	January 24	The rocket development division of the German Army successfully launched, for the first time, an A-9 vehicle.
1946	May	RAND authorities determined that it was feasible to design a capsule with wings for manned space flight.
1952	April 17	The Bell Aircraft Company offered a proposal to the Wright Air Development Center for a manned bomber-missile, known as Bomi.
1954	April 1	The Air Force and the Bell Aircraft Company arranged a contract for the study of an advanced, bomber-reconnaissance weapon system.
1955	January 4	ARDC headquarters issued System Requirement 12, which called for studies of a reconnaissance aircraft or missile possessing a range of 3,000 nautical miles and capable of reaching 100,000 feet.
	May 12	Air Force headquarters announced General Operational Requirement 12 for a piloted, high-altitude, reconnaissance weapon system available by 1959.
	September 21	The Bomi contract of the Bell Aircraft Company was extended as a study for the Special Reconnaissance System 118P.
	December 19	The Air Force requested the aviation industry to investigate the feasibility of developing a manned, hypersonic, rocket-powered, bombardment and reconnaissance weapon system.
1956	March	The Research and Target Systems Division of ARDC headquarters completed an abbreviated development plan for a glide-rocket, research system, designated Hywards.
	March 20	The Air Force and Bell Aircraft Company completed negotiations for a study contract involving Reconnaissance System 459L, Brass Bell.
1956	June 12	ARDC headquarters issued System Requirement 126, outlining the requirements for a rocket-bomber, named Robo.
	November 6	ARDC headquarters issued System Requirement 131, requesting information from Air Force agencies for the preparation of a Hywards abbreviated development plan.
1957	April 30	Air Force headquarters directed the Air Research and Development Command to formulate a development plan encompassing all hypersonic weapon systems.
	June 20	A committee, with representation from ARDC headquarters, the Wright Air Development Center, the Cambridge Air Force Research Center, and the Air Materiel Command, was formed to evaluate contractor studies on Robo.
	October 10	ARDC headquarters consolidated Hywards, Brass Bell, and Robo studies into a three-step abbreviated development plan for System 464L, Dyna-Soar.

November 15 Air Force headquarters approved the abbreviated development plan for Dyna-Soar.

25 Air Force headquarters issued Development Directive 94, which allocated $3 million of fiscal year 1958 funds for Dyna-Soar.

December 21 ARDC headquarters issued System Development Directive 464L, directing the implementation of the Dyna-Soar program.

1958 May   20 The Air Force and the National Advisory Committee for Aeronautics signed an agreement for NACA participation in the Dyna-Soar program.

June   16 The Air Force announced that the Boeing Airplane Company and the Martin Company had been chosen to compete, during a period of 12 to 18 months, for the Dyna-Soar contract.

September 30 Air Force headquarters informed Detachment One of ARDC headquarters that the $10 million procurement fund for fiscal year 1959 had been canceled from the Dyna-Soar program.

November   The Dyna-Soar project office completed a preliminary development plan, involving a two-step program: the development of a research vehicle and then a weapon system.

1959 January   7 Deputy Secretary of Defense, D. A. Quarles reinstated the $10 million of fiscal year 1959 funds for the Dyna-Soar program.

February 17 Air Force headquarters revised General Operational Requirement 12. Instead of a high-altitude reconnaissance system, ARDC was to develop a bombardment system.

April   13 Dr. H. F. York, the Director of Defense for Research and Engineering, established the primary objective of the Dyna-Soar program as the suborbital exploration of hypersonic flight.

May   7 ARDC headquarters issued System Requirement 201. The purpose of the Dyna-Soar vehicle was to determine the military potential of a boost-glide weapon system and provide research data on flight characteristics up to and including global flight.

June   The Dyna-Soar source selection board completed its evaluation of the proposals of the Boeing Airplane Company and the Martin Company. The board recommended the development of the Boeing glider but also favored the employment of the orbital Titan C booster offered by Martin.

November 1 In a development plan, the Dyna-Soar project office formulated a new three-step approach, involving the development of a suborbital glider, an orbital system, and an operational weapon system.

9 The Secretary of the Air Force announced that the Boeing Airplane Company was the system contractor, while the Martin Company would be an associate contractor for booster development.

24 Dr. J. V. Charyk, Assistant Secretary of the Air Force for Research and Development, directed a Phase Alpha study to determine the validity of the Dyna-Soar approach to manned, orbital flight.

December 11   The Air Force and the Boeing Airplane Company completed contractual arrangements for the Phase Alpha study.

1960 January   27   The Vice Commander of the Wright Air Development Division directed the formation of an Air Force committee to evaluate the contractor studies for Phase Alpha.

February   8   Lieutenant General B. A. Schriever, ARDC commander, and Lieutenant General S. E. Anderson, AMC commander, signed an agreement which delineated the responsibilities of BMD and AMC in the Dyna-Soar program.

March   The Air Force committee concluded from the Phase Alpha study that a glider with medium lift-to-drag ratio, such as Dyna-Soar, would be the most feasible approach for an investigation of manned re-entry.

April   1   The Dyna-Soar project office completed another development plan, detailing the three-step approach first offered in the November 1959 development plan.

8   Professor C. D. Perkins, Assistant Secretary of the Air Force for Research and Development, approved the Phase Alpha results and the development plan and directed implementation of the suborbital Step I.

11-14   The Air Force and the National Aeronautics and Space Administration held a joint conference at the Langley Research Center, Virginia, to provide industry and government agencies with a progress report concerning manned hypervelocity and re-entry vehicles.

19   The Assistant Secretary of the Air Force for Materiel, P. B. Taylor, authorized the negotiation of fiscal year 1961 contracts for the Step I program.

22   The Department of Defense endorsed the Dyna-Soar program and permitted the release of $16.2 million of fiscal year 1960 funds.

27   The Air Force and the Boeing Airplane Company negotiated a letter contract for Step I of Dyna-Soar.

June   8   The Air Force gave the Martin Company responsibility for the development of the Dyna-Soar booster airframe.

9   The Air Force completed arrangements with the Aero-space Corporation to provide technical services for the Step I program.

27   The Air Force authorized the Aero-Jet General Corporation to develop booster engines for the Dyna-Soar system.

July   21   Air Force headquarters issued System Development Requirement 19, which sanctioned the three-step approach.

August   4   ARDC headquarters directed that the conduct of flight testing be firmly placed in the control of the project offices.

October   12   Air Force headquarters issued Development Directive 411, which gave approval to Step II and III studies.

Air Force headquarters requested the project office formulate a "stand-by" plan for accelerating the orbital flight date of the Dyna-Soar program.

November 28   The Assistant Secretary of the Air Force requested ARDC to examine the feasibility of employing Titan II instead of Titan I for Dyna-Soar suborbital flights.

December   The Dyna Soar office completed a "stand-by" plan which would accelerate the program by employing the same booster for both suborbital and orbital flights.

6   ARDC headquarters issued a system study directive, -which allotted $250,000 for a Step III study.

The Air Force granted authority to the Minneapolis-Honeywell Regulator Company to develop the primary guidance subsystem.

16   The Air Force completed negotiations with the Radio Corporation of America for the development of the communication and data link subsystem.

1961 January   12   Air Force headquarters announced that Titan II would be the suborbital Step I booster.

February 3   Air Force headquarters informed the Dyna-Soar office that the fiscal year 1962 funding level had been set at $70 million.

14   The Air Force and the Boeing Airplane Company completed negotiations for Step IIA and IIB studies.

March   28   Air Force headquarters announced that the Department of Defense had decided to raise fiscal year 1962 funds for Dyna-Soar to $100 million.

April   24   Dr. J. V. Charyk, Under Secretary of the Air Force, authorized the negotiation of contracts for the entire Step I program.

26   The Dyna-Soar program office completed a system package program, further elaborating the three-step approach.

May   4   The Boeing Company offered a "streamline" approach for accelerating the Dyna-Soar program by the elimination of suborbital flights.

12   A Dyna-Soar technical evaluation board recommended the Martin C plan for a Step IIA booster.

29   The Space Systems Division completed two development plans for an Advanced Re-entry Technology program and a SAINT II program.

July   11   The Dyna-Soar Directorate of the Space Systems Division recommended employment of the Phoenix A388 space launch system for the Step IIA booster.

August   The Dyna-Soar program was placed under the jurisdiction of the Designated Systems Management Group of Air Force Headquarters.

General B. A. Schriever, AFSC commander, directed a study for a Manned, Military Space, Capability Vehicle.

September 11-22   Air Force and NASA officials conducted a mock-up inspection of the Dyna-Soar system at the Boeing Company facilities in Seattle, Washington.

28   The Air Force completed the study of the Manned, Military, Space, Capability Vehicle.

October   7   The Dyna-Soar program office completed an abbreviated development plan for a Dyna-Soar military system.

13   The Department of Defense approved the Titan III as the space launch system for the Air Force.

November 16   The Deputy Commander for Aerospace Systems completed a development plan for the Dyna-Soar program, which characterized the program as a manned, orbital research system.

December 11   Air Force headquarters approved the November 1961 development plan.

27   Air Force headquarters issued System Program Directive 4, which formalized the objectives of the November 1961 development plan.

1962   January   8   AFSC headquarters halted any further consideration of a Step III study.

31   General B. A. Schriever, AFSC commander, rescinded the 4 August 1960 test policy and directed that Air Force test wings and centers prepare and implement test plans and appoint local test directors for the conduct of AFSC flight tests.

February   21   Air Force headquarters amended System Development Requirement 19, by deleting references to suborbital flight and the development of military subsystems.

23   Secretary of Defense, Robert S. McNamara, officially limited the objective of the Dyna-Soar program to the development of an orbital, research system.

1962   May   14   The Dyna-Soar program office completed a new system package program, which included multi-orbital flights.

June   26   The Department of Defense officially designated the Dyna-Soar glider as the X- 20.

30   The Boeing Company completed the Step IIA and IIB studies.

July   13   Air Force headquarters informed ARDC headquarters that the Department of Defense had given qualified approval of the May 1962 system package program.

October   10   The Dyna-Soar program completed a system package program, which made X-20 flight dates compatible with projected Titan IIIC schedules.

15   Air Force headquarters issued System Program Directive 9, authorizing research and development of Titan III, System 624A.

16   The function of the ASD Field Test Office was transferred to the 6555th Aerospace Test Wing of the Ballistic Systems Division.

November   The Department of Defense set $130 million for fiscal year 1963 and $125 million for 1964 as the allotment for the Dyna-Soar program.

5-7 The Dyna-Soar Symposium was held at Wright Field to insure dissemination of information to industry and government agencies concerning progress in Dyna-Soar technology.

26 The X-20 office completed the "Westward-Ho" plan, which proposed consolidation of the flight control centers at Edwards Air Force Base.

December 19 The Vice Commander of AFSC directed the establishment of a manned, space flight, review group to examine all aspects of the X-20 test program.

1963 January 11 The Dyna-Soar program completed a system package program, which incorporated the "Westward-Ho" proposal.

18 The Secretary of Defense directed a review of the Dyna-Soar program.

19 The Secretary of Defense directed a review of the Titan III program and the Gemini program of NASA.

21 The Department of Defense and the National Aeronautics and Space Administration completed an agreement for defense department participation in the Gemini program.

March 15 The Secretary of Defense directed the Air Force to conduct a comparison of the military potentials of Dyna-Soar and Gemini.

30 Lieutenant General H. M. Estes, AFSC vice commander, forwarded four funding alternatives for the X-20 program to Air Force headquarters.

April 12 USAF headquarters approved $130 million and $135 million as the most feasible funding level for the Dyna-Soar program in fiscal years 1963 and 1964.

May 9 General Schriever assigned responsibility for X-20 orbital test direction to the Space Systems Division and placed the flight control center at Satellite Test Center, Sunnyvale, California.

10 Officials of the Space Systems Division and the Aeronautical Systems Division completed their joint response to Secretary McNamara's request for the military potentialities of Dyna-Soar and Gemini.

27 Based on an anticipated funding level of $130 million for fiscal year 1963 and $135 million for 1964, the Dyna-Soar office completed a system package program which acknowledged a two month delay in the flight schedules.

June 8 The Secretary of the Air Force approved the 27 May system package program.

July 3 AFSC headquarters informed the X-20 office that the Department of Defense would only allow $125 million for fiscal year 1964.

31 General Schriever assigned responsibility for X-20 air-launch program and pilot training to the Space Systems Division.

CHAPTER I
SEVEN FROM PEENEMÜNDE

By March 1945, the Allies had overrun the possible launching areas for the A-4 rocket, commonly known as the V-2. Germany had been denied further employment of its more advanced vengeance weapon. Awaiting the end and under military guard, Lieutenant General Walter R. Dornberger, director of guided missile development for the German ministry of munitions, retired on 6 April with his Peenemünde band of rocket experts to the recesses of southern Germany. The A-4 had prematurely reached operational status in September 1944, and nearly 3,000 missiles were fired against targets. Further development of this and other advanced rocket weapons, however, was hopeless.[1] While employed too late to alter the apparent outcome of the war, the A-4 not only radically changed the concept of weapon delivery but offered the promise of rapidly extending the speed, range, and altitude of manned flight.

The first application of the A-4 rocket engine, capable of delivering over 50,000 pounds of thrust, to the problem of extending the regime of piloted flight was the formulation of the boost-glide concept. Here, a winged-vehicle would be propelled by a rocket booster to a sufficient altitude where, after fuel in the rocket stages had been expended, the craft would perform a gliding flight and then execute a conventional landing. The first intense effort at the refinement of this concept began in 1943, under the direction of General Dornberger, at the German Army's research facilities in the east sector of Peenemünde.

Dr. Wernher von Braun, Dornberger's assistant in charge of planning and design, reasoned that by merely attaching wings to the A-4 airframe, the range of this vehicle, now designated the A-9, could be extended from 230 to 360 miles. The director of preliminary planning went further and considered the possibility of placing the A-9 vehicle on a proposed A-10 booster, capable of producing 440,000 pounds of thrust and accelerating the craft to a velocity of 4,000 feet per second. With this arrangement, the A-9 could traverse a distance of 3,000 miles in 17 minutes. Further in the future, the rocket expert planned a multi-stage engine which could boost the A-9 to orbital velocities.

By mid-1943, preliminary designs had been completed, trajectories calculated, guidance systems investigated, and wind tunnel data gathered for the development of the A-9. Priority, however, demanded full effort on the A-4, and General Dornberger halted the work on the A-9. Late in 1944, greater range for the A-4 was demanded and development of the A-9 was resumed. After two unsuccessful launchings, this advanced vehicle reached a height of 50 miles and a speed of 4,000 feet per second on 24 January 1945.[2]

Independent of the Peenemünde group, Dr. Eugen Sänger and his assistant, Dr. Irene Bredt, were pursuing similar investigations for the German air ministry's Research Establishment for Gliders. By August 1944, they had completed their elaborate calculations for a manned, rocket bomber. The winged-rocket was to have a length of 92 feet, a span of 50 feet, and a takeoff weight of 110 tons. Unlike von Braun, Sänger preferred horizontal launch. For 11 seconds, a rocket sled would propel the bomber along tracks, two miles in length, until a takeoff velocity of 1,640 feet per second was attained. Under power of its own rocket engine, the vehicle would then climb to an altitude varying from 30 to 60 miles. At the end of ascent, the bomber would proceed in an oscillating, gliding flight, conceivably circumnavigating the Earth.

Sänger was intent on explaining the military value of his proposed system and detailed possible modes of attack. To achieve a strike on a specific point, the vehicle would be accelerated only until it acquired enough velocity to reach the target. After releasing its bomb, the vehicle would turn at the lowest possible speed, ignite its engine, and then return to its original base. For greater distances and bomb loads, the possession of an auxiliary landing site near the target was necessary. If such a site were not available, the rocket bomber would have to be sacrificed. An attack on a larger area, however, did not necessitate a low velocity over the target, and, consequently, there was more likelihood that the bomber could circumnavigate the globe.

The drawbacks to Sänger's proposal were obvious, and, consequently, the German military did not give serious consideration to the rocket bomber. The difficulties inherent in turning the rocket bomber at hypersonic speeds only increased the desirability for an antipodal landing site. To depend on the possibility of possessing friendly landing areas so near a target was unrealistic. Even if a fleet of rocket bombers could circle the Earth, a bomb capacity of about 8,000 pounds per vehicle, as estimated by Sänger, could not have changed the course of conflict.[3]

Apparently Russian military officials obtained copies of Sänger's analysis at the end of the War and became interested in the possibilities of boost-glide flight. In 1958, an article which appeared in a Soviet aviation journal referred to a Russian glide-bombing system, capable of attaining an altitude of 295,000 feet and striking a target at a distance of 3,500 nautical miles. Later, an American aviation periodical reported that Russian scientists were developing an antipodal, glide-missile, designated the T-4A. By March 1960, the Assistant Chief of Staff for Intelligence, USAF headquarters, estimated that the Soviets were at least conducting research directed towards the development of a boost-glide vehicle. Such a system could lead to the development of a craft capable of performing reconnaissance and bombing missions. Air Force intelligence analysts believed that limited flight tests of the manned stage could begin in 1962 and an operational system could be available by 1967.[4]

Soon after the war, American military officials also exhibited interest in the possibilities of a boost-glide vehicle. In 1946, the Army Air Force, under a contract with the Douglas Aircraft Company, sheltered a group of American scientists and specialists in various social science areas in an effort to provide analyses and recommendations relating to air warfare. One of the first studies completed under the new Project RAND centered on the design of an orbital vehicle.

Basing their analysis on the technological developments of the Peenemünde scientists, RAND experts considered that it was possible, by employing either a four-stage, alcohol-oxygen, or a three-stage, hydrogen-oxygen booster, to place a 500 pound capsule in orbit at an altitude of 300 miles. The initial objective was to provide an orbiting, scientific laboratory, nevertheless, RAND authorities stated that it was feasible to design a capsule with wings for future manned flight.[5] In 1948, RAND made a few more studies investigating the technological difficulties involved in flight beyond the atmosphere; however, the next step was taken by the Bell Aircraft Company.

Dr. von Braun did not become associated with any American efforts in refining the boost-glide concept but, from 1945 through 1950, served as a technical adviser for the Army Ordnance Department at the White Sands Proving Grounds, New Mexico. Dornberger, on the other hand, was held in England until 1947 when he became a consultant on guided missiles for the Air Materiel Command at Wright-Patterson Air Force Base, Ohio. In 1950, he left the Air Force and became a consultant for Bell Aircraft.

Perhaps the German missile expert was influential in persuading this contractor to undertake a study of boost-glide technology, for, on 17 April 1952, Bell officials approached the Wright Air Development Center (WADC) with a proposal for a manned bomber-missile, abbreviated to Bomi. Bell's glide-vehicle was to be boosted by a two-stage rocket and was to be capable of operating at altitudes above 100,000 feet, at speeds over mach 4.0, and at a range of 3,000 nautical miles. A month later, Bell submitted a proposal to Wright center for the initiation of a feasibility study. The contractor believed that the study would cost $398,459 and would take 12 months.[6]

By 28 November, the Air Research and Development Command (ARDC) headquarters had completed a review of the Bomi project. While Bell's proposal duplicated parts of the Atlas intercontinental ballistic missile and the Feedback satellite reconnaissance programs, command headquarters considered that some phases of Bomi would advance the Air Force's technical knowledge. Consequently, ARDC headquarters requested WADC to evaluate the proposal with the view of utilizing the concept both as a manned bomber and as a reconnaissance vehicle.[7]

Wright center officials completed their evaluation by 10 April 1953 and listed several reasons for not accepting the Bell proposal. A range of 3,000 nautical miles was too short for intercontinental operations. It was difficult to conceive how the vehicle could be adequately cooled, nor was there sufficient information concerning stability, control, and aeroelasticity at the proposed speeds. Furthermore, Bell's estimated lift-to-drag ratio was far too optimistic. Since it was to operate under an extreme environment, there was also the question of the value of providing a piloted vehicle. Before undertaking such a project, Wright engineers emphasized that the cost and military worth of such a system first had to be established. Center officials added that some doubt existed concerning the ability of the contractor to complete the program successfully.[8]

Bell Aircraft, however, was persistent, and, on 22 September, its representatives briefed ARDC headquarters on the Bomi strategic weapon system. Brigadier General F. B. Wood, Deputy Chief of Staff for Development, did think the proposal "somewhat radical" but stated that it could not be considered "outside the realm of possibilities." General Wood then requested WADC to give further consideration to Bell's proposal.[9] Apparently, Wright center officials reconsidered their first evaluation of Bomi, for, in their reply to ARDC headquarters on 23 November, they assumed a more favorable position.

Wright engineers considered that the Atlas ballistic missile and the Navaho cruise missile programs offered more promise of successful development than Bomi. The Bell proposal, however, appeared to present a reconnaissance ability far in advance of the Feedback program. Furthermore, Wright officials reasoned that the Bomi vehicle would provide a test craft for several unexplored flight regimes and would offer a guide for the development of manned, hypersonic, military systems. Because of the lack of information, Wright authorities did not recommend the initiation of development but thought that the potential reconnaissance value of Bomi necessitated a two-year study program. Specifically, Wright officials recommended that Bell be offered a $250,000 contract for one year with the possibility of extending the study for an additional year. This investigation should determine whether the piloted, Bomi vehicle was more advantageous than an unmanned version and whether a reconnaissance mission would compromise the strategic striking ability of the system.[10]

ARDC headquarters agreed and approved Wright center's recommendation. Brigadier General L. I. Davis, acting Deputy Chief of Staff for Development, emphasized that the strategic requirements for an intercontinental vehicle, with a range up to 25,000 nautical miles, should be considered. General Davis stated that development of a program such as Bomi would not be undertaken until other contractors could offer competitive concepts. In accordance, the acting deputy chief of staff requested that the Boeing Airplane Company include in its efforts for Project MX-2145 (Design Studies for an Advanced Strategic Weapon System) investigations of a manned, glide-rocket system.[11]

Boeing had undertaken MX-2145 in May 1953 in order to determine the characteristics of a high performance bomber which could succeed the B-58 Hustler and be capable of delivering nuclear weapons over intercontinental ranges by 1960. Later, as directed by ARDC headquarters, Boeing briefly considered the possibility of a manned, reconnaissance glide-rocket. The contractor regarded the method of traveling an intermediate distance and then reversing direction to return to the point of origin as impractical. Rather, Boeing emphasized that it would be more feasible to orbit the Earth. The contractor, however, pointed to the difficulties of devising structures to withstand high temperature and equipment for reconnaissance. Yet, because of the military potential of such a system, the contractor thought that further investigations were indicated.[12]

On 1 April 1954, Wright center completed a contract with the Bell Aircraft Corporation for a design study of an advanced, bomber-reconnaissance weapon system. The contractor was to define the various problem areas and detail the requirements for future programs. Bell had to focus on such problems as the necessity for a manned vehicle, the profiles of possible missions, performance at high temperatures, and the feasibility of various guidance systems.[13]

Bell Aircraft now envisaged a three-stage system, with each stage riding pickaback. This system would total more than 800,000 pounds. Bomi, now designated as MX-2276, would be launched vertically, and the three rocket engines would be fired simultaneously, delivering 1.2 million pounds of thrust. Bell proposed manning the booster stage in order to achieve recovery by use of aerodynamic surfaces. The third-stage would also be piloted and would carry navigation, reconnaissance, and bombardment equipment. Bomi would be capable of reaching an altitude of 259,000 feet, attaining a speed of 22,000 feet per second, and possessing a range of 10,600 nautical miles.

The contractor believed that a piloted system such as Bomi held several advantages over an unmanned version. Reliability of the system would be increased, bombing precision augmented, and reconnaissance information easily recovered. Furthermore, operational flexibility would be enhanced with the possibility of selecting alternate targets. Unmanned instrumentation certainly could not provide for all the necessary contingencies.[14]

With the completion of the initial study in May 1955, the contract expired, but Bell continued its efforts without government funds or direction. On 1 June, WADC personnel discussed with the contractor the possibility of officially extending its work. The purpose of the Air Force in considering an extension was to investigate the feasibility of adapting the Bomi concept to Special Reconnaissance System 118P.

On 4 January 1955, ARDC headquarters had issued System Requirement 12, which called for studies of a reconnaissance aircraft or missile possessing a range of 3,000 nautical miles and an operational altitude of more than 100,000 feet. Wright center officials established System 118P, and several contractors investigated the adaptability of boost-glide rockets and vehicles using air-breathing engines to the system requirement. To bring Bell into these efforts, ARDC headquarters gave assurance, in June, that $125,000 would be released for the purpose of extending Bell's Bomi contract, and by 21 September 1955, contract negotiations were completed. Bell's efforts would continue.[15]

At the request of the Assistant Secretary of the Air Force for Research and Development, Trevor Gardner, personnel from the Bombardment Aircraft Division of AFDC headquarters and Bell Aircraft gave several presentations to ARDC and USAF headquarters in November, where the Bomi concept was received with approval.[16]* Meanwhile, officials from the laboratories of Wright center, the laboratories of the National Advisory Committee for Aeronautics (NACA), and the Directorate of Weapon Systems in ARDC headquarters had evaluated the results of the Bomi study and had drawn several conclusions.

Representatives from the three organizations thought that Bell's concept was theoretically practicable and promising, and that the Bomi program should be continued to determine the feasibility of such a weapon system. Emphasis, however, should be placed on a test program to validate Bell's analysis. The members considered that the most advantageous procedure for Bomi would be a three-step program with the development of a 5,000 nautical mile, a 10,000 nautical mile, and a global system.[17]

By 1 December 1955, Bell had completed its final engineering report for the supplementary contract and had expanded a total of $420,000 for the Bomi studies. For System 118P, Bell's design had included a two-stage rocket to boost a vehicle to 165,000 feet at a velocity of mach 15. The contractor, however, was once again out of funds. Brigadier General H. M. Estes, Jr., Assistant Deputy Commander for Weapon Systems, AFDC headquarters, estimated that about $4 million more would be required for the next 12 to 18 months. General Estes then requested the Deputy Commander for Weapon Systems at ARDC headquarters to allocate $1 million for fiscal year 1956 and to grant authority for the continuation of the program.[18]

While the question of future funding was being debated, officials from the New Development Weapon Systems Office of ARDC headquarters and Bell Aircraft visited Langley Air Force Base, Virginia, in December 1955, to obtain the views of NACA on the Bomi concept. The advisory committee had first become interested in the boost-glide concept when it undertook a preliminary study in 1953 to determine the feasibility of manned, hypersonic flight. On 30 September 1955, Dr. I. H. Abbott, Assistant Director for

---

*   On 1 August 1955, the management of weapon systems development was transferred from the Wright Air Development Center to ARDC headquarters. Detachment One of the Directorate of Systems Management, which included the Bombardment Aircraft Division, however, was located at Wright-Patterson Air Force Base.

Research, NACA, thought that more data was required before a development program could be initiated for Bomi. Dr. Abbott hoped that the Air Force would continue to inform NACA on the future progress of the program in order that its laboratories could contribute to the research program. The conference in December resulted in an invitation to NACA for participation in the validation testing for Bomi.[19]

Early in January 1956, the Intelligence and Reconnaissance Division of ARDC headquarters informed the New Development Weapon Systems Office that $800,000 had been allocated for continuation of Bomi. The Air Force, however, considered that the Bell program should now be directed towards the fulfillment of the General Operational Requirement 12, which had been issued on 12 May 1955. This directive called for a piloted, high-altitude, reconnaissance weapon system which was to be available by 1959. Accordingly, the Air Force concluded a contract with Bell on 20 March 1956, totaling $746,500, for Reconnaissance System 459L, commonly known as Brass Bell. In October, the contract was extended to 31 August 1957, bringing total expenditures to approximately $1 million. Later in 1956, Bell was awarded an additional $200,000 and four more months to complete its work.[20]

By December 1956, Bell Aircraft had conceived of a manned, two-stage system which would be propelled over 5,500 nautical miles at a velocity of 18,000 feet per second to an altitude of 170,000 feet by Atlas thrust chambers. With the addition of another stage, Bell engineers reasoned that the range could be extended to 10,000 nautical miles with a maximum speed of 22,000 feet per second.[21]

While the Air Force had channeled Bell's work towards the eventual development of a boost-glide, reconnaissance system, it had not abandoned the application of this concept to the development of a bombardment vehicle. On 19 December 1955, the Air Force had sent a request to the aircraft industry for a study which would incorporate analytical investigations, proposed test programs, and design approaches for a manned, hypersonic, rocket-powered, bombardment and reconnaissance weapon system. Boeing, the Republic Aircraft Company, the McDonnell Aircraft Corporation, the Convair Division of the General Dynamics Corporation, Douglas, and North American Aviation responded to the request. Study contracts, amounting to $860,000 were awarded to the latter three for investigations extending from May through December 1956. Later, the Martin Company, Lockheed Aircraft, and Bell joined in the study. By the end of fiscal year 1957, an additional $3.2 million was expended by Boeing, Convair, North American, Republic, Douglas, and Bell from their own funds.[22]

On 12 June 1956, ARDC headquarters outlined the conditions for the rocket-bomber study, now designated as Robo, in its System Requirement 126. The purpose of the study was to determine the feasibility of a manned, hypersonic, bombardment and reconnaissance system for intercontinental operation by 1965. The main requirement of the proposed system was the ability to circumnavigate the globe and yet operate at a minimum altitude of 100,000 feet. Furthermore, the vehicle would not only have to perform strategic strike missions but, in addition, fulfill a reconnaissance role. The contractors would also have to determine the effects of carrying weapons, ranging in weight from 1,500 to 25,000 pounds, on vehicle design and investigate the feasibility of launching air-to-surface missiles.[23]

The importance of advanced systems such as Brass Bell and Robo was given added emphasis by ARDC commander, Lieutenant General T. S. Power, at his conference on "radical" configurations, held on 15 February 1956. General Power stated that the Air Force should stop considering new and novel configurations and should start developing them. Speeds to any conceivable extent and operation of manned, ballistic rockets beyond the atmosphere should be investigated.[24]

Encouraged by General Power's statement, Major G. D. Colchagoff of the Research and Target Systems Division, ARDC headquarters, considered that one of the promising proposed programs was the manned, glide-rocket, research system. This was to be a vehicle similar to Brass Bell and Robo and would be used to obtain scientific data rather than fulfill a military role. The research and target division prepared an abbreviated development plan for the test system and submitted it to Air Force headquarters in March. On 29 June, headquarters approved the proposal but requested a full development plan.[25] Research and target managers, however, had already encountered funding difficulties.

In April 1956, the research and target division had estimated that $4 million was required for the manned glide-rocket, and a total of $33.7 million was needed for the research-vehicle program, which included the X-13, the X-14 the XB-47D, the X-15, and a vertical-takeoff-and-landing (VTOL) aircraft. Air Force headquarters, however, had set a ceiling of $8.5 million for all of these programs. The research and target division then undertook negotiations with the Air Materiel Command to determine a method of funding to alleviate this deficiency. If this attempt failed, the division warned USAF headquarters that the Air Force would not have a research-vehicle program.[26]

Air Force headquarters, however, drastically reduced the budget for fiscal year 1957, allocating no funds for the manned glide-rocket. General Power warned that this reduction would postpone his bold research program for at least one year. He cautioned headquarters that this action would seriously jeopardize America's qualitative lead over Russia.[27]

In spite of inadequate funding, ARDC issued System Requirement 131 on 6 November 1956, which requested information from the ARDC director of systems management, Wright center, the flight test center and the Cambridge research center for the preparation of an abbreviated system development plan. The manned, glide-rocket, research program was now titled Hypersonic Weapons Research and Development Supporting System (Hywards) and was classified as System 455L. By 28 December, the ARDC Directorate of Systems Plans had completed a development plan for Hywards.[28]

The purpose of the Hywards vehicle was to provide research data on aerodynamic, structural, human factor, and component problems and was to serve as a test craft for development of subsystems to be employed in future boost-glide systems. The research and target division considered three propulsion choices as satisfactory for boosting Hywards. The 35,000 pound thrust chambers, employing fluorine-ammonia fuel, which Bell had under development, was one possibility. The 55,500 and 60,000 pound thrust sustainer engines for the Atlas and Titan systems comprised another. The 50,000 pound thrust XLR-99 engine, employed in the X-15 vehicle, was the third option. One of these rocket systems would propel the Hywards craft to a velocity of 12,000 feet per second and an altitude of 360,000 feet. The initial flight test program was to employ the airdrop technique, similar to the X-15 launch, while later testing would use a rocket-boosted, ground-launch method. The research and target division emphasized that by appropriate modifications to Hywards, increased velocities and orbital flight could be attained to provide continuing test support for the Air Force's technological advances.[29]

On 27 February 1957, the development plans for both Hywards and Brass Bell were presented to USAF headquarters, where it was decided that the two programs were complementary, and, therefore, should be consolidated. Funding, however, proved more difficult. For fiscal year 1958, ARDC headquarters had requested $5 million for Hywards and $4.5 million for Brass Bell. Air Force headquarters, however, reduced these requests to a total of $5.5 million. Lieutenant General D. L. Putt, Deputy Chief of Staff for Development, USAF headquarters, hesitated endorsing the boost-glide programs. The lack of Air Force funds necessitated giving priority to the advanced satellite reconnaissance system, 117L, rather than to Hywards or Brass Bell. Furthermore, the X-15 program would provide a more dependable source of research data than the boost-glide programs. Major General R. P. Swofford, Director of Research and Development, USAF headquarters, did recommend that $1 million be allocated for the boost-glide systems, but, on 30 April, Air Force headquarters informed ARDC headquarters that the two development plans were disapproved and that a new plan, encompassing all hypersonic weapon systems should be prepared.[30]

Before the new development plan for Brass Bell and Hywards was completed, additional investigations for the Robo programs were accomplished. On 20 June 1957, an ad hoc committee, consisting of representatives from ARDC headquarters, Wright Air Development Center, the Cambridge Air Force Research Center, and the Air Materiel Command, was formed to evaluate the Robo studies of the contractors. Advisory personnel from the Strategic Air Command, the National Advisory Committee for Aeronautics, and the Office of Scientific Research were also present.

During the first three days of the conference, the contractors working on System Requirement 126 presented their proposals, most of which centered on the feasibility of manned vehicles. Both Bell and Douglas favored a three-stage, boost-glide vehicle, the former employing fluorine and the latter, an oxygen propellant. The Convair Division also proposed a three-stage system, using fluorine fuel, but its concept differed from the previous two in that a control rocket and turbojet engine were placed in the glider. While North American advanced a two-stage vehicle, using conventional rocket fuel, Republic advocated an unmanned vehicle, powered by a hypersonic cruise, ramjet engine, and boosted by a single-stage rocket. Republic's proposal also involved an unmanned, satellite, guidance station, which was to be placed in orbit by a three-stage booster. Finally, Boeing favored an unmanned version and advanced an intercontinental glide-missile. In the opinion of Boeing officials, a manned vehicle would involve a longer development cycle and would not possess any great advantage over a missile.

After the presentation of the contractors' proposals, the committee spent the next two days evaluating the concepts. While Wright officials thought that the boost-glide concept was feasible and would offer the promise of an operational weapon system by 1970, they also pointed to several problems confronting the Air Force. The details of configuration design were yet unknown. The status of research in the area of materials was not sufficiently advanced. Lack of hypersonic test facilities would delay ramjet development until 1962. Rocket engines were not reliable enough to allow an adequate safety factor for manned vehicles during launch. Finally, center officials pointed to the difficulty of providing a suitable physiological environment for a piloted craft.

Officials of the Cambridge Research Center focused on a different set of problems. All the proposals employed an inertial, auto-navigating system, and Cambridge officials pointed out that these systems required detailed gravitational and geodetical information in order to strike a target accurately. The effect of the Earth's rotational motion became extremely important at hypersonic speeds, and, consequently, this factor would have to be considered in determining the accuracy of the guidance systems. Research center scientists also emphasized that an ion sheath would be created as the vehicle penetrated the atmosphere during re-entry; this phenomenon would hinder communication. There were other difficulties that required investigation. The thermal properties of the atmosphere would have to be studied in order to determine the extent of aerodynamic heating. Adequate data on the effect of wind turbulence and the impact of meteor dust on the vehicle would have to be determined. Officials of the Cambridge center added one more problem: the presence of ionization trails, infrared radiation, and vehicle contrails could facilitate hostile detection of the vehicle.

It was apparent to the representatives of the Air Materiel Command that the development of either a manned or unmanned system would be feasible only with increased and coordinated efforts of six to eight years of basic research. More detailed knowledge was required of the system design in order that a determination could be made of various logistical problems and the complexity of the launching area. Viewing the development costs for the ballistic missile programs, materiel officials estimated that the cost for Robo would be extremely high. In order that the Robo program could be continued, air materiel officials recommended that the participating contractors be given specific research projects. A contracting source for the conceptual vehicle should then be chosen, and, after approximately six years, competition for the weapon system development should be held.

After surveying the contractors' proposals and the analyses of Wright center, the materiel command, and the Cambridge center, the ad hoc committee concluded that a boost-glide weapon system was technically feasible, in spite of the numerous problems inherent in the development of such a system. With moderate funding, an experimental vehicle could be tested in 1965, a glide-missile in 1968, and Robo in 1974. The committee emphasized that the promise of boost-glide vehicles to be employed either for scientific research or as weapon systems was necessity enough for the undertaking. The members of the committee went beyond the scope of the Robo proposals and recommended that ARDC headquarters submit a preliminary development plan to USAF headquarters, covering the entire complex of boost-glide vehicles.[31]

By 10 October 1957, the Director of Systems Plans, ARDC headquarters, had completed consolidating the details of the Hywards, Brass Bell, and Robo programs into a three-step, abbreviated, development plan for the new Dyna-Soar (a compound of dynamic soaring) program. Like Hywards, the first phase of System 464L involved the development of a manned, hypersonic, test vehicle which would obtain data in a flight regime significantly beyond the reach of the X-15 and would provide a means to evaluate military subsystems. To avoid further confusion between the purpose of Dyna-Soar and the X-15 vehicle, the directorate made a clear distinction between a research vehicle and a conceptual test vehicle. Both vehicles were designed to obtain flight data in a regime which had not been sufficiently well defined; however, the latter was to obtain information for the development of a specific system. The initial objectives of the Step I vehicle would be a speed of approximately 18,000 feet per second and altitudes of 350,000 feet and would be attained by use of one of the three engines considered for Hywards.

The Brass Bell program assumed the position of Step II in the Dyna-Soar plan. A two-stage rocket booster would propel the reconnaissance vehicle to a speed of 18,000 feet per second and an altitude of about 170,000 feet. The vehicle would then glide over a range of 5,000 nautical miles. The system would have to be capable of providing high quality photographic, radar, and intelligence information. The vehicle would also have to possess the ability of performing strategic bombing missions. The Director of Systems Plans considered that the liquid rocket Titan sustainer appeared usable; however, investigations under Step I could prove the fluorine engine more valuable.

Step III incorporated the Robo plans, and encompassed a more sophisticated vehicle which would be boosted to 300,000 feet and 25,000 feet per second and would be capable of orbital flight. Like the earlier phase, this vehicle would be able to execute bombardment or reconnaissance missions.

Because of insufficient data, the directorate reasoned that the Dyna-Soar program could not be immediately initiated. A two-phase program for preliminary investigations had to come first. Phase one would involve validation of various assumptions, theory, and data gathered from previous boost-glide studies, provide design data, and determine the optimum flight profile for the conceptual vehicle. The second part would refine vehicle design, establish performance, and define subsystems and research instrumentation. While this two-phase preliminary program would consume 12 to 18 months, preliminary studies for the Brass Bell and Robo phases of Dyna-Soar could be started. Following this procedure, flight testing at near satellite speeds for the conceptual test vehicle would begin in 1966. The estimated operational date for Dyna-Soar II was set in 1969, and for Dyna-Soar III in 1974.

The Director of Systems Plans argued that the hypersonic, boost-glide vehicle offered a considerable extension of speed, range, and altitude over conventional Air Force systems. Furthermore, this concept represented a major step towards manned space flight. It could not be safely assumed, the systems plans directorate reasoned, that the intercontinental ballistic missile would destroy all the required targets in the decade of the 1970's. Difficulties in penetrating hostile territory by air-breathing vehicles further enhanced the necessity for a manned, boost-glide vehicle. Additionally, the proposed reconnaissance ability of Dyna-Soar could provide more detailed and accurate intelligence data than other Air Force reconnaissance systems then under development. The director warned that time could not be economically bought. If the boost-glide weapon system were necessary, it was imperative to initiate the Dyna-Soar program by allowing a funding level $3 million for fiscal year 1958.[32]

On 17 October 1957, Lieutenant Colonel C. G. Strathy of the Research and Target Systems Division presented the Dyna-Soar plan to Air Force headquarters. Brigadier General D. Z. Zimmerman, Deputy Director of Development Planning, USAF headquarters, gave enthusiastic endorsement but thought that ARDC headquarters should take a more courageous approach. Command headquarters, he stated, should immediately consider what could be accomplished with greater funding than had been requested. Also present at the briefing was Dr. J. W. Crowley, Associate Director for Research of NACA. He pointed out that the national advisory committee was strongly in favor of initiating the conceptual vehicle program as a logical extension of the X-15 program. He emphasized that his organization was directing its research towards the refinement of the boost-glide concept and was planning new facilities for future research.[33]

Brigadier General H. A. Boushey, Deputy Director of Research and Development, USAF headquarters, informed ARDC headquarters, on 15 November, that the Dyna-Soar abbreviated development plan had been approved. General Boushey's office then issued, on 25 November, Development Directive 94, which allocated $3 million of fiscal year 1958 funds for the hypersonic, glide-rocket weapon system. The boost-glide concept offered the promise of a rapid extension of the manned flight regime, and, following General Zimmerman's reasoning, the deputy director stated that the philosophy of minimum risk and minimum rate of expenditure must be abandoned. If the concept appeared feasible after expenditure of fiscal year 1958 and 1959 funds, the boost-glide program should definitely be accelerated. Not certain of the feasibility of piloted flight, Air Force headquarters directed that the study of manned and unmanned reconnaissance and bombardment weapon systems should be pursued with equal determination. A decision on whether the vehicle was to be piloted would be made in the future and based on substantial analysis. Finally, USAF headquarters stressed that the only objective of the conceptual test vehicle was to obtain data on the boost-glide flight regime. Early and clear test results from this system must be obtained.[34]

While Dr. Sänger had elaborated the theoretical foundation, Dornberger's Peenemünde group demonstrated the practicability of boost-glide flight by launching a winged-precursor, the A-9. The Air Force, however, refined the concept with the Bomi, Brass Bell, Robo, and Hywards study programs. These steps advanced the proposal towards a clearly delineated development program for an orbital, military vehicle – Dyna-Soar.

CHAPTER I
REFERENCES

1.   W. R. Dornberger, V-2 (New York: The Viking Press, 1954), p. 268; W. F. Craven and J. L. Cate (eds.), The Army Air Forces in World War II, Vol. III: Europe: Argument to V-E Day, January 1944 to May 1945 (Chicago: The University of Chicago Press, 1951), p. 542.

2.   Dornberger, V-2, pp. 140-143, 250-251, 271.

3.   Col. L. E. Simon, Report of German Scientific Establishments (Washington: U.S. Army, 1945), pp. 156-158, 200-201; Drs. Eugen Sänger and Irene Bredt, A Rocket Drive for Long Range Bombers (Santa Barbara, Calif.: Robert Gornog, 1952), pp. 6, 50-51, 58, 85, 122, 129, 136-138; Willy Ley, Rockets, Missiles, and Space Travel (2nd. ed. rev., New York: The Viking Press, 1961), pp. 4, 8-10, 429-434.

4.   Ley, Rockets, Missiles, and Space Travel, pp. 429-430; Missiles and Rockets, 11 Jan. 1960, p. 13; ltr., Col. J. F. Berry, Dep. Dir., Warning and Threat Assessments Ofc., ACS/Intel., Hqs. USAF to Dir., Adv. Tech., DCS/Dev., Hqs. USAF, 18 Mar. 1960, subj: Estimate of USSR Boost-Glide Activity. Unless otherwise stated, all documents cited in this monograph which are not included in the appended document volumes are found in the files of the Dyna-Soar System Program Office, ASD.

5.   Douglas Aircraft Company, Preliminary Design of an Experimental World-Circling Spaceship (Santa Monica, Calif.: Douglas Aircraft Co., 1946), pp. ii-viii, 192-193, 203, 211.

6.   Ltr., R. J. Sandstrom, Vice Pres. of Eng., Bell Aircraft Corp. to Hqs. WADC, 17 Apr. 1952, subj.: Proposed Glide Bomber Study, Doc. 1; ltr., J. W. Rane, Jr., Asst. Dir/Contracts, WADC, 26 May 1952, subj.: Rocket Bomber Development Program – Proposal for Feasibility Study, Doc. 2.

7.   Ltr., Lt. Col. H. Heaton, Acting Dir., Aero. & Propul. Ofc., Dep/Dev., Hqs. ARDC to Hqs. WADC, 28 Nov. 1952, subj: Rocket Bomber Feasibility Studies, Doc. 3.

8.   Ltr., Hqs. WADC to Hqs. ARDC, 10 Apr. 1953, subj.: Rocket Bomber Feasibility, Doc. 4.

9.   Ltr., Brig. Gen. F. B. Wood, DCS/Dev., Hqs. ARDC to Hqs. WADC, 1 Oct. 1953, subj: Bell Aircraft Corporation Strategic Weapon System D-143 Bomi, Doc. 5.

10. Ltr., Col. M. C. Demler, Vice Cmdr., WADC to Cmdr., ARDC, 23 Nov. 1953, subj.: Bell Aircraft Corporation Strategic Weapon System D-143 Bomi, Doc. 6.

11. Ltr., Brig. Gen. L. I. Davis, Acting DCS/Dev., Hqs. ARDC to Hqs. WADC, 23 Dec. 1953, subj.: Bell Aircraft Corporation Strategic Weapon System D-143 Bomi, Doc. 7.

12. Rpt., Boeing Airplane Company, 22 Apr. 1954, subj.: Preliminary Study of Design Characteristics of Hypersonic Bombardment Rockets, pp. 24-26; rpt., Boeing, 1 July 1953, subj.: Planning Report, pp. 1-2; History of Wright Air Development Center, 1 July - 31 December 1953, II, 275-277.

13. C. H. Uyehara, Dyna-Soar Antecedents (unpublished manuscript, ASD Hist. Div. files, 1960), p. 14.

14. Ibid., pp. 14, 19, 22.

15. Directorate of Systems Management Weekly Activity Report, 7 and 14 June 1955, 22 Sept. 1955, hereafter cited as DSM WAR.

16. DSM WAR, 25 Nov. 1955, 15 Dec. 1955.

17. Ltr., Brig. Gen. H. M. Estes, Jr., Asst. Dep. Cmdr/Weap. Sys., Hqs. ARDC to Dep. Cmdr/Weap. Sys., Hqs. ARDC, 21 Dec. 1955., subj.: Evaluation of Project Bomi, Doc. 9.

18. Ibid.; DSM WAR, 15 Dec. 1955, 5 Apr. 1956.

19. Ltr., Dr. I. R. Abbott, Asst. Dir/Res., NACA to Hqs. WADC, 30 Sept. 1955, subj.: Requested Comments on Project MX-2276; DSM WAR, 22 Dec. 1955.

20. GOR 12, Hqs. USAF, 12 May 1955., subj.: Piloted Very High Altitude Reconnaissance Weapon Systems, Doc. 8; DMS WAR, 12 Jan. 1956; Uyehara, Antecedents, pp. 35-36; R&D Proj. Card, System 459L, Intel. & Recon. Sys. Div., Hqs. ARDC, 31 Dec. 1956, subj.: Brass Bell.

21. R&D Proj. Card, System 459L, 31 Dec. 1956, subj.: Brass Bell.

22. Uyehara, Antecedents, pp. 44-45; rpt., Ad Hoc Committee, 1 Aug. 1957, subj.: Evaluation Report for ARDC System Requirement No. 126, Robo, Doc. 13.

23. SR 126, Hqs. ARDC, 12 June 1956, subj.: Rocket Bomber Study, Doc. 11.

24. Memo., Maj. G. D. Colchagoff, Res. & Target Sys. Div., Dep. Cmdr/Weap. Sys., Hqs. ARDC to Lt. Col. R. C. Anderson, Res. & Target Sys. Div., 16 Feb. 1956, subj.: New Research Systems, Doc. 10.

25. Ibid.; Uyehara, Antecedents, p. 51.

26. Memo., Res. & Target Sys. Div., Hqs. ARDC, 16 Apr. 1956, subj.: Research Vehicle Program.

27. Ltr., Lt. Gen. T. S. Powers, Cmdr., ARDC to Hqs. USAF, 16 July, 1956, subj.: Fiscal Year 1957 Funds.

28. SP 131, Hqs. ARDC, 6 Nov. 1956, subj.: Abbreviated Development Plan for System 455L, Doc. 12; Chronology, DSM, Hqs. ARDC, 14 Mar. 1959, subj.: Significant Events, Dyna-Soar Background, p. 15.

29. R&D Proj. Card, System 455L, Res. & Target Sys. Div., Hqs. ARDC, 28 Dec. 1956, subj.: Hypersonic Glide Rocket Research System.

30. Uyehara, Antecedents, pp. 55-56, 57.

31. Rpt., Ad Hoc Committee, 1 Aug. 1957, subj.: Evaluation Report of the Ad Hoc Committee for ARDC System Requirement Number 126, Robo, Doc. 13; rpt., Ad Hoc Committee, 1 Oct. 1957, Subj.: Supplement 1 to the Evaluation Report, Doc. 14.

32. Abbreviated System Development Plan, System 464L, Hqs. ARDC, 10 Oct. 1957, subj.: Hypersonic Strategic Weapon System.

33. Trip rpt., Lt. W. C. Walter, New Dev. Weap. Sys. Ofc., Hqs. ARDC to R. D. Hodge, Ch., New Dev. Weap. Sys. Ofc., 21 Oct. 1957, subj.: Visit to Hqs. USAF on Project Dyna-Soar, Doc. 15; ltr., Brig. Gen. H. A. Boushey, Dep. Dir., Res. & Dev., DCS/Dev., Hqs. USAF to Cmdr., ARDC, 15 Nov. 1957, subj.: Approval of Abbreviated Development Plan for System 464L, Doc. 16.

34. DD 94, Hqs. USAF, 25 Nov. 1957, subj.: Hypersonic Glide Rocket Weapon System, Doc. 17.

CHAPTER II

SYSTEM 464L

With the approval of the abbreviated development plan, the direction of the Dyna-Soar program appeared clearly marked. An experimental glider, a reconnaissance vehicle, and a bombardment system comprised a three-step progression. During the existence of System 464L, however, officials in the Department of Defense subjected the program to severe criticism. The necessity of orbital flight and the feasibility of a boost-glide weapon system were points frequently questioned. By November 1959, the project Office had to undertake an exacting investigation of the Dyna-Soar approach to manned, space flight. Certainty of program objectives had momentarily disappeared.

On 21 December 1957, ARDC headquarters issued System Development Directive 464L, which stipulated that the mission of the conceptual test vehicle, Dyna-Soar I, was to obtain data on the boost-glide flight regime in support of future weapon system development. Headquarters suggested that a system development plan for Dyna-Soar I and the recommended weapon system programs be completed on 31 October 1958 and set July 1962 as the date for the first flight of the conceptual test vehicle. Finally, ARDC headquarters approved immediate initiation of the program by directing the source selection process to begin.[1]

By 25 January 1958, a task group of the source selection board had screened a list of 111 contractors to determine potential bidders for the Phase I design. The working group considered that Bell, Boeing, Chance-Vought Aircraft, Convair, General Electric Company, Douglas, Lockheed, Martin, North American, and Western Electric Company would be able to carry out the development. Later, the list was amended to include McDonnell Aircraft, Northrop Aircraft, and Republic Aviation.[2]

The source selection board had received, by March 1958, proposals from nine contractor teams. Essentially, two approaches were taken in considering the development of Dyna-Soar I. In the satelloid concept, a glider would be boosted to an orbital velocity of 25,500 feet per second to an altitude of 400,000 feet, thereby achieving global range as a satellite. In the flexible boost-glide proposal, however, the projected vehicle would follow a glide-trajectory after expenditure of the booster. With a high lift-to-drag ratio at a velocity of 25,000 feet per second and an altitude of 300,000 feet, the glider could circumnavigate the Earth.

Three contractors offered the first approach, the satelloid concept, as the most feasible. Republic conceived of a 16,000 pound, delta-wing glider, boosted by three, solid propellant stages. The vehicle, along with a 6,450 pound space-to-earth missile, would be propelled to a velocity of 25,700 feet per second and an altitude of 400,000 feet. Lockheed considered a 5,000 pound glider similar in design to that of Republic. This vehicle could operate as a satelloid, however, the contractor suggested a modified Atlas booster which lacked sufficient thrust for global range. A 15,000 pound vehicle similar to the X-15 craft comprised the proposal of North American. The booster was to consist of a one-and-a-half stage liquid propellant unit with an additional stage in the glider. Operated by a two-man crew, the vehicle was also to have two, small, liquid engines for maneuvering and landing. The glider was to be propelled to a velocity of 25,600 feet per second and an altitude of 400,000 feet and would operate as a satelloid.

Six contractors concentrated on the flexible boost-glide concept. Douglas considered a 13,000 pound, arrow-wing glider which was to be boosted by three, modified solid propellant stages of the Minuteman system. An additional stage would provide a booster for advanced versions of Dyna-Soar. McDonnell offered a design similar to that of Douglas but proposed, instead, the employment of a modified Atlas unit. A delta-wing glider, weighing 11,300 pounds, was recommended by Convair. This contractor did not consider the various possibilities for the booster system but did incorporate a turbojet engine to facilitate landing maneuvers. Martin and Bell joined to propose a two-man, delta-wing vehicle, weighing 13,300 pounds, which would be propelled by a modified Titan engine. Employing Minuteman solid propellant units, Boeing offered a smaller glider, weighing 6,500 pounds. Finally, Northrop proposed a 14,200 pound, delta-wing glider, which was to be boosted by a combination liquid and solid propellant engine.

The task group of the source selection board, after reviewing the proposals, pointed out that with the exception of the North American vehicle all of the contractors' proposed configurations were based on a delta-wing design. The size of the proposed vehicles was also small in comparison with current fighter aircraft such as the F-106. McDonnell and Republic offered vehicles which could carry the biggest payload, yet they in turn required the largest boosters. At the other extreme was BOEING'S proposal which could carry only 500 pounds, including the weight of the pilot. The task group also emphasized that of the three contractors proposing the satelloid concept Lockheed's vehicle fell short of a global range. Of the six contractors offering the flexible boost-glide approach, only the Martin-Bell team and Boeing proposed a first-step vehicle capable of achieving orbital velocities. The other four considered a global range in advanced versions.[3]

By the beginning of April, the working group had completed its evaluation of the contractors' proposals, and, on 16 June 1958, Air Force headquarters announced that the Martin Company and the Boeing Airplane Company both had been selected for the development of Dyna-Soar I.[4] Major General R. P. Swofford, Jr., then Acting Deputy Chief of Staff for Development, USAF headquarters, clarified the selection of two contractors. A competitive period between Martin and Boeing would extend from 12 to 18 months at which time selection of a single contractor would be made. General Swofford anticipated that $3 million would be available from fiscal year 1958 funds and $15 million would be set for 1959. The decision as to whether Dyna-Soar I would operate as a boost-glide or a satelloid system was left open, as well as the determination of a piloted or unmanned system. The acting deputy directed that both contractors should proceed as far as possible with available funds towards the completion of an experimental test vehicle. The design, however, should approximate the configuration of a Dyna-Soar weapon system.[5]

Apparently some questioning concerning the validity of the Dyna-Soar program occurred at Air Force headquarters, for, on 11 July, Major General J. W. Sessums, Jr., Vice Commander of ARDC, stated to Lieutenant General R. C. Wilson, USAF Deputy Chief of Staff for Development, that Air Staff personnel should stop doubting the necessity for Dyna-Soar. Once a new project had been sanctioned by headquarters, General Sessums considered, support should be given for its completion.[6] In reply, General Wilson assured General Sessums that the Air Staff held the conviction that Dyna-Soar was an important project. However, due to the interest of the Advanced Research Projects Agency (ARPA) and the National Aeronautics and Space Administration (NASA) and their undetermined responsibilities in the development of systems such as Dyna-Soar, the Air Force firmly had to defend its projects to the Department of Defense.* General Wilson closed by reassuring General Sessums of his full endorsement of the Dyna-Soar program.[7]

While the Dyna-Soar program had the verbal support of USAF headquarters, Lieutenant General S. E. Anderson, ARDC commander, considered that the program required additional funds. He reminded General Wilson that ARDC headquarters, with the efforts of only one contractor in mind, had requested $32.5 million for fiscal year 1959. The Air Staff had limited this amount to $15 million for the contributions of both Boeing and Martin. Consequently, $52 million was now required for the 1959 Dyna-Soar program.

The ARDC commander emphasized that if System 464L were to represent a major step in manned, space flight, then the delay inherent in the reduced funding must be recognized and accepted by Air Force headquarters.[9] General Wilson agreed with General Anderson's estimation and stated that the approved funding level for fiscal year 1959 would undoubtedly delay the program by one year. The stipulated $18 million for both fiscal years 1958 and 1959, although a minimum amount, would permit the final contractor selection. General Wilson did assure the ARDC commander that the Air Staff would try to alleviate the situation and thought there was a possibility for increasing fiscal year 1959 funding.[10]

Major General V. R. Haugen, Assistant Deputy Commander for Weapon Systems, Detachment One, made another plea to the Deputy Chief of Staff for Development. He estimated that inadequate funding would push the flight date for the research vehicle back by eight months. Such austerity would hinder the developmental test program and cause excessive design modification. General Haugen strongly urged the augmentation of fiscal year 1959 funding to $52 million. Besides this, it was important that the full release of the planned $15 million be immediately made.[11]

On 4 September, Colonel J. L. Martin, Jr., Acting Director of Advanced Technology, USAF headquarters, offered additional clarification of the funding situation to Detachment One. He stated that the two separate efforts by Boeing and Martin should only be maintained until study results pointed to a single, superior approach. It was possible for this effort to be terminated within 12 months. Colonel Martin pointed out that the Air Staff was aware that the $18 million level would cause delays; these funds, however, would provide the necessary information for contractor selection. He did announce that release of the $15 million had been made. Lastly, Colonel Martin directed that the term "conceptual test vehicle" would no longer be used to refer to Dyna-Soar I and, in its place, suggested the words "experimental prototype."[12]

The Dyna-Soar project office replied that the competitive period could be terminated by April instead of July 1959; however, additional funding could be effectively utilized.[13] These efforts to increase the Dyna-Soar allotment had no effect, for, on 30 September 1958, USAF headquarters now informed Detachment One that the $10 million procurement funds for fiscal year 1959 had been canceled. All that remained for development of Dyna-Soar was $3 million from fiscal year 1958, with $5 million for 1959. In his 12 August letter to General Anderson, General Wilson mentioned the possibility of increased funding

---

*   Previously, considerable discussion within the Air Force had taken place concerning the role which the National Aeronautics and Space Administration, earlier designated the National Advisory Committee for Aeronautics, was going to play in the Dyna-Soar program. On 31 January 1958, Lieutenant General D. Z. Putt, Deputy Chief of Staff for Development, USAF headquarters, asked NACA to join with the Air Force in developing a manned, orbiting, research vehicle. He further stated that the program should be managed and funded along the lines of the X-15 program. It appeared that General Putt was proposing a Dyna-Soar I program under the direction of NACA. ARDC headquarters strongly recommended against this contingency on the grounds that Dyna-Soar would eventually be directed towards a weapon system development. By 20 May, General T. D. White, Air Force Chief of Staff, and Dr. H. L. Dryden, NACA director, signed an agreement for NACA participation in System 464L. With the technical advice and assistance of NACA, the Air Force would direct and fund Dyna-Soar development. On 14 November 1958, the Air Force and NASA reaffirmed this agreement.[8]

for fiscal year 1959. Apparently a figure of $14.5 million was being considered; however, Air Force headquarters also informed ARDC that this proposed increase would not be made. Headquarters further directed that expenditure rates by the contractors be adjusted in order that the $8 million would prolong their efforts through 1 January 1959.[14]

From 20 through 24 October 1958, Mr. W. E. Lamar, in the Deputy for Research Vehicles and Advanced Systems, and Lieutenant Colonel R. M. Herrington Jr., chief of the Dyna-Soar project office, briefed Air Force headquarters on the necessity of releasing funds for the Dyna-Soar program. The discussions resulted in several conclusions. The objectives of the program would remain unchanged, but further justification would have to be given to Department of Defense officials. The position of NASA in the program was reaffirmed, and it was further stipulated that ARPA would participate in system studies relating to Dyna-Soar.[15] These decisions, however, did not offer immediate hope for increased funding.

Early in November 1958, Colonel Herrington and Mr. Lama briefed officials of both ARDC and USAF headquarters on the question of Dyna-Soar funding. General Anderson, after hearing the presentation, stated that he supported the program but thought that references to space operation should be deleted in the presentations to the Air Staff. Later, during a briefing to General Wilson, USAF officials decided that suborbital aspects and possibilities of a military prototype system should be emphasized. With the sanction of the Air Force Vice Chief of Staff, General C. E. LeMay, the Dyna-Soar presentation was given to Mr. R. C. Horner, the Air Force Assistant Secretary for Research and Development. The latter emphasized that if a strong weapon system program were offered to Department of Defense officials, Dyna-Soar would probably be terminated.

Rather, Secretary Horner suggested that the program be slanted towards the development of a military, research system. He stated that a memorandum would be sent to the defense secretary requesting release of additional funds for Dyna-Soar.[16] While Colonel Herrington and Mr. Lamar achieved their funding objectives, it was also apparent that the final goal of the Dyna-Soar program – the development of an operational weapon system – was somewhat in jeopardy.

In accordance with ARDC System Development Directive 464L, the Dyna-Soar project office had completed, in November, a preliminary development plan which supplanted the abbreviated plan of October 1957. Instead of the three-step approach, the Dyna-Soar program would follow a two-phase development. Since the military test vehicle would be exploring a flight regime which was significantly more severe than that of existing Air Force systems, the first phase would involve a vehicle whose function was to evaluate aerodynamic characteristics, pilot performance, and subsystem operation. Dyna-Soar I was to be a manned glider with a highly-swept, triangular-planform wing, weighing between 7,000 and 13,000 pounds. A combination of Minuteman solid rockets could lift the vehicle, at a weight of 10,000 pounds, to a velocity of 25,000 feet per second and an altitude of 300,000 feet. By employing a liquid rocket such as the Titan system, a 13,000 pound vehicle could be propelled to a similar speed and height. The project office stipulated that a retro-rocket system to decelerate the glider and an engine to provide maneuverability for landing procedures would be necessary.

Assuming a March 1959 approval for the preliminary development plan, the Dyna-Soar office reasoned that the air-drop tests could begin in January 1962, the suborbital, manned, ground-launch tests in July 1962, and the first, piloted, global flight in October 1963. While this first phase was under development, weapon system studies would be conducted concurrently, with the earliest operational date for a weapon system set for 1967. This Dyna-Soar weapon could perform reconnaissance, air defense, space defense, and strategic bombardment missions.[17] The problem of obtaining funds to continue the program, not an outline of Dyna-Soar objectives, was still, however, of immediate importance.

On 4 December 1958, the Secretary of the Air Force requested the Secretary of Defense to release $10 million for the Dyna-Soar program. Apparently the defense department did not act immediately, for, on 30 December, Air Force headquarters informed Detachment One that release of these funds could not be expected until January 1959.[18] The project office urgently requested that procurement authorizations be immediately issued.[19] Finally, on 7 January, the Deputy Secretary of Defense, D. A. Quarles, issued a memorandum to the Secretary of the Air Force, which approved the release of $10 million for the Dyna-Soar program. The deputy secretary emphasized that this was only an approval for a research and development project and did not constitute recognition of Dyna-Soar as a weapon system. The stipulated increase of $14.5 million was not to be released until a decision was made concerning the Boeing-Martin competition.[20]

Air Force headquarters, on 14 January 1959, requested the Dyna-Soar office to provide a detailed program schedule. Concerning the Dyna-Soar I military test system, planning should be based on the following projected funding: $3 million for fiscal year 1958, $29.5 million for 1959, and $35 million for 1960. Headquarters further directed that the competitive period for the contractors would end by 1 April with a final selection announced by 1 July 1959. While emphasis on a weapon system would be minimized, joint Air Force and ARPA weapon system studies would proceed under separate agreement with Dyna-Soar contractors. The project office was also directed to consider two other developmental approaches. The first would assume that Dyna-Soar objectives had definitely been changed to center on a research vehicle, similar to the X-15 craft, and planning would be based on a projected funding of $78 million for fiscal year 1961, $80 million for 1962, $80 million for 1963 and $40 million for 1964. In the second approach, the Dyna-soar program would include weapon system objectives, and a funding total of $650 million extending from fiscal year 1961 through 1967 would be assumed. The next day, Air Force headquarters partially revised its directions by stipulating that the source selection process should be completed by 1 May 1959.[21]

On 6 February 1959, the Dyna-Soar project office pointed out that the 1 May date was impracticable, but the office did anticipate a presentation on source selection to the Air Council by 1 June. The project office went on to emphasize that the funding forecasts were incompatible with the flight dates which had been specified to the contractors. It was apparent to the project office that only heavy expenditures during the beginning of phase two could result in the questioned flight dates. The Dyna-Soar office, consequently, requested Air Force headquarters to provide a more realistic funding schedule.

In mid-February, the Dyna-Soar office further clarified its position. The approval of only $5 million in development funds for fiscal year 1959 (the release of $10 million had been for procurement), instead of a revised request of $28 million, had a serious effect on the program by reducing the applied research and development program. Furthermore, the project office had originally requested $187 million for fiscal year 1960, an estimate that was predicated on more extensive effort during fiscal year 1959 than was actually taking place under the reduced funding level. Air Force headquarters had only projected $35 million for fiscal year 1960. The result would be a prolongation of the program.[23] This statement of the project office had some impact on headquarters, for, on 17 February, the Air Staff requested the project office to provide additional information on the program based on fiscal year 1960 funding levels of either $50 million or $70 million.[24]

The depreciation of Dyna-Soar as a weapon system by the defense department, as exemplified by the Secretary Quarles' memorandum of 7 January, did not alter the necessity, in the opinion of the Air Force, for a boost-glide weapon. On 17 February 1959, Air Force headquarters revised its General Operation Requirement 92, previously issued on 12 May 1955. Instead of referring to a high-altitude reconnaissance system, the Air Force now concentrated on a bombardment system. USAF headquarters stated that this system, capable of target destruction, was expected to operate at the fastest attainable hypersonic speed, within and above the stratosphere, and could complete at least one circumnavigation of the Earth. This projected system would be capable of operation from 1966 to 1970.[25]

On 13 April 1959, Dr. H. F. York, Director of Defense for Research and Engineering, firmly established the objectives for Dyna-Soar I. The primary goal was the non-orbital exploration of hypersonic flight up to a velocity of 22,000 feet per second. Launched by a booster already in production or planned for the national ballistic missile and space programs, the vehicle would be manned, maneuverable, and capable of controlled landings. Secondary objectives were the testing of military subsystems and the attainment of orbital velocities. The Department of Defense instructed that the accomplishment of these last objectives should only be implemented if there were no adverse effects on the primary objective. The additional $14.5 million was now authorized for fiscal year 1959, giving a total of $29.5 million for that year. The Department of Defense inquired whether this figure plus a proposed $35 million for fiscal year 1960 would be sufficient. to carry out the program. If the Air Force did not consider this feasible, then an alternate program should be submitted for review.[26]

Command headquarters was not in accord with these directions. In an effort to fulfill the conditions established by General Operational Requirement 92, the research and development command issued, on 7 May 1959, ARDC System Requirement 201. The Dyna-Soar I vehicle was to be a military test system developed under the direction of the Air Force with technical assistance from the National Aeronautics and Space Administration. The purpose of this system would be to determine the military potential of a boost glide weapon system and provide research data on flight characteristics up to and including global flight. Concurrently, studies would be made concerning a weapon system based on this type of hypersonic vehicle. Headquarters then directed its Detachment One to prepare a development plan for Dyna-Soar by 1 November 1959.[27]*

Major General Haugen, in reply to the directions of Dr. York, "strongly recommended" that the attainment of orbital velocities and the testing of military subsystems should be a primary, not a secondary objective. He further stated that Dyna-Soar was the only manned vehicle program which could determine the military potential in the near-space regime. It was "extremely important," the systems management director stated, that the accomplishment of the Dyna-Soar mission not be compromised by restrictions which limited safety, reliability, and growth potential in deference to short-term monetary savings.[28]

General Haugen's organization then drew up a position paper substantiating these recommendations. The directorate firmly believed that both the primary and secondary objectives had to be achieved. Concentration on the first set of objectives would prevent investigation of re-entry from orbit and the adequate testing of military subsystems. The directorate then recommended a program involving the fabrication of eight unmanned vehicles, eight manned vehicles, and 27 boosters, all to be employed in a total of 25 launchings. This would cost a total of $665 million. While modification of this program to conform with only the primary objectives would reduce the cost by $110 million, it would seriously lessen the possibility of evolving a weapon system from Dyna-Soar I.[29]

Excluding $18 million expended during contract competition, the Deputy Chief of Staff for Development in Air Force headquarters established, on 28 May, $665 million as the maximum total of the Dyna-Soar program. For planning purposes $77 million was set for fiscal year 1960. On 11 June 1959, [30] the Air Force Council considered this last figure to be excessive, and the deputy chief of staff had to recant: $35 million was to be used in place of the $77 million.[31]

During a briefing on 23 June 1959, officials of the project office and Dr. J. V. Charyk, Assistant Secretary of the Air Force for Research and Development, further discussed the questions of Dyna-Soar funding and objectives. Apparently, Dr. Charyk, at this point, was not in full agreement with Dr. York's position. The assistant secretary considered that the over-all purpose of the program was to exploit the potentialities of boost-glide technology, and, consequently, he implied that orbital velocities should be attained early in the program. For fiscal year 1960, he favored $77 million instead of $35 million but raised the question of how much a total funding level of $300 million to $500 million would compromise the program.** Dr. Charyk then reported to the project officials that Dr. York appeared quite concerned over the effort necessary for modification of a proposed Dyna-Soar booster.[32]

---

*   By January 1959, the preliminary development plan of November 1958 had been forwarded to ARDC and USAF headquarters, however, apparently neither headquarters gave it official sanction.

** The documentary source, as cited in reference 32, for Dr. Charyk's comments referred to the $77 million and $35 million as projected figures for fiscal year 1959. Placed in context of the funding discussions concerning the Dyna-Soar program, these estimates obviously applied to fiscal year 1960 and not 1959.

The Air Force source selection board had already appraised the Boeing and Martin proposals. Although both contractors offered similar delta-wing designs, they differed in their selection of boosters. While Boeing only considered an orbital Atlas-Centaur combination, Martin officials offered a surborbital Titan A (later renamed the Titan I) and an orbital Titan C. The board deemed the Boeing glider superior but also recommended use of Martin's orbital booster. The Secretary of the Air Force, J. H. Douglas, did not agree. Development of a new booster, capable of orbital velocities, was clearly not in accord with Dr. York's direction. The secretary recommended further study of the configuration and size of the vehicle to determine whether the glider could be modified to permit compatibility with a basic, suborbital, Titan system. Furthermore, Secretary Douglas was concerned about the total cost of the program. He did not think that funding should be increased by attempting to configure a vehicle which conformed to an anticipated weapon system. Consequently, the Secretary of the Air Force directed a reassessment of the Dyna-Soar program, with the ultimate objective of reducing the over-all expense. Accordingly, USAF headquarters directed Detachment One to examine the possibilities for a lighter vehicle and to analyze a development program based on a total cost of not more than $500 million.[33]

Designation of the booster, management of booster development and procurement, and most important, the purpose of the program, were problems that became intertwined in the series of discussions following Secretary Douglas' instructions. After a 14 July meeting with Dr. Charyk, General Boushey, Colonel W. L. Moore, Jr., and Lieutenant Colonel Ferer, General Haugen directed systems management to prepare a presentation designed to answer the questions raised by Secretary Douglas and also to outline the participation of the Ballistic Missiles Division (BMD) in the Dyna-Soar program.* After reviewing this briefing on 22 July 1959, Lieutenant General B. A. Schriever, now ARDC commander, instructed General Haugen's directorate to prepare a detailed management plan for booster development.[34]** Dr. York, however, on 27 July, placed a new complication in this planning effort by requesting the Air Force secretary and the director of ARPA to investigate the possibility of a common development of a Dyna-Soar booster and a second stage for the Saturn booster of NASA. The Director of Defense for Research and Engineering stated that no commitments for the propulsion system would be made until this proposal had been considered. Dr. York apparently had in mind reviving consideration of the Titan C for System 464L and modifying this booster for use in the Saturn program.[35]

On 28 and 29 July, General Haugen and Brigadier General. O. J. Ritland, BMD commander, completed a tentative agreement concerning the management of Dyna-Soar booster development. During a series of meetings on 11 and 13 August, however, General Schriever and General Anderson, AMC commander, could not agree on a method of booster procurement. With the exception of the parts pertaining to BMD participation in the Dyna-Soar program. Mr. Lamar then gave the Dyna-Soar presentation to Dr. Charyk, with Generals Wilson, Ferguson, and Haugen attending. After preliminary data was given on Titan C and the Saturn second stage, Dr. Charyk was asked to recommend to the defense department that a contractor source selection be made for Dyna-Soar. He declined: subcontractor selection had not been adequately competitive and the proposed Dyna-Soar funding was too high.[36]

By the middle of August, the Ballistic Missiles Division had completed its evaluation of possible Dyna-Soar boosters. Largely because of serious stability and control problems, an Atlas-Centaur combination was rejected in favor of the Titan C. Concerning Dr. York's proposal, west coast officials believed that it was impractical to employ a precisely identical booster stage for both the Dyna-Soar and Saturn projects. Since Titan C was essentially a cluster of four LR87-AJ-3 engines, ballistic division engineers did recommend employing two of these propulsive units as a Saturn second stage.[37] Discussions between Dr. Charyk, Dr. York, and ballistic division officials concerning selection of the Dyna-Soar booster followed. Finally, while a booster was not designated, Dr. Charyk, Generals Wilson, Ferguson, and Boushey decided, on 25 September, that Titan C would not be employed in the Program.[38]

---

*   Colonel Moore succeeded Colonel R. M. Herrington, Jr., as chief of the Dyna-Soar Weapon System Project Office early in July-1959.
** On 10 March 1959, Lieutenant General S. E. Anderson, previously ARDC commander, became commander of the Air Materiel Command. Lieutenant General B. A. Schriever, on 25 April 1959, assumed command of ARDC.

On 23 September, Lieutenant General W. F. McKee, AMC vice commander, took up the question of booster procurement and proposed to General Schriever a management plan, based on discussions between ARDC and AMC personnel, for the Dyna-Soar program. Because of the wide participation of government agencies and industry, control of Dyna-Soar had to be centralized in a specific organization. While the system was to be procured under two contracts, one for the glider and one for the propulsion unit, the contractor responsible for the manufacture of the vehicle would be given responsibility for integration of the entire system and would act as weapon system contractor. Over-all management would be vested in a joint ARDC and AMC project office located at Wright-Patterson Air Force Base. Concerning the procurement authority of the Aeronautical Systems Center (ASG) and the Ballistic Missiles Center (BMC), both of the materiel command, General McKee suggested that the aeronautical center negotiate the two contracts, utilizing the experience available at the ballistic center. The Aeronautical Systems Center, however, would delegate authority to the ballistic center to contractually cover engineering changes. This delegation would be limited to actions not affecting over-all cost, compatibility between booster and vehicle, and system performance. General McKee closed by recommending that ARDC and AMC forward a message to Air Force headquarters outlining this proposal.[39]

General Schriever, on 2 October, informed AMC officials that he agreed with General McKee's proposed message to USAF headquarters. He did wish to point out, however, that the plan did not adequately reflect the increased role that ARDC agencies at Wright Field were intending to play. General Schriever further stated that ARDC was going to establish a single agency for all booster research and development which would incorporate the use of BMD and BMC.[40] General Anderson replied that he did not understand the ARDC commander's statement concerning increased management responsibility of Wright agencies. He stated that the AMC plan stressed this aspect. General Anderson further emphasized that the materiel command recognized BMD's technical responsibility for the Dyna-Soar booster and had agreed to delegate necessary procurement authority. The AMC commander did not think it was necessary, however, to delegate authority to negotiate contracts. This authority, along with over-all technical management should rest in the ARDC and ASC weapon system project offices.[41]

On 29 October, General Boushey re-examined the Dyna-Soar requirements established by the 13 April memorandum of Dr. York. Orbital flight and testing of military subsystems could only be permitted, Dr. York insisted, if these efforts did not adversely affect the central objective of non-orbital, hypersonic flight. General Boushey reiterated the opinion of USAF headquarters: both sets of objectives should be definitely achieved. Assuming a total funding of $665 million, ARDC was directed to formulate a two-phase development approach for a 9,000 to 10,000 pound glider.[42]

By 1 November 1959, the Dyna-Soar office completed an abbreviated development plan in fulfillment of ARDC System Requirement 201. As suggested by the Office of the Secretary of Defense, the project office once again structured the program in a three-step approach. In Step I, a manned glider, ranging in weight from 6,570 to 9,410 pounds would be propelled to suborbital velocities by a modified Titan booster. Step II encompassed manned orbital flight of the basic glider and interim military operations. A weapon system, founded on technology from the previous steps, comprised Step III. The project office anticipated 19 air-drop tests to begin in April 1962; the first of eight unmanned, suborbital flights to occur in July 1963; and the first of eight piloted, suborbital launches to take place in May 1964. The first, manned, global flight of Step II was scheduled for August 1965. To accomplish this program, the project office estimated the development cost to total $623.6 million from fiscal year 1960 through 1966.[43] On 2 November, the Weapons Board of Air Force headquarters approved the revised Dyna-Soar plan. The Air Council, in addition to sanctioning the three-step program, also approved of an ARDC and AMC arrangement concerning booster procurement.[44]

Generals Schriever and Anderson, on 4 November, forwarded a joint ARDC and AMC letter to USAF headquarters. After detailing the essentials of the program, the two commanders outlined their agreement on booster procurement: the project office would utilize the "experience" of the ballistic division in obtaining a booster for Dyna-Soar. They further stated that the proposed program would make full use of existing national booster programs, essentially satisfying Dr. York's requirement, and would also attain Air

Force objectives by achieving orbital velocities. General Schriever and General Anderson closed by urging the source selection process to be completed.[45]

Following this advice, the Secretary of the Air Force, on 9 November 1959, announced the Dyna-Soar contracting sources. The Boeing Airplane Company had won the competition and was awarded the systems contract. The Martin Company, however, was named associate contractor with the responsibility for booster development.[46] On 17 November, Air Force headquarters directed the research and development command to implement Step I and to begin planning for Step II of the Dyna-Soar program.[47] Three days later, Dr. Charyk gave the Air Force authority to negotiate Step I contracts for fiscal year 1960. There was, however, an obstruction. The assistant secretary instructed the Deputy Chief of Staff for Development that, prior to obligating any funds for the Dyna-Soar program, now designated System 620A Dr. Charyk's office would have to be given financial plans and adequate work statements. No commitments could be made before the Air Force had a concise understanding of the direction of the project.[48]

In an effort to obtain approval to obligate funds for fiscal years 1959 and 1960, General Boushey and some of his staff met with Dr. Charyk on 24 November, and Dr. Charyk made it clear that he did not wish to release any funds for Dyna-Soar at that time. Instead, he was going to institute Phase Alpha, the purpose of which would be to examine the step-approach, the proposed booster, the vehicle size, and the flight test objectives. Dr. Charyk stated that no funds would be obligated until the Alpha exercise was completed. Once Dyna-Soar was implemented, the assistant secretary wanted to review the program step-by-step and release funds as the program proceeded.[49] To cover the work carried on under Phase Alpha, the Air Force released a total of $1 million. Pending further approval by Dr. Charyk, obligations could not exceed this amount.[50]

CHAPTER II

REFERENCES

1.   SDD 464L, Hqs. ARDC, 12 Dec. 1957, Doc. 18.

2.   Ltr., Col. H. M. Harlow, Dir., F-106 WSPO, ASC, AMC to Source Selection Bd., 25 Jan. 1958, subj.: Report of Working Group for System 464L; mangt. rpt., W. E. Lamar, Asst. Ch., Bomb. Ac. Div., Det. 1, Hqs. ARDC, 4 Feb. 1958, subj.: Hypersonic Strategic Weapon System, Doc. 21.

3.   Rpt., Source Selection Bd., ident. no. 58RDZ-9197 n.d., subj.: Summary of Contractor Proposals.

4.   DSM WAR, 10 Apr.1958; TWX, AFDRD-AN-32317, Hqs. USAF to Cmdr., Det. 1, Hqs. ARDC, 16 June 1958, Doc. 25.

5.   Ltr., Maj. Gen. R. P. Swofford, Jr., Acting DCS/Dev., Hqs. USAF to Cmdr., ARDC, 25 June 1958, subj.: Selection of Contractor for WS 464L (Dyna-Soar) Development, Doc. 27.

6.   Ltr., Maj. Gen. J. W. Sessums, Jr., Vice Cmdr., ARDC to Lt. Gen. R. C. Wilson, DCS/Dev., Hqs. USAF, 11 July 1958, subj.: Dyna-Soar Program, Doc. 28.

7.   Ltr., Lt. Gen. R. C. Wilson, DCS/Dev., Hqs. USAF to Maj. Gen. J. W. Sessums Jr., Vice Cmdr., ARDC, 23 July 1958, subj.: Dyna-Soar Program, Doc. 29.

8.   Ltr., Lt. Gen. D. L. Putt, DCS/Dev. Hqs. USAF to Dr. H. L. Dryden, Dir., NACA, 31 Jan. 1958, subj.: Dyna-Soar Program, Doc. 20; TWX, RDZPR-2-9-E, Hqs. ARDC to Cmdr., Det. 1, Hqs. ARDC, 13 Feb. 1958, Doc. 22; memo., Gen. T. D. White, CS/AF and Dr. H. L. Dryden, Dir., NACA, 20 May 1958, subj.: Principles for Participation of NACA in the Development and Testing of the "Air Force System 464L Hypersonic Boost Glide Vehicle (Dyna-Soar I)," Doc. 24A; memo., Gen. White and T. K. Glennan, Administrator, NASA, 14 Nov. 1958, subj.: Principles for Participation of NASA in the Development and Testing of the "Air Force System 464L Hypersonic Boost Glide Vehicle (Dyna-Soar I)," Doc. 40.

9.   Ltr., Lt. Gen. S. E. Anderson, Cmdr., ARDC to Lt. Gen. R. C. Wilson, DCS/Dev., Has. USAF, 24 July 1958, subj.: Fiscal Year 1959 Dyna-Soar Funding, Doc- 30.

10.   Ltr., Lt. Gen. R. C. Wilson, Hqs. USAF to Lt. Gen. S. E. Anderson, Cmdr., ARDC, 12 Aug. 1958, subj.: Dyna-Soar Funding, Doc. 32.

11.   Ltr., Maj. Gen. V. R. Haugen, Asst. Dep. Cmdr/Weap. Sys., Det. 1, Hqs. ARDC to DCS/Dev. Hqs. USAF, 6 Aug. 1958, subj.: Status of the Dyna-Soar Program, Doc 31.

12.   Ltr., Col. J. L. Martin, Jr., Acting Dir., Adv. Tech., Hqs. USAF to Cmdr., Det. 1, Hqs. ARDC, 4 Sept. 1958, subj.: Action on Dyna-Soar, Doc 33.

13.   TWX RDZSXB-3117, Cmdr., Det. 1, Hqs. ARDC to Hqs. USAF, 16 Sept. 29583 Doc. 34.

14.   TWX, AFDAT-58885, Hqs. USAF to Cmdr., Det. 1, Hqs. ARDC, 30 Sept. 1958, Doc. 35.

15. Chronology, Dyna-Soar WSPO, ident. no. 60WWZR-14775 Z n.d., subj.: Dyna-Soar Program, August 1957 to June 1960, hereafter cited as Dyna-Soar Chronology; DSM WAR, 31 Oct. 1958.

16. DSH WAR, 14 Nov. 1958; Dyna-Soar Chronology.

17. Preliminary Development Plan, System 464L, Det. 1, Hqs. ARDC, Nov. 1958, pp. 2, 8, 11, 32.

18. Dyna-Soar Chronology; TWX, AFDAT-54520, Hqs. USAF to Cmdr., Det. 1, Hqs. ARDC, 30 Dec. 1958, Doc. 41.

19. TWX, RDZSXB-12-31619-E, Cmdr., Det. 1, Hqs. ARDC to Hqs. USAF, 31 Dec. 1958, Doc. 42.

20. Memo., Dep. Secy. of Def. to SAF, 7 Jan. 1959, subj.: Release of Dyna-Soar Funds, Doc. 44.

21. TWX, AFDAT-55061, Hqs. USAF to Cmdr., Det. 1, Hqs. ARDC, 14 Jan. 1959, Doc. 47; TWX, AFDAT-48049, Hqs. USAF to Cmdr., Det. 1, Hqs. ARDC, 16 Jan. 1959, Doc- 49.

22. TWX, RDZXSB-30164-E, Cmdr., Det. 1, Hqs. ARDC to Hqs. USAF, 6 Feb. 1959, Doc. 51.

23. TWX, Cmdr., Det. 1, Hqs. ARDC to Hqs. USAF, 13 Feb. 1959, Doc. 53.

24. TWX, AFDAT-56643, Hqs. USAF to Cmdr., Det. 1, Hqs. ARDC, 17 Feb. 1959, Doc. 54.

25. GOR 92, revised, Hqs. USAF, 17 Feb. 1959, Doc. 55.

26. TWX, AFDAT-59299, Hqs. USAF to Cmdr., ARDC, 24 Apr. 1959, Doc. 61.

27. SR 201, Hqs. ARDC, 7 May 1959, Doc. 64.

28. Ltr., Maj. Gen. V. R. Haugen, Asst. Dep. Cmdr/Weap. Sys., Det. 1, Hqs. ARDC to Hqs. USAF, 15 May 1959, subj.: Dyna-Soar I Program Guidance, Doc. 65.

29. Rpt., DSM, Det. 1, Hqs. ARDC n.d., subj.: Dyna-Soar, Substantiation of RDZ Program, Doc. 66.

30. TWX, AFDAT-51437, Hqs. USAF to Cmde., Det. 1, Hqs. ARDC, 28 May 1959, Doc. 67.

31. DSM WAR, 12 June 1959; Dyna-Soar Chronology.

32. Presn., W. E. Lamar, Dep. Dir/Dev., Dyna-Soar SPO, ASD to Hqs. AFSC, 20 Feb. 1963, subj.: History of Dyna-Soar (X-20) to Present; memo., Lt. Col. B. H. Ferer, Asst. Boost-Glide Sys. DCS/Dev., Hqs. USAF, 24 June 1959, subj.: Dyna-Soar Briefing to Dr. Charyk, Tuesday, 23 June 1959, Doc. 69.

33. TWX, AFDAT-53166, Hqs. USAF to Cmdr., Det. 1, Hqs. ARDC, 10 July 1959, Doc. 70.

34. TWX, RDG-23-7-5-E, Hqs. ARDC to Hqs. BMD, 23 July 1959, Doc. 71; rpt., DSM, Det. 1, Hqs. ARDC, ident. no. 59RDZ-23128 n.d., subj.: Dyna-Soar Meeting, July-August 1959, Doc. 76.

35. Memo.; Dr. H. F. York, DDR&E, DOD to SAF and Dir., ARPA, DOD, 27 July 1959, subj.: Saturn, Dyna-Boar Propulsion, Doc. 72.

36. Dyna-Soar Chronology.

37. Ltr., Brig. Gen. C. H. Terhune, Jr., Vice Cmdr., BMD to Det. 1, Hqs. ARDC, 13 Aug. 1959, subj.: Dyna-Soar Booster Evaluation, Doc. 75.

38. Ltr., Brig. Gen. H. A. Boushey, Dir., Adv. Tech., DCS/Dev., Hqs. USAF to Hqs. ARDC, 29 Oct. 1959, subj.: Required Action on Dyna-Soar, Doc. 83.

39. Ltr., Lt. Gen. W. F. McKee, Vice Cmdr., AMC to Cmdr., ARDC, 23 Sept. 1959, subj.: The Dyna-Soar Program, Doc. 78.

40. TWX, Lt. Gen. B. A. Schriever, Cmdr., ARDC to Lt. Gen. S. E. Anderson, Cmdr., AMC, 12 Oct. 1959, Doc. 79.

41. TWX, Lt. Gen. S. E. Anderson, Cmdr., AMC to Lt. Gen. B. A. Schriever, Cmdr., ARDC, 14 Oct. 1959, Doc. 80.

42. Ltr., Brig. Gen. Boushey to Hqs. ARDC, 29 Oct. 1959, subj.: Required Action on Dynes-Soar, Doc. 83.

43. Abbreviated Development Plan, System 464L, Dynes-Soar WSPO, Det. 1, Hqs. ARDC, 1 Nov. 1959.

44. DSM WAR, 6 Nov. 1959.

45. Ltr., Lt. Gen. B. A. Schriever, Cmdr., ARDC and Lt. Gen. S. E. Anderson, Cmdr., AMC to Hqs. USAF, 4 Nov. 1959, subj.: Dyna-Soar Selection, Doc. 84.

46. DSM WAR, 13 Nov. 1959; TWX, RDZSXB-31253-E, Det. 1, Hqs. ARDC to Hqs. BMD, 10 Nov. 1959, Doc. 87; TWX, RDZSXB-31261-E, DSM, Hqs. ARDC to Hqs. SAC, 13 Nov. 1959, Doc. 88.

47. TWX, AFDAT-90938, Hqs. USAF to Cmdr., ARDC, 17 Nov. 1959, Doc. 90.

48. Memo., Dr. J. V. Charyk, ASAF/Res. & Dev. to DCS/Dev., Hqs. USAF, 20 Nov. 1959, subj.: D&F 50-11-284 Dyna-Soar program, Doc. 92.

49. Memo., Lt. Col. B. H. Ferer, Asst/Boost-Glide Sys., DCS/Dev., Hqs. USAF, 24 Nov. 1959, subj.: Assistant Secretary Air Force R&D Restrictions on Dyna-Soar Determinations and Findings, Doc. 93.

50. TWX, AFDAT-95166, Hqs. USAF to Cmdr., ARDC, 7 Dec. 1959, Doc. 96.

CHAPTER III

ALPHA TO ONE

Before the Dyna-Soar Weapon System Project Office could undertake the suborbital Step I of the program, the Air Force had to institute Phase Alpha and appraise the Dyna-Soar approach to eventual manned, orbital flight. Early in December 1959, the Aero and Space Vehicles Panel of the Scientific Advisory Board offered some recommendations concerning the objectives of this study. The panel pointed to the inadequacy of technical knowledge in the areas of aerodynamics and structures and, consequently, considered that development test programs to alleviate these deficiencies should be formulated during the study. Concerning the entire program, the scientific advisory group strongly supported the Dyna-Soar approach. While the program could be severely limited by a restricted budget and the absence of a high military priority, the Aero and Space Vehicles Panel insisted that Dyna-Soar was important because, if properly directed, it could yield significant information in the broad research areas of science and engineering.[1]

Dr. J. V. Charyk, Assistant Secretary of the Air Force for Research and Development, concurred with the position of the panel. In Alpha, emphasis would be placed on the identification and solutions of technical problems, and the objective of Step I would be the development of a test vehicle rather than a weapon system. Dr. Charyk then authorized the release of an additional $2.5 million for this study.[2]

On 11 December 1959, the Air Force and the Boeing Airplane Company had already signed a contract for the Alpha study, but the Air Force was undecided as to which contractors or Air Force agencies would provide Boeing with booster analyses. By the end of January 1960, the Dyna-Soar office recommended that the Ballistic Missile Division and the Space Technology Laboratories provide the booster studies. Since Alpha had to be completed in March 1960, the project office did not consider that there was sufficient time to complete a contract with Martin for the Alpha study.[3] The Aeronautical Systems Center objected and maintained that the existing contracts with Boeing could not be extended to allow participation in booster studies.[4] Command headquarters disagreed and resolved the issue on 3 February: the Ballistic Missiles Center would arrange contracts with the space laboratories and the Martin Company and the Aeronautical Systems Center would extend the Boeing contract.[5]

Booster information for Alpha was not the only problem; ARDC headquarters still had to settle the question of booster procurement for the entire Dyna-Soar program. Lieutenant General B. A. Schriever, Commander of ARDC, and Lieutenant General S. E. Anderson, Commander of AMC, had apparently delineated the authority of their respective commands in their 4 November 1959 letter, but a formal agreement had not been reached. Early in December 1959, General Schriever had completed an agreement within his command which assigned technical responsibility for booster development to the Ballistic Missiles Division. General Schriever hoped that General Anderson also intended to delegate commensurate contractual authority to the Ballistic Missiles Center.[6] General Anderson was essentially in agreement with General Schriever's position, but he objected to an agreement made between the ARDC project office and the ballistic division without participation of AMC elements. Consequently, the air materiel commander urged that the two commands complete a joint agreement concerning the development of the Dyna-Soar booster.[7]

On 8 February 1960, Generals Schriever and Anderson reached such an understanding which detailed the position of the west coast complex in the Dyna-Soar program. While management and financial authority for the entire program rested in the weapon system project office, the ballistic division and center, with the approval of the system office, would define the statements of work and complete contractual arrangements for booster development. All changes in the booster program which significantly altered performance, configuration, cost, or schedules, however, would necessitate concurrence of the project office.[8]

In the middle of January 1960, Brigadier General H. A. Boushey, Assistant for Advanced Technology in Air Force headquarters, gave more specific instructions concerning the direction of the Phase Alpha study. The objective of this review was to examine selected configurations for controlled, manned re-entry, to determine the technical risks involved in each, and to define a development test program for Step I.[9] In order to evaluate the efforts of Boeing, Martin, the ballistic division, and the space laboratories in this study, Colonel W. R. Grohs, Vice Commander of the Wright Aeronautical Development Division (WARD), then directed the formation of an ad hoc committee.[10]*

This group was established early in February with representation not only from the Wright division but also from the Air Force Flight Test Center, the Air Force Missile Test Center, the Air Materiel Command, and the National Aeronautics and Space Administration. The central objective of this committee was to determine the kind of research vehicle the Air Force required to solve the problems involving manned re-entry from orbital flight. Consequently, the ad hoc committee contracted with several companies, which were placed under the direction of Boeing, to investigate the potentialities of several categories of configurations. Variable geometric shapes such as the drag brake of the AVCO Manufacturing Corporation, a folding-wing glider of Lockheed Aircraft, and an inflatable device of Goodyear Aircraft were all examined. The committee also analyzed ballistic shapes such as a modified Mercury Capsule of McDonnell and lifting body configurations offered by the ad hoc committee itself and General Electric. Finally, gliders with varying lift-to-drag ratios were also proposed by the committee, Bell Aircraft, Boeing, and Chance-Vought Aircraft.

After examining these various configurations, the ad hoc group concluded that the development and fabrication of a ballistic shape or a lifting body configuration with a lift-to-drag ratio up to 0.5 would only duplicate the findings of the National Aeronautics and Space Administration in its Mercury program. Conversely a glider with a high lift-to-drag ratio of 3.0 would not only provide a maximum amount of information on re-entry but would also demonstrate the greatest maneuverability in the atmosphere and allow the widest selection of landing sites. Such a glider, however, presented the most difficult design problems. Consequently, the ad hoc committee decided that a medium lift-to-drag glider, in the range of 1.5 to 2.5, offered the most feasible approach for advancing knowledge of re-entry problems.[11]

At the end of March 1960, the Aero and Space Vehicles panel again reviewed the Dyna-Soar program with emphasis on the results of the Alpha study. If the overriding requirement were to orbit the greatest weight in the shortest development time, the panel reasoned that the modified ballistic approach was preferable. However, the members noted that gliders would advance technical knowledge of structures and would provide the greatest operational flexibility. The vehicles panel further emphasized the importance of attaining early orbital flight and, consequently, suggested a re-examination of the need for a sub-orbital Step I and more precise planning for the orbital Step II.[12]

The Dyna-Soar glider, as conceived by the Alpha group and the project office, was to be a low-wing, delta-shape vehicle, weighing about 10,000 pounds. To undergo the heating conditions during re-entry, the framework was to be composed of René 41 braces which would withstand a temperature of 1,800 degrees Fahrenheit. The upper surface of the glider was to be fabricated of René 41 panels, where the temperature was expected to range from 500 to 1,900 degrees. The lower surface was to be a heat shield, designed for a maximum temperature of 2,700 degrees and was to consist of molybdenum sheets attached to insulated René 41 panels. The leading edge of the wings would have to withstand similar heat conditions and was to be composed of coated molybdenum segments.

The severest temperature, ranging from 3,600 to 4,300 degrees, would be endured by the nose cap, which was to be constructed of graphite with zirconia rods.[13]

In conjunction with the ad hoc group, the Dyna-Soar project office completed, by 1 April 1960, a new development plan which further elaborated the three-step program presented in the November 1959 approach. Step I was directed towards the achievement of four objectives: exploration of the maximum heating regions of the flight regime, investigation of maneuverability during re-entry, demonstration of

---

*   With the formation of the Wright Air Development Division, on 15 December 1959, the management of weapon system development was transferred from ARDC headquarters to the Wright complex.

conventional landing, and evaluation of the ability of man to function usefully in hypersonic flight. While Step I was limited to suborbital flight, the purpose of Step IIA was to gather data on orbital velocities and to test military subsystems, such as high resolution radar, photographic and infrared sensors, advanced bombing and navigation systems, advanced flight data systems, air-to-surface missiles, rendezvous equipment, and the requisite guidance and control systems While Step IIB would provide an interim military system capable of reconnaissance and satellite inspection missions, the objective of Step III was a fully operational weapon system.

Whereas the last two steps were only outlined, the main consideration of the project office was the suborbital Step I. In order to demonstrate the flying characteristics of the glider up to speeds of mach 2, the Dyna-Soar office scheduled a program of 20 air-drop tests from a B-52 carrier to begin in July 1963.* Beginning in November 1963, five unmanned flights were to be conducted to Mayaguana in the Bahama Islands and Fortaleza, Brazil, with velocities ranging from 9,000 to 19,000 feet per second. Eleven piloted flights, scheduled to start in November 1964, would then follow, progressively increasing the velocity to the maximum 19,000 feet per second and employing landing sites in Mayaguana, Santa Lucia in the Leeward Islands, and, finally, near Fortaleza.

To accomplish this Step I program, the Dyna-Soar office estimated that $74.9 million would be required for fiscal year 1961, $150.9 million for 1962, $124.7 million for 1963, $73.6 million for 1964, $46.8 million for 1965, and $9.9 million for 1966. Including $12.8 million for 1960, these figures totaled $493.6 million for the suborbital program.

During the first week in April 1960, officials of the Dyna-Soar project office presented the new development plan and the results of Phase Alpha to Generals Schriever, Anderson, and Boushey, and the Strategic Air Panel and the Weapons Board of Air Force headquarters. On 8 April, Dyna-Soar representatives explained the program to the Assistant Secretary of the Air Force for Research and Development, now Professor C. D. Perkins, and received his approval to begin work on the suborbital Step I.[15] On 19 April, the Assistant Secretary of the Air Force for Materiel, P. B. Taylor authorized negotiations of fiscal year 1961 contracts for this phase of the program.** The Department of Defense, on 22 April, endorsed the new program and permitted the release of $16.2 million of fiscal year 1960 funds.[16] Consequently, on 27 April, the Air Force completed a letter contract with the Boeing Airplane Company as system contractor. Source selection procedures had previously been initiated for the award of two associate contracts. On 6 December 1960, the Air Force granted authority to the Minneapolis-Honeywell Regulator Company for the primary guidance subsystem, and, on 16 December, the Air Force gave responsibility to the Radio Corporation of America for the communication and data link subsystem.***

Air Force headquarters, on 21 July 1960, further recognized the three step program by issuing System Development Requirement 19. With the segmented approach, the Air Force could develop a manned glider capable of demonstrating orbital flight, maneuverability during hypersonic glide, and controlled landings. Furthermore, Dyna-Soar could lead to a military system able to fulfill missions of space maneuver and rendezvous, satellite inspection, and reconnaissance. Headquarters looked forward to the first manned, suborbital launch which was to occur in 1964.[17]

While the Step I program was approved and funded, the Dyna-Soar project office firmly thought that studies for the advanced phases of the program should also be initiated. In early August 1960, the project

---

*   For the air-drop program, the Dyna-Soar office was considering employment of either the XLR-11 or the AR-1 liquid rocket engines to propel the glider to specified speeds. Late in 1960, however, the project office decided to use a solid acceleration rocket not only for abort during launch but also for the air-drop tests.

**   On 24 April 1961, Dr. Charyk, then Under Secretary of the Air Force, permitted contractual arrangements for the entire Step I program rather than for only particular fiscal years.

***   The Air Force granted three other associate contracts for the Dyna-Soar program. On 8 June 1960, the Martin Company received responsibility for the booster airframe, while, on 27 June, the Air Force authorized the Aero-Jet General Corporation to develop the booster engines. Previously, on 9 June, the Air Force made arrangements with the Aerospace Corporation to provide technical services for the Step I program.

office recommended to ARDC headquarters that $2.32 million should be made available through fiscal year 1962 for this purpose. If these funds were released immediately, the project office anticipated completion of preliminary program plans for Steps IIA, IIB, and III by December 1961, January 1962, and June 1962, respectively.[18] Later in the month, the Dyna-Soar office again reminded command headquarters of the urgency in releasing these funds.[19]

The apparent source of delay was that the authority to negotiate contracts, issued by Assistant Secretary Taylor on 19 April 1960, referred specifically to Step I of the program. Colonel E. A. Kiessling, Director of Aeronautical Systems in ARDC headquarters, met with Professor Perkins on 22 and 23 September, and the assistant secretary agreed that this authority did not prohibit Step II and III studies. The restraint only applied to the expenditure of fiscal year 1961 funds for the purchase of equipment for the advanced phases.[20]* This decision was confirmed on 12 October when Air Force headquarters approved Steps II and III studies by issuing Development Directive 41122**. ARDC headquarters then issued, on 6 December, a system study directive for Step III and allotted $250,000 for this work.[24] By the middle of 1961, however, it was questionable whether the Air Force would continue the three-step approach. The Air Staff consequently postponed the Step III investigation, and, early in 1962, command headquarters canceled the study.[25]

In the April 1960 development plan, the Dyna-Soar office had proposed the employment of Titan I as the Step I booster. The first stage of this system was powered by the LR87-AJ-3 engine, capable of developing 300,000 pounds of thrust, while the second stage, an LR91-AJ-3 engine, could produce 80,000 pounds of thrust. This booster would be able to propel the Dyna-Soar glider to a velocity of 19,000 feet per second on a suborbital flight from Cape Canaveral to Fortaleza, Brazil. Professor Perkins, however, considered this booster marginal for Step I flights and, on 28 November 1960, requested the Air Force to examine the feasibility of employing Titan II for the suborbital step and a combination Titan II first stage and a Centaur-derivative upper stage for the orbital phase.[26] The Titan II was a two-stage liquid rocket and, unlike the Titan I, employed hypergolic, storable propellants. The first stage consisted of an XLR87-AJ-5 engine, capable of producing 430,000 pounds of thrust, while the second stage was an XLR91-AJ-5 unit, capable of delivering 100,000 pounds of thrust.

Late in December 1960, Mr. R. C. Johnston of the Dyna-Soar office and Major G. S. Halvorsen of the Ballistic Missiles Division presented the advantages of Titan II to ARDC headquarters, and the proposal to employ the advanced Titan received the endorsement of General Schriever. A presentation to Air Force headquarters followed. Assistant Secretary Perkins appeared satisfied with the recommendation but stated that Department of Defense approval would probably not be given unless the booster change was considered in conjunction with an anticipated funding level of $70 million for fiscal year 1962, instead of the requested $150 million.[27]

A few days later, the project office protested the $70 million level and insisted that it would result in serious delays to the program. Regardless of the funding arrangements, The Dyna-Soar office urged approval of Titan II.[28] Colonel Kiessling concurred with this position and appealed to USAF headquarters. Even with the proposed low funding level, the Director of Aeronautical Systems stated, employment of the Titan II promised a substantially improved Dyna-Soar program and this booster change should be immediately approved.[29]

* Colonel T. T. Omohundro, Deputy Director for Aeronautical Systems, ARDC headquarters, informed the Dyna-Soar office, on 4 October 1960, that Air Force headquarters would probably have to issue a new authority to negotiate contracts for Step II and III studies before funds could be released. Apparently, Colonel Kiessling had not told his deputy of Professor Perkins' previous decision.[21]

** On 14 February 1961, the Air Force and Boeing completed a contract for Step IIA and IIB studies with an effective date of 9 November 1960. Boeing was allotted $1.33 million and given until 30 June 1962 to complete the studies. With the assumption that a new orbital booster would provide Step II propulsion, Boeing concluded that it was feasible for the Dyna-Soar glider to perform military missions such as reconnaissance, satellite interception and inspection, space logistics, and bombardment. The last mission, however, the contractor considered could be performed with less expense by intercontinental ballistic missiles.[23]

Mr. Johnston and Major Halvorsen again went to Air Force headquarters. After receiving the approval of Major General M. C. Demler, Director of Aerospace Systems, the Dyna-Soar representatives informed the Strategic Air Panel of the attributes of Titan II. Discussion of the panel centered on the availability of the new booster for Step I flights, limitations of the combination Titan II and Centaur-derivative for the orbital booster, and the apparent inadequate funding level for fiscal year 1962. In spite some doubts, the panel approved the proposed booster for Dyna-Soar I and further recommended that approximately $150 million should be allocated for fiscal year 1962.[30]

At the request of Assistant Secretary Perkins, General Demler had prepared a summary on the advantages of Titan II over the earlier version. The Director of Aerospace Systems insisted that Titan I was barely sufficient for achieving the objectives of Step I and, furthermore, could not be modified to provide orbital velocities for the glider. The April 1960 development plan had stipulated that with Titan I the first unmanned ground-launch would occur in November 1963, while employment of the more powerful Titan II would only push this date back to January 1964. General Demler pointed out that if the program were limited to $70 million, October 1964 would be the date for the first unmanned ground-launch with Titan I, while December 1964 would be the date for Titan II. The aerospace director estimated that with a $150 million level for fiscal year 1962, the development of Titan II would cost an additional $33 million, while the cost would still be $26 million with the $70 million funding level. General Demler considered that the total booster cost for Step I and II employing the Titan I and then a Titan II-Centaur combination would be $320.3 million. If Titan II were immediately used for Step I, the booster cost would be $324.3 million. Thus the additional cost for using the more powerful booster in the first phase of the Dyna-Soar program only amounted to $4.2 million. The conclusion was obvious, however, General Demler refrained from making recommendations.[31]

Following the briefing to the Strategic Air Panel, Mr. Johnston and Major Halvorsen gave the Titan II presentations to the Weapons Board. The members were familiar with the logic of General Demler's summary, and, while expressing interest in the early attainment of orbital flight, they endorsed the change to Titan II. The board recommended that Air Force headquarters immediately instruct ARDC to adopt the new booster.[32]

However, Major General V. R, Haugen and Colonel B. H. Ferer, both in the office of the Deputy Chief of Staff for Development, decided to seek the approval of the Department of Defense. The Titan II presentations were then given to Mr. J. H, Rubel, Deputy Director of Defense for Research and Engineering. While reiterating the necessity of a $70 million budget, Mr. Rubel, agreed to the technical merits of Titan II. On 12 January 1961, Air Force headquarters announced approval of this booster for Step I flights.[33]

During these discussions over Titan II, it was apparent that the Department of Defense was seriously considering limiting the fiscal year 1962 figure to $70 million. This financial restriction was confirmed on 3 February, when Air Force headquarters directed the Dyna-Soar office to reorient the Step I program to conform with this lower funding level.[34] By the end of the month, the project office and the Dyna-Soar contractors had evaluated the impact of this reduction on the program. It was clear that flight schedules would be set back almost one year.[35]

Apparently Department of Defense officials relented, for, on 28 March 1961, Air Force headquarters announced that the fiscal year 1962 level would be set at $100 million. The following day, Colonel W. L. Moore, Dyna-Soar director, and his Deputy Director for Development, W. E. Lamar, reported on the status of the program to Air Force headquarters. Both Dr. Charyk and Major General Haugen directed that the program be established on a "reasonable" funding level. Colonel Moore noted that a definition of this statement was not offered.[36] Finally, on 4 April, headquarters of the Air Force Systems Command (AFSC) officially instructed the program office to redirect Dyna-Soar to a $100 million level for fiscal year 1962.[37]*

---

*   On 1 April, 1961, the Air Research and Development Command, by acquiring the procurement and production functions from the Air Materiel Command, was reorganized as the Air Force Systems Command. At Wright-Patterson Air Force Base, the Wright Air Development Division combined with the Aeronautical Systems Center to become the Aeronautical Systems Division (ASD).

By 26 April 1961, the Dyna-Soar office had completed a system package program. This plan further elaborated the familiar three-step approach. Step I would involve suborbital missions of the Dyna-Soar glider boosted by the Titan II. For the research and development of this program, the Dyna-Soar office stated that $100 million was required for fiscal year 1962, $143.3 million for 1963, $114.6 million for 1964, $70.7 million for 1965, $51.1 million for 1966, and $9.2 million for 1967. If these funds were allotted, the first air-drop would take place in January 1964, the first unmanned ground-launch in August 1964, and the first manned ground-launch in April 1965.

The objective of Step IIA was to demonstrate orbital flight of the Dyna-Soar vehicle on around-the-world missions from Cape Canaveral to Edwards Air Force Base. The program office proposed the testing, on these flights, of various military subsystems such as weapon delivery and reconnaissance subsystems. Because of high cost, the Dyna-Soar office did not recommend the evaluation of a space maneuvering engine, space-to-earth missiles, or space-to-space weapons during Step IIA flights.

For fiscal years 1963 through 1968, the program office estimated that this phase of Step II would total $467.8 million and, assuming the selection of the orbital booster by the beginning of fiscal year 1962, reasoned that the first manned, orbital flight could be conducted in April 1966.

In Step IIB, the Dyna-Soar vehicle would provide an interim operational system capable of fulfilling reconnaissance, satellite interception, space logistics, and bombardment missions. With the exception of $300,000 necessary for an additional Step IIB study, the Dyna-Soar office did not detail the financial requirements for this phase, however, it did anticipate a Step IIB vehicle operating by October 1967. The program office looked further in the future and maintained that $250,000 would be necessary for each fiscal year through 1964 for studies on a Step III weapon system, which could be available by late 1971.[38]

In the April 1961 system package program, the Dyna-Soar office outlined an extensive Category I program, consisting of structural and environmental, design, and aerothermodynamic testing, which was necessary for the development of the glider. In order to verify information obtained from this laboratory testing, the system office recommended participation in another test program which would place Dyna-Soar models in a free-flight trajectory.[39] The first approach which the Dyna-Soar office considered was System 609A of the Ballistic Missiles Division.

During the March 1960 review, the Aero and Space Vehicles Panel emphasized the difficulty in predicting behavior of structures utilizing coated heat shields and recommended Dyna-Soar participation in the 609A program.[40] The system office agreed and decided to place full-scale sections of the glider nose on four hyper-environmental flights.[41]*

Although subsequent planning reduced the number to two flights, command headquarters refused to release funds for such tests, and, consequently, Colonel Moore terminated Dyna-Soar flights in the System 609A test program on 5 October. The project director gave several reasons for this decision: low probability of obtaining sufficient data with only two flights, insufficient velocity of the boosters, and high cost for Dyna-Soar participation.[43]

Air Force headquarters was concerned over this cancellation and emphasized to ARDC headquarters that the absence of a free-flight test program for Dyna-Soar failed to carry out assurances previously given to the Department of Defense.[44] The National Aeronautics and Space Administration had another approach which it had been proposing since May 1960. Dyna-Soar models constructed by both NASA and the Air

---

*   Models of the AVCO drag brake were also scheduled to ride 609A launches. In February 1960, Air Force headquarters had transferred the management of this project from the Directorate of Advanced Systems Technology, WADD, to the Dyna-Soar Weapon System Project Office. In March, the Air Force granted AVCO a study contract, and, in July, ARDC headquarters approved a development program for the drag brake. Air Force headquarters was reluctant to authorize funds, and the program was terminated in December. Nevertheless, in February 1961, Major General J. R. Holzapple, WADD Commander, reinstated research on certain technical areas of the drag brake program.[42]

Force would be placed on RVX-2A re-entry vehicles and boosted by Atlas or Titan systems. Project office engineers could thereby obtain data on heat transfer and aerodynamic characteristics. By November 1960, the Dyna-Soar office was seriously considering verification of laboratory data by this RVX-2A program.[45]

In May 1961, Major General W. A. Davis, ASD Commander, emphasized to AFSC headquarters the requirements for RVX-2A tests: funds and space on Titan II launches. [45] After two more appeals by the program office, Major General M. F. Cooper, Deputy Chief of Staff for Research and Engineering gave the position of AFSC headquarters. Placing a re-entry vehicle with Dyna-Soar models on the Titan II would impose several limitations on the test schedule of the booster, requiring several modifications to the airframe and the launch facilities. General Cooper further stated that the $10 million estimated by NASA officials for the RVX-2A program would necessitate approval by Air Force headquarters. Consequently, General Cooper intended to incorporate this program in a future Dyna-Soar development plan. The RVX-2A proposal was included in a 7 October 1961 plan for the development of a Dyna-Soar weapon system; however, this program did not receive the approval of USAF headquarters.[48] The attempt by the Dyna-Soar office to provide a specific program for free-flight verification of its laboratory test data ended at that point.

The April 1961 system package program also reflected changes in the Dyna-Soar flight plan. While 20 air-drop tests were still scheduled, only two unmanned ground-launches, instead of the previously planned four, were to be conducted. On the first flight, the Titan II would accelerate the glider to a velocity of 16,000 feet per second, reaching Santa Lucia. During the second unmanned launch, the vehicle would attain a velocity of 21,000 feet per second and land near Fortaleza. Twelve manned flights were then planned with velocities ranging from 16,000 to 22,000 feet per second. If the two additional vehicles for unmanned launches were not expended, additional piloted flights would then take place.[50]

The scheduling of flights to Fortaleza, however, was becoming academic. As early as June 1960, Air Force headquarters notified ARDC headquarters that the State Department was concerned over the problem of renewing an agreement with Brazil for American military use of its territory.[51] This subject reappeared in May 1961 when the acting Director of Defense for Research and Engineering, J. H. Rubel, informed the Department of the Air Force that discussions with State Department officials indicated the difficulty, if not the impossibility, of obtaining a landing site for Dyna-Soar in Brazil.[52] Unless Air Force headquarters would tolerate increased costs, reduced flight test objectives, or employment of a new booster, the Dyna-Soar office thought that a landing field in Brazil was essential. The program office stated that employment of alternative landing sites would seriously affect the conduct of Category II flights and would probably prevent attainment of important research objectives.[53] Although Dr. Brockway McMillan, Assistant Secretary of the Air Force for Research and Development, reiterated this position to the Department of Defense, the subject of a Fortaleza landing site did not assume a greater significance because the Air Force was already seriously questioning the need for suborbital flight.[54]

From January 1960 through April 1961, the Dyna-Soar program office had defined the three-step program and had implemented the suborbital phase. While Air Force headquarters had approved the April 1960 development plan, it had not sanctioned the more detailed April 1961 system package program. The reason for this suspended action was apparent. The Dyna-Soar office was engaged in a study which promised to eliminate suborbital flight, accelerate the date for the first manned orbital launch, and, consequently, radically alter the three-step approach.

CHAPTER III

REFERENCES

1.   Rpt., Aero & Space Vehicles Panel, SAB, Jan. 1960, subj.: Dyna-Soar.

2.   Memo., J. V. Charyk, ASAF/Res. & Dev., to DCS/Dev., Hqs. USAF, 4 Jan. 1960, subj.: Dyna-Soar "Phase Alpha" Program, Doc. 103.

3.   DSM WAR, 18 Dec. 1959, 15 Jan. 1960; TWX, RDZSXB-30061-E, Hqs. WADD to Hqs. ARDC, 20 Jan. 1960, Doc. 109.

4. TWX, RDZSXB-1-1420-E, Hqs. WARD to Hqs. ARDC, 28 Jan. 1960, Doc. 110.

5. D5M WAR, 5 Feb. 1960.

6. Ltr., Lt. Gen. B. A. Schriever, Cmdr., ARDC to Lt. Gen. S. E. Anderson, Cmdr., AMC, 5 Dec. 1959, subj.: Management of Dyna-Soar Project, Doc. 95; memo., Hqs. ARDC and Hqs. BMD n.d., subj.: Statement of Agreement, System 620A, Dyna-Soar I, for the Air Force Ballistic Missiles Division, ARDC, Doc. 95A.

7. Ltr., Lt. Gen. S. E. Anderson, Cmdr., AMC to Lt. Gen. B. A. Schriever, Cmdr., ARDC, 11 Dec. 1959, subj.: Management of the Dyna-Soar Program, Doc. 98.

8. Memo., Lt. Gen. S. E. Anderson, Cmdr., AMC and Lt. Gen. B. A. Schriever, Cmdr., ARDC, subj.: AMC and ARDC Management Procedures for the Dyna-Soar Program, 8 Feb. 1960, Doc. 114.

9. Ltr., Brig. Gen. H. A. Boushey, Asst/Adv. Tech., DCS/Dev., Hqs. USAF to Hqs. ARDC, 15 Jan. 1960, Doc. 107.

10. Ltr., Col. W. R. Grohs, Vice Cmdr., WADD to Dep/Adv. Sys., DSM, WARD, 27 Jan. 1960, subj.: Formation of Ad Hoc Group for Dyna-Soar.

11. DSM WAR, 5 Feb. 1960; interview, C. B. Hargis, Asst. Dep. Dir/Dev., Dyna-Soar SPO, ASD by C. J. Geiger, Hist. Div., ASD, 10 May 1963; presn., W. E. Lamar, Dep. Dir/Dev., Dyna-Soar SPO, ASD to Hqs. AFSC, 20 Feb. 1963, subj.: History of Dyna-Soar (X-20) to Present.

12. Rpt., Aero & Space Vehicles Panel, SAB, 15 Apr. 1960, subj.: Dyna-Soar and Phase Alpha Review.

13. System Development Plan, Dyna-Soar (Step I) Program, 620A, Dyna-Soar WSPO, WADD, 1 Apr. 1960, pp. III-8 to III-12, III-49 to III-52, III-65 to III-72, III-195 to III-207; System Package Program, System 620A, Dyna-Soar SPO, ASD, 26 Apr. 1961, pp. 84-92, 132-139, 182-187.

14. System Development Plan, Dyna-Soar (Step I) Program, 620A, Dyna-Soar WSPO, WADD, 1 Apr. 1960, pp. III-1 to III-2, III-267 to III-272, IV-1 to IV-2, V-i, VIII-1; IX-1.

15. DSM WAR, 8, 16 Apr. 1960.

16. TWX, AFDAT-89082, Hqs. USAF to Hqs. ARDC, 26 Apr. 1960, Doc. 124.

17. SDR 19, Hqs. USAF, 21 July 1960, Doc. 132.

18. Ltr., Dyna-Soar WSPO, WADD, to Hqs. ARDC, subj.: Dyna-Soar Step II and Step III Funding, 2 Aug. 1960, Doc. 136.

19. TWX, WWZRM-8-1041-E, Hqs. WADD to Hqs. ARDC, 23 Aug. 1960, Doc. 144.

20. Memo., Col. E. A. Kiessling, Dir., Aero. Sys., Hqs. ARDC, 27 Sept. 1960, subj.: Dyna-Soar Step II and Step III Funding.

21. Ltr., Col. Kiessling, Hqs. ARDC to Hqs. WADD, subj.: WADD Dyna-Soar Step II and Step III Funding, 4 Oct. 1960, Doc. 152.

22. DD 411, Hqs. USAF, 12 Oct. 1960, Doc. 159.

23. Rpt., Dyna-Soar SPO, ASD, Dec. 1962, subj.: Evaluation of the Dyna-Soar Step IIA and Step IIB Studies, pp. 1, 11.

24. SSD, Hqs. ARDC, 6 Dec. 1960, subj.: Dyna-Soar Step III Study, Doc. 176.

25. TWX, AFDAP-78950, Hqs. USAF to Hqs. AFSC, 16 June 1961, Doc. 234; TWX, SCLDA-8-1-2-E, Hqs. AFSC to Hqs. ASD, 8 Jan. 1962.

26. TWX, AFDSD—66668, Hqs. USAF to Hqs. ARDC, 5 Dec. 1960, Doc. 175.

27. Memo., R. C. Johnston, Ch., Booster Br., Dyna-Soar WSPO, WADD, 27 Dec. 1960, subj.: Record of Presentations on Proposed Use of Titan II as a Dyna-Soar Booster, Doc. 187.

28. TWX, WWZR-5-1-1081, Hqs. WADD to Hqs. ARDC, 5 Jan. 1961, Doc. 190.

29. TWX, RDRAS-6-1 8, Hqs. ARDC to Hqs. USAF, 6 Jan. 1961.

30. Memo., R. C. Johnston, Ch., Booster Br., Dyna-Soar WSPO, WADD, 12 Jan. 1961, subj.: Trip Report to Hqs. USAF, 9-11 Jan. 1961, Titan II Booster Presentation, Doc. 193.

31. Rpt., Maj. Gen. M. C. Demler, Dir., Aerospace Sys. Dev., DCS/Dev., Hqs. USAF n.d., subj.: System 620A, Dyna-Soar (Step I) Booster, Doc. 191.

32. Memo., R. C. Johnston, Ch., Booster Br., Dyna-Soar WSPO, WADD, 12 Jan. 1961, subj.: Trip Report to Hqs. USAF, 9-11 Jan. 1961, Titan II Booster Presentation, Doc. 193.

33. Ibid.; TWX, AFDSD-76644, Hqs. USAF to Hqs. ARDC, 12 Jan. 1961, Doc. 194.

34. TWX, AFDSD-82107, Hqs. USAF to Hqs. WADD, 3 Feb. 1961.

35. DSM WAR, 3 Mar. 1961.

36. Trip Rpt., Col. W. L. Moore, Jr., Dir., Dyna-Soar SPO, WADD to Brig. Gen. A. T. Culbertson, Dir., Sys. Mangt., WADD, 31 Mar. 1961, subj.: Presentation on Status of Dyna-Soar Program, Doc. 219.

37. TWX, SCRAS-3-4-2, Hqs. AFSC to Hqs. ASD, 4 Apr. 1961, Doc. 220.

38. System Package Program, System 620A, Dyna-Soar SPO, ASD, 26 Apr. 1961, pp- 42, 411, 446-447, 454, 460-474, 487a, 489-490, 502, 510, 519, 522.

39. Ibid., p. 189.

40. Rpt., Aero & Space Vehicles Panel, SAB, 15 Apr. 1960, subj.: Dyna-Soar and Phase Alpha Review.

4.1 Memo., W. E. Lamar, Ch., Dir. Sys. Engg., Dyna-Soar WSPO, WADD, n.d., subj.: Conference at NASA Regarding RVX-2 Test for Dyna-Soar, Doc. 165; System Development Plan, Dyna-Soar (Step I) Program, 620A, Dyna-Soar WSPO, WADD, 1 Apr. 1960, p. III-77; TWX, WWZRB-23-8-1017, Hq. WADD to Hqs. BMD, 23 Aug. 1960, Doc. 143.

42. Ltr., Col. W. L. Moore, Jr., Ch., Dyna-Soar WSPO, WARD to J. A. Boykin, DSM, WADD, 1 Feb. 1960, subj.: AVCO Drag Brake Project, Doc. 111; TWX, WWZRB-6-4-E, Hqs. WADD to Hqs. BMD, 14 June 1960, Doc. 129; memo., J. B. Trenholm, Jr., Asst. Dir., Dyna-Soar SPO, WADD, 21 Feb. 1961, subj.: AVCO Drag Brake-Project 6065, Doc. 206; ltr., C. B. Hargis, Jr., Asst. Ch., Dyna-Soar WSPO, WADD to Hqs. BMD, 1 Apr. 1960, subj.: AVCO Drag Brake Program, Booster Data, and Availability Requirements, Doc. 122; DSM WAR, 1 Apr. 1960, 1 July 1960, 2 Aug. 1960, 2 and 16 Sept. 1960, 6 Jan. 1961; TWX, RDRAS-14-12-26, Hqs. ARDC to Hqs. WADD, 14. Dec. 1962, Doc. 179.

43. Ltr., Col. W. L. Moore, Jr., Ch., Dyna-Soar WSPO, WADD to Hqs. ARDC, 5 Oct. 1960, subj.: Dyna-Soar HETS Program, Doc. 155.

44. TWX, AFDSD-AS/0-93477, Hqs. USAF to Hqs. ARDC, 13 Oct. 1960, Doc. 163.

45. Memo., W. E. Lamar, Ch., Dir. Sys. Engg., Dyna-Soar WSPO, WADD n.d., subj.: Conference at NASA Regarding RVX-2 Test for Dyna-Soar, Doc. 165; memo., NASA, 4 Nov. 1960, subj.: The Atlas-RVX-2 and Communications Experiments Program Proposed by NASA, Doc. 169; TWX, WWDR-7-11-2, Hqs. WADD to Hqs. ARDC, 8 Nov. 1960, Doc. 170.

46. Ltr., Maj. Gen. W. A. Davis, Cmdr., ASD to Hqs. AFSC, subj.: Aeronautical Systems Division Free Flight Environmental Test Program, 2 May 1961, Doc. 226.

47. Ltr., Maj. Gen. M. F. Cooper, DCS/Res. & Engg., Hqs. AFSC to Dr. I. H. Abbott, Dir., Adv. Res. Prog., NASA, 28 June 1961, subj.: Dyna-Soar Re-entry Research Program, Doc. 240.

48. Abbreviated Development Plan, System 620A, Dyna-Soar SPO, ASD, 7 Oct. 1961, subj.: Manned Military Space Capability, p. 64.

49. Memo., W. M. Ritchey, Asst. Test Force Dir., Dyna-Soar WSPO, WADD, 1 Nov. 1960, subj.: Meeting to Consider System Development Test Approach, Doc. 166; DSM WAR, 4 Nov. 1960.

50. System Package Program, System 620A, Dyna-Soar SPO, ASD, 26 Apr. 1961, pp. 306-307.

51. TWX, RDROF-9-6-4, Hqs. ARDC to Hqs. WADD, 9 June 1960, Doc. 127.

52. Memo., J. H. Rubel, Acting DDR&E, DOD to SAFUS, 3 May 1961, subj.: Dyna-Soar Landing Site in Brazil, in files Dyna-Soar Sys. Staff Ofc., DCS/Sys. & Log., Hqs. USAF.

53. TWX, WWZRT-25-5-1079, Hqs. ASD to Hqs. USAF, 25 May 1961, Doc. 231.

54. Memo., Brockway McMillan., ASAF/Res. & Dev. to DDR&E. DOD, 12 June 1961, subj.: Dyna-Soar Landing Site in Brazil, in files of Dyna-Soar Sys. Staff Ofc., DCS/Sys. & Log., Hqs. USAF.

CHAPTER IV

REDIRECTION

When Brigadier General M. B. Adams, Deputy Director of Systems Development in Air Force headquarters, forwarded Development Directive 411 in October 1960, he initiated a series of studies which eventually resulted in a redirection of the Dyna-Soar program. General Adams instructed the Air Research and Development Command to formulate a "stand-by" plan for achieving orbital flight with the Step I glider at the earliest possible date.[1] In December, the Dyna-Soar office was ready with such a proposal. By merging Steps I and IIA into a continuous development and employing an orbital booster for both suborbital and orbital flights, the time for the first, manned, orbital launch could be accelerated by as much as 17 months over the three-step schedules.[2]

Depending on either a March 1961 or a November 1961 approval date, Dyna-Soar officials estimated that by using a Titan II in combination with a Centaur derivative, the program would cost either $726 million or $748 million. If Saturn C-1 was designated, the figures would be $892 million or $899 million. The total, however, for a separate suborbital Step I and an orbital Step IIA would approximate $982.6 million. This financial difference between "stand-by" and the three-step approach stemmed from the employment of the same booster for both suborbital and orbital flights. The Dyna-Soar office favored this accelerated approach and recommended that ARDC headquarters immediately approve "stand-by."[3] Command headquarters did not agree and took the position that "stand-by" would only be approved when the international situation necessitated a higher priority and additional funds for Dyna-Soar.[4]

The logic of employing the same booster for Steps I and IIA pointed to a further conclusion. On 4 May 1961, Boeing officials proposed another plan for acceleration. This "streamline" approach encompassed the elimination of suborbital flight, temporary employment of available subsystems, and the use of Saturn C-1.

Assuming a June 1961 approval date, Boeing representatives anticipated the first unmanned, orbital flight to occur in April 1963, instead of August 1964 as scheduled in the three-step approach.[5]

Temporary subsystems would only decrease system reliability, the program office reasoned, and, consequently, BOEING'S proposal was not entirely acceptable. Dyna-Soar officials considered that the key to accelerating the orbital flight date was not only the question of booster availability, but also the time required to develop the various glider subsystems. If funding for fiscal year 1962 were increased, it would be possible to accelerate the glider schedules and advance the orbital flight date.

By the end of June, the program office had refined BOEING'S original plan. The first phase, "streamline," involved the development of an orbital research vehicle. The purpose of the second phase was the development and testing of military subsystems with the final phase resulting in an operational weapon system. Either a modified Saturn booster, a Titan II with a hydrogen-oxygen second stage, or a Titan II augmented by solid propellant engines, was acceptable for the "streamline" phase. The program office now estimated that this phase would cost a total of $967.6 millions with the first unmanned, orbital flight occurring in November 1963.[6]

While the Dyna-Soar office was considering ways to accelerate the orbital flight date of its glider, the newly established Space Systems Division (SSD) completed, on 29 May 1961, two development plans for demonstrating orbital and far-earth orbital flight of a lifting body design. Essentially, the objective of the Advanced Re-entry Technology program (ART) was to determine whether ablative or radiative heat protection was more feasible for lifting re-entry.[7] The second program advanced by SSD was a manned, satellite, inspector proposal, SAINT II.

The space division had under its cognizance a SAINT I program, the purpose of which was the development of an unmanned, prototype, inspector vehicle. The SAINT II proposal involved the development of a manned vehicle, capable of achieving precise orbital rendezvous and fulfilling space logistic missions. This lifting body would be able to maneuver during re-entry and accomplish conventional landing at a pre-selected site. Officials of the space division listed several reasons why the Dyna-Soar configuration could not, in their opinion, accomplish SAINT II missions. The re-entry velocity of Dyna-Soar could not be significantly increased because of the inadapability of this configuration to ablative heat protection. Furthermore, winged-configurations did not permit sufficient payload weights and incurred structural penalties to the booster. Finally, rendezvous and logistic missions would require prohibitive modifications to the Dyna-Soar glider.

The proposed SAINT II demonstration vehicle was to be a two-man, lifting, re-entry craft, launched by a Titan II and Chariot combination. This Chariot upper stage would employ fluorine and hydrazine propellants and would produce 35,000 pounds of thrust. The vehicle would be limited to 12,000 pounds, but, with approval of an Air Force space launch system, the weight could be increased to 20,000 pounds. Twelve orbital demonstration launches were scheduled, with the first unmanned flight occurring early in 1964 and the initial manned launches taking place later that year. From fiscal year 1962 through 1965 this program would require $413.9 million.[8]

After examining the space division proposal and the Dyna-Soar plan for acceleration, General B. A. Schriever, AFSC commander, deferred a decision on Dyna-Soar until the relationship between "streamline" and SAINT II was clarified. Moreover, further analysis of an orbital booster for Dyna-Soar would have to be accomplished.[9]

From 1 through 12 May 1961, a Dyna-Soar technical evaluation board, composed of representatives from the Air Force Systems Command, the Air Force Logistics Command (AFLC), and the National Aeronautics and Space Administration, had considered 13 proposals for orbital boosters from the Convair Division, the Martin Company and NASA. The evaluation board decided that the Martin C plan was the most feasible approach. The first stage of this liquid booster consisted of an LR87-AJ-5 engine, capable of producing 430,000 pounds of thrust, while the second stage, with a J-2 engine, could deliver 200,000 pounds of thrust.[10]

The Dyna-Soar Directorate of the Space Systems Division, having the responsibility for developing boosters for System 620A, also made a recommendation on the Step IIA propulsion. On 11 July, Colonel Joseph Pellegrini informed the Dyna-Soar office that his directorate favored employment of the projected Space Launch System A388. This proposal was an outgrowth of an SSD study on a Phoenix series of varying combinations of solid and liquid boosters to be used in several Air Force space missions. Phoenix A388 was to have a solid first stage, which could produce 750,000 pounds of thrust, and a liquid propellant second stage, using the J-2 engine.[11]

On 3 and 4 August 1961, Colonel Walter L. Moore, Jr., director of the Dyna-Soar program, brought the "streamline" proposal before the Strategic Air Panel, the Systems Review Board, and the Vice Chief of Staff. The program director pointed out that by eliminating suborbital flight the first air-drop would occur in mid-1963; the first, unmanned, orbital flight in 1964; and the first, piloted, orbital launch in early 1965. In comparison, the first, piloted, Step IIA flight had been scheduled for January 1967. Not only would the orbital flight date be accelerated but considerable financial savings would also accrue. Colonel Moore now estimated that the combined cost of Steps I and IIA was projected at 1.201 billion, while the figure for "streamline would run $1.026 billion. The director concluded by emphasizing that Dyna-Soar provided the most effective solution to an Air Force, manned, space program, and "streamline" was the most expeditious approach to piloted, orbital flight.[12]

Officials from SSD and the Aerospace Corporation presented their considerations for a "streamline" booster. At this point, it was clear that previous SSD evaluations for a Step IIA booster were simply incorporated in the "streamliner" analysis. The first choice of Aerospace and SSD officials was again their proposed Phoenix space launch system. Assuming a November 1961 approval date, Phoenix A388 allowed the first, unmanned launch to occur in July 1964, and, based on an 18-flight Dyna-Soar program, the cost for Phoenix development from fiscal year 1962 through 1966 would total $183.3 million. The second option was the Soltan, derived by attaching two 100-inch diameter solid propellant engines to the Titan II. The projected Soltan schedule permitted the same launch date as the Phoenix, but the cost was estimated at $325.4 million. Although the Saturn C-1 allowed an unmanned launch date in November 1963 and the cost would total $267.2 million, this booster was the third choice, largely because it was deemed less reliable. The space division representatives then concluded their part of the presentation by discussing the merits of ART and SAINT II.[13]

The Assistant Secretary of the Air Force for Research and Development, Dr. Brockway McMillan, was not as enthusiastic for acceptance of the Phoenix system. While he did not recommend use of the Saturn, Dr. McMillan thought that the Air Force should seriously consider the fact that the big NASA booster would provide the earliest launch date for Dyna-Soar. The assistant secretary believed, however, that an Atlas-Centaur combination would be the most feasible space launch vehicle for 10,000 pound payloads through 1965. After this time period, Dr. McMillan favored Soltan.[14]

Prior to these briefings, General Schriever was already convinced that Dyna-Soar had to be accelerated. He further believed that the best selection for the booster was Phoenix A388.[15] On 11 August, he informed ASD, SSD and his Deputy Commander for Aerospace Systems, Lieutenant General H. M. Estes, Jr., that "streamline" had the approval of AFSC headquarters and had to be "vigorously supported" by all elements of the command. Yet, the acceleration of Dyna-Soar was not that simple. Then AFSC commander was still concerned over the duplication of the manned, SAINT proposal and an orbital Dyna-Soar. He stated that these plans constituted a complex, and, at that point, an indefinable approach to military space flight which could not be presented to USAF headquarters. Consequently, General Schriever directed that a Manned, Military, Space, Capability, Vehicle study be completed by September. This proposed program would consist of "streamline," and a Phase Beta study which would determine vehicle configuration, boosters, military subsystems, and missions for an operational system which would follow Dyna-Soar. General Schriever also directed that the applied research programs of his command be reviewed to assure contributions to Dyna-Soar and far-earth orbital flights.[16]

During an August 1961 meeting of the Designated Systems Management Group, the Secretary of the Air

Force, Eugene M. Zuckert, commented on the question of Dyna-Soar acceleration.* He directed the three-step approach to continue until the position of Dyna-Soar in a manned, military space program was determined. Within the confines of the $100 million fiscal year 1962 budget, the secretary stated that action could be taken to facilitate the transition from a Step 1 to a "streamline" program. Finally, he requested a study on various approaches to manned, military, orbital flight.[18]

Under the direction of General Estes, a committee was formed in mid-August 1961 with representation from the Air Force Systems Command, RAND, MITRE, and the Scientific Advisory Board for the purpose of formulating a manned, military, space plan. The work of the committee was completed by the end of September with diverse sets of recommendations.

One of the working groups, chaired by a representative from the Aerospace Corporation, favored terminating the Dyna-Soar program and redirecting BOEING'S efforts to the development of a lifting body. Such an approach would cost $2 billion. A second alternative was to accelerate a suborbital Dyna-Soar program, cancel the orbital phases and initiate studies for far-earth, orbital flights. This proposal would total $2.6 billion. The least feasible approach, this group considered, was implement "streamline," and initiate a Phase Beta. Such a program would be the most expensive, totaling $2.8 billion.[19]

The opposite position was assumed by a panel of Scientific Advisory Board members, chaired by Professor C. D. Perkins, which strongly supported the last alternative of the Aerospace group. The Perkins group thought that military applications of a lifting body approach did not offer more promise than Dyna-Soar. To emphasize this point, the group questioned the control characteristics of a lifting body design which could make the execution of conventional landings hazardous. The group further argued that "streamline" should be directed towards defining military space objectives and insisted that a Phase Beta and an applied research program should be undertaken before considering an advanced Dyna-Soar vehicle.[20]

General Estes reached his own conclusions about a manned, military, space study. "Streamline" should receive Air Force approval; however, it should have unquestionable military applications, namely satellite inspection and interception missions. The deputy commander doubted that a Dyna-Soar vehicle could accomplish far-earth orbital flights and undergo the resulting re-entry velocity, ranging from 35,000 to 37,000 feet per second, and, consequently, he firmly stated that a Phase Beta study, conducted by Boeing, was necessary to determine a super-orbital design for Dyna-Soar.[21]

Secretary of Defense Robert S. McNamara also made a pronouncement on Dyna-Soar. After hearing presentations on the program and the military, space proposal of SSD, the secretary seriously questioned whether Dyna-Soar represented the best expenditure of national resources.[22] From this encounter with the defense department, the Air Staff derived a concept which was to dominate the Dyna-Soar program. Before military applications could be considered, the Air Force would have to demonstrate manned, orbital flight and safe recovery.[23]

During a meeting of the Designated Systems Management Group in early October 1961, it was very clear that the Air Force had decided in favor of "streamline." The management group had severely criticized SAINT II, by insisting that the projected number of flight tests and the proposed funding levels were too

---

* In early April 1961, Lieutenant General R. C. Wilson, Deputy Chief of Staff for Development, appeared concerned with the management of Air Force headquarters over the Dyna-Soar program. Although the Air Staff had devoted considerable attention to this program, it had not always been successful in affecting the decisions of the Secretary of the Air Force or the Secretary of Defense. General Wilson indicated to General C. E. LeMay, the Vice Chief of Staff, that this situation could be alleviated if the program were placed under the management of the Air Force Ballistic Missile and Space Committee. General LeMay, on 5 May, concurred and pointed out that the Department of the Air Force would have to place increasing emphasis on Dyna-Soar because it was a system leading to manned space flight. Dr. J. V. Charyk, the Air Force under secretary, disagreed and thought that since Dyna-Soar was primarily a research project, transfer of the management in the department should be deferred until a Dyna-Soar weapon system was under development. On 25 July, the Secretary of the Air Force replaced the ballistic and space committee with the Designated Systems Management Group. Composed of important officials in the Department of the Air Force, this group was to assist the Secretary of the Air Force in managing significant programs. By 1 August 1961, the Dyna-Soar program was listed as one of the systems under the jurisdiction of the designated management group.[17]

unrealistic. As a result of this review, the Department of the Air Force prohibited further use of the SAINT designation.[24]

Dyna-Soar officials completed, on 7 October 1961, an abbreviated development plan for a manned, military, space, capability program. The plan consisted of "streamline"; a Phase Beta study, which would determine approaches to the design of a super-orbital Dyna-Soar vehicle; supporting technological test programs; and an applied research program. The objectives of the proposed Dyna-Soar plan were to provide a technological basis for manned, maneuverable, orbital systems; determine the optimum configuration for super-orbital missions, and demonstrate the military capability of both orbital and super-orbital vehicles.

The program office considered the Phoenix system acceptable but derived, instead, a new two-step program based on the employment of Titan III, which differed from Soltan by using two 120-inch diameter solid propellant engines. While Dyna-Soar I would encompass the "streamline" proposal, Dyna-Soar II would involve the development of a far-earth, orbital vehicle. The program office anticipated the first, unmanned, orbital flight in November 1964, and the first, piloted flight in May 1956. The next five flights would be piloted with the purpose of accomplishing multiorbital missions. The ninth flight test, occurring in June 1966, however, would be an unmanned exploration of super-orbital velocities. The remaining nine flight tests would be piloted, with the purpose of demonstrating military missions of satellite interception and reconnaissance. The flight test program was to terminate by December 1967.

To accomplish this program, the Dyna-Soar office considered that $162.5 million would be reared for fiscal year 1962, $211.7 million for 1963, $167.4 million for 1964, $168.6 million for 1965, $99.0 million for 1966, $21.0 million for 1967, and $2.4 million for 1968. With $88.2 million expended prior to fiscal year 1962, these figures would total $921 million for the development of a manned, military, Dyna-Soar vehicle.[25]

On 15 October 1961, Colonel B. H. Ferer of the Dyna-Soar system staff office, USAF headquarters, requested W. E. Lamar, Deputy Director for Development in the Dyna-Soar office, to brief Dr. Brockway McMillan and a military, manned, spacecraft panel, convened to advise the Secretary of Defense. Mr. Lamar gave a comprehensive narrative of the history of the Dyna-Soar and its current status to the assistant secretary. While Dr. McMillan approved the briefing as suitable for the spacecraft panel, he requested Mr. Lamar not to emphasize military applications at that time. The briefing to the panel followed, but Colonel Ferer once again called Lamar. The deputy for development was rescheduled to brief Dr. L. L. Kavanau, Special Assistant on Space in the Department of Defense. Dr. Kavanau appeared quite interested in the various alternatives to accelerating Dyna-Soar and finally stated that it was sensible to go directly to an orbital booster.[26]

Based on the October proposal, General Estes prepared another development plan for Dyna-Soar. This approach was presented in a series of briefings to systems command headquarters, the Air Staff, and, on 14 November, to the Designated Systems Management Group.[27] The central objective was to develop a manned, maneuverable vehicle, capable of obtaining basic research data, demonstrating re-entry, testing subsystems, and exploring man's military function in space. These objectives were to be achieved by adapting the Dyna-Soar glider to a Titan III booster, in place of the previously approved, suborbital Titan II.*

The Dyna-Soar office considered two alternate funding plans. Plan A adhered to the established $100 million ceiling for fiscal year 1962, set $156 million for 1963, and required $305.7 million from 1964 through 1967. Total development funds would amount to $653.4 million and would permit the first, unmanned ground-launch by November 1964. Plan B followed the ceilings of $100 million for fiscal year 1962 and $125 million for fiscal year 1963. Under this approach, $420.2 million would be required from 1964 through 1968, totaling $736.9 million. This latter plan established April 1965 as the earliest date for

---

*   While accepting the standard space launch concept, the Department of Defense decided against the employment of a Phoenix
       system and, on 13 October, informed Dr. McMillan that Titan III was to be the Air Force space booster.[28]

the first, unmanned ground-launch. Regardless of which approach was taken, the proposed program would substantially accelerate the first, manned, orbital flight from 1967 to 1965.[29]

On 11 December 1961, Air Force headquarters informed the systems command that the Secretary of the Air Force had agreed to accelerate the Dyna-Soar program. The suborbital phase of the old three-step program was eliminated, and the central objective was the early attainment of orbital flight, with the Titan III booster. Plan B of the November 1961 development plan was accepted, and $100 million for fiscal year 1962 and $125 million for 1963 was stipulated. Finally, the Air Staff instructed the Dyna-Soar office to present a new system package program to headquarters by early March 1962.[30]

Colonel Moore set the following tentative target dates to be considered in reorienting the program: the first air-launch in July 1964; the first, unmanned, orbital ground-launch in February 1965; and the first, manned, orbital ground-launch in August 1965. The program director commented that the advancement of the program to an orbital status represented a large step toward meeting the over-all objectives of Dyna-Soar.[31]

The program office then issued instructions to its contractors, the Boeing Company, the Minneapolis-Honeywell Regulator Company, and the Radio Corporation of America, pertaining to the redirected program. The tentative dates offered by Colonel Moore were to be used as guidelines for establishing attainable schedules. The Dyna-Soar glider was to be capable of completing one orbit with all flights terminating at Edwards Air Force Base, California. The system office informed the contractors that no requirements existed for maneuvering in space nor for the development of military subsystems. The contractors were to make only a minimum number of changes to the glider and the transition section in order to adapt the airframe to the Titan IIIC. To conform to budget restrictions, a serious reduction in program scope was necessary. Certain wind tunnel tests would have to be suspended. The air-launch program would consist of only 15 drops from a B-52 and would terminate in April 1965. The first two ground-launches were to be unmanned, and the 32 remaining eight were to be piloted.[32]

On 27 December 1961, the Deputy Chief of Staff for Systems and Logistics, USAF headquarters, issued System Program Directive 4, which reiterated the program objective announced in the November 1961 development plan. The deputy chief of staff emphasized the Air Force view that man would be required to perform missions essential to national security in space. The Dyna-Soar program would provide a vehicle which offered an economical and flexible means to return to a specific landing site, and, consequently, would fulfill a vital military need not covered in the national space program. The directive specified that Titan IIIC was to be the booster, and that only single orbits were contemplated for each ground-launch, Although Air Force headquarters chose the low funding level of Plan B, $100 million for fiscal year 1962 and $115 million for 1963, headquarters also insisted on the accelerated flight dates of Plan A.* The deputy chief of staff would accept later flight dates only if an examination bit the systems command revealed the impossibility of achieving such a schedule. Lastly, a new system package program had to be completed by March 1962.[33]**

---

* The flight schedule of Plan A in the November 1961 development plan stipulated April 1964 for the air-launch program, November 1964 for the unmanned ground-launch, and May 1965 for the manned ground-launch.

** Major General W. A. Davis, ASD commander, protested that the March 1962 date was an arbitrary limitation and did not allow the system office enough time to reshape the program. Air Force headquarters apparently received this recommendation favorably because, on 2 February 1962, the Deputy Chief of Staff for Systems and Logistics issued an amendment to the system program directive of 27 December 1961, extending the completion date of a new system package program to the middle of May 1962.[34]

To give further legal sanction to the redirected program, Air Force headquarters, on 21 February 1962, issued an amendment to the advanced development objective, dated 21 July 1960.* This amendment deleted references to suborbital flights and to the development of military subsystems. Air Force headquarters, however, did state that a reliable method for routine recovery of space vehicles would make military missions practical. The amendment further stipulated that the program was oriented to single orbital flights, with the first, unmanned ground-launch occurring in November 1964.[35]

In a memorandum of 23 February 1962, Secretary McNamara officially endorsed the redirection of the Dyna-Soar program. He directed the termination of the suborbital program and the attainment of orbital flight, by employment of the Titan IIIC booster. The funding level was limited to $100 million in fiscal year 1962 and $115 million in 1963. Finally, Secretary McNamara insisted on a redesignation of the Dyna-Soar program to a nomenclature more suitable for a research vehicle.[36]

By the end of February, a draft version of the system package program was completed, and, in the middle of March, the program office offered the preliminary outlines to AFSC and Air Force headquarters. The central point of this briefing was that the $115 million fiscal year 1963 ceiling would endanger the attainment of desired system reliability and would also limit the flight profile of the glider. As a result of these presentations, Air Force headquarters instructed the systems command to prepare a briefing for the Department of Defense.[37]

On 17 April, officials of the Dyna-Soar office made a presentation to Dr. Harold Brown, Director of Defense for Research and Engineering. The program office wanted approval of a $12.2 million increase for fiscal year 1963 and, also, an additional $16.7 million to realize an unmanned ground-launch date of May 1965. Dr. Brown offered to give both proposals further consideration and requested the Dyna-Soar office to present alternative funding levels to meet a May or July 1965 unmanned launch date.[38]

By 23 April 1962, the system package program was completed. The objective of the new Dyna-Soar program had been clearly announced by the November 1961 development plan and was reiterated in this more elaborate proposal. Dyna-Soar was a research and development program for a military, test system to explore and demonstrate maneuverable re-entry of a piloted, orbital glider which could execute conventional landing at a pre-selected site. For the Dyna-Soar office, the new program represented a fundamental step towards the attainment of future, piloted, military, space flight.

Prior to redirection in December 1961, the Dyna-Soar system office had final authority over the Step I booster being developed by the space division. Under the new program, however, the Dyna-Soar glider would only be one of the payloads for the standard space launch system, designated 624A. Titan IIIA formed the standard core and was essentially a modified Titan II with a transtage composed of an additional propulsive unit and a control module. This version of the standard launch system, although it had no assigned payload, as yet, was capable of placing 7,000 pounds into an orbit of 100 nautical miles. The Dyna-Soar glider, however, was scheduled to ride the Titan IIIC booster. This launch system was derived from the standard core with an attached first-stage of two, four-segment, solid, rocket motors, capable of delivering a total of 1,760,000 pounds of thrust.** The second and third stages were liquid propulsive units and would produce 474,000 and 100,000 pounds of thrust, respectively. Titan IIIC could place a maximum of 25,000 pounds in low-earth orbit, however, for the particular Dyna-Soar trajectory and conditions, the payload capability was 21,000 pounds.[40]

---

*   This advanced development objective had been previously designated System Development Requirement 19, issued on 21 July 1960.

**   Late in May 1962, the Assistant Secretary McMillan requested the Dyna-Soar office to investigate the impact of employing a five-segment Titan IIIC on the program. Although this change would necessitate glider modifications amounting to $5.4 million, the program office recommended that the five-segment configuration be selected for Dyna-Soar, and command headquarters concurred on 25 July.[39]

The flight test program was defined in three phases. One Dyna-Soar glider was now scheduled to accomplish 20 air-launches from a B-52C aircraft to determine glider approach and landing characteristics, obtain data on lift-to-drag ratio and flight characteristics at low supersonic velocities, and accumulate information on the operation of the glider subsystems. On four of the air-launches, the acceleration rocket would power the glider to a speed of Mach 1.4 and a height of 70,000 feet.

Figure 1. Titan IIIC and Dyna-Soar prior to launch (artist's drawing).

Figure 2. Dyna-Soar flight during solid-stage burning.

Figure 3. Second-stage ignition.

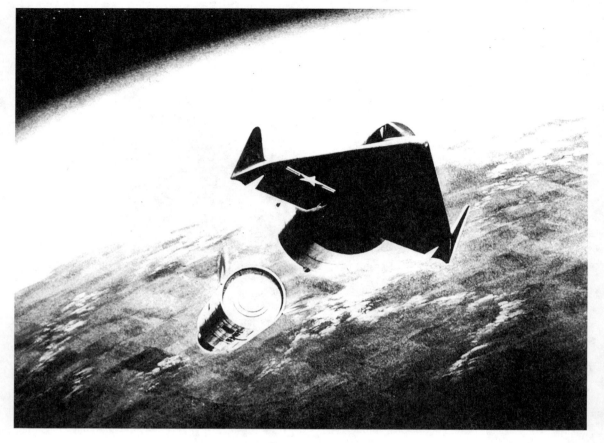

Figure 4. Transtage separation.

Figure 5. Transition section separation.

Figure 6. Beginning of re-entry glide.

Figure 7. Re-entry glide during high temperature regime.

Figure 8. Landing.

Following the air-launch program, two, unmanned, orbital launches could occur. The purpose was to verify the booster-glider system as a total vehicle for piloted flight, and demonstrate glider-design for hypersonic velocities. The Titan IIIC would propel the glider to a velocity of 24,490 feet per second, and after fulfilling its orbital mission, the vehicle would land at Edwards Air Force Base by employment of the drone-landing techniques. Eight, piloted, orbital flights were to follow, further exploring and defining the Dyna-Soar flight corridor.

According to the reasoning of the Dyna-Soar office, the first air-launch would occur in September 1964, with the final drop taking place in July 1965. The first, unmanned ground-launch was to be conducted in May 1965, with the second, unmanned flight occurring in August 1965. The first, piloted flight was scheduled for November 1965 and the last, manned, orbital mission for the beginning of 1967. The Dyna-Soar office had hopefully attempted to obtain the earliest possible launch dates and still remain within the $115 million fiscal year 1963 ceiling set by USAF headquarters on 27 December 1963.[41]

On 25 April 1962, General Davis forwarded the system package program the approval of AFSC headquarters. In line with Dr. Brown's request for alternative funding proposals, the Dyna-Soar office submitted a more realistic funding schedule. To meet a May 1965 schedule for the first, unmanned launch, $144.8 million was required for fiscal year 1963 and $133.1 million for 1964. If the first, unmanned launch was to occur in May 1965, then $127.2 million was needed for fiscal year 1963 and $133.1 million for 1964.[42]*

---

* General Davis also pointed out that the Pacific Missile Range of the Department of the Navy had issued a financial requirement of $100 million for the construction of four vessels which would be employed in the Dyna-Soar program. The ASD commander emphasized that other space programs would eventually use these facilities, and, consequently, this cost should not be fully attributed to System 620A. Pacific range officials lowered the requirement to three new ships and modification of an existing vessel, totaling $69 million. By the middle of May, Navy officials agreed that ship costs of $36 million and a total range requirement of $49 million were directly related to the Dyna-Soar program. Because of subsequent revisions to the program, range officials then submitted an increased estimate of $69 million for both the 10 October 1962 and the 11 January 1963 system package programs. The Dyna-Soar office did not concur with this figure, however, total range costs relating to System 620A were agreeably reduced to $48.888 million in May 1963.[43]

Following completion of the system package program, a series of presentations were made to elements of AFSC headquarters, Air Force headquarters and the Department of Defense. To remain within the $115 million fiscal year 1963 ceiling, the Dyna-Soar office was forced to reduce the development test program, thereby decreasing the reliability of the glider system and limiting the scope of the flight test program. During one of the briefings to the Department of Defense, Dr. Brown recommended significant changes to the Dyna-Soar program. Additional funds would be allotted for further development testing, and most important, the Dyna-Soar glider was to fulfill multiorbit missions.[44]

On 14 May, the program office had completed a revision of its system package. The wind tunnel program was expanded. Glider and panel flutter tests were added. Work to increase the heat resistant ability of certain sections of the glider was contemplated. Refinement of the glider design and dynamic analysis of the air vehicle vibration were additional tasks. The program office further scheduled additional testing of the reaction control, the environmental control, and the guidance systems. A more comprehensive reliability program for the glider and the communication and tracking systems was to be inaugurated, and an analysis of a means to reduce the weight of the glider subsystems was to take place.

For the Dyna-Soar office, multiorbital missions were a logical and relatively inexpensive addition to the basic program and would probably be scheduled for the fifth or sixth ground-launch. Such a demonstration, in the opinion of the Dyna-Soar office, was a prerequisite to more extensive exploration of the military function in piloted, space flight. Multiorbital missions, however, necessitated modification of the guidance system, increased reliability of all subsystems, and the addition of a de-orbiting unit.

Previously, a single-orbit, Dyna-Soar mission did not require the employment of a de-orbiting system, largely because the flight profile was only an around-the-world, ballistic trajectory. The Dyna-Soar office considered two alternatives for equipping the glider with a de-orbiting ability. One possibility was to place a system in the transition section of the glider. Another approach, actually chosen, was to employ the transtage of the Titan IIIC vehicle. This fourth stage would permit accurate orbital injection of the glider and would remain attached to the transition section to provide de-orbiting propulsion.

Along with these additions to the system package program, the Dyna-Soar office submitted a new funding schedule. The requirement was $152.6 million for fiscal year 1963, $145.2 million for 1964, $113.7 million for 1965, $78.3 million for 1966, and $17.7 million for 1967. This proposal would set the total cost for the Dyna-Soar program at $682.1 million.[45]

Before the Department of Defense acted on these revisions, the system office and Air Force headquarters had to determine a new designation for Dyna-Soar, more accurately reflecting the experimental nature of the program. In his February memorandum, Secretary of Defense McNamara directed Secretary Zuckert to replace the name "Dyna-Soar" with a numerical designation, such as the X-19. Mr. J. B. Trenholm, Jr., assistant director of the program office, requested his director for program control to derive a new nomenclature for Dyna-Soar. The assistant director added that the program office should officially request retention of "Dyna-Soar" as the popular name. Whatever the designation, Air Force headquarters required it by April.[46]

Following Air Force regulations, the director for program control reluctantly submitted ARDC form 81A, offering the designation, XJN-1 and, at the same time, requested use of "Dyna-Soar." Colonel Ferer at USAF headquarters did not concur with the XJN-1 label but offered instead XMS-1, designating experimental-manned-spacecraft. Other elements in Air Force headquarters and in the Department of Defense objected to both designations. Finally, on 19 June 1962, USAF headquarters derived and approved the designation, X-20.[47] On 26 June, a Department of Defense news release explained that this new designation described the experimental character of the program.[48] By the middle of July, Air Force headquarters allowed the word, "Dyna-Soar," to stand with X-20.[49]

On 13 July 1962, USAF headquarters informed the systems command that the Secretary of Defense conditionally approved the 14 May revision of the system package program. Instead of the requested

$152.6 million for fiscal year 1963, Secretary McNamara authorized $135 million and insisted that future funding would not exceed this level. He further stipulated that Dyna-Soar schedules would have to be compatible with Titan IIIC milestones and that technical confidence and data acquisition in the X-20 program would have precedence over flight schedules. Air Force headquarters then directed the program office to make appropriate changes to the system package as soon as possible.[50]

In spite of the fact that the Dyna-Soar program had been redirected, funds and approval were still lacking for System 624A, Titan III. Since the X-20 was scheduled to ride the fourth development shot of Titan IIIC, flight dates for Dyna-Soar could not be set until the Titan schedule was determined. On 31 August 1962, the space division informed the X-20 office that calendar dates for booster launchings could not be furnished until funding had been released. This was expected by November, with program development beginning in December 1962. The first Titan IIIC launch would occur 29 months later, and the fourth shot (the first, Dyna-Soar, unmanned launch) would take place 36 months after program "go-ahead."[51]

Based on this Titan IIIC scheduling assumption, the X-20 system office completed, on 10 October, another system package program. Twenty air-drop tests were to be conducted from January through October 1965. Two unmanned, orbital launches were to occur in November 1965 and February 1966. The first of eight, piloted flights was to take place in May 1966, with a possible multiorbit launch occurring in November 1967.* The Dyna-Soar office stipulated that $135 million would be required in fiscal year 1963, $135 million in 1964, $102.78 million in 1965, $107.51 million in 1966, $66.74 million in 1967, and $10 million in 1968. The program would require $766.23 million for the development of the orbital X-20 vehicle.[52] Major General R. G. Ruegg, ASD commander, submitted this system package program to AFSC headquarters on 12 October 1962, however, it never received command endorsement.

While the X-20 office was concerned with Titan III schedules and approval of a new package program, AFSC headquarters directed a change in the organization of ASD which had possible significance for the Dyna-Soar program. On 28 September 1962, the systems command directed that the function of the ASD Field Test Office at Patrick Air Force Base, Florida, be transferred to the 6555th Aerospace Test Wing of the Ballistic Systems Division.[53]

Previously ARDC headquarters had established, on 4 August 1960, a general policy on test procedures which firmly placed control of system testing in the various project offices rather than the test centers.[54] With headquarters approval, the Dyna-Soar office appointed a test director for the entire Category II program and directed that the Air Force Flight Test Center provide a Deputy Director for Air-Launch and the WADD Field Test Office at Patrick Air Force Base, a Deputy Director for Ground-Launch.[55] The test centers, however, objected to giving the project offices full authority, largely because such a policy did not fully utilize their ability to conduct flight test progress. Consequently, on 31 January 1962, General Schriever rescinded the August 1960 policy and directed that, while over-all authority still rested in the program offices, the centers and test wings would prepare and implement the test plans and appoint local test directors.[56] While the purpose of this new policy was to give the test centers more authority in the test program, it did not result in any significant changes to the structure of the Dyna-Soar test force. Under this new arrangement, the program office appointed a Deputy System Program Director for Test, while the flight test center provided the Air-Launch Test Force Director and the Patrick field office, the Ground-Launch Test Force Director.[57]

Throughout these changes in the Dyna-Soar test structure, the 6555th Aerospace Test Wing of the Ballistic Systems Division had authority only during the operation of the booster. With the transfer of the functions of the ASD field office to this test group, however, the aerospace wing became, in effect, the director of the orbital flight tests. This test group was responsible to the commander of BSD., who, in the instance of conflicting requirements of various assignments, would determine priorities for the operations of his test wing.[58]

---

* These X-20 schedules proved compatible with the Titan III schedules, for on 15 October 1962, Air Force headquarters issued System Program Directive 9. This authorized research and development of the space booster to begin on 1 December 1962 with a total of $745.5 million from fiscal year 1962 through 1966.

In an effort to conserve program funds, the X-20 office formulated a flight test program, the "Westward-Ho" proposal, which would eliminate the necessity for the construction of several control centers and multiple flight simulators. Previous planning had located a flight control center at Edwards Air Force Base for the conduct of the air-launch tests. The ground-launch program required a launch center and a flight control center, both at Cape Canaveral, and also a recovery center at Edwards Air Force Base. "Westward-Ho" simply proposed the consolidation of the flight control centers for both the air-drop and ground-launch tests at Edwards, leaving only a launch control center at the Cape. The Air Force Flight Test Center would provide a test director for both the air-drop and orbital flight tests, who would be responsible in turn to the X-20 program office. By establishing one flight control center and employing only one flight simulator, the Dyna-Soar office estimated a savings of at least $3 million.[59]

The "Westward-Ho" logic of the X-20 office was not apparent to AFSC headquarters. On 19 December, the AFSC vice commander, Lieutenant General Estes, directed the establishment of a manned, space flight, review group, for the purpose of examining all aspects of the X-20 test program including the relationships of the various AFSC agencies. Brigadier General O. J. Glasser of the Electronic Systems Division was named chairman of this group, which was to be composed of representatives from AFSC headquarters, the aeronautical division, the space division, the missile test center and the missile deployment center.[60]

Colonel Moore noted that the Air Force Flight Test Center, the key agency in "Westward-Ho" had not been permitted representation at this review. Furthermore, he had offered to familiarize the committee with a presentation on the Dyna-Soar test requirement, but this proposal was rejected.[61] The significance of the coming review was not entirely clear to the X-20 program office.

General Glasser's committee formally convened on 3 and 23 January and 5 February 1963. While no decisions were made at these meetings, the members discussed several critical points of the Dyna-Soar program. Although the Test Support Panel seemed to favor the location of a single flight control center at Edwards Air Force Base, it was clear that "Westward-Ho" impinged on the interests of the Air Force Missile Development Center, the Space Systems Division, and the Air Force Missile Test Center. General Glasser, however, emphasized the central problem confronting the Dyna-Soar program: the open conflict between the Space Systems Division and the Aeronautical Systems Division for control of the only Air Force, manned, space program. The Organization and Management Panel offered some solutions to this problem. First, management of the program by AFSC headquarters would have to be altered. Like the Titan III program, the Dyna-Soar system should be placed under the guidance of the Deputy to the Commander for Manned Space Flight instead of the Deputy Chief of Staff for Systems. More important, the panel strongly recommended that the entire program be reassigned to the Space Systems Division. General Glasser did not favor such a radical solution but thought that a single AFSC division should be made the arbiter for both the Titan III and X-20 programs.[62]

While designating his deputy for manned space flight as a headquarters point of contact for the Dyna-Soar program, General Schriever, on 9 May 1963, altered the structure of the X-20 test force. He directed that the Space Systems Division would name the director for X-20 orbital flights, with the flight control center being located at the Satellite Test Center, Sunnyvale, California. The commander of AFSC did emphasize, however, that the Aeronautical Systems Division was responsible for the development of the X-20.[63] At the end of July, General Schriever also assigned responsibility for the air-launch program and pilot training to the space division.[64]

Although the Air Force had undertaken a manned, military, space study in 1961, the Department of Defense still had not determined a military, space mission for the Air Force. While the 1961 study had essentially compared the Dyna-Soar glider with a SAINT II lifting body, Secretary McNamara was also interested in the military potentialities of the two-man, Gemini capsule of NASA. In his 23 February 1962 memorandum, the Secretary of Defense expressed interest in participating in this program with the National Aeronautics and Space Administration for the purpose of demonstrating manned rendezvous.[65] On 18 and 19 January 1963, Secretary McNamara directed that a comparison study between the X-20 glider and the Gemini vehicle be made which would determine the more feasible approach to a military capability. He also asked for an evaluation of the Titan III and various alternative launch vehicles.[66]

A few days later, Gemini became even more significant to the Air Force, for the Department of Defense completed an agreement with the National Aeronautics and Space Administration, which permitted Air Force participation in the program. A planning board, chaired by the Assistant Secretary of the Air Force for Research and Development and the Associate Administrator of NASA, was to be established for the purpose of setting the requirements of the program. The agreement stipulated that the Department of Defense would not only participate in the program but would also financially assist in the attainment of Gemini objectives.[67]

At the end of January, Major General O. J. Ritland, Deputy to the Commander for Manned Space Flight, emphasized to the commanders of ASD and SSD, that Secretary McNamara intended to focus on the X-20, Gemini and Titan III programs with the ultimate objective of developing a manned, military, space system. General Ritland warned that once a decision was made, it would be difficult for the Air Force to alter it. Consequently, command headquarters, the space division and the aeronautical division would have to prepare a comprehensive response to the secretary's request. General Ritland then gave the Space Systems Division the responsibility for providing statements of the Air Force, manned, space mission and for defining space system requirements, tests, and operations.[68]

By the end of February 1963, AFSC headquarters had compiled a position paper on the X-20 program. Six alternative programs were considered: maintain the present program, reorient to a lower budget through fiscal year 1964, accelerate the flight test program, reinstate a suborbital phase, expand the program further exploring technological and military objectives, and, finally, terminate the X 20 program. The conclusion of command headquarters was to continue the present X-20 and Titan III programs.[69]

Early in March, General LeMay offered his thoughts on the coming review by the Secretary of the Air Force. He firmly stated that continuation of Titan III was absolutely necessary and, most important, the current X-20 program should definitely proceed. The Air Force Chief of Staff emphasized that the Dyna-Soar vehicle would provide major extensions to areas of technology important to the development of future military systems and, consequently, the Air Force should not consider termination of the X-20 program or delay of schedules for the approval of an alternative space program. General LeMay insisted that the purpose of Air Force participation in the Gemini program was limited to obtaining experience and information concerning manned space flight. The Chief of Staff underlined that the interest of the Air Force in the NASA program was strictly on the basis of an effort in addition to the Dyna-Soar program.[70]

After hearing presentations of the X-20, Gemini, and Titan III programs in the middle of March, Secretary McNamara reached several conclusions which seemed to reverse his previous position on the experimental nature of the Dyna-Soar program. He stated that the Air Force had been placing too much emphasis on controlled re-entry when it did not have any real objectives for orbital flight. Rather, the sequence should be the missions which could be performed in orbit, the methods to accomplish them, and only then the most feasible approach to re-entry. Dr. Brown, however, pointed out that the Air Force could not detail orbital missions unless it could perform controlled re-entry. Furthermore, the Director of Defense for Research and Engineering, stated that the widest lateral mobility, such as possessed by the X-20, during landing was necessary in performing military missions.

Dr. McMillan surmised that Secretary McNamara did not favor immediate termination of the X-20 program.[71] Secretary McNamara did request, however, further comparison between Dyna-Soar and Gemini in the light of four military missions: satellite inspection, satellite defense, reconnaissance in space, and the orbiting of offensive weapon systems.[72]

On 10 May 1963, a committee composed of officials from the aeronautical and space divisions completed their response to Secretary McNamara's direction. The committee was aware that the Dyna-Soar glider had sufficient payload capacity for testing a large number of military components and that the X-20 demonstration of flexible re-entry would be an important result of the flight test program. Concerning

Gemini, the committee also recognized that this program would enhance knowledge relating to maneuverability during orbit and consequently recommended the incorporation of a series of experiments leading to the testing of military subsystems. Further in the future, both vehicles could be adapted to serve as test craft for military subsystems; however, neither could, without modification, become a fully qualified weapon system for any of the missions specified by Secretary McNamara. With the employment of Titan III instead of Titan II and the incorporation of a mission module, this Gemini system could provide greater orbital maneuverability and payload capacity than the X-20. The Dyna-Soar vehicle, however, would provide greater flexibility during re-entry and, unlike Gemini, could return the military subsystems to Earth for examination and re-use.[73]

General Ritland forwarded this report to Air Force headquarters a few days later. The deputy for manned space flight recommended that the X-20 program be continued because of the contribution that a high lift-to-drag ratio re-entry vehicle could make for possible military missions. Air Force participation in the Gemini program, however, should be confined to establishing a small field office at the NASA Manned Space Center and seeing that military experiments were part of the program.[74]

While the Department of Defense had not made a final determination concerning the X-20 and Gemini, General Estes cautioned the Dyna-Soar office at the end of June that the Secretary of Defense was still studying the military potential of both approaches. The vice commander stated that the system office had to maintain a position which would permit continuation of the program, while at the same time restricting contractor actions to assure minimum liability in event of cancellation.[75]

While the X-20 and Gemini approaches to orbital flight were under examination, the Dyna-Soar office was also confronted with an adjustment to the program because of a pending budget reduction. In November 1962, it had been apparent that the Department of Defense was considering restriction of fiscal year 1963 and 1964 funds to $130 million and $125 million instead of the previously stipulated level of $135 million for both years.[76] Colonel Moore pointed out to AFSC headquarters that only through aggressive efforts would $135 million be sufficient for fiscal year 1963 and any proposed reduction would be based on a lack of understanding of the Dyna-Soar requirements. Furthermore, an increase in fiscal year 1964 funds was necessary, raising the figure to $147.652 million.[77] Later, the system office informed General LeMay that schedules could not be maintained if funding were reduced and that $135 million and $145 million would be required for fiscal years 1963 and 1964.[78]

During March 1963, the X-20 office prepared four funding alternatives, which General Estes submitted to Air Force headquarters at the end of the month. The most desirable approach was to maintain the program schedules as offered in the 10 October 1962 system package program by increasing the funding. The X-20 office estimated that $135 million was required for fiscal year 1963, $145 million for 1964, and $114 million for 1965, which gave a total program cost of $795 million. The second alternative was to authorize a ceiling of $792 million, with $135 million allotted for 1963, $135 million for 1964, and $120 million for 1965. This reduction could be accomplished by deferring the multiorbit flight date by six months. The third option required $130 million for 1963, $135 million for 1964, and $130 million for 1965, with a program total of $807 million. Such a funding arrangement would delay the entire program by two months and defer the multiorbit flight from the fifth to the seventh ground-launch. The least desirable approach was to delay the entire program by six months, authorizing $130 million for 1963, $125 million for 1964, and $125 million for 1965. Under this alternative, the program would total $828 million.[79]

On 12 April 1963, Air Force headquarters accepted the third alternative. A funding level of $130 million was established for 1963 and the system office was directed to plan for $135 million in 1964. Headquarters stipulated that program schedules could not be delayed by more than two months and that a new system package program had to be submitted by 20 May.[80]

On 15 January 1963, the Dyna-Soar office had completed a tentative package program, which included the same funding and flight schedules as the 10 October 1962 proposal. The central difference was that the latter program incorporated the "Westward-Ho" proposal.[81] This system package program, however, was not submitted to AFSC headquarters for approval. In accordance with the 12 April 1963 instruction, the X-20 office completed another system package program on 6 May which was distributed to the various program participants for their comments. On 9 May, however, General Schriever assigned the orbital test responsibility to the Space Systems Division, and, consequently, AFSC headquarters again instructed the Dyna-Soar office to revise the X-20 system package program by 13 May.[82]

In the 13 May system package program, the X-20 office estimated that $130 million was required for fiscal year 1963, $135 million for 1964, $130 million for 1965, $110 million for 1966, and $73 million for 1967. The air-launch program was to extend from March 1965 through January 1966, with the two unmanned, ground-launches occurring in January and April 1966. The first piloted flight would take place in July 1966, with the first multiorbit flight occurring in May 1967. The eighth and final piloted flight was to be conducted in November 1967.[83] Brigadier General D. M. Jones, acting commander of ASD, informed AFSC headquarters that there had been insufficient time to incorporate the details of the new test organization in the program package. Furthermore, a funding level of $130 million and $135 million for fiscal years 1963 and 1964 could delay Dyna-Soar flights by more than the two months anticipated in the 12 April direction of USAF headquarters.[84]

On 27 May, another system package program was completed. The same funding rates as the 13 May proposal were retained but the flight schedule was revised in order to conform with firm contractor estimates. The air-launch program was to extend from May 1965 through May 1966. The two unmanned launches were to take place in January and April 1966, and the first piloted launch was to occur in July 1966. Recognizing the necessity for a four month interval between single and multiorbit flights, the X-20 office set August instead of May 1967 for the first multiorbit launch. The Dyna-Soar flight test program was to terminate in February 1968 with the eighth orbital launch.[85]

The Secretary of the Air Force gave his approval to this system package program on 8 June 1963; however, the Department of Defense did not accept the recommended funding. On 3 July, AFSC headquarters informed the Dyna-Soar office that attempts to secure additional funding had failed. The funding level for fiscal year 1964 was $125 million.[86] By September, it was clear to the Dyna-Soar office that the consequence of this reduced funding level would be to delay multiorbital flight from the seventh to the ninth ground-launch.[87]

While final approval by the Department of Defense of the Dyna-Soar system package program was still pending in the middle of 1963, the impact of the December 1961 redirection on the Dyna-Soar program was apparent. The first Dyna-Soar development plan of October 1957 had definite military objectives leading to the development of orbital reconnaissance and bombardment vehicles. In April 1959, Dr. York, then Director of Defense for Research and Engineering, altered these goals and placed major emphasis on the development of a suborbital research vehicle. In spite of intensive comparative studies with manned SAINT and Gemini vehicles, the central purpose, as established by Dr. York, had not changed. While the system program directive of December 1961 and Secretary McNamara's memorandum of February 1962 elevated Dyna-Soar to an orbital vehicle, the glider was officially described as an experimental system.

Conceivably the redirected program could appear as a reversal of the three-step approach which was aimed at the development of a suborbital system, an orbital glider with interim military ability, and an operational weapon system. Yet, under this old development plan, the real Dyna-Soar program had only consisted of a glider which would perform suborbital fight. Consequently, Department of Defense sanction of the new program marked an advancement over the three-step approach in that orbital and even multiorbital flights of the X-20 glider were now established objectives of Dyna-Soar.

CHAPTER IV

REFERENCES

1. Ltr., Brig. Gen. M. B. Adams, Dep. Dir., Sys. Dev., Hqs. USAF to Cmdr., ARDC, 12 Oct. 1960, subj.: Development Directive No. 411, Doc. 161.

2. Ltr., Col. W. L. Moore, Jr., Dir., Dyna-Soar SPO, ASD, to Cmdr., ARDC, 23 Feb. 1961, subj.: Preliminary Results of Dyna-Soar "Stand-by" Program Study, Doc. 207.

3. Ibid.; TWX, AFDSD—AS-81749, Hqs. USAF to Hqs. ARDC, 31 Jan. 1961; System Package Program, System 620A, Dyna-Soar SPO, ASD, 26 Apr. 1961, pp. 411, 487A.

4. Memo., Col. W. L. Moore, Jr., Dir., Dyna-Soar SPO, ASD, 20 Mar. 1961, subj.: "Stand -by" Program, Doc. 216.

5. Presn., Boeing Co. to Dyna-Soar SPO, ASD, 4 May 1961, subj.: Project Streamline, pp. 1, 29.

6. Rpt., Dyna-Soar SPO, ASD, 1 Sept. 1961, subj.: Description; System Package Program, System 620A, Dyna-Soar SPO, ASD, 26 June 1961, pp. 395-399.

7. Proposed Development Plan for an Advanced Re-entry Technology Program, Hqs. SSD, 29 May 1961; p. 1-5.

8. System Package Program, Expanded SAINT Program, Hqs. SSD; 29 May 1961, pp. 1-1 to 1-8, V-63, VII-2.

9. Ltr., Col. W. L. Moore, Jr., Dir., Dyna-Soar SPO to Brig. Gen. A. T. Culbertson, Dir., Sys. Mangt., ASD, 28 June 1961, subj.: Dyna-Soar, SSD Advanced Re-entry Technology Program and SAINT II Proposed Conflict, Doc. 239.

10. Minutes, Capt. J. W. Launderville, Dyna-Soar Dir., SSD, 12 May 1961, subj.: The Ninth Meeting of the Technical Evaluation Board.

11. TWX, SSVS-10-7-3, Hqs. SSD to Hqs. ASD, 11 July 1961, Doc. 244.

12. Presn., Col. W. L. Moore, Jr., Dir., Dyna-Soar SPO, to Strategic Air Panel, Hqs. USAF, 3 Aug. 1961, subj.: Project Streamline
13. Ibid.

14. Memo., Brockway McMillan, ASAF/Res. & Dev., 18 Aug, 1961, subj.: Standardized Space Booster Program, in the files of the Dyna-Soar Sys. Staff Ofc., DCS/Sys. & Log., Hqs. USAF.

15. Ltr., Gen. B. A. Schriever, Cmdr., AFSC to Gen. C. E. LeMay, CS/AF, 1 Aug. 1961, subj.: Acceleration of Dynes-Soar Program, in the files of the Dyna-Soar Sys. Staff Ofc., DCS/Sys. & Log., Hqs. USAF.

16. TWX, SCGV-11-8-18, Hqs. AFSC to DCAS, Hqs. AFSC, 11 Aug. 1961, Doc. 249.

17. Memo., Lt.. Gen. R. C. Wilson, DCS/Dev., Hqs. USAF to CS/AF, 11 Apr. 1961, subj.: Dyna-Soar Management, in files of Dyna-Soar Sys. Staff Ofc., DCS/Sys. & Log., Hqs. USAF; ltr., Gen. C. E. LeMay, VCS/AF to SAF, 5 May 1961, subj.: Dyna-Soar Program, in files of Dyna-Soar Sys. Staff Ofc.; memo., J. V. Charyk, SAFUS to CS/AF n.d., subj.: Dyna-Soar Program, in files of Dyna-Soar Sys. Staff Ofc.; memo., SAF, 25 July 1961, subj.: Designated Management Group; memo., Maj. Gen. R. M. Montgomery, Asst. VCS/AF, i Aug. 1961, subj.: Designated Systems Management Group and Systems Review Board.

18. Ltr., Col. T. H. Runyon, Ec. Secy., DSMG, Hqs. USAF to DCS/Sys. & Log., Hqs. USAF, 5 Sept. 1961, subj.: DSMG Meeting on Dyna-Soar, in files of the Dyna-Soar Sys. Staff Ofc., DCS/Sys. & Log., Hqs. USAF.

19. Ltr., Lt. Gen. H. M. Estes, Jr., DCAS, Hqs. AFSC to Gen. B. A. Schriever, Cmdr, AFSC, 28 Sept. 1961, subj.: MMSCV Study, Doc. 256.

20. Ltr., Ad Hoc Committee, SAB to Gen. B. A. Schriever, Cmdr., AFSC, 28 Sept. 1961, subj.: SAB Recommendations Concerning Dyna-Soar; rpt., SAB, 12 Dec, 1961, subj.: A Report on Operations During 1961, p. 15.

21. Ltr., Lt. Gen. Estes to Gen. Schriever, 28 Sept. 1961, Doc. 256.

22. Memo., SAFUS to CS/AF, 28 Sept. 1961, subj.: West Coast Trip by Secretary McNamara.

23. TWX, AFSDC-67166, Hqs. USAF to Hqs. AFSC, 3 Oct. 1961.

24. TWX, AFSDC-S-67777, Hqs. USAF to Hqs. AFSC, 5 Oct. 1961.

25. Abbreviated Development Plan, System 620A, Dyna-Soar SPO, ASD, 7 Oct. 1961, subj.: Manned, Military, Space Capability, pp. 1-2, 6-7, 17-25, 41-42, fig. 2.

26. Trip rpt., W. E. Lamar, Dep. Dir/Dev., Dyna-Soar SPO to Col. W. L. Moore, Jr., Dir., Dyna-Soar SPO, ASD, 18 Oct. 1961, subj.: Briefing to Military, Manned, Spacecraft, Study Panel on Dyna-Soar; trip rpt.. Lamar to Maj. Gen. W. A. Davis, Cmdr., ASD, 1

Nov. 1961, subj.: Briefing to Dr. Kavanau, DDR&E, 25 Oct. 1961.

27.   Interview, W. M. Lyon, Asst. Ch., Ops. Div., Dyna-Soar SPO, by C. J. Geiger, Hist. Div., ASD, 6 Oct. 1961.

28.   Memo., Dep., DDR&E, DOD to ASAF/Res. & Dev., 1,3 Oct, 1961, subj.: Titan III Launch Vehicle Family.

29.   Development Plan, Dyna-Soar, DCAS, Hqs. AFSC, 16 Nov, 1961, pp. 7-21, 44-52.

30.   TWX, AFSDC-85-081, Hqs. USAF to Hqs. AFSC, 11 Dec. 1961.

31.   Ltr., Col. Floyd Finberg, Asst. Dep. Dir/Prog. Control, Dyna-Soar SPQ, to W. E. Lamar, Dep. Dir/Dev., Dyna-Soar SPO, ASD, 12 Dec. 1961, subj.: Redirected Program; memo., Col. W. L. Moore, Jr., Dir., Dyna-Soar SPO, ASD, 15 Dec. 1961, subj.: Program Redirection.

32.   Memo., Col. J. E. Christensen, Dep. Dir/Proc. & Prod., Dyna-Soar SPO, ASD, 15 Dec. 1961, subj.: Program Reorientation, Bulletin No. 10.

33.   SPD 4, DCS/Sys. & Log., Hqs. USAF, 27 Dec. 1961, Doc. 257.

34.   TWX, ASG-27-12-9, Maj. Gen. W. A. Davis, Cmdr., ASD to Hqs. AFSC, 27 Dec. 1961; amend. 1 to SPD 4, DCS/Sys. & Log., Hqs. USAF, 2 Feb. 1962, Doc. 259.

35.   Amend. 1 to ADO 19, DCS/Res. & Tech., Hqs. USAF, 21 Feb. 1962, Doc. 260.

36.   Memo., Secy. of Def. to SAF, 23 Feb. 1962, subj.: The Air Force Manned, Military, Space Program, Doc. 261.

37.   Daily Activity Rpt., Dyna-Soar SPO, ASD, 19 Mar. 1962; Program Summary, System 620A, Dyna-Soar SPO, 14 May 1962, pp. 2-3.

38.   Daily Activity Rpt., Dyna-Soar SPO, ASD, 19 Apr. 1962; ltr., Maj. Gen. W. A. Davis, Cmdr., ASD to Maj. Gen. M. F. Cooper, DCS/Sys., Hqs. AFSC, 25 Apr. 1962, subj.: System 620A Package Program, Doc. 264.

39.   TWX, AFSDC-S-93823, Hqs. USAF to Hqs. AFSC, 5 June 1962; TWX, ASZR-17-7-1014, Hqs. ASD to Hqs. AFSC, 18 June 1962; TWX, LSCSAD-23-7-18, Hqs. AFSC to Hqs. ASD, 25 July 1962, Doc. 272.

40.   Proposed System Package Plan, System 624A, Hqs. SSD, 30 Apr. 1962, Vol. I, p. 1-2.

41.   System Package Program, System 620A, Dyna-Soar SPO, ASD, 14 May 1962, pp. 1-1, 2-21, 5-3, 5-5, 6-10.

42.   Ltr., Maj. Gen. W. A. Davis, Cmdr., ASD to Maj. Gen, M. F. Cooper, DCS/Sys., Hqs. AFSC, 25 Apr. 1962, subj.: System 620A Package.

43.   Ibid; System Package Program, System 620A, Dyna-Soar SPO, ASD, 26 Apr. 1962, p. 11-13; System Package Program, System 620A, 14 May 1962, p. 11-14; System Package Program, System 620A, 11 Oct. 1962, p. 11-15; System Package Program, System 620A, 11 Jan. 1963, p. 11-15, System Package Program, System 620A, 13 May 1963, p. 11-7.

44.   Interview, W. M. Lyon, Asst. Ch., Ops. Div., Dyna-Soar SPO, ASD by C. J. Geiger, Hist. Div., ASD, 6 Oct. 1962.

45.   System Package Program, System 620A, Dyna-Soar SPO, ASD, 14 May 1962 pp. 6-1, 6-54, 6-55, fig. 1-1.

46.   Ltr., J. B. Trenholm, Jr., Asst. Dir., Dyna-Soar SPO, to Dep. Dir/Prog. Control, Dyna-Soar, SPO, ASD 20 Mar. 1962, subj.: Item for Action.

47.   Interview, R. C. Johnston, Ch., Ops. Div., Dyna-Soar SPO, ASD by C. J. Geiger, Hist. Div., ASD, 12 Oct. 1962.

48.   DOD News Release, 26 June 1962, subj.: Dyna-Soar Designated X-20 by the Air Force.

49.   Memo., Maj. J. C. Buckner, Ops. Div., Dyna-Soar SPO, ASD, 17 July 1962, subj.: X-20 Designation for Dyna-Soar.

50.   TWX, AFSDC-5-9-65922, Hqs. USAF to Hqs. AFSC, 13 July 1962, Doc. 269.

51.   TWX, SSBT-31-8-26, Hqs. SSD to Hqs. ASD, 31 Aug. 1962.

52.   System Package Program, System 620A, Dyna-Soar SPO, ASD, 10 Oct. 1962. pp. 2-6, 11-5.

53.   TWX, ASFP-5-10-1, ASD Fld. Test Ofc. to Hqs. ASD, 5 Oct. 1962.

54.   Ltr., Maj. Gen. M. F. Cooper, DCS/Res. & Engg., Hqs. ARDC to Hqs. WADD, 15 Aug. 1950, ARDC Policy on Testing.

55.   TWX, RDRA-30-9-43, Hqs. ARDC to Hqs. WADD, 30 Sept. 1960.

56.   Interview, Lt. Col. C. Svimonoff, Dep. Sys. Prog. Dir/Test, Dyna-Soar SPO, ASD by C. J. Geiger, Hist. Div., ASD, 30 Apr. 1963; memo., Gen, B. A. Schriever, Cmdr., AFSC, 31 Jan. 1962, subj.: AFSC System Test Policy Procedures.

57. System Package Program, System 620A, Dyna-Soar SPO, ASD, 14 May 1962, pp- 5-15, 5-19.

58. Memo., Maj. Gen. W. A. Davis, Cmdr., BSD and Maj. R. G. Ruegg, Cmdr., ASD, 16 Oct. 1962, subj. Support of Test Programs by the 6555th ATW, Patrick AFB, Florida, Doc. 277.

59. Rpt., Dyna-Soar SPO, ASD, 26 Oct 1962, subj.: A Re-examination of the X-20 Test Operation concept, pp. 2, 4, 11-12, 16, 26.

60. TWX, SCSS-19-21-1, Hqs. AFSC to Hqs. ASD, 19 Dec. 1962.

61. Ltr., Col. W. L. Moore, Jr., Dir., Dyna-Soar SPO, ASD to Maj. Gen. R. G. Ruegg, Cmdr.; ASD, 2 Jan. 1963, subj.: Manned Space Flight Review Group.

62. Minutes (unofficial), Dyna-Soar SPO, ASD, 3, 23 Jan. 1963, 5 Feb. 1963, subj.: Manned Space Flight Review Group.

63. Ltr., Gen. B. A. Schriever, Cmdr., AFSC to Cmdr. SSD, 9 May 1963, subj.: X-20 Program Management, Doc. 295.

64. Ltr., Gen. Schriever to Cmdr. ASD, 31 July 1963, subj.: X-20 Program, Doc. 303.

65. Memo., Secy. of Def. to SAF, 23 Feb. 1962, subj.: The Air Force, Manned, Military, Space Program, Doc. 261.

66. Memo., Secy. of Def. to DDR&E, DOD, 18 Jan. 1963, subj.: Review of Dyna-Soar, Doc. 278; memo., Secy. of Def. to DDR&E, 19. Jan. 1963, subj.: Review of Gemini, Doc. 279; memo., Secy. of Def. to DDR&E, 19 Jan 1963, Review of Titan III, Doc. 280.

67. Memo., J. E. Webb, Administrator, NASA and R. S. McNamara, Secy. of Def., 21 Jan 1963, subj.: Agreement Between the National Aeronautics and Space Administration and the Department of Defense Concerning the Gemini Program, Doc. 281.

68. Ltr., Maj. Gen. O. J. Ritland, Dep. to Cmdr/Manned Space Flight, Hqs. AFSC to Cmdr., SSD, 31 Jan. 1963, subj.: Gemini-Dyna-Soar Task Groups, Doc. 283.

69. Rpt., Hqs. AFSC, 26 Feb. 1963, subj.: The X-20 Program.

70. Ltr., Gen. C. E. LeMay, CS/AF to SAF, 2 Mar. 1963, subj.: DOD Review of USAF Space Program, Doc. 289.

71. Memo., Brockway McMillan, ASAF/Res. & Dev. to SAF, 15 Mar. 1963, subj.: Secretary McNamara's Trip to Seattle, Doc. 290.

72. Memo., Secy. of Def. to SAF, 15 Mar. 1963, subj.: X-20, Gemini, and Military Missions, Doc. 291.

73. Rpt., Dep/Tech., SSD, 10 May 1963, subj.: Response to Secretary McNamara's 15 Mar. 1963, Questions, I, viii-xiii.

74. Ltr., Maj. Gen. O. J. Ritland, Dep. to Cmdr/Manned Space Flight, Hqs. AFSC to Hqs. USAF, 22 May 1963, subj.: Response to Secretary McNamara's 15 Mar. 1963 Questions, Doc. 300.

75. Lt. Gen. H. M. Estes, Jr., Vice Cmdr., AFSC to Maj. Gen. R. G. Ruegg, Cmdr., ASD, 21 June 1963, subj.: X-20 Program Guidance for FY 1964, Doc. 302.

76. Ltr., Maj. Gen. M. F. Cooper, DCS/Sys., Hqs. AFSC to Lt. Gen. H. M. Estes, Jr., Vice Cmdr., AFSC, 16 Nov 1962, subj.: X-20 (Dyna-Soar) System Package Program.

77 Ltr., Col. W. L. Moore, Jr., Dir., Dyna-Soar SPO to Hqs. AFSC, 13 Dec. 1962, subj.: Redline Presto Follow-on Information; ltr., Col. Moore to Hqs. AFSC, 26 Dec. 1962, subj.: Redline Presto Follow-on information.

78. TWX, AFRAE-69512, CSAF to Hqs. AFSC, 18 Jan. 1963; TWX, SCSAS-28-1-59, Hqs. AFSC to CSAF, 28 Jan. 1963.

79. Lt. Gen. H. M. Estes, Jr., Vice Cmdr., AFSC to Hqs. USAF, 30 Mar. 1963, subj.: X-20 Program, Doc, 295.

80. TWX, AFRDC—92965, Hqs. USAF to Hqs. AFSC, 12 Apr. 1963, Doc. 296.

81. Tentative System Package Program, System 620A, Dyna-Soar SPO, 15 Jan. 1963, pp. 2-27, 11-5.

82. TWX, MSFA-9-5-17, Hqs. AFSC to Hqs. ASD, 9 May 1963, Doc. 298.

83. System Package Program, System 620A, Dyna-Soar SPO, 13 May 1963, p. 1-5.

84. Ltr., Brig. Gen. D. M. Jones, Acting Cmdr., ASD to Hqs. AFSC, 13 May 1963, subj.: Revised X-20A System Package Program.

85. System Package Program, System 620A, Dyna-Soar SPO, 27 May 1963, p. 1-5.

86. TWX, AFRDDG-68949, Hqs. USAF to Hqs. AFSC, 10 June 1963, Doc. 301; TWX, SCCP3-7-2, Hqs. AFSC to Hqs. ASD, 3 July 1963.

87. System Package Program, System 620A, Dyna-Soar SPO, 3 Sept. 1963, p. 1-5.

DEPARTMENT OF DEFENSE NEWS BRIEFING
with

HON. ROBERT S. McNAMARA
SECRETARY OF DEFENSE
THE PENTAGON

December 10, 1963
12:30 p.m.

MR. SYLVESTER: This is on the record, Mr. Secretary.

SECRETARY McNAMARA: It has been a long time since we have met together. Not the least of the reasons for the intervals has been the fact that we have been extraordinarily busy the last several days and weeks, and we are working on the Defense budget and military program for fiscal 1965.

My purpose today is not to comment, upon that in any detail, but rather, to announce the cancellation of the Dyna-soar project. You will recall that during my testimony before the Congressional committees earlier this year in connection with the fiscal 1964 military program, on several occasions I indicated that there were questions regarding the desirability of completing the Dyna-soar program. I indicated that the Air Force was studying alternatives to that program and that later this year we would review the program further and consider whether it should be continued or cancelled. We have done so. We believe that it should be cancelled.

We propose to substitute for it an expanded Asset program, which I will describe in a moment, plus a new program which we are calling a manned orbital laboratory, sometimes known as Gemini X, plus a laboratory – Gemini X referring to a modification of the Gemini capsule that would be used in connection with what might be thought of as a trailer to follow behind it, which itself would be the manned orbiting laboratory.

We expect that the cancellation of Dyna-soar, plus the expansion of Assets, plus the Gemini X program, inclusive of the manned orbital laboratory, will result in expenditure of savings of approximately $100 million during the next 18 months, and this will be true, we believe, despite a substantial expansion in the usefulness of the program itself.

The Dyna-soar, as many of you know, was designed primarily to explore re-entry techniques with the hope that it would be possible to develop means of controlling re-entry more precisely, in order that a body from orbit could be moved to a precise landing field on the surface of the earth, but the Dyna-soar program, as it was originally conceived, was not intended to develop a capability for ferrying vehicles or personnel or equipment into orbit, nor was it intended that the Dyna-soar would provide a capability for extended stay in orbit, nor was it intended that it would provide a capability for placing substantial payloads, useful payloads, in orbit, and hence, it had a limited objective. It was very expensive.

It is true that there has been expended on it some $400 million at present or through the cancellation period, but there are hundreds of millions left to be spent to achieve a very narrow objective. We have broadened the objective; the objective of the Gemini X manned orbiting laboratory program will be to explore operations in space using equipment and personnel which may have some military purpose, and the objective of the Asset program will be to explore using unmanned equipment, alternative forms of re-entry, and re-entry techniques and a combination of these two, we believe, will substantially expand our knowledge of both re-entry and orbital operations, and as I say, at an expenditure savings in the next 18 months of approximately $100 million.

Now I have, I think, distributed to you, a press statement outlining substantially what I have said. I will be happy to try to answer your questions you have on this..

QUESTION: Mr. Secretary, will the astronauts who will be involved in this program, will they be Air Force only or Air Force and NASA?

SECRETARY McNAMARA: This entire program will be Air Force managed, and, therefore, the choice of personnel will be the Air Force's.

It is possible, and I think probable and desirable that NASA may, in a sense, ask the Air Force to undertake certain experiments using these vehicles for the account of NASA in the same way that the Air Force today is arranging for NASA, under its management in the pure Gemini program, to carry out certain experiments for the account of the Air Force, the Air Fore paying for those.

The important point to emphasize is that the Gemini program, insofar as it is associated with and related to the lunar landing program, is entirely under the management of NASA, and the extent to which the Air Force participates in that program is up to NASA, and done with Air Force financing, but done under NASA's management.

This program is exactly the reverse. This Gemini X and manned orbiting laboratory program is an Air Force managed program, financed by the Air Force, and to the extent any experiments are done for NASA, it would be done under Air Force management with NASA paying the incremental cost.

Now, there was a question over here.

QUESTION: You mentioned something called Acid?

SECRETARY McNAMARA: Asset: Asset is a program involving unmanned vehicles to be launched into orbit to re-enter the atmosphere to check alternative shapes and alternative re-entry techniques to obtain very much the same type of information as was the objective of Dyna-soar to obtain, but to do so with unmanned vehicles instead of manned vehicles, and this will permit the exploration of a much wider range of shapes and techniques, because we needn't be so careful as we would have to be if a man's life were at stake.

QUESTION: Can we use this rundown as a quote from you?

SECRETARY McNAMARA: Absolutely.

QUESTION: Is that A-s-s-e-t?

SECRETARY McNAMARA: A-s-s-e-t; yes.

QUESTION: How did it get the name?

SECRETARY McNAMARA: I can't answer your question.

QUESTION: Mr. Secretary, is there a requirement to get authorization through from the Houses and the Senate and House Armed Services Committee for a new project like this?

SECRETARY McNAMARA: Oh, yes; we will have the Gemini X program as a major part of our fiscal '65 program.

QUESTION: How do you get your $100 million saving in 18 months? You are adding a new program and subtracting an old program?

SECRETARY McNAMARA: That is right. And a net difference is in this 18 month period, approximately $100 million — I can't give you the whole thing from memory. I do recall the Dyna-soar numbers in particular, the fiscal '64 budget for Dyna-soar was $125.4 million and the fiscal '65 program was $146 million. The $146 million is cancelled entirely, the fiscal '65 — the fiscal '64 program of $125.4 will be cancelled effective the 15th of December. There will be the runout of the current commitments, which will have to be settled, and there will be contract cancellation costs involved.

But the resultant savings associated with the cancellation of Dyna-soar is offset by the hope for authorizations of Congress for the Gemini X will result in an adjustment for the expanded Asset program will result in the
$100 million expenditure savings.

QUESTION: Did you mention that so far the Dyna-soar had cost $400 million? Can you tell us what -

SECRETARY McNAMARA: I. said approximately $400 million.

QUESTION: Can you tell us what you have to show for that expenditure? Is there anything in the way of knowledge or anything?

SECRETARY McNAMARA: There is; yes. There is knowledge of materials and of re-entry vehicle designs. There have been substantial tests, wind tunnel and other tests, and substantial metallurgical developments carried on and this will be useful in other defense and NASA space programs.

QUESTION: Sir, what do you estimate the total eventual cost of this development program will be for MOL?

SECRETARY McNAMARA: I think that is an excellent question. It would have been difficult to answer in the same way, it would be difficult to answer what the ultimate cost of Dyna-soar would be. The current Dyna-soar program; with its very limited objective, was estimated to cost about $860 million. It seemed probable that that would overrun about $150 million, so I think you would say that the probable cost of this — what I will call the limited Dyna-soar program — was about $1 billion.

That limited Dyna-soar program provided for only approximately 12 launches, and the orbital life of the capsule was very short. The first launch would have an orbital life of about 1½ hours. It was a one orbital launch – the 12th launch would have an orbital life of something on the order of three orbits, roughly 4½ hours.

This is because the objective of Dyna-soar was not extended life in orbit, but experiments associated with precise control in re-entry, Dyna-soar has a potential for expansion of the orbital life beyond the periods I have given to you, but that expansion of the orbital life would require redesign of the capsule itself and perhaps the addition of a trailer. Both of those would result in substantial cost, the exact amount none of us can say because the program never was developed to that point.

It's that type of program that we are planning on in Gemini X, plus the manned orbiting laboratory. The total cost of this will be in the hundreds of millions of dollars, but I can't be more precise at this time. We will be before we take it to Congress in fiscal '65. It's too early to give you precise figures on that. We will include those in our fiscal '65 program to the Congress.

Needless to say, there will be a number of both unmanned and manned launches associated with Gemini X and the manned orbiting laboratory.

QUESTION: Mr. Secretary, while you are saying hundreds of millions of dollars would be the cost MOL, could you compare it with the – and I hesitate to give you the exact number – could you compare it with the $1 billion that you expected to spend on Dyna-soar?

SECRETARY McNAMARA: No, I don't wish to be more precise in this, Jack. I don't think it is wise at this time.

QUESTION: Could you tell us what decision or what knowledge you have now of this canceling of the Dyna-soar program that you did not have, say, last year?

SECRETARY McNAMARA: Yes, we have a much better understanding now on the size of equipment that we may wish to experiment with in orbit, of the support systems required to sustain life in orbit for a specific period of time, and of the total payload capacity in terms of cubic feet required for that purpose and beyond that we have a better understanding now of what can be done with unmanned vehicles such as the Asset program to experiment with both materials and techniques of precision re-entry.

All of that information has been accumulated in the past several months and has been the foundation for the decision to cancel Dyna-soar.

QUESTION: In other words, you just know a lot more now?

SECRETARY McNAMARA: We do, indeed, and we will know a lot more, of course, as we go in the future, and I think, therefore, one must expect these space programs to change over a period of years as our knowledge expands.

QUESTION: What is the military application of the MOL? You mentioned this briefly.

SECRETARY McNAMARA: The military purposes include potential navigation aids, meteorological projects, and other classified projects.

QUESTION: Inspection?

SECRETARY McNAMARA: Just other classified projects.

QUESTION: Mr. Secretary, there has been some discussion about the Air Force's role in space, and their needs. Mr. Gilpatrick has often mentioned the need to develop the ability to contact a hostile vehicle in orbit to neutralize that, if necessary. Is this a step in that direction?

SECRETARY McNAMARA: No. It is in a sense, but we will learn more about man's capabilities in space by this program, but this is an experimental program, not related to a specific military mission. I have said many times in the past that the potential requirements for manned operations in space for military purposes are not clear. But that, despite the fact that they are not clear, we will undertake a carefully controlled and carefully scheduled program of developing the techniques which would be required were we to ever suddenly be confronted with a military mission in space, and it is for that reason that you recall we proposed and the Congress approved the Titan III program. It is for that reason that we are proposing the manned orbiting laboratory; not for a precise, clearly defined, well recognized military mission, but because we feel that we must develop certain of the technology that would be the foundation for manned military operations in space should the specific need for those ever become clear and apparent.

In effect, what we are doing is shortening the lead time associated with manned military operations in space by developing a large-thrust launch vehicle and by developing a capsule and the capability for sustained manned operations in that capsule, these being the long-lead components of potential military operations in space.

QUESTION: How much money was spent on Dyna-soar, sir?

SECRETARY McNAMARA: Roughly $400 million. There was a question over here.
QUESTION: Isn't NASA supposed to be doing this same basic thing — exploring man's capabilities to exist and function as a capable human being in space?

SECRETARY McNAMARA: Their program is related to the lunar program. Gemini, the Gemini project, is a project specifically designed as a component of the lunar landing program.

QUESTION: But they are doing other things in space other than the lunar landing?

SECRETARY McNAMARA: There are none. To the best of my knowledge – you will have to ask them about this. They have no near-earth orbit manned operations planned comparable to this. Our requirements are directly related to potential military requirements. I emphasize "potential" because, as I said earlier, we haven't precisely defined manned military missions and I think we must be very careful that we don't overly anticipate those and undertake programs so large in cost as to be wasteful.

What we have constantly tried to do is hold back on particular space projects until we can be completely clear that that space project has a pure military purpose such as a transit satellite, unmanned, or is so closely related to the technology that it would be the foundation for manned military operations in space if and when those become required, that we can say we must carry them out as an insurance program. That as the character of Titan III and that is the character of Gemini X.

QUESTION: I gather from this that you do feel, so far as we know what man's military role in space may be, that it will be somewhere in the near-earth orbit?

SECRETARY McNAMARA: Yes, I think that is a very good point. That is correct. And secondly, we believe it will require some substantial cubic space in which to operate. Therefore, we must have either a capsule and/or a trailer, if you will, with sufficient cubic content to allow man or men, plus equipment, to function in space, and this project is designed with those cubic contents in mind.

QUESTION: Mr. Secretary, have you said what is the capability?

SECRETARY McNAMARA: I haven't said, but it will have a multi-man capability.

QUESTION: Would it be more than two?

SECRETARY McNAMARA: It would have more than two, although we wouldn't necessarily utilize all of them.
QUESTION. Would you have to redefine the Gemini, however, to take more than two?

SECRETARY McNAMARA: The laboratory, orbiting laboratory, is part of this whole system and the combination of the two would have a multi-man capability. Yes?

QUESTION: What order payload and what order orbit - and functioning altitude?

SECRETARY McNAMARA: Near-earth orbit, say, roughly -

QUESTION: Say 350 miles out?

SECRETARY McNAMARA: Yes, or less.

QUESTION: Mr. Secretary, do you have a formal agreement with Mr. Webb concerning this?

SECRETARY McNAMARA: Well, in the sense of — we have discussed the project and we are both in agreement that this is a wise move from the point of view of the Nation.

QUESTION: Of much less significant nature, you have that formal agreement —

SECRETARY McNAMARA: I would say we have a formal agreement in the sense that we both agree that this is a wise move for the Defense Department and we agreed with him, as we have on the Gemini, that we will, although this project is under the Air Force management, we will recognize NASA's request for participation in it to the extent that that does not compromise the Air Force mission, in the same way that Gemini has recognized the Air Force request for piggyback payloads, if you will, to the extent it doesn't compromise the lunar landing priority and requirement.

QUESTION: Is there any national decision taken that this will be the near-earth orbital project?

SECRETARY McNAMARA: Yes, I think it is first —

QUESTION: — at the Air Force's request last year for some kind of a manned orbital laboratory, and how far have we gone in the study in that direction? Has there been any drawings or any laboratory studies actually started on this particular thing?

SECRETARY McNAMARA: There has been a tremendous amount of work done over a period of time both in NASA and in the Defense Department, including the Air Force, on this whole problem of manned operations in space and there were tens of studies done on it. So the answer is yes, there has been a lot of work done on it.

QUESTION: Is this part of the program or was this decided rather quickly?

SECRETARY McNAMARA: Oh, no. We have been working on it for months.

QUESTION: Mr. Secretary, now, apparently you are going to send this thing up — the MOL capsule itself, with the Gemini coupled to it?

SECRETARY McNAMARA: That is right.

QUESTION: This indicates to me that you are trying to do this to get rid of the necessity for the rendezvous. In the first instance, they will go up together so that there will be an exchange. Rendezvous in this particular drill would come in only in the event you send a second Gemini up there.

SECRETARY McNAMARA: That is entirely right, a second Gemini X. You are quite right in your understanding of this, and the implication to what you are saying is that rendezvous, per se, is not an objective of this program, and that is correct. But rendezvous, as a capability, is a definite potential for the program.

We don't propose to develop rendezvous unless we can see a clearer requirement for it than we have at the present time. As you suggested, it would be proposed to launch these two units together. Rendezvous wouldn't be necessary unless it was desired to obtain a ferry capability.

QUESTION: Or to replace personnel?

SECRETARY McNAMARA: Right, as the ferry capability being desirable, if at all, in order to extend the life of the experiment in orbit for longer than the initial crew or initial equipment could sustain.

QUESTION: On that part, Mr. Secretary, what would the estimated life of the orbiting laboratory be in orbit?

SECRETARY McNAMARA: I would say two to four weeks.

QUESTION: Mr. Secretary, would this announcement have been made regardless of the change in Administration? Is this something that you had planned to announce at this time?

SECRETARY McNAMARA: Well, I can't say. It was President Johnson who approved it.

QUESTION: Had you been working toward this?

SECRETARY McNAMARA: Oh, we have been working on it. We have been working in this direction, as I suggested, for several months, as my testimony before Congress would indicate. Put I think it is important to recognize the emphasis that President Johnson has placed on the economy in the Defense budget, with, however, the same emphasis that President Kennedy placed on first achieving the military structure in force levels required as a proper foundation for our foreign policy, and then; exercising the utmost in economy in order to accomplish those at the lowest possible cost.

QUESTION: Mr. Secretary, first, when do you visualize this first launch might be?

SECRETARY McNAMARA: Well, I think it is too early to make any definitive statement on it, but it is estimated that the first launch of the Gemini X capsule would be an unmanned in the first half of 1966, and manned in the second half of '66, and the laboratory launches associated with Gemini X roughly a year later. In the case of the unmanned and, well, probably more than a year later, late 1967 for the manned orbiting laboratory, unmanned, and first half of '68 manned.

QUESTION: Mr. Secretary, I have a question I would like to ask. You told us about the savings in the next 18 months.

SECRETARY McNAMARA: Yes.

QUESTION: How do you visualize this as far as the following fiscal year? Is it a savings or –

SECRETARY McNAMARA: Let me say this: I personally believe this will be a very substantial savings over an extended period of years, but this gets you back into the question of what would happen to Dyna-soar. You see, the Dyna-soar was programmed at roughly $1 billion for 12 launches to lead to a very narrowly conceived objective, which clearly is not an adequate program for the future, and, therefore, what we can, I think, call Dyna-soar Phase I, the $1 billion phase, would certainly have been succeeded by Dyna-soar Phase II, and the question of the concept of Dyna-soar Phase II and the cost thereof are questions that no one has an answer to.

I began discussing them with Boeing almost two years ago, and have carried on several conversations since that time, but we have never really developed a Phase II Dyna-soar program with costs and content and, therefore, it is not possible to compare this much broader program of Gemini X plus a manned orbiting laboratory with a total Dyna-soar Phase I and II program, which would be not totally comparable, but more comparable in purpose and extent to this Gemini X manned laboratory. There is no question in my mind what these substantial savings would be, however –

Yes, this gentleman.

QUESTION: Sir, you talked about savings. Now let's talk about spending. Now, you have got $125 million to $146 million for Dyna-soar.

SECRETARY McNAMARA: That is right.

QUESTION: How much do you expect to spend on both Asset and the MOL?

SECRETARY McNAMARA; At the present time, because we haven't taken our program to the Congress, I don't want to give you the exact figures, but it will be just about $100 million less: than we would have spent on the Dyna soar.

QUESTION: Do you know, then, how much of the $125 million you have spent so far this first six months?

SECRETARY McNAMARA: We estimate that about $80 million out of the $125 million will be required to close out the contract. This is what has been spent, plus what will be spent during the cancellation phase.

QUESTION: How, is Asset a going program now?

SECRETARY McNAMARA: Yes, but on a very limited scale, and what we are proposing to do therefore, is to expand the scope of Asset to increase, the number of launches, to increase the shape, and materials studied, and substantially expand our knowledge of precision reentry gained through this unmanned program.

QUESTION: Is it envisioned, sir, that with each of the Gemini/MOL launches, you will carry the trailer with you?

SECRETARY McNAMARA: No. There will be a larger number of capsules launched than trailers launched.

QUESTION: What about the rendezvous, then?

SECRETARY McNAMARA: No, this is simply in order to test the modified capsule before we hook the trailer behind.

QUESTION: But when you put the manned laboratory up there, when you go into the manned phase of it, will each launch carry the trailer with it ?

SECRETARY McNAMARA: No, this is getting far into the future, but I think probably not. I think probably they will experiment with a man in the modified capsule before we hook the laboratory on it.

QUESTION: Mr. Secretary, as I understand it, one of the objectives of the Administration is to avoid creating any unnecessary instabilities in the military relation between the United States and the Soviet Union. Does this change from the Dyna-soar program to the MOL have any relationship whatever to that?

SECRETARY McNAMARA: No, no, not in the slightest. We have both, as you know, supported the United Nations resolution stating it is not our purpose to utilize space as an environment from which to – in which to develop the capability for use of large-scale mass destruction weapons.

QUESTION: Would the MOL be less in its experimental project and perhaps less provocative than the Dyna-soar?

SECRETARY McNAMARA: I don't think in terms of provocation it is either more or less provocative.

QUESTION: Would you anticipate that any preliminary work with industry would be done in fiscal '64?

SECRETARY McNAMARA: Yes, I do. Yes?

QUESTION: Would you indicate that that might be, programming –

SECRETARY McNAMARA: No. It will be essentially a program definition period.

QUESTION: Connected with the Gemini X?

SECRETARY McNAMARA: The Gemini X and the manned orbiting laboratory together.

QUESTION: Mr. Secretary, how long do you hope to remain Secretary of Defense?

SECRETARY McNAMARA: Well, I don't really think how long I hope to remain is as pertinent as how long the President will ask me to remain.

QUESTION: Do you want to remain any longer?

SECRETARY McNAMARA: I will stay the Secretary of Defense as long as the President believes I can effectively carry out his policies, and guess that pretty well completes the Dyna-soar-Gemini X program.

Gentlemen, it has been nice to meet with you. I will try to insure that it won't be as long an interval in the future as it has been in the past, since we last met. I think we have had enough on it fellows, as the weeks and months go by. You will have many more questions on it, and we will meet again. Thank you.

QUESTION: Do you have anything on what it will look like?

SECRETARY MCNAMARA: No, we are not in the advertising business, so we don't have any.

THE PRESS: Thank you, Mr. Secretary.

Date: Dec 1963   Title: DOD Press Release about MOL

DOD PRESS RELEASE
DATED December 10, 1963,

#1556-63

AIR FORCE TO DEVELOP MANNED ORBITAL LABORATORY

Secretary of Defense, Robert McNamara, today assigned to the Air Force a new program for the development of a near earth manned orbiting laboratory (MOL).

The MOL program, which will consist of an orbiting pressurized cylinder approximately the size of a small house trailer, will increase the Defense Department effort to determine military usefulness of man in space. This program, while increasing this effort, will permit savings of approximately $100 million over present 1964-1965 military space program expenditures.

MOL will be designed so that astronauts can move about freely in it without a space suit and conduct observations and experiments in the laboratory over a period of up to a month. The first manned flight of the MOL is expected late in 1967 or early in 1968.

In initiating the MOL program, it was decided to terminate the Dyna-Soar (X-20) program because the current requirement is for a program aimed directly at the basic question of man's utility in space, rather than a program limited to finding means to control the return of man from space, the Dyna-Soar project was designed to do the latter.

The Dyna-Soar vehicle is a one man-spacecraft, launched from a Titan III booster. It was designed to test the feasibility of maneuverability during reentry, thus allowing the pilot to choose a landing site and land in a manner similar to a conventional aircraft.

The MOL will be attached to a modified Gemini capsule and lifted into orbit by a Titan III booster. The Gemini capsule is being developed by NASA for use in the Apollo moon shot program. The Titan III is being developed as a standardized space booster by the Air Force.

Astronauts will be seated in the modified Gemini capsule during launch, and will move to the lab after injection into orbit. After completion of their tasks in space, the astronauts will return to the capsule, which will then be detached from the lab to return to earth.

The design of the MOL vehicle will permit rendezvous in space between the orbiting lab and the second Gemini capsule, so that relief crews could replace original crews in the lab. Such an operation would be undertaken as man's utility in a space environment will demonstrate and long operations in the space lab were needed.

The MOL program will make use of the existing NASA controls facilities. These include the tracking facilities which have been set up for the Gemini and other space flight programs of NASA and of the DOD throughout the world.

The lab itself will conduct military experiments involving manned use of equipment and instrumentation in orbit and, if desired by NASA, for scientific and civilian purposes. Preliminary ground or aircraft simulation will be made in all cases before full commitment to space experimentation.

The problem of reentry conditions, materials and techniques can be studied at substantially lower costs without actually using a manned vehicle like Dyna-Soar. The MOL program will permit much more extensive exploration of in-flight capabilities of the manned space vehicle. If results of the MOL and the unmanned reentry programs warrant, a new and more advanced ferry vehicle program may be initiated some years in the future.

Dictated by Willoughby-DOD Press to blp.

Date: Feb 1964   Title: Termination of the Dyna-Soar (History)

AFSC Historical Publication Series

64–51–3

History of the
AERONAUTICAL SYSTEMS DIVISION

July - December 1963
Volume III
Termination of the X-20A Dyna-Soar
(Narrative)

64 ASE–39

TERMINATION OF THE X-20A DYNA-SOAR
by
Clarence J. Geiger

Historical Division Information Office
Aeronautical Systems Division
Air Force Systems Command

Frontispiece. The X-20 vehicle in orbit (artist's drawing).

TABLE OF CONTENTS

VOLUME III
    List of Illustrations
    Chronology
    TERMINATION OF THE X-20A DYNA-SOAR
    Glossary of Abbreviations

LIST OF ILLUSTRATIONS

CHRONOLOGY

1961 December 11 The Air Force eliminated suborbital launches of the Dyna-Soar vehicle and directed early attainment of orbital flight. The objectives of the program were to obtain research data on maneuverable re-entry and demonstrate conventional landing at pre-selected sites.

1962 February 23 The Secretary of Defense, R. S. McNamara, confirmed the redirection of the Dyna-Soar program and stated that the establishment of the necessary technology and experience for manned space missions were the immediate goals of the military space program.

1963 January 18 Secretary McNamara directed a review of the X-20 program.

    19 The Secretary of Defense instructed the Air Force to re-examine the Titan III program and the Gemini program of the National Aeronautics and Space Administration.

    21 The Department of Defense completed an agreement with NASA for Air Force participation in the Gemini program.

    March 15 Secretary McNamara directed the Air Force to conduct a comparison of the military potentials of the X-20 and Gemini programs.

    May 9 The Commander of the Air Force Systems Command, General B. A. Schriever, assigned the X-20 orbital test program to the Space-Systems Division. The mission control center was to be located at the Satellite Test Center instead of Cape Canaveral.

    10 AFSC completed a report comparing the X-20 and Gemini and recommended the addition of military experiments to the Gemini program and possible further flights of the X-20.

    22 Major General O. J. Ritland, Deputy to the Commander for Manned Space Flight, AFSC headquarters, recommended to Air Force headquarters the continuation of the X-20 program and the limitation of Air Force participation in the Gemini program to a series of military experiments.

May 27 Based on an anticipated funding level of $135 million for fiscal year 1964 and firm contractor estimates of flight schedules, the X-20 office completed another revision of the system package program.

June 1 The Dyna-Soar System Program Office completed a study concerning the use of the X-20 for anti-satellite missions.

5 The Assistant Secretary of the Air Force for Research and Development, Dr. Brockway McMillan, recommended to the Secretary of Defense that the X-20 program be continued.

July 3 AFSC headquarters informed the X-20 office that the defense department would only allow $125 million instead of $135 million for fiscal year 1964.

12 The Secretary of the Air Force, E. M. Zuckert, directed that AFSC study the operational applications of the X-20 vehicle.

22 Vice President Lyndon B. Johnson requested the Secretary of the Defense to prepare a statement on the importance to national security of a space station.

31 The Commander of AFSC assigned the responsibility for the X-20 air-launch and pilot training programs to the Space Systems Division.

August 9 In his reply to the Vice President, Secretary McNamara stressed the necessity of multi-manned orbital flights of long duration.

30 The Director of Defense for Research and Engineering approved a study program for a military, orbiting, space station.

September 3 The X-20 office completed a system package program based on a funding level of $125 million for fiscal year 1964, with the first multiorbital flight delayed from August 1967 to December 1967.

12 The President's Scientific Advisory Committee requested a briefing from the Air Force on possible military space missions, biomedical experiments to be performed in space, and the capability of Gemini, Apollo, and the X-20 vehicles to execute these requirements.

23 The Dyna-Soar office completed Revision A to the system package program which detailed financial adjustments to the program if the mission control center remained at Cape Canaveral.

October 7-8 Dr. A. C. Hall, Deputy Director for Space in the Office of the Director for Research and Engineering, and Dr. A. H. Flax, Assistant Secretary of the Air Force for Research and Development, visited the Boeing facilities in Seattle, Washington, for a status briefing on the X-20 program.

23 Secretary McNamara was briefed on the Titan III and Dyna-Soar programs at the Martin Company facilities in Denver, Colorado.

November 14 The Director of Defense for Research and Engineering recommended to the Secretary of Defense cancellation of the X-20 program and initiation of a space station program.

18 With the assistance of the Boeing Company; the Minneapolis-Honeywell Regulator Company, and the Air Force Aerospace Medical Division, the X-20 office completed a report for SSD on the use of Dyna-Soar for satellite inspection missions.

29 AFSC headquarters informed the X-20 office that USAF headquarters had approved three of the proposed four military capability studies relating to Dyna-Soar.

30 Largely because of NASA objections to the space station proposal, Dr. Brown suggested to the Secretary of Defense an orbiting laboratory program, employing a Gemini capsule and a 1,500 cubic foot test module.

December 4 In a memorandum to the Secretary of the Air Force, Dr. Flax disagreed with Dr. Brown's space station proposal and argued against the cancellation of the X-20.

4 Secretary Zuckert informed the Secretary of Defense that he supported the position of Dr. Flax.

4 Major General J. K. Hester, Assistant Vice Chief of Staff, offered a space station program which employed the X-20.

5 Secretary Zuckert forwarded General Hester's proposal to the Secretary of Defense and stated that there was no reason to omit the X-20 from consideration as part of a space station program.

10 The Secretary of Defense announced the termination of the Dyna-Soar program and the initiation of the Manned, Orbiting, Laboratory program.

10 The X-20 office directed the Dyna-Soar contractors and various Air Force agencies to stop all efforts involving X-20 funds.

11 The Secretary of the Air Force directed that X-20 efforts important to other space programs be continued.

13 The X-20 System Program Office completed the first phase-out plan, and the X-20 Engineering Office compiled a list of useful efforts for continuation.

16 AFSC headquarters canceled two studies relating to the military applications of the X-20.

19-20 Representatives from various government agencies met at the system program office to determine the allocation of X-20 hardware.

20 Both the system program office and engineering office completed revisions to the termination plan and the list of efforts for possible continuation.

27 The program office again revised its termination plan.

1964 January 3 Further revisions were made to the termination plan and the list of efforts for continuation.

23 A final edition of the program office's termination plan was completed.

23 USAF headquarters informed AFSC that the Secretary of the Air Force had approved 36 tasks for continuation.

29 The X-20 Engineering Office completed a management plan for the continuation of useful X-20 efforts.

TERMINATION OF THE X-20A DYNA-SOAR

In 1963 the Department of Defense was again seriously questioning the necessity for the Dyna-Soar program. It appeared that the alternatives for the X-20 had been severely narrowed: direct the program towards achieving military goals or terminate it in lieu of another approach to a manned, military, space system. During the Phase Alpha studies of 1960 and the Manned, Military, Space, Capability Vehicle studies of 1961, the re-entry approach of the Dyna-Soar glider was critically compared with other re-entry proposals and systems. On these two occasions, both the Air Force and the Department of Defense deemed the Dyna-Soar as the most feasible. The X-20 program, however, was not as fortunate in the 1963 evaluations.

In December 1961, Air Force headquarters had eliminated suborbital launches of the Dyna-Soar vehicle and had directed the early attainment of orbital flight. The objectives were to obtain research data on maneuverable re-entry and demonstrate conventional landing at a pre-selected site.[1] Secretary of Defense Robert S. McNamara later confirmed this redirection and identified the purposes of the military space program. He stated that the establishment of the necessary technology and experience for manned space missions were the immediate goals. The Secretary placed emphasis on acquiring the ability to rendezvous with uncooperative targets, to maneuver during orbital flight and re-entry, to achieve precise recovery, and to re-use the vehicles with minimum refurbishment. In order to realize these ends, Secretary McNamara offered three programs. The orbital, research, Dyna-Soar program would provide a necessary technological basis. A cooperative effort with the National Aeronautics and Space Administration in its Gemini program would give experience in manned rendezvous. Lastly the defense secretary stated that a manned space laboratory to conduct sustained tests of military systems could be useful.[2]

It was not until January 1963 that Secretary McNamara took another significant step in defining a military space program. He directed a comparison between the Dyna-Soar program and the Gemini program of NASA to determine which would be of more military value.[3] Gemini became even more important a few days later when the Department of Defense completed an agreement with the national aeronautics administration for Air Force participation. Following a review in the middle of March of the Dyna-Soar program, Secretary McNamara further clarified his directions concerning the Gemini and X-20 study. He considered that the Air Force had placed too much emphasis on controlled re-entry and not on the missions which could be performed in orbit. Inspection, reconnaissance, defense of space vehicles, and the introduction of offensive weapons in space were all significant. He suggested that the Air Force take as long as six months to determine the most practicable test vehicle for these military space missions. The Secretary of Defense then suggested that a space station serviced by a ferry vehicle could be the most feasible approach.[4] Air Force headquarters directed the Air Force Systems Command to organize studies concerning X-20 and Gemini contributions to these four missions.[5]

By 10 May, a committee, under the leadership of the Space Systems Division and composed of representatives from the Aerospace Corporation, Air Force Systems Command headquarters, and the Aeronautical Systems Division, completed a comparison of Gemini and the X-20. The committee considered that the current X-20 program could be rapidly, and with relative economy, adapted for testing of military subsystems and military operations. There were several reasons. The Dyna-Soar glider had a payload volume of 75 cubic feet, sufficient power, and enough cooling capacity to accommodate subsystems required for military missions. Furthermore, the orbital duration of the vehicle could be extended to 24 hours or longer.

Concerning reconnaissance missions, the committee thought that the X-20 program could develop low, orbital, operational techniques and ground recognition ability. The research data from the program would also be applicable for the verification of the feasibility, design and employment of glide bombs. The fact that the X-20 would develop maneuvering techniques and quick return methods made the program valuable for the development satellite defensive missions. Since deceleration occurred slowly during lifting re-entry, such an approach would provide a safe physiological environment for transfer of personnel from space stations and for other logistical missions. Lastly, significant information for the development of future maneuvering re-entry spacecraft would be obtained from the X-20 program.

The committee then detailed the necessary modifications to the X-20 glider in order to allow the incorporation of weather reconnaissance or satellite inspection equipment. A test program of four X-20A flights, and two demonstration flights, would total $206 million from fiscal years 1964 through 1968. The same type of program, this time for the testing and demonstration of inspection subsystems, would total $228 million.[6]

In contrast, the technology being developed by the Gemini program of NASA related to the ability to rendezvous and orbit for long duration. The committee estimated that to incorporate a series of military experiments into the current NASA program with only minor equipment and operational flight changes would total about $16.1 million from fiscal years 1964 through 1966. If the Department of Defense conducted two Gemini launches and employed the same booster as NASA, the Titan II, the cost for inspection and reconnaissance experiments would total $129 million from fiscal years 1964 through 1967. If six Department of Defense flights were conducted, the total would be $458 million. The committee then considered a series of Gemini launches conducted by the Department of Defense, this time using the Titan IIIC. Because the 5,000 pound Gemini capsule only had a limited payload capacity of 10 cubic feet, the committee considered the addition of a mission module, which would have to be discarded in space, to the Gemini capsule. The largest test module which was considered had a volume of 700 cubic feet. The committee then examined the applicability of such a test system to reconnaissance and inspection missions. Considering a six flight program beginning in July 1966, with the following flights at five month intervals, an inspection test flight program would total $509 million and a reconnaissance flight test program would cost $474 million.[7]

The committee concluded that the main advantage of the Gemini vehicle was that it was lighter than the X-20 and consequently could carry more fuel for orbital maneuverability or have a larger payload. The inherent advantage of the X-20 was its maneuverability during re-entry, which meant that it could land quicker and with more landing site options. The committee recommended that a series of military experiments should be implemented in the NASA Gemini program and that additional flights of the X-20 might be warranted. Both systems could be modified to perform reconnaissance, inspection, satellite defense, and logistical missions; however, neither would directly provide a means of introducing offensive weapons into earth orbit.[8]

On 22 May, Major General O. J. Ritland, Deputy to the Commander for Manned Space Flight, AFSC headquarters, forwarded the report to Air Force headquarters with the recommendation that the X-20 program be continued because of the contribution a high lift-to-drag ratio vehicle could make to future military systems. Air Force participation in the Gemini program should be limited to incorporating a series of military experiments into the NASA program.[9]* A few weeks later, Brockway McMillan, the Assistant Secretary of the Air Force for Research and Development, summarized the report in a memorandum to the Secretary of Defense. The assistant secretary recommended that the X-20 program be energetically continued. He suggested that further examination of the military applications of the X-20 and Gemini be extended under various study programs.[10]

At the request of AFSC headquarters, the program office then completed a study concerning the use of the X-20 in anti-satellite missions. The Dyna-Soar office proposed an X-20B which would have an interim operational capability of satellite inspection and negation. The program office suggested that the last six flights of the current X-20A program be altered to carry inspection sensors and additional fuel for space maneuver demonstration. Two additional flights would be added to demonstrate an interim operational capability. This would necessitate a weight reduction to the X-20 glider of 700 pounds which could be achieved through a series of design changes. Such a program would total $227 million from fiscal years 1964 through 1968. To conduct a 50 flight operational program following the completion of the two demonstration flights would cost $1.229 billion from fiscal years 1965 through 1972.[11]

---

*   Secretary McNamara approved the incorporation of Air Force experiments in the NASA Gemini program on 20 June 1963.

Near the end of June 1963, the Space Systems Division requested the X-20 office to conduct, as part of the 706 Phase 0 studies, an analysis which would show the capability of the Dyna-Soar vehicle and modified and versions to fulfill satellite inspection missions.[12] With the assistance of the Boeing Company, the system contractor, the Minneapolis-Honeywell Regulator Company, an associate contractor, and the Air Force Aerospace Medical Division, the Dyna-Soar office completed its report by the middle of November. This study offered an inspection vehicle, the X-20X, which could have provisions for a one or two-man crew, permit orbital flight for 14 days, and be capable of inspecting targets as high as 1,000 nautical miles. The Dyna-Soar office estimated a first flight date of the X-20X in September 1967 and a probable funding requirement, depending upon the extent of modifications, ranging from $324 million to $364.2 million for fiscal years 1965 through 1971.[13]

Since the completion of the Step IIA and IIB studies by Boeing in June 1962, the Dyna-Soar office had on several occasions, requested funds for intensive military application studies, and, on 8 July 1963, W. E. Lamar, Director of the X-20 Engineering Office, reiterated this request during a presentation to the Secretary of the Air Force, E. M. Zuckert.[14] A few days later, Secretary Zuckert, attending a meeting of the Designated Systems Management Group, directed studies of the operational applications of Dyna-Soar. He stated that the X-20 program would probably prove to be invaluable to the national military space program.[15]

Before the purpose of these studies was clarified, the future of the Dyna-Soar became tied to a projected space station program. On 22 July, Vice President Lyndon B. Johnson raised the question of the importance of space stations to national security and requested the Secretary of Defense to prepare a statement on this subject.[16] Secretary McNamara replied a few days later and stressed a factor which the Air Force now had to consider: multi-manned orbital flights of long duration. The Secretary outlined some premises upon which America's manned, military, space program was to be based. He stated that the investigation of the military role in space was important to national security. Because there was no clearly defined military space mission, present efforts should be directed towards the establishment of the necessary technological base and experience in the event that such missions were determined. The Secretary of Defense pointed out that Air Force participation in the Gemini program would provide much of this technological base. He considered that an orbital space station could prove useful in conducting experiments to improve capability in every type of military mission. Such a system could even evolve into an operational military vehicle. Secretary McNamara informed Vice President Johnson that he hoped to have the characteristics of an orbital space station delineated by early 1964.[17]

In September, a subcommittee of the President's Scientific Advisory Committee Space Vehicle Panel was formed to review the available data relative to a manned orbiting station. The President's Office of Science and Technology requested the Air Force to brief the subcommittee on possible military space missions, biomedical experiments which could be performed in space, and the capability of Gemini, Apollo, and the X-20 vehicles to execute these possible future requirements.[18]

Additional instructions concerning the briefing to the President's Scientific Advisory Committee were relayed from the Director of Defense for Research and Engineering by Air Force headquarters to the Aeronautical Systems Division. Considerations such as modifications of the X-20 and discussion of an orbital space station should be emphasized. Air Force headquarters pointed out that the Department of Defense was not convinced that an orbital space station was needed. Rather a study of the requirements to test military equipment in space was necessary to answer questions such as equipment characteristics and the usefulness of man in space.[19]

A few days later, Dr. Lester Lees, chairman of the subcommittee, gave additional information to Mr. Lamar about the coming presentation. Emphasis was to be on specific, meaningful experiments which the Air Force could conduct with either Gemini, Apollo, or the X-20, in order to provide a technological basis for future military space missions. Dr. Lees pointed out that it would be necessary to convince a number of governmental officials that military man had a definite mission in space. The usual arguments for manned space flight such as decision-making and flexibility were inadequate. The subcommittee chairman stated that more specific reasons must be given or it was unlikely that extensive funds would be available for the development of manned space systems.[20]

The briefings to the President's Scientific Advisory Committee on 10 October essentially covered the findings concerning Gemini and the X-20 in the earlier 10 May report of the Air Force to Secretary McNamara. More detail, however, was presented on the use of the X-20 as a shuttle vehicle capable of rendezvous and docking. A configuration of the X-20 with an orbital development laboratory was also considered.[21] After completion of the presentations, Dr. Lees commented to Mr. Lamar that although he had previously been against the continuation of the Dyna-Soar program he now saw a definite need for the X-20. He would no longer oppose the program. [22]

By the end of October, the purpose of the Dyna-Soar capability studies, which Secretary Zuckert had agreed to in July, were clarified, Following the instructions of Air Force headquarters, Lieutenant General H. M. Estes, AFSC Vice Commander, informed Major General R. G. Ruegg, ASD Commander, that the purpose of the first study was to formulate a program of military space experiments involving only engineering changes to the X-20 subsystems. The Vice Commander added that this program of experiments should be compared to a similar one employing the Gemini vehicle to ensure that the Dyna-Soar approach offered the most economical and effective means of accomplishment. A second study would integrate the findings of various other studies and establish a series of mission models for reconnaissance, surveillance, satellite inspection, and also logistical support of a space station. A third study was to examine the future operational potential of re-entry vehicles having a lift-to-drag ratio greater than the X-20. A final study would examine the economic implications of various modes of recovering space vehicles from near-earth orbit.[23] At the end of November, AFSC headquarters informed the X-20 office that Air Force headquarters had approved all but the second proposal which had just been submitted.[24]*

Early in October 1963, General B. A. Schriever, AFSC Commander, informed ASD and SSD that the Secretary of Defense intended to visit the Martin Company facilities at Denver, Colorado, to receive briefings on the status of the X-20 and Titan III programs.[25] Colonel W. L. Moore, X-20 program director, later noted that the directions were somewhat in error because it became apparent during these presentations that Secretary McNamara desired far more than a status briefing.[26]

Prior to these briefings, there were numerous indications that the future of the Dyna-Soar program was uncertain. Several X-20 displays and activities had been planned for the Air Force Association convention which was to be held in the middle of September. One of the proposed events involved the continuous showing of a brief film on the nature and objectives of the Dyna-Soar program. Although this film was an updated version of one previously unclassified and released, the Office of the Secretary of Defense refused its clearance for the convention.[27] Furthermore, neither Dr. A. C. Hall, Deputy Director for Space in the Office of the Director of Defense for Research and Engineering, Dr. A. H. Flax, now Assistant Secretary of the Air Force Research and Development, indicated agreement to a briefing by the Air Force Plant Representative at Boeing on the necessity for manned, military, space flight.[28] It was reported that some X-20 Boeing officials became concerned over the future of the program after this visit.[29] In addition, the Director of Defense for Research and Engineering, Dr. Harold Brown, had not approved the release of funds for X-20 range requirements. The AFSC Vice Commander was concerned and considered that the range operational date of October 1965 for the Dyna-Soar program was certainly in jeopardy.[30] Lastly, Dr. Brown, in a speech before the United Aircraft Corporate Systems Center at Farmington, Connecticut, appeared critical of the Air Force, manned, space programs. He stated that both the Gemini and X-20 programs had very limited ability to answer the question of what man could do in space. Unless an affirmative answer were found, there would be no successor to these programs.[31]

A few days later, on 23 October, Secretary McNamara, accompanied by R. L. Gilpatric, Deputy Secretary of Defense, Harold Brown, and Brockway McMillan, now Under Secretary of the Air Force, were briefed by Titan III and X-20 officials. At the conclusion of his presentation, Colonel Moore stated that it would be

* On 16 December, AFSC headquarters canceled the first two studies, both of which dealt directly with the Dyna-Soar program.

desirable to have the Department of Defense publicly state its confidence in the Dyna-Soar program. The X-20 director then asked if there were any questions.[32]

Both Secretary McNamara and Dr. Brown asked a series of questions directed towards obtaining information on the necessity of manned, military, space systems. Secretary McNamara stated that the X-20 office had been authorized to study this problem since March 1963. He emphasized that he considered this the most important part of the X-20 program. The Secretary of Defense wanted to know what was planned for the Dyna-Soar program after maneuverable re-entry had been demonstrated. He insisted that he could not justify the expenditure of about $1 billion for a program which had no ultimate purpose. He was not interested in further expenditures until he had an understanding of the possible space missions. Only then would the department give a vote of confidence to the X-20 program. Secretary McNamara then directed Dr. McMillan to get the answers.[33]

Some of the participants arrived at varying conclusions concerning the reaction of Secretary McNamara to the briefing. Mr. J. H. Goldie, BOEING'S X-20 chief engineer, thought that the Secretary of Defense did not appear to be firmly against the X-20 nor in favor of Gemini. Rather, Secretary McNamara seemed willing to allow the Air Force to use the X-20 as a test craft and a military system if a case could be adequately made for a manned, military, space system.[34] Mr. Lamar concluded that the Secretary of Defense was not satisfied with the response and that "drastic consequences" were likely if an adequate reply were not made.[35] Colonel Moore prophetically stated that Secretary McNamara "probably will not ask us again."[36]

Just as serious as Secretary McNamara's reception of the X-20 briefing was the refusal of the Department of Defense to sanction a revision of the system package program. From May through September 1963, several changes involving the test organization and funding were made to the X-20 program. On 9 May 1963, General Schriever had directed that the Dyna-Soar orbital test program be assigned to the Space Systems Division. The AFSC commander further ordered that the mission control center be located at the Satellite Test Center in Sunnyvale, California, instead of the Air Force Missile Test Center.[37] The 27 May 1963 system package program reflected this change in the test program and registered a requirement of $135 million for fiscal year 1964.

While Air Force headquarters approved this system package program in June, the Department of Defense would only allow $125 million for fiscal, year 1964. On 3 July, the Air Force Systems Command headquarters informed the X-20 office that attempts to obtain the higher funding level had failed.[38] The Director of Defense for Research and Engineering considered that the primary purpose of the program was to acquire data on maneuverable re-entry. Incorporation of multiorbital flight was only of secondary importance, and the X-20 office could defer the first multiorbital flight date to remain within budget limitations.[39] AFSC headquarters then directed that a revised system package program be completed by early September.[40] Before this could be accomplished, General Schriever transferred not only orbital test direction to the space division but also responsibility for the air-drop program and the training of X-20 pilots.[41] These additional changes would also have to be incorporated into the revised system package program.

The 3 September program package presented the adjusted financial estimates and flight schedules. Considering that $125 million had been authorized for fiscal year 1964 and a total of $339.20 million had previously been expended, the program office estimated that $139 million would be required for 1965, $135.12 million for 1966, $93.85 million for 1967, $31.85 million for 1968, and $3 million for 1969. The total cost for the Dyna-Soar program would amount to $867.02 million. The reduction of fiscal year 1964 funds was absorbed by delaying the necessary modifications for multiorbital flight and deferring the date of the ninth ground-launch (the first multiorbital flight) from August 1967 to December 1967. The 20 air-launches were to occur from May 1965 through may 1966, and the two unmanned ground-launches were to take place in January 1966 and April 1966. The first piloted ground-launch was to occur in July 1966, and the last piloted flight was to be conducted in February 1968.[42]

Soon after the issuing of this program package, there was some concern over the expense involved in locating the mission control center at Sunnyvale. Colonel Moore estimated that this relocation would increase program costs by several million dollars. [43] Major General L. I. Davis, a special assistant to the AFSC Vice Commander, supported this argument by stating to General Schriever that many of the functions necessary for launch control were also necessary for mission control. It would be less expensive to keep both control centers at the Air Force Missile Test Center.[44]

At the request of AFSC headquarters, the X-20 office forwarded, on 23 September, a revision of the 3 September system package program, which detailed adjustments to program costs if the mission control center remained at Cape Canaveral. The X-20 office estimated that $138.13 million would be required for fiscal year 1965, $130.66 million for 1966, $88.34 million for 1967 and $31.09 million for 1968. The total program cost would amount to $853.23 million instead of the previously estimated $867.02 million.[45] On 17 October 1963, AFSC headquarters forwarded the system package program to the Air Staff, informing them that it was more feasible to locate the mission control center at the missile test center.[46] This program package did not receive the endorsement of either headquarters. As late as 21 November, the X-20 assistant director, J. B. Trenholm, reminded AFSC headquarters that it would be beneficial to the program if the systems command would approve of the program package.[47]

It had been reported that, on the day following the 23 October 1963 briefing to Secretary McNamara, Dr. Brown had offered a manned, orbiting, laboratory program to the Air Force in exchange for Air Force agreement to terminate the X-20 program. General C. E. LeMay, the Air Force Chief of Staff, did not agree and directed an Air Force group to prepare a rebuttal to such a proposal.[48] Previously, in August, Dr. Brown had approved an Air Force request to conduct a study of an orbital space station. He authorized the expenditure of $1 million for fiscal year 1964. The Air Force was to focus on the reconnaissance mission with the objective of assessing the utility of man for military purposes in space. In determining the characteristics of such a station, the Air Force should consider the use of such programs as the X-15, the X-20, Mercury, Gemini, and Apollo. This study had to be concluded by early 1964.[49]

Before the completion of this space station study, however, Dr. Brown recommended a program for such an effort to Secretary McNamara in a 14 November 1963 memorandum. The Director of Defense for Research and Engineering analyzed varying sizes of space station systems which would incorporate either the Gemini or Apollo capsules as ferry vehicles and would employ either the Titan II, the Titan IIIC, or the Saturn IB booster. Two of the approaches were suitable. One would involve the use of the Lunar Excursion Module (LEM) adapter as a space station and the Saturn IB as the booster. The Apollo command module and the Titan IIIC would perform the logistics function. Dr. Brown estimated that this approach would cost $1.286 billion from fiscal years 1964 through 1969. The first, manned, ferry launch could take place in late 1966, and active station tests could be conducted by late 1967.

The alternative which the Director of Defense for Research and Engineering preferred was to develop a space station with provisions for four men, use the Gemini capsule as a ferry vehicle, and separately launch both the station and capsule with a Titan IIIC booster. From fiscal years 1964 through 1968, this approach would total $983 million. The first, manned, ferry launch could occur in the middle of 1966, and active space station tests could begin in the middle of 1967.

Dr. Brown, however, was concerned because both of the recommended approaches would employ primitive landing methods, and, consequently, he suggested the development of a low lift-to-drag ratio vehicle which could perform maneuverable re-entry and conventional landing. The Director of Defense for Research and Engineering suggested that models of such a craft be tested in the Aerothermodynamic, Structural Systems, Environmental, Test program (ASSET) during 1964 and 1965, and he estimated that an improved ferry vehicle could be available for later station tests. The total for this more sophisticated vehicle program would amount to $443 million for fiscal years 1964 through 1968.

Dr. Brown's recommendation to Secretary McNamara was brief: cancel the X-20 program and initiate the Gemini approach to a manned, military, space station. Management of the Gemini program should be transferred from NASA to the Department of Defense by October 1965.[50]

Discussions between National Aeronautics and Space Administration and Department of Defense officials made it clear that the space agency would agree to a coordinated, military, space program, but it was not prepared to support a space station program. Instead NASA suggested a program for an orbiting military laboratory which did not involve ferrying, docking, and resupplying. On 30 November, Dr. Brown, in another memorandum to Secretary McNamara, analyzed an approach more agreeable to NASA. This alternative would involve the orbiting by a Titan IIIC booster of a Gemini capsule and a 1,500 cubic foot test module, capable of supporting two to four men for 30 days. Dr. Brown maintained that such an approach could easily be converted into the Gemini alternative he had recommended on 14 November. This simplified approach would total $730 million from fiscal year 1964 through 1968, and the manned, orbital, test program could be conducted in late 1967. Dr. Brown, however, advised the Secretary of Defense that the space station proposal of 14 November was still the most feasible and should be initiated.[51]

While NASA had suggested a simplified Gemini approach, it by no means concurred with the proposed termination of the X-20 program. The Associate Administrator for Advanced Research and Technology, Dr. R. L. Bisplinghoff, pointed out that advanced flight system studies had repeatedly shown the importance of developing the technology of maneuverable hypersonic vehicles with high-temperature, radiation-cooled, metal structures. Test facilities were unable to simulate this lifting re-entry environment, and, consequently, X-20 flights were necessary to provide such data. NASA had always supported the Dyna-Soar program and should it be canceled the space agency would have to initiate a substitute program.[52]

In order to achieve the objective of obtaining data on re-entry, Dr. Bisplinghoff recommended some changes to the Dyna-Soar program. After completion of an adequate air-drop program and a satisfactory unmanned ground-launch flight, a piloted orbital flight should be conducted.[53] Dr. Brown requested Dr. Flax to examine such an alternative for the X-20.[54] With the assistance of the X-20 program office and AFSC headquarters, Dr. Flax completed his reply on 4 December. He estimated that such a curtailed program would reduce the total cost by $174.4 million through fiscal year 1969. He pointed out, however, that such an approach would result in the loss of technical data which would be disproportionate to the financial savings.[55]

On the same day, in another memorandum to the Secretary of the Air Force, Dr. Flax firmly disagreed with the recommendations of Dr. Brown's 14 November memorandum. The Assistant Secretary pointed out that the X-20 had not been given serious consideration as an element in any of the space station proposals. He emphasized that major modifications were necessary to both the Gemini and the X-20 if either were to be employed in an orbital station program. Furthermore, the Dyna-Soar approach possessed several advantages: the vehicle could make emergency landings without the costly deployment of air and sea elements and there would be a more tolerable force of vehicle deceleration during-re-entry. Dr. Flax continued by emphasizing the importance of the X-20 program. Its technology not only supported the development of re-entry vehicles, including Dr. Brown's improved ferry vehicle, but also an entire class of hypersonic winged-vehicles. Since about $400 million had already been expended on the X-20 program, the Assistant Secretary severely questioned the proposal to cancel Dyna-Soar and initiate a new program with similar objectives. While he endorsed the purposes of the space station program, Dr. Flax believed that the decision to begin such a program was independent of the question to terminate the X-20.[56]

On the same day, Secretary of the Air Force Zuckert forwarded Dr. Flax's memorandum to Secretary of Defense McNamara with the statement that it represented the best technical advice available in the Air Force. The Secretary of the Air Force added that both he and Dr. Brockway McMillan were in accord with Dr. Flax's position. Secretary Zuckert further stated that he did not wish to see the Air Force abandon a program such as Dyna-Soar and start a new program which perhaps had been projected upon optimistic schedules and costs.[57]

As an Air Force reply to Dr. Brown's 14 November memorandum, Major General J. K. Hester, the Assistant Vice Chief of Staff, suggested to the Secretary of the Air Force several alternatives for varying sizes of space stations, all of which employed the X-20 vehicle. The first alternative offered an extended

X-20 transition section which would provide a module of 700 cubic feet. This would be a two-man station employing an X-20 launched by a Titan IIIC. The second approach comprised a separately launched two-room station by the Titan II. This would have 1,000 cubic feet of volume and would be serviced by an X-20 shuttle vehicle boosted with a Titan IIIC. The third alternative, recommended by General Hester as the most feasible, involved a five man station, launched by Titan IIIC and capable of orbiting for one year. This approach would require $978.4 million from fiscal years 1964 through 1969 for the development of a space station and the X-20 ferry vehicle. The Assistant Vice Chief of Staff considered that the first space station launch could take place by the middle of 1967. With an X-20 approach to a space station program, it was not necessary to have a separate program for an improved ferry vehicle. Rather, only an annual funding level of $6.4 million for the ASSET program was necessary to advance space technology. General Hester, therefore, recommended the initiation of a space station program employing the X-20 and, if economy were essential, the cancellation of the Gemini program.[58]

On the next day, Secretary Zuckert forwarded General Hester's memorandum to Secretary McNamara. The Air Force Secretary stated that the Air Staff study clearly indicated that there was no definite reason for omitting the X-20 from consideration as a re-entry vehicle for an orbital space station or orbital laboratory program. This was particularly important because of safety and cost advantages which the X-20 offered for long duration orbital missions. Secretary Zuckert believed that the X-20 alternative deserved serious consideration.[59]

On 8 December, a rumor circulated in Air Force headquarters that the Defense Department had reduced X-20 fiscal year 1964 funds from $125 million to $80 million and had not allocated any money for fiscal year 1965.[60] The next day, defense officials conferred with President Johnson. Apparently, Secretary McNamara recommended the termination of Dyna-Soar, and the President agreed.[61] On 10 December, the Secretary of Defense announced the cancellation of the X-20 project. The program had been reviewed, alternatives studied, and the decision made. In its place would be a manned orbital laboratory (the NASA proposal which Dr. Brown explained in his 30 November 1963 memorandum). The Secretary of Defense also stated that there would be an expanded ASSET program (the improved ferry vehicle program which Dr. Brown offered in his 14 November memorandum) to explore a wide range of re-entry shapes and techniques. By taking the Gemini approach to a space program, Secretary McNamara estimated that $100 million would be saved in the following 18 months.

The Secretary of Defense explained his reasons for canceling the X-20. He stated that the purpose of the program had been to demonstrate maneuverable re-entry and landing at a precise point. The Dyna-Soar vehicle was not intended to develop a capability for carrying on space logistics operations. Furthermore, the X-20 was not intended to place substantial payloads into space, nor fulfill extended orbital missions. The Secretary of Defense stated that about $400 million had already been expended on a program which still required several hundred million dollars more to achieve a very narrow objective.[62]

A few days after the termination announcement, Dr. Brown, in a memorandum to the Secretary of the Air Force, replied to the arguments of Dr. Flax and General Hester. Dr. Brown stated that before reaching a decision the Air Force alternatives were carefully considered. There were three objections. The Air Force recommended program involved construction of a space station and a new and larger X-20. The Department of Defense considered that such a large step was not justified and a test module and Gemini vehicle were chosen as the logical first step. Furthermore, the Air Force suggestion to cancel Gemini was not within the power of the Department of Defense since this was a NASA program. Lastly, the Air Force recommendation involved a greater degree of schedule risk than the chosen program. The Air Force proposal could not be accepted as a feasible substitute for the Manned, Orbiting, Laboratory program.[63]

Following Secretary McNamara's news conference on 10 December, Air Force headquarters informed all of its commands of the termination of the X-20 and the initiation of an orbital laboratory program.[64] On the same day, General Schriever met with some of his staff to discuss the new space approach. He stated that both the orbiting laboratory and the expanded ASSET programs would be placed under the management of the Space Systems Division.[65] Later, General Schriever requested the Commander of the

Research and Technology Division, Major General Marvin C. Demler, to aid the space division in the preparation of a new ASSET development plan. The objective of this program as first announced by Dr. Brown remained unchanged: the development of an advanced ferry vehicle.[66]

Although official instructions were not received from AFSC headquarters until 17 December, the X-20 program office instructed the Dyna-Soar contractors and various Air Force agencies on 10 December to stop all activities involving the expenditure of X-20 funds.[67] On the next day, Secretary Zuckert authorized the Air Force to terminate the X-20 program; however, it was to continue certain X-20 efforts which were deemed important to other space programs. A preliminary report was due no later than 16 December.[68] The day following this direction, the ASD program office recommended the continuation of ten activities: studies of pilot control of booster trajectories, fabrication of the Dyna-Soar heat protection system, construction of the full pressure suit, fabrication and testing of the high temperature elevon bearings, final development testing of the nose cap, flight testing on the ASSET vehicle of coated molybdenum panels, final acceptance testing of the test instrumentation subsystem ground station, development of the very high frequency (VHF) search and rescue receiver and transmitter, employment of existing Boeing simulator crew station and flight instruments for further research, and development of certain sensing and transducing equipment for telemetry instrumentation.[69]

On 18 December, Air Force headquarters informed the program office that the Secretary of the Air Force had approved the ten items, and funding for continuation of these contracts would be limited to $200,000 a month.[70]

The X-20 engineering office, however, had recommended a list of several items for reinstatement which were in addition to the ten efforts continued by the program director. The X-20 Program Director had not supported the engineering office items either because he did not consider them of sufficiently wide applicability or he could not adequately establish their merit.[71] This list, however, was revised on 14 December by representatives from AFSC headquarters, the Space Systems Division, the Aeronautical Systems Division, and the Research and Technology Division. The officials decided to identify the items not only by technical area, as originally presented by the engineering office, but also by four categories. Category A involved efforts whose cost for completion would be equal to the termination expense. Category B comprised items which were applicable to various space programs. Category C included items which would contribute to the advancement of the state-of-the-art. The final classification, Category D, contained efforts which possessed a potential future use.[72]

On 20 December 1963, a revision of this list had been completed and coordinated with the laboratories of the Research and Technology Division. The items were classified both by technical area and the suggested categories. At the end of the month, officials from USAF headquarters, AFSC headquarters, ASD, and RTD again reviewed proposed items for continuation, and this time a new classification was suggested. Category I included items which would advance the state-of-the-art. Category II involved items which only required feasibility demonstration or design verification. Category III comprised equipment which was nearly completed, and Category IV were efforts which necessitated further justification.[73]

By 3 January 1964, a last revision of the proposed useful efforts had been completed. A Category V was added which included items that had been suggested for continuation by various organizations but were considered unacceptable by the X-20 engineering office. Essentially, the engineering office recommended for continuation the 38 efforts which comprised Categories I, II, and III. Included in these were the ten items which were being continued by the program office itself. A few days later, General Estes requested from USAF headquarters authority to retain sufficient funds for program termination, which would include $3.1 million for the completion of the first three categories.[74] On 23 January, USAF headquarters informed AFSC that the Secretary of the Air Force had approved, with the exception of two items, all the efforts listed under the first three categories. The Air Force would allow an expenditure of $70 million from fiscal year 1964 funds for the Dyna-Soar program, $2.09 million of which would be directed towards completing the three categories.[75]* The Research and Technology Division was then assigned authority to formulate a management plan for completion of this work.[76] The X-20 engineering office completed a plan at the end of January, recommending that separate contracts be negotiated for the three categories of items which had not been already reinstated. These contracts would be administrated by the Research and Technology Division except for two which were to be transferred to the Air Force Missile Development Center and the Air Force Flight Test Center.[77] While Air Force headquarters did not give an official approval, this plan was put into operation.

The Air Force calculated that Boeing had completed 41.74 percent of its tasks. The Minneapolis-Honeywell Regulator Company, the associate contractor for the primary guidance subsystem, had finished 58 percent, and the Radio Corporation of America, the associate contractor for the communication and tracking subsystem, had completed 59 percent of its work. At the time of Secretary McNamara's announcement, Boeing had 6,475 people involved in the X-20 program, while Minneapolis-Honeywell had 630 and RCA, 565. The governmental expenditure for these contracts amounted to $410 million.[78]

While it had only approximately reached mid-point, the Dyna-Soar program definitely advanced the technology of radiation-cooled structures. Thirty-six X-20 tasks were continued and would directly contribute to other Air Force space efforts. Also significant was the initiation of an expanded ASSET program directed towards the development of a lifting, re-entry, shuttle vehicle. Paradoxically, the cancellation of X-20 development apparently made the maneuverable re-entry concept far more acceptable to the Department of Defense and some elements of the Air Force than it had been during the existence of the Dyna-Soar program.

*   For a list of the 36 items which were continued, see document 107.

REFERENCES

1.   TWX, AFSDC-85081, Hq. USAF to Hq. AFSC, 11 Dec. 1961; SPD, DCS/Sys. & Log., Hq. USAF, 27 Dec. 1961.

2.   Memo., R. S. McNamara, Secy. of Def. to E. M. Zuckert, SAF, 23 Feb. 1962, subj.: The Air Force Manned Military Space Program.

3.   Memo., Secy. of Def. to Harold Brown, DDR&E, 18 Jan. 1963, subj.: Gemini and Dyna-Soar.

4.   Memo., Brockway McMillan, ASAF/R&D to SAF, 15 Mar. 1963, subj.: Dyna-Soar Briefing to Secretary McNamara.

5.   Memo., ASAF/R&D to SAF, 18 Mar. 1963, subj.: Response to McNamara's Questions.

6.   Rpt., Dep/Tech., SSD, 10 May 1963, subj.: Response to Secretary McNamara's 15 March 1963 Questions, I, 2-19, to 2-23, 2-27 to 2-29, 2-37, 2-47c, 2-47h.

7.   Ibid., 2-54c to 2-54j, 2-55 to 2-57, 2-63, 2-79, 2-87.

8.   Ibid., ix-xiii.

9.   Ltr., Maj. Gen. O. J. Ritland, Dep. to the Cmdr/Manned Space Flight, Hq. AFSC to Hq. USAF, 22 May 1963, subj.: Response to Secretary McNamara's 15 March 1963 Questions.

10.   Memo, ASAF/R&D to Secy. of Def., 5 June 1963, subj.: Review of Air Force Space R&D Program, Doc. 7.

11.   Rpt., X-20 SPO, ASD, 1 June 1963, subj.: X-20 Anti-Satellite Mission, pp. 7, 8, 12, 21.

12.   Ltr., Col. W. D. Brady, Acting Dep/Tech., SSD to Col. W. L. Moore, Prog. Dir., X-20 SPO, 25 June 1963, subj.: Program 706, Phase 0 Study, Doc. 12.

13.   Rpt., X-20 SPO, 18 Nov. 1963, subj.: Phase 0 Study of X- 20 Program 706, 1, 2, 14, 15.

14.   Presn., W. E. Lamar, Dir., X-20 Engg. Ofc., RTD to SAF, 8 July 1963, subj.: X-20 Status Report.

15.   Minutes, Col. C. R. Tosti, Exec. Secy., DSMG, Hq. USAF, 23 July 1963, subj.: Sixty-fifth Meeting, Designated Systems Management Group, 12 July 1963, Doc. 19.

16.   Memo., L. B. Johnson, Vice Pres. to R. S. McNamara, Secy. of Def., 22 July 1963, subj.: Space Stations, Doc. 18.

17.   Memo., McNamara to Johnson, 9 Aug. 1963, subj.: Orbital Space Station, Doc. 24.

18.   Memo., N. E. Golovin, Ofc., Science and Tech., Exec. Ofc. of the Pres. to A. C. Hall, Dep. DDR&E, 12 Sept. 1963, subj.: Briefing

for Lees Subcommittee of PSAC Space Vehicle Panel, Doc. 30.

19. Ltr., Maj. Gen. W. B. Keese, DCS/Plans, Hq. USAF to Hq. ASD, 19 Sept. 1963, subj.: Briefing to PSAC Panel on 10 October 1963, Doc. 32.

20. Trip rpt., Lamar, 23 Sept. 1963, subj.: Briefing for President's Advisory Committee – Space Vehicle Panel, Doc. 36.

21. Rpt. to PSAC, Hq. USAF, 10 Oct. 1963, subj.: Manned Orbiting Station and Alternatives, Vol. IV.

22. Interview, Lamar by C. J. Geiger, Hist. Div., ASD, 14 Jan. 1964.

23. Memo., A. H. Flax, ASAF/R&D, Hq. USAF, 18 Sept. 1963, subj.: Dyna-Soar Operational Capability Study, Doc. 31; ltr., Lt. Gen. H. M. Estes, Vice Cmdr., AFSC to Hq. ASD, 25 Oct. 1963, subj.: Space Planning Studies, Doc- 51.

24. TWX, SCLDS-27-11-3, Hq. AFSC to Hq. ASD, 29 Nov. 1963.

25. TWX, SCG-0-10-6, Hq. AFSC to Hq. ASD, 10 Oct. 1963.

26. Memo., Col. Moore, 30 Oct. 1963, subj.: Record Memorandum of the X-20 Presentation to Secretary of Defense McNamara, 23 October 1963, and Pertinent Background, Doc. 53.

27. Hist. Rpt., Dir., Dev., DCS/R&D, Hq. USAF, July-Dec. 1963, p. 77, Doc. 102.

28. Ltr., Col. W. H. Price, AFPR, Boeing Co. to Hq. AFSC, 10 Oct. 1963, subj.: Report of Visit, Assistant Secretary of the Air Force, Doc. 4.2.

29. Interview, Lamar by Geiger, 27 Feb. 1964.

30. Hist. Rpt., Dir., Dev., DCS/R&D, Hq. USAF, July-Dec. 1963, p. 77, Doc. 102; ltr., Lt. Gen. Estes to Lt. Gen. James Ferguson, DCS/R&D, Hq.. USAF, 17 Oct. 1963, subj.: X-20A System Package Program, Doc. 43.

31. Speech, Harold Brown, DDR&E, 17 Oct. 1963, subj.: National Space Program, Doc. 44.

32. Memo., Col. Moore, 30 Oct. 1963, subj.: Record Memorandum of X-20 Presentation to Secretary of Defense McNamara, 23 October 1963, and Pertinent Background, Doc. 53.

33. Trip rpt., Lamar n.d., subj.: Paraphrased Transcript of Discussion After X-20 Status Briefing to Mr. McNamara by Col. Moore in Denver, 23 October 1963, Doc. 46.

34. Trip rpt., J. H. Goldie, Boeing Co., 24 Oct. 1963, subj.: Questions Comments, and Impressions from McNamara Briefing in Denver, 23 October 1963, Doc. 49.

35. Trip rpt., Lamar n.d., subj.: Paraphrased Transcript, Doc. 46.

36. Memo., Col. Moore, 24 Oct. 1963, subj.: X-20A Status Report to Secretary McNamara, 23 October 1963, Doc. 50.

37. Ltr., Gen. B. A. Schriever, Cmdr., AFSC to Hq. ASD, 9 May 1963, subj.: X-20 Program Management.

38. TWX, SCCP-3-7-2, Hq. AFSC to Hq. ASD, 3 July 1963.

39. Memo., DDR&E n.d., subj.: Rationale of FY 64 RDT&E Program Adjustments by OSD, Doc. 21.

40. TWX, MSFA-30-7-47, Hq. AFSC to Hq. ASD, 30 July 1963.

41. Ltr., Gen. Schriever to Cmdr., ASD, 31 July 1963, subj.: X-20 Program.

42. System Package Program, System 620A, X-20 SPO, 3 Sept. 1963, pp. 11-1, 11-3 , 11-5.

43. Memo., Col. Moore, 9 Sept. 1963, subj.: X-20 Test Program, Doc. 28.

44. Ltr., Maj. Gen. L. I. Davis, Spec. Asst. to Vice Cmdr., AFSC to Gen. Schriever, 19 Sept. 1963, subj.: X-20 Test Program, Doc. 35.

45. Revision A, System Package Program, System 620A, X-20 SPO, 3 Sept. 1963, p. 11-3.

46. Ltr., Lt. Gen. Estes to Lt. Gen. Ferguson, 17 Oct. 1963, subj.: X-20A System Package Program, Doc. 143.

47. Ltr., J. B. Trenholm, Asst. Dir., X-20 SPO to Lt. Col. C. L. Scoville, Dir., Mil. Space Prog., Hq. AFSC, 21 Nov. 1963, Doc. 59.

48. Memo., Lt. Col. Scoville n.d., subj.: Events Concerning the X-20 Cancellation, Doc. 100.

49. Memo., DDR&E to SAF, 30 Aug. 1963, subj.: Military Orbiting Space Station, Doc. 27.

50. Memo., DDR&E to Secy. of Def., 14 Nov. 1963, subj.: Approaches to a Manned Military Space Program, Doc. 57.

51. Memo., DDR&E to Secy. of Def., 30 Nov. 1963, subj.: Evaluation of an Orbital Test Module, Doc. 64.

52. Ltr., R. L. Bisplimghoff, Assoc. Adm/Adv. Res. % Tech., NASA to Assoc. Adm., NASA, 22 Nov. 1963, subj.: X-20 Program, Doc. 60.

53. Ibid.

54. Memo., DDR&E to ASAF/R&D, 29 Nov. 1963, subj,: X-20 Program, Doc. 63.

55. Memo., Col. Moore, 3 Dec. 1963, subj.: Telephone Request from Hq. AFSC, Doc. 65; memo., ASAF/R&D to DDR&E, 4 Dec. 1963, subj.: X-20A Program, Doc. 67.

56. Memo., ASAF/R&D to SAF, 4 Dec. 1963, subj.: Manned Military Space Program, Doc. 68.

57. Memo., SAF to Secy. of Def., 4 Dec. 1963, subj.: Manned Military Space Program, as noted by Max Rosenberg, Hist. Div., Liaison Ofc., Hq. USAF.

58. Memo., Maj. Gen. J. K. Hester, Asst. VCS/AF to SAF, 4 Dec. 1963, subj.: Approaches to a Manned Military Space Program, Doc. 69.

59. Memo., SAF to Secy. of Def., 5 Dec. 1963, subj.s Manned Military Space Program, as noted by Rosenberg.

60. Hist. Rpt., Dir., Dev., DCS/R&D, Hq. USAF, July-Dec. 1963, p. 81, Doc. 102.

61. New York Times, 10 Dec. 1963.

62. News Briefing, Secy of Def., 10 Dec. 1963, subj.: Cancellation of the X-20 Program, Doc. 71.

63. Memo., DDR&E to SAF, 12 Dec. 1963, subj.: X-20 Program, Doc. 83.

64. TWX, AFCVC-1918/63, Hq. USAF to All Commands, 10 Dec. 1963, Doc. 73.

65. Memo., Lt. Col. Scoville n.d. subj.: Events Concerning the X-20 Cancellation, Doc. 100.

66. Ltr., Gen. Schriever to Maj. Gen. Marvin C. Demler, Cmdr., RTD, 16 Dec. 1963, subj.: Manned Space Program, Doc. 88.

67. TWX, ASZR-10-12-1011, Hq. ASD to Hq. SSD, 10 Dec. 1963, Doc. 74; TWX, ASZRK-12-10-249, Hq. ASD to AFPR, Boeing Co., 10 Dec. 1963, Doc. 75.

68. TWX, SAF to Hq. AFSC, 11 Dec. 1963, Doc. 78; memo., SAF to CS/AF, 12 Dec. 1963, subj.: Dyna-Soar Termination, Doc. 82.

69. X-20 Phase-out Plan, X-20 SPO, 13 Dec. 1963, pp. IV-1 to IV-11.

70. TWX, AFRDD-79094, CS/AF to Hq. AFSC, 18 Dec. 1963, Doc. 99.

71. X-20 Phase-out Plan, X-20 SPO, 13 Dec. 1963, p. IV-1.

72. TWX, MSF-17-12-45, Hq. AFSC to Hq. ASD, 17 Dec. 1963, Doc. 95.

73. X-20 Detailed Termination Plan, X-20 SPO, 3 Jan. 1964, p. IV-18.

74. Ltr., Lt. Gen. Estes to Hq. USAF, 8 Jan. 1964, subj.: X-20A Technology, Doc. 103.

75. TWX, AFRDDG-86985, Hq. USAF to Hq. AFSC, 23 Jan. 1964, Doc. 106.

76. TWX, MSFAM-16-1-38, Hq. AFSC to Hq. ASD, 16 Jan. 1964, Doc. 105.

77. Management Plan for X-20 Continuation Tasks, X-20 Engg. Ofc., RTD, 31 Jan. 1964, pp- 4-5, Doc. 107.

78. Ltr., P. J. DiSalvo, Dep. Dir/Procurement, X-20 SPO to AFPR, Boeing Co., 12 Mar. 1964, subj.: AF33(657)-7132, The Boeing Company Percentage of Completion and SPO Recommendations for Final Settlement, Doc. 109; ltr., DiSalvo to AFPR, Minneapolis-Honeywell Regulator Co., 27 Feb. 1964, subj.: AF33(657)-7133, Minneapolis-Honeywell Regulator Company Percentage of Completion and SPO Recommendation for Final Settlement, Doc. 108; ltr., DiSalvo to AFPR, RCA, 30 Mar. 1964, subj.: Contract AF33(657)-7134, Radio Corporation of America, Percentage of Completion and SPO Recommendation for Final Settlement, Doc. 110; X-20 Detailed Termination Plan, X-20 SPO, 23 Jan. 1964, p. III-2; interview, DiSalvo, Acting Ch., Hitting Msl. SPO, by Geiger, 5 Aug. 1964.

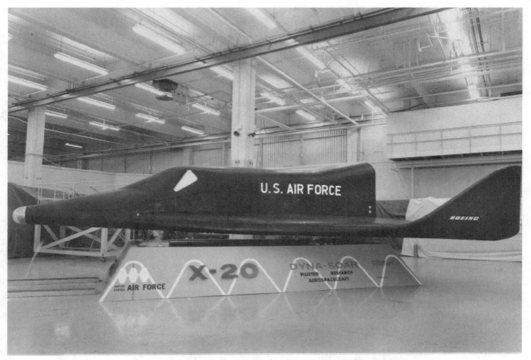

Figure 1. A Mock-up of the X-20 glider in April 1963.

Figure 2. An X-20 right wing model in June 1963.

Figure 3. A leading edge segment being fabricated in July 1963.

Figure 4. The fuselage structure assembly of the X-20 in December 1963.

Figure 5. A weld assembly of the X-20 pilot's compartment in September 1963.

Figure 6. The X-20 manufacturing assembly model in July 1963.

GLOSSARY OF ABBREVIATIONS

Adm.	Administrator		Ltr.	Letter
Adv.	Advanced		Maj.	Major
AF	Air Force		Memo.	Memorandum
AFFTC	Air Force Flight Test Center		MMSCV	Manned, Military, Space, Capability Vehicle
AFMDC	Air Force Missile Development Center		MOL	Manned, Orbiting Laboratory
AFMTC	Air Force Missile Test Center		Msl.	Missile
AFPR	Air Force Plant Representative		NASA	National Aeronautics and Space
AFSC	Air Force Systems Command			Administration
ASAF	Assistant Secretary of the Air Force		N. D.	No Date
ASD	Aeronautical Systems Division		No.	Number
ASSET	Aerothermodynamic, Structural Systems,		Ofc.	Office
	Environmental Test		P., PP.	Page, Pages
Assoc.	Associate		Pres.	President
Asst.	Assistant		Presn.	Presentation
Cmdr.	Commander		Prog.	Program
Co.	Company		PSAC	President's Scientific Advisory Committee
Col.	Colonel		RCA	Radio Corporation of America
CS	Chief of Staff		R&D	Research and Development
DCS	Deputy Chief of Staff		RDT&E	Research, Development, Testing, and
DDR&E	Director of Defense for Research and			Evaluation
	Engineering		Res.	Research
Def.	Defense		Rpt.	Report
Dep.	Deputy		RTD	Research and Technology Division
Dev.	Development		SAF	Secretary of the Air Force
Dir.	Directorate		SAFUS	Under Secretary of the Air Force
Div.	Division		Secy.	Secretary
Doc.	Document		SPD	System Program Directive
DOD	Department of Defense		Spec.	Special
DSMG	Designated Systems Management Group		SPO	System Program Office
Engg.	Engineering		SSD	Space Systems Division
Exec.	Executive		Subj.	Subject
Fig.	Figure		Sys.	Systems(s)
FY	Fiscal Year		Tech.	Technology
Gen.	General		TWX	Teletypewriter Exchange Message
Hist.	Historical		U. S.	United States
Hq.	Headquarters		USAF	United States Air Force
LEM	Lunar Excursion Module		VCS	Vice Chief of Staff
Log.	Logistics		VHF	Very High Frequency
Lt.	Lieutenant		Vol.	Volume

Date: Apr 1964   Title: AFSC News Release Dyna-Soar Research

NEWS SERVICE
For Release: April 23, 1964, P.M.

UNITED STATES AIR FORCE
ANDREWS AIR FORCE BASE,
WASHINGTON 25, D. C.

43-10-44 21/4/64

RESEARCH FROM DYNA-SOAR WILL APPLY TO-MOL PROGRAM

ANDREWS AFB, Md.—The full pressure suit developed for the X-20 (Dyna-Soar) pilots now will be utilized in support of the Air Force Manned Orbiting Laboratory (MOL) program, the Air Force Systems Command (AFSC) announced today.

The X-20 program was cancelled by the Department of Defense on December 10, 1963. At that point, $410 million had been spent, approximately half of the estimated total cost of the program.

The suit is one of 36 program activities identified in a study conducted by AFSC's Aeronautical Systems and Research and Technology Divisions which could be brought to the point where they would provide an advanced technical payoff, demonstrate feasibility, verify design, or mark the completion and delivery of hardware. An additional $3,000,000 has been authorized to pursue these projects. Nine other projects still are being studied. Research and testing equipment and raw materials already bought and paid for will be converted to other uses.

One of the largest intangibles in the program will be impossible to relate to dollars and cents. This is the research conducted by contractors and government agencies that went into development of the Dyna-Soar. All research findings are being documented and forwarded to the Defense Documentation Center where they will be available to all government agencies, contractors, and universities for use on future projects.

Since the program was cancelled, the X-20 System Program Office (SPO) at the Aeronautical Systems Division, Wright-Patterson AFB, Ohio, has been terminating contracts and surveying other government agencies for their interest in the technology and equipment developed under the program.

Shortly after the termination, a joint ASD/RTD group compiled a list of program activities to be considered for continuation and categorized them as follows:

(1) Provide state-of-the-art payoff;
(2) demonstrate feasibility or verify design;
(3) finish and deliver hardware;
(4) require further justification by user; and
(5) unaccepted.

The group recommended that 36 of 66 projects be continued since they clearly provided technological benefits sufficient to justify their cost. All were in the first three categories and approval for their continuation was, given by Headquarters USAF. Illustration of the projects earmarked for continuation are these 10:

The X-20 full pressure suit.
    The contract is complete except for helmets for the last five suits. Of the ten suits that will be complete when the helmets are delivered, three will go to Air Force Flight Test Center, Edwards AFB, Calif., for Manned Orbiting Laboratory (MOL) mockup support; three to Aerospace Medical Research Laboratories, Wright-Patterson AFB, for continuation of tests, now 85 per cent completed; two to Systems Engineering Group at Wright-Patterson for support of any future systems that might need them; one to the Manned Spacecraft Center, NASA, Houston, Tex., for possible use in the Apollo program; and one, designed for a NASA X-20 pilot to be used by him in the flight tests of a lifting body vehicle at Edwards AFB.

Search and Rescue Receiver/Transmitter.
    This new high power (5 watt) UHF solid state personnel rescue beacon was being developed to enhance the chances of rescue for a pilot who has to bail out. The higher power will provide three to five times greater range than existing rescue beacons. The standard Air Force rescue beacon (AN-URC-10) has a power output of 200 milliwatts and yields an ADF range of 180 to 220 miles as

compared with 60 to 70 miles of the URC-10. Redesign is required to replace discontinued transistor types with silicon transistors, thereby extending its temperature range so it will have general application to the Air Force. Upon completion it will be made available to ASD, NASA and Space Systems Division, Los Angeles, Calif., for possible use in C-141, KC-135, B-47, F-111, and B-52 aircraft, Gemini, Apollo, MOL and other manned vehicles.

Studies of Pilot Control of Booster Trajectories (PIBOL).

The feasibility of a pilot contributing to the guidance and control function of a launch vehicle has been investigated as part of the X-20 program. Underlying these studies is an attempt to take advantage of the pilot's reasoning and control capabilities to back up the launch vehicle guidance system, thereby increasing overall system reliability. Studies by the Boeing Company, Seattle, Wash., Martin-Marietta, Baltimore, Md., ASD and others have shown that a pilot is capable of controlling; a large, aerodynamically unstable boost vehicle during its entire ascent phase, including initial azimuth orientation and orbital injection cutoff. Studies by Boeing resulted in detailing the hardware mechanization required to implement PIBOL in the X-20 glider. At the termination of the X-20 program, Martin-Marietta, Denver, Colo., was working under an SSD contract to determine the mechanization and hardware modifications required to implement PIBOL in the Titan III booster. Possible uses of PIBOL include the Air Force MOL project and various NASA manned space projects.

Nose Cap.
The Ling-Temco-Vought nose cap concept currently is being subjected to a series of ground tests simulating the X-20A vehicle mission environment. Approximately one-half of the scheduled tests have been accomplished. Although this nose cap concept was successfully demonstrated in a previous test, design changes make further testing desirable. Successful completion of the tests would provide a proved radiation cooled nose cap concept for future re-entry vehicles.

X-20 Heat Protection System.
Verification of thermal analysis methods for variable temperature and pressure conditions encountered during glide re-entry of a radiation-cooled vehicle is not available. By fabricating an 18-inch square section of a complete thermal protection element, including external skin, insulation, corrugated panels, and water wall, and subjecting it to a realistic re-entry thermal condition in a vacuum chamber, the opportunity to correlate theoretical methods of analysis with experimental results would be afforded. The panel will be constructed from previously fabricated X-20A hardware and tested. This work will be valuable to future re-entry and orbital vehicle structural technology.

Coated Molybdenum Panels.
To demonstrate coating performance under actual re-entry conditions, complete coating of X-20 panels for flight testing on ASSET test vehicles has been accomplished. One panel was flown on ASSET vehicle ASV No. 1, but the loss of the vehicle precluded post flight inspection. Another panel has been coated and will be installed on the No. 3 vehicle for a future test. A third panel will be flown on Vehicle No. 4.

High Temperature Bearings.
Control surface bearing applications for future Air Force and NASA re-entry vehicles will require high temperature and load applications. Complete fabrication and high temperature testing of the X-20 elevon bearings will be accomplished, providing proved material and design concepts.

Sensors/Transducers.
Presently there are no sensors with adequate accuracy suitable for obtaining research data, such as low pressure, flutter, density, etc., at extremely high temperatures. Boeing has been evaluating a number of sensor/transducer developments and techniques in the 1,800-3,000-degree range. Completion of these evaluations will provide considerable background and experience in obtaining test results and conducting data analysis for application to other re-entry programs. At the completion of the Boeing evaluation, it is expected that follow-on research will be accomplished by Air Force transferred to development laboratories.

Test Instrumentation Subsystem Ground Station.
This ground station has the capability to recover telemetry data of both PCM and analog nature. Lacking only acceptance tests, the ground station will be tested by personnel familiar with the equipment. After acceptance, it will be transferred from Seattle, where it was built for the Boeing Company by Electro Mechanical Research to the AFFTC, Edwards, to support the Titan III or X-15 programs.

Flight Simulation Instruments and Controls.
A development simulator cockpit and associated consoles and hardware will be shipped from the Boeing plant in Seattle to AFFTC. There it will be used in human engineering research, in space vehicle display and control techniques and also in advanced pilot training.

UNISPHERE

NEW YORK WORLD'S FAIR 1964-1965 CORPORATION
INTERNATIONAL EXPOSITION
AT FLUSHING MEADOW PARK , FLUSHING 52, NY.

TELEPHONE-AREA CODE 212.WF 4-1964
CABLE ADDRESS "WORLDSFAIR"

ROBERT MOSES PRESIDENT

NEWS: TUESDAY AUGUST 18, 1964

MARTIN COMPANY SPACE VEHICLES AT NEW YORK WORLD'S FAIR

N.Y. WORLD'S FAIR — AUGUST 18 - - - SPECTATORS AT THE WORLD'S FAIR GOT AN UNEXPECTED GLIMPSE OF THE FUTURE TODAY AS THESE TWO FULL-SCALE SPACE SHIPS WERE LIFTED 130-FEET INTO THE AIR AND DROPPED INTO THE NEW HALL OF SCIENCE BUILDING, MARTIN COMPANY, A MAJOR AEROSPACE FIRM, IS PRODUCING A SPACE SPECTACULAR "RENDEZVOUS IN SPACE" FOR THE CITY OF NEW YORK'S NEW PERMANENT SCIENCE MUSEUM WHICH OPENS TO THE PUBLIC THE WEEK OF SEPTEMBER 9TH. THE SHOW, DIRECTED BY ACADEMY AWARD WINNER FRANK CAPRA, FEATURES A WIDE-SCREEN MOTION PICTURE INTEGRATED WITH THE FULL-SIZED SPACE VEHICLES WHICH MANEUVER HIGH OVER THE HEADS OF THE AUDIENCE THE TWO SHIPS ARE AUTHENTIC CREATIONS BASED ON THIS NATION'S ADVANCED PLANS TO ORBIT A SPACE LABORATORY AND ITS RE-SUPPLY VEHICLES.

MARTIN COMPANY'S "RENDEZVOUS IN SPACE"

To portray a critical phase in the operations of an orbital laboratory — the vehicle planned for the first comprehensive research by and on man in space — the Martin Company has produced "Rendezvous in Space," for the Hall of Science at the World's Fair.

Flights of fancy into outer space are hardly unique. But "Rendezvous in Space" is not fancy: it is projected scientific fact. Feasibility of this phase of space operations has been firmly established, and it should be an actuality within a decade. It is one logical extension of Project Gemini, which this year is expected to orbit two astronauts for as long as two weeks, carry out investigations made possible by their prolonged participation, including extravehicular experiments, and develop techniques for orbital rendezvous and docking.

The "Rendezvous in Space" concept if founded on the fact that man is both the most precise and the most productive instrument for scientific space investigation. By the spontaneous use of his judgment he is able to make observations, change procedures in mid-experiment and reach conclusions unanticipated in pre-launch programming. And with the orbital laboratory, on which the exhibit focuses, man becomes more than ever not only the experimenter but the experiment.

Before man can live and travel in space for any length of time, there are pressing scientific questions about how this utterly alien environment will affect him and the tools he uses. Because the research that will provide the answers can only be pursued in the environment under scrutiny, an orbital laboratory has presented itself as the most efficient, productive and safe means to carry through the investigations.

To produce the information required for protracted future flights in space such as a manned expedition to

Mars, a series of laboratories with an orbital life ultimately approaching three years is contemplated. Even the initial laboratories, of a span of a month or two, would necessitate periodic resupplying and changes of crew as man learns for how long a time he can live weightlessly in space. This would be accomplished by a regular earth-to-laboratory shuttle by space supply vehicles, or "taxis. " Serving also as space "lifeboats," they would be an advanced type of re-entry vehicle, designed to be launched into the orbital zone of the laboratory, maneuvered for rendezvous and transfer, flown down from orbit and, much like an ordinary airplane, landed at a chosen airport. The vital attribute putting them a significant step ahead of the Mercury-Gemini-Apollo re-entry vehicles would be their maneuverability. The pilot of the space taxi portrayed in the Martin production would be able to select a landing site anywhere within an area about the size of the United States. And he could land, gently, on any jet landing strip. The laws of ballistics, which bind contemporary reentry pilots closely, prescribe a minute landing area under any circumstances. In a sudden emergency such as the eruption of solar radiation suggested in the exhibit film, the pre-determined trajectory might land the vehicle in a polar ocean or a tropical jungle.

"Rendezvous in Space" is a pre-enactment of a regular supply mission. The full-sized, manned spacecraft which rendezvous at the production's climax are scrupulously faithful adaptations in design, dimension and operation of the actual vehicles. Hall of Science audiences will see models of astronauts move through a tunnel in the nose of the taxi into the laboratory after the docking operations are completed.

The laboratory is conceived as a pressurized and double-walled aluminum cylinder powered by heat from radioisotopes. The outer skin would serve as a thermal and meteoritic shield, and the inner skin would provide a constant pressure wall on which interior equipment would be mounted. The walls would be separated by insulation, enclosing a warning system to detect meteorite impact.

The laboratory will contain four sections housing elaborate equipment for a wide variety of material and observation tests in zero gravity, or weightlessness, and an environmental control system designed to reduce human fatigue and enhance a sense of security (a green floor and a blue ceiling, for instance, provide references to counter the weightless "where's up?" feeling). The living compartments, which contain sleeping quarters and dining and recreation facilities amid biological testing equipment, will be the headquarters of man's scientific observation of himself in space. The station control and maintenance center, where the nerve ends of the laboratory come together, will have the instruments necessary to direct and monitor the vehicle's position, the rendezvous, the thermal and atmospheric sub-systems, communications and testing equipment, electrical power and the general warning system.
As weightlessness is the critical unknown in protracted space exploration, it must become known how man will react to many weeks and perhaps months of this state. It is impossible to find out in any condition but zero gravity. So the orbiting crewmen will be intensively studied as they eat and sleep, work and relax. Not only their physiological reactions but also their sustained capacity as equipment operators and scientific observers will be in the balance.

A variety of experiments is contemplated: Astronomical observations of planets and stars can be made without the distortions imposed by the earth's atmosphere.

Tests can be run on the long-term effects of space on the laboratory's systems and metals and other materials that will make up the systems for the longer probes of the future. The reactions of a variety of organisms can be evaluated as they live, reproduce and die. Techniques of purifying wastes can be refined as man moves closer to the space ideal of a closed ecology.

The cylindrical laboratory, 30 feet long and 12 feet in diameter, is supported by a lightweight frame of 1, 500 feet of aluminum tubing covered with more than 15,000 square feet of fiberglass panels. The escape and shuttle vehicles are 24 feet long, 16 feet wide and nearly six feet in diameter. They are covered with a total of over 5,000 board feet of a new lightweight foam material called dystron, laminated with polyester resins and overlaid with a thin layer of fiberglass skin for durability. Masonite templates give the overall external configuration. Luminous paint was sprayed over all three vehicles to create the illusion of flight in space by tending to make the monorail, on which the vehicles operate approximately 80 feet above the heads of the spectators, invisible.

Date: Feb 1965   **Title:**Dornberger Aerospace-plane Article

WITHIN the last few years, space has proven to be a medium in which manned and unmanned devices can operate quite successfully. The first steps have been taken to unveil the secrets of the universe.

Future decades will see an ever increasing effort by leading nations of the world to make the best possible use of this new environment for the benefit of mankind. More and more satellites will orbit the earth. Manned space stations will be established, supplied and maintained, and manned and unmanned spacecraft will explore our planetary system. Operation in space will become a routine everyday event.

With this technically feasible future in mind, I think it is high time to have a hard look at our present prohibitively expensive approach to space operation and really do something about it.

To be ready in time in the most economical manner should be our guide line. We must explore, conquer and utilize space without endangering our national economy.

A start in the right direction would be primarily the development of boosters, not only recoverable but reusable, without delay and long overhaul. The day of the one-way, large ballistic, non-reusable booster is coming to an end. Since the birth of the space age in 1957, we have pushed too far in one direction by concentrating on development of the big ballistic boosters. Certainly, the more problems we experience in the launching of the big, one-way boosters, the more the economy of space exploration will be investigated.

A landing on the moon and accessible neighboring planets is a must. What might have been considered is how such a goal could have been attained in a reasonable time most economically while at the same time advancing future progress in space.

One possible way to accomplish this would have required the establishment of earth-orbiting space stations this side of the Van Allen radiation belt and the development of a recoverable, reusable space transporter to supply them.

In 1955 I predicted that such a transporter could be available ten years hence "if our research were sufficiently pushed." But that hoped-for "push" has in the interim been little more than a slight shove. It now appears to be another fifteen years in the future at least. The cancellation of the Dyna-Soar space glider program will prove one day to have been a serious set-back. We now have no experimental manned spacecraft capable of accomplishing tasks required in the predictable future. What has to be accomplished between now and the time this new mode of transportation becomes available is a task, certainly, of tremendous magnitude.

Future space bases in earth orbit may one day be used as re-fueling and transfer stations, as laboratories and assembly facilities circling the earth at an altitude of 400 to 500 miles.

Supplying these space bases may require 20 to 30 recoverable space transporters, each with about 20 to 30 tons payload, taking off in fast sequences from regular airport runways. At the space bases themselves, many types of spacecraft can be tested and launched for a variety of missions in outer space and returned to the orbiting base, parked and maintained.

If a recoverable space transporter and a fully equipped space station would have been our first goal in space, lunar orbit bases could have been established by a multitude of spacecraft, taking off from the earth-orbiting space stations together. Established 150 miles above the moon's surface, landing craft could then have landed and returned similarly to the proposed Apollo rendezvous technique. It is easily recognizable that by use of this method, more men, more spacecraft and more payload would have been available in lunar orbit at the same time during a lunar landing.

By landing enough men on the moon's surface, a lunar station for scientific research could have been established faster and more efficiently.

Only rocket-powered spacecraft with rocket engines delivering about 20,000 lbs. of thrust, but with a long burning time, would have been needed for this trip from earth orbit to moon orbit.

The propellant required could have been transferred from the space transporter into tanks at the earth-orbiting space bases. Additional tanks could have been attached to the spacecraft for lengthening the burning time and could have been dropped after the propellant had been consumed.

Even now, using such a system, the U.S. could gain a space installation of tremendous value and potentiality for future missions in space.

There are two concepts of special interest presently under study for achieving shuttle service between earth and orbiting space stations: the space plane and the aerospace plane. Looking at them and realizing their usefulness, one thing is certain. If we had invested only a small part of the energies and funds expended on ballistic booster development into the space plane as a space transporter, we would not be heading today down one of the most expensive dead-end roads, as some experts have put it, and would be much better prepared for the future.

The space plane in some versions is conceived as a large, heavy, winged, one-stage design of two to four million pounds take-off weight. This plane takes off from the ground vertically, ascends through the atmosphere, operates in space, re-enters the atmosphere and lands horizontally in the conventional manner. After lift-off by rocket power it will use air-breathing engines for further ascent within the atmosphere. During the atmospheric flight, so-called air separators are in operation, scooping up large quantities of air to be cooled down and liquefied by hydrogen, carried on board, and used as fuel. The oxygen is separated and pumped into the tanks of the space plane. As a liquid it is then used as an oxidizer for the rocket drive during operation in space.

The more appealing aerospace plane is a winged two-stage design. It takes off horizontally from a conventional airport runway. The first stage is equipped with air-breathing hypersonic ramjets. When it achieves speeds up to Mach 8, the second, winged rocket-powered stage separates from the first to operate in space. Both stages are recoverable and reusable after landing on any conventional runway.

The first stage is equipped with an air separator, similar to the one used in the space plane, but the liquid oxygen is used to fill the empty tanks in the upper stage. The take-off weight of the aerospace plane is about 500,000 to 750,000 pounds, much lighter than that of the spaceplane, since all oxidizer tanks are empty at take-off and aerodynamic lift is used from the beginning of the flight up to the fringe of the atmosphere, where the second stage is released for its flight to the space station.

To improve the economy of such a development, it may well be that some day the first stage of this aerospace plane will also be used to bring hypersonic commercial airplanes up to the border of the atmosphere, for flights over continents to their destination.

Date:1944   Title: Sänger-Bredt Report Bibliography

Gaedicke, Der gefahrlose Menschenflug (Safe Manned Flight). Hephästos-Verlag, Hamburg, 1911.

Pelterie, Considérations sur les résultats de l'allegément indéfini des moteurs. (Considerations of the Results of the Undefined Tractive Effort of Motors), Journal de physique, 1913.

Ziolkowsky [Tsiolkovsky], Erforschung der Welträume mittels Reaktionsraumschiffen (Exploration of Space by Reaction Spaceships). Kaluga-Leningrad, 1914.

Goddard, A Method of Reaching Extreme Altitudes. Smithsonian Institute, Washington, 1919.

Ziolkowsky [Tsiolkovskiy], Rine Rakete in den kosmischen Raum
(A Rocket in Space). I. Kaluga, 1924.

Hohmann, Die Erreichbarkeit der Himmelskörper (The Attainability of Celestial bodies) R. Oldenbourg, Munchen, 1925.

Oberth, Die Rakete zu den Planetenräumen (Rocket Into Interplanetary Space). R. Oldenbourg, München, 1928.

Pelterie, L'exploration par fusées de la très haute atmosphère et la possibilité des voyages interplanétaires (Rocket Exploration of the Upper Atmosphere and the Possibility of Interplanetary Travel). Paris, 1927.

Ley, Die Möglichkeit der Weltraumfahrt (The Possibility of Space Travel). Hachmeister und Thal, Leipzig, 1928.

Ziolkowsky, Erste praktische Vorversuche mit Reaktionsraumschiffen (First Practical Preliminary Experiment With Reaction Spaceships). VI. Kaluga, 1928.

Scherschefsky, Die Rakete für Fahrt und Flug (The Rocket for Travel and Flight). Volckmann, Berlin, 1929.

Noordung, Das Problem der Befahrung des Weltraumes (The Problem of Space Navigation). R. C. Schmidt, Berlin, 1929.

Oberth, Wege zur Raumschiffahrt (Ways to Space Flight). R. Oldenbourg, München, 1929.

Rakete, Zeitschrift des Vereins für Raumschiffahrt. (Journal of Society for Space Flight) Breslau, 1929.

Rynin, Weltraumfahrten, Träume, Legenden and Phantasien (Spaceflight: Dreams, Legends and Fantasies). Leningrad, 1928.

Rynin, Die Raumschiffahrt in der zeitgenössischen Belletristik (Spaceflight in Coutemporary Literature). Soikin Press, Leningrad, 1928.

Kondratjuk [Kondrat'yuk], Die Eroberung der Planetenräume (The Conquest of Interplanetary Space). Novosibirsk, 1929.

Perelmann, Weltraumfahrten (Astronautics). VI. Aufl. Moskau, 1929.

Rynin, Theorie der Bewegung durch direkten Rückstoss (Theory of Direct-Reaction Movement). T. H., Leningrad, 1929.

Ziolkowsky, Fernflug- und Mehrfachraketen (Long-Distance and Multiple Rockets). Kaluga, 1929.

Rynin, Raketen and Vortriebsmittel direkter Reaktion (Rockets and Direct-Reaction Propellants). Soikin Press, Leningrad, 1929.

Ziolkovsky, Ziele der Raumschiffahrt (Goals of Space Travel). Kaluga, 1929.

Ziolkovsky, Den Sternfahrern (Travelers to the Stars). Kaluga, 1930.

Ziolkovsky, Das neue Flugzeug (The New Airplane). Kaluga, 1930.

Biermann, Weltraumschiffahrt (Space Travel). Bremen, 1931.

Rynin, Sternnavigation, Zwischenplanetenverkehr (Astronautics and Interplanetary Travel). Acad. Sci. USSR Press, Leningrad, 1932.

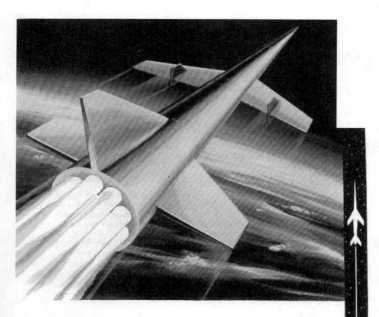

NOW...

**Bell harnesses Fluorine
to give man more power
for his push into space**

Elemental liquid fluorine—so flammable it will
set asbestos or water on fire—has been tamed by
Bell Aircraft for use as an oxidizer with rocket
fuels.

Special techniques in metal treatment, handling and welding had to be developed—tough
storage, transportation and operational problems
had to be overcome—before Bell was able to
accomplish the first large-scale rocket thrust
chamber firings using this ultimate in liquid
rocket oxidizers.

Fluorine will step up the power output of today's rocket engines 22% to 40% using present
fuels. It will make possible a 70% increase in
the payload of existing missiles and new space
vehicles now in development stages.

Thus, with this final major breakthrough in
chemical rockets, Bell Aircraft engineers have
added another first to an impressive list of contributions to aviation progress and national defense. This development brings chemical rocket
engines close to their ultimate capability. Only
nuclear fuels and even more exotic forms of
propulsion can offer further significant advances
in years to come. Perhaps you have rocket propulsion problems which Bell's highly qualified
engineers can help solve.

ROCKETS DIVISION **BELL** *Aircraft corp*

BUFFALO 5, N. Y.

BELL AIRCRAFT CORPORATION IS A MEMBER OF THE MARTIN-
BELL INDUSTRY TEAM DEVELOPING THE DYNA-SOAR HYPER-
SONIC GLIDER FOR THE U. S. AIR FORCE. OTHER TEAM MEMBERS
ARE THE MARTIN COMPANY, BENDIX AVIATION CORPORATION,
MINNEAPOLIS-HONEYWELL REGULATOR COMPANY, GOODYEAR
AIRCRAFT COMPANY AND AMERICAN MACHINE AND FOUNDRY
COMPANY.

DYNA SOAR
*Manned
Maneuverable
Space Flight*

Honeywell Inertial Navi-
gation System and Flight
Control Subsystem Elec-
tronics for the spacecraft
and Inertial Reference
system on Titan booster
for launch guidance.

Your Space-Age OPPORTUNITIES with Dyna-Soar

Bell Aircraft Corporation, a member of the Martin-Bell industry
team developing the Dyna-soar
hypersonic glider for the U. S. Air
Force, offers high-level openings on
this challenging project. These
positions will appeal particularly to
experienced engineers who desire an
opportunity for rewarding progress
in advanced research, analysis,
design and development of space
vehicles affording full scope to their
creative ingenuity with unusual
opportunities for rapid advancement and professional recognition.

**NAVIGATION SYSTEMS SPE-
CIALIST** to investigate and establish
requirements for inertial, radar
radio and air data sensing systems
with responsibility for setting up
and planning tests as required to
verify theories. BSEE or BSME
with 6 years' experience in related
fields required.

**FLIGHT CONTROL SYSTEMS
SPECIALIST** to determine need for
the most suitable methods of obtaining automatic and manual flight
control, stability augmentation,
power boost pilot displays and controls. BSEE or BSME with 8
years' experience in related fields.

**AERODYNAMIC HEAT TRANS-
FER GROUP LEADER** to direct and
lead heat transfer group in performing analysis of aerodynamic heating
of aircraft through complete speed
range from subsonic to hypersonic,
and to develop and apply new
methods for performing these analyses. PhD or MS in Aero, Physics
or Applied Math with 3-5 years'
experience.

GASDYNAMICS SPECIALIST to
initiate and perform fundamental
and applied theoretical research in
fields of gasdynamics particularly at
hypersonic speeds, and to initiate,
perform and/or monitor basic applied experimental research on
fundamental hypersonic flow and
physical phenomena. PhD or MS
in Aero Engineering, Physics or
Applied Math with 3 years' experience.

*Salaries are commensurate with
your background. Good living
and working conditions prevail
with liberal benefits. Write:*

**Supervisor, Engineering Employment,
Dept. Y-63**

BELL AIRCRAFT CORPORATION
BUFFALO 5, NEW YORK

SPACE GLIDER. Drawing of Dyna-Soar space glider, which will combine extreme speed of a ballistic missile with controlled and accurate flight of a manned aircraft. Designed to be rocketed into space, where it could travel at speeds approaching 18,000 mph, Dyna-Soar will be able to re-enter earth's atmosphere and make conventional pilot-controlled landing. Boeing is system contractor for Dyna-Soar, now being developed by U. S. Air Force with cooperation of National Aeronautics and Space Administration.

Capability has many faces at Boeing

THREE - ENGINE JET. Scale model of America's first short-range jetliner, the Boeing 727. Already, 117 Boeing 727s have been ordered by American, Eastern, Lufthansa and United airlines for delivery beginning late in 1963.

PLASMA PHYSICS. Boeing Scientific Research Laboratories scientist has verified experimentally, for the first time, a theory concerning ionized gas—important in future harnessing of thermonuclear power.

AUTOMATIC SKY FIGHTER. Supersonic Boeing Bomarc missile, now operational, is the United States Air Force's push-button defense weapon against airborne missiles and attacking bombers. New Bomarc "B" models have scored test intercepts up to 446 miles from base at altitudes of more than 100,000 feet.

BOEING

HOW DO YOU FEED A DYNA-SOAR?

CONFIDENTLY . . . fast, cold, safe . . . and without spilling a drop! The X-20's cryogenic "spoon" is a manually applied, two part quick-disconnect assembly. ■ Both the reusable ground sub-assembly and the airborne assembly are fabricated of aluminum and stainless steel, measure 10 by 15 by 10 inches, and weigh only 46 pounds complete . . . the airborne assembly weighing less than 11 pounds. ■ The six fill-and-vent connectors are locked to the bird with a patented ball-locking positive seal. ■ Normal pyrotechnic separation is used. A lanyard for emergency disconnect may be supplied. ■ We help feed all sorts of commercial and military birds which require state-of-the-art technology and highest reliability. An expert staff of fueling specialists and complete facilities are maintained for design, fabrication and on-site testing for systems requiring kerosene, JP-4, JP-5, Aviation Gasoline, HEF, IRFNA, UDMH and Cryogenics. ■ Inquiries invited.

SCHULZ TOOL & MANUFACTURING COMPANY
Fueling equipment for the aircraft and aerospace industries
425 South Pine Street, San Gabriel, California

SALES OFFICES: EAST COAST ENGINEERING SALES AND SERVICE CO., ROSLYN, NEW YORK AND ST. LOUIS, MISSOURI
D AND D SALES, INC., RENTON, WASHINGTON □ TRADING COMPANY AVIO-DIEPEN N. V., THE HAGUE, HOLLAND

INDEX

The enclosed DVD Video disc is Region Free NTSC.

Contents

USAF 1961-1962 Dyna-Soar Progress Report

USAF Documentary "*The Story of Dyna-Soar*"

Press Conference X-20 Roll Out September 1962
   Including: Captain Albert Crews, Major Henry Gordon, Captain William
   Knight, Major Russell Rogers, General Bernard Schreiver, Milton
   Thompson, Major James Wood, Undersecretary of the Air Force Dr.
   Joseph Charyk, Secretary of the Air Force Eugene Zuckert

Silent USAF construction footage

Silent Pressure suit and cockpit tests
   Including: Neil Armstrong, Henry Gordon, Gus Grissom, William Knight,
   Wally Schirra, Milt Thompson, James Wood.

Over an hour and half of extremely rare color footage of America's first experimental space shuttle.